A COLEOPTERIST'S HANDBOOK

(4th Edition)

by J. Cooter
and
M.V.L. Barclay

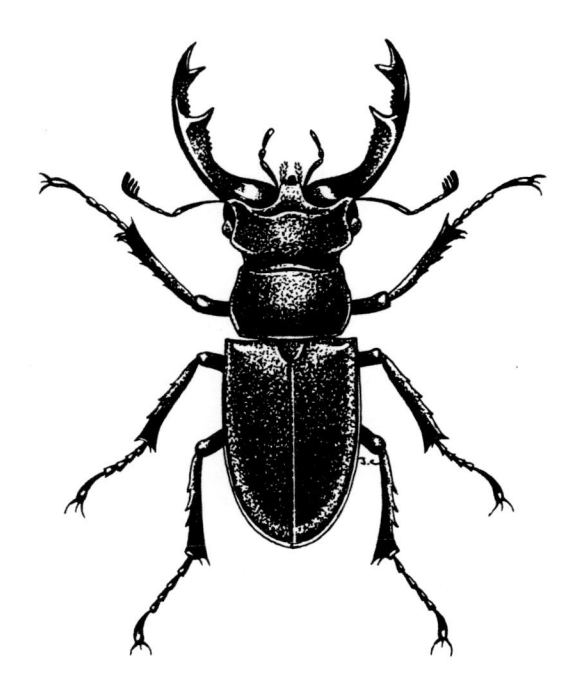

THE AMATEUR ENTOMOLOGIST
VOLUME 11

2006

The moral rights of J.Cooter and M.V.L.Barclay
as joint editors of and contributing authors to this work
and the moral rights of all other contributing authors have been asserted.

First edition 1954
Second edition 1974
Third edition 1991
Fourth edition 2006

Copyright © The Amateur Entomologists' Society 2006
Registered Charity 267430
ORPINGTON, KENT, ENGLAND

General Editor Fiona Merrion-Vass

Printed by Cravitz Printing Company Ltd.
1 Tower Hill, Brentwood, Essex CM14 4TA
E-mail: cravitzprinting@btconnect.com

ISBN 0 900054 70 0

*Dedicated to our families
past and present for encouraging,
supporting and at times putting up
with our interests in entomology*

Front cover illustration:
Saperda scalaris (L.) (Cerambycidae)
© Michal Hoskovec.

CONTENTS

The Beetle Families 1
Collecting
    Equipment 201
    Autocatcher 214
    Vacuum samplers 215
    Sample habitats
        winter 219
        spring 222
        summer 227
        autumn 237
    Collecting abroad 239
    Beetles associated with ants 248
    Stored product species 258
    Vascular plants and beetles 270
        Beetles and their host plants plants 270
        Host plants and their beetles 290

Curatorial
    Killing and relaxing 311
    Mounting equipment 315
    Mounting procedure 320
    Data and other labels 322
    Storage 328
    Pest control 331
    Spirit preservation 332
    Organising a collection 333
    Degreasing 335
    Nomenclature 336
    Identification 340
        microscopes 342
        lighting 343
        identification literature 344
    Chromosome differences 346
    Genitalia, systematic importance 352
        equipment required 353
        dissecting and mounting 356
        male 362
        female 364
    van Doesburg method 366

Breeding
- Terrestrial predatory larvae 369
- Aquatic predatory larvae 370
- Aquatic phytophagous larvae 374
- Dung inhabiting larvae 374
- Soil-inhabiting root feeders 375
- Terrestrial phytophagous larvae 376
- Wood and bark inhabiting species 377
- Detritus and fungal feeders 378

Conservation 379

Creating a database 397

Recording 403

Glossary 413

Beetle index 431

PREFACE
To the fourth edition (almost fully revised)
By J.Cooter

It is doubtless a personal trait, but I find revising the *Handbook* a very daunting business with occasional but ephemeral moments of pleasure. This despite copious help given freely by numerous friends who have contributed specialist chapters and knowing the *Handbook* fills a niche and is generally well received. I say 'generally' as I recall a reviewer of the third edition took me to task for my suggestion of killing off the contents of a wasp nest before investigating it for Coleoptera; evidently a conservation issue in the mind of the reviewer, or as another reviewer put it 'stop carping and be grateful someone has sat down and written such a useful book.'

Whilst working on this revised fourth edition, I was told independently by two 'Young Turks' of entomology, that they had their youthful enthusiasm nurtured by the *Handbook*, it was to them necessary reading and a huge help in making the transition from 'bug hunting' to 'entomology.' Both are now employed as entomologists in our top two entomological museums. Similarly, on my first visit to the International Insekten Börse in Prague in 1994, Czech friends mentioned they found the *Handbook* useful. Hearing that feed-back years after publication of the third edition was a great boost for me, the effort had, after all, been worthwhile (wasp nests aside).

As with previous editions, I have received great help and encouragement from numerous contributors, each a specialist in their field. Each has met my deadlines and not harangued me for then taking such a long time to get the text finalised, collated and ready for the printer. Likewise, Fiona Merrion-Vass, the Society's Publications Officer, has been supportive in not chivvying me along and demanding text by this or that date. To all involved I offer my sincere thanks.

Special thanks go to Max Barclay who, as well as contributing specialist text, has read through the entire final draft checking for consistent use of the hybrid taxonomy adopted for this edition (which will probably set tongues wagging), writing new text, revising some of the third edition chapters, checking and correcting spelling, grammar and checking references, making sure all measurements are metric, dates correct and other almost nit-picking points which, if left unchecked, would annoy the reader (if indeed the Dear Reader spotted them or had nothing better to do).

I also tender sincere thanks to Darren Mann for regularly checking references, supplying photocopies from a range of non-British journals and promptly answering numerous e-mails on a range of matters, many of which were of an obscure and 'nit-picky' nature on text throughout the book. All this in addition to supplying the revision of the Scarabaeoidea chapter, and other text.

The new edition has a brief chapter 'Collecting Abroad' possibly written in journalistic style but one which I hope will be useful. British coleopterists, compared to our colleagues across the Channel on mainland Europe, have a very parochial outlook and it is rare to find a British colleague with an interest in anything but our own very restricted and relatively well documented beetle fauna. I recommend a visit to the International Insekten Börse held on the first weekend of March and October each year in Prague. Here you will find an entomological hot house. I have just returned from the 2005 March event where I met up with friends from Poland, Ukraine, Austria (2), Germany (2), Italy (2), France, Czech Republic (7), Slovakia (2), Netherlands, Russia (2) and three from England. I received boxes of beetles to identify for friends and handed out several boxes from my last Chinese trip; this material will have been identified ready for return at the next (October) Börse. Evenings are spent enjoying a meal and a few drinks with friends and colleagues and there's even time for sight-seeing.

I hope this edition will help and encourage at least one or two putative coleopterists to get immersed in the study of beetles; it is important there is a 'next generation.' Despite the sentiments expressed in the first paragraph, I strive to do my best and never think of the job as 'completed' until I am totally satisfied and actually proud of the amassed text, all 'i's' dotted and 't's' crossed.

Returning to my opening theme, and in no way thinking to compare this small effort to such a great time-enduring work, I quote from the Preface of Rev Fowler's 1886 *Coleoptera of the British Islands* it is simply that I see parallels between my methods and hopes and those of The Great Man:

"I may perhaps be allowed in conclusion to add that it has been written at odd times, often amidst constant interruptions, and during the intervals of close ordinary work. I hope, however, that it may, at all events, prove of some help towards encouraging the study of our native Coleoptera."

Jonathan Cooter
Hereford, March, 2005.

PREFACE TO THE THIRD EDITION

During January 1974 the second edition of this *Handbook* was published. Finances dictated its format, with new material added as appendices. The result was adequate but not entirely satisfactory and it was obvious that, once the second edition became out of print, a completely revised edition would be required. With stocks low, I was again approached by Peter Cribb and set about the work. A new format was devised and the help of many specialists became a required necessity; my own knowledge, possibly in common with many other coleopterists, is selectively gleaned to suit personal interests and bias. The whole exercise has taught me four lessons:

- My own knowledge of British Coleoptera is patchy.
- Use of a word-processor would have made life easier and seen an earlier publication of this book.
- To think hard before volunteering for anything!
- Dead-lines really are quite flexible.

With regard to the latter I must point out that very often the flexibility was my own and not that of (the bulk of) the individual contributors. With this in mind I tender my apologies to those assiduous specialists who promptly prepared their valued contributions. In all, delays included, I hope this edition meets with approval and will be a useful text for the novice and more experienced coleopterist alike. In revising the *Handbook* it soon became evident that some of the original material (in the 1954 and 1974 editions) needed so little alteration, often only the revising of nomenclature, that to re-write would be not only a waste of time but something akin to plagiarism.

With so many individual contributors, there may be some inconsistency in terminology and style. Efforts have been made to bring a degree of uniformity, but doubtless some instances remain.

I offer my thanks to all those persons that have contributed and are named in the text. My sincere thanks are also offered to Dr Lena Ward and David Spalding (Institute of Terrestrial Ecology) and Dr M. Majerus. Howard Mendel and Alex Williams, with a naivety born of friendship, kindly offered to read through the assembled manuscript; both made valuable comments and suggestions, the majority of which have been incorporated. I am pleased to report that burdensome task had no adverse effect on our friendship. Peter Cribb once again has taken on the onerous job of not only giving continual encouragement and not complaining too strongly about my missed dead-lines but for editing the manuscript and seeing it through the publication stages. John Read very kindly produced the excellent drawings of beetles featured in the *Handbook*; to all four gentlemen I tender my heartfelt thanks.

In a book that proposes to give advice and guidance perhaps the 'Preface' is the place to mention the best piece of advice that was ever given to me; it has helped me to maintain a sense of perspective over the years. At my very first 'Verrall Supper' when about 18 years old, I found myself in conversation with Horace Last and Rev. E.J. Pearce, two authorities with whom I had corresponded for several years, but never previously met in person. Towards the end of our chat Horace (now a close family friend and personal mentor) took me to one side and confided to me 'Now remember, Jonathan, there's more to life than beetles.'

Jonathan Cooter
1990

PREFACE TO THE SECOND EDITION

Since the first edition of this *Handbook* published in 1954 many advances have been made in the study of the Coleoptera. Some of the original contributors are alas no longer with us and it is to the original editors and associate authors that I would like to dedicate this partly revised and extended edition.

Although published twenty years ago, the *Handbook* is still the most useful and informative work of its kind available, surely a reflection of the thoroughness and knowledge of its authors. It is hoped that additions made to this edition have increased the value of the book. The chapter on photography has been omitted as the Society hopes to publish soon a new leaflet on insect photography. Minor changes and the additions have been incorporated in appendices and a new chapter covering certain aspects of the extraction, preservation and uses of the genitalia has been included. Here the Society is indebted to Dr. R.A. Crowson for contributing the section on the systematic use of the genitalia in the Coleoptera. The reader may be baffled by the absence of Fig. 23. This has been left out and the numbering of the following figures and references to them have not been changed. The blocks of the original cover picture have long been lost or destroyed, hence the more formal illustration I have made of the Stag Beetle.

I would like to thank Dr. Crowson for his help with the section on the genitalia, Peter Hammond of the British Museum (NH) for the advice and encouragement given during the preparation of the revisions and Peter Cribb, without whose helpful editorial advice this work would not have been finished. I also thank the editors of the *Entomologist's Monthly Magazine* and the Royal Entomological Society of London for permission to make use of drawings appearing in their publications, and the Joint Committee for the Conservation of British Insects for the inclusion of the Code for Collectors. My father, John Cooter, kindly assisted in producing the drawings for the Glossary.

Stoneleigh. January, 1974.
Jonathan Cooter.

PREFACE TO THE FIRST EDITION

The Council of the Amateur Entomologists' Society, having in mind the success of its former Handbooks for the Hymenoptera and the Silk Moth Rearer, both of which are sold out, felt that it could make a further contribution to practical entomology by the production of *A Coleopterist's Handbook*. They have high hopes of it, for the response to their request for advance orders was satisfyingly widespread.

Their thanks are due to each of the contributors who wrote specially for this *Handbook*, and to those others who permitted useful quotations from their writings elsewhere; to the photographers* who have provided an essential part of the illustrations; and to the artists, particularly Mr. K.H. Poole, who drew for the three-coloured cover; but especially are they due to the editors, Messrs G.B. Walsh and J.R. Dibb, upon the former of whom the brunt of the work has fallen.

A handbook of this kind, printed in small numbers and outside the general flow of book-publishing, is necessarily a relatively costly venture and this fact is bound to be reflected in the price. It would have been a venture beyond an amateur body's slender resources but for the munificence of a grant of £100 through the Royal Society, supplemented by substantial donations from the Honourable Robert Gerard and Professor W.A.F. Balfour-Browne, F.R.S.E., which provided the plates. The Council are very grateful for this support. They believe that the reproductions from Schiödte's rare work, *De metamorphosi eleutheratorum observationes* (Plates VI to XII) will be particularly welcomed.

In accordance with the Society's policy of helping the beginner to overcome the practical difficulties of entomological work and of emphasising the attraction of life-studies and rearing as well as those of making a good collection, this *Handbook* gives a good deal of space to the relationships of the living insects and the appearance of their pre-adult stages, as well as to the day-to-day arts and wiles which enable them to be found and studied.

Entomology is, fortunately, one of the fields of Science in which the amateur observer has still a valuable part to play, both in amassing accurate data on the habits and distribution of insects and in throwing light upon obscurities of development, behaviour or habitat. If a few of the readers of this *Handbook* are able to profit by the experience of those who have contributed to the symposium, their labour of love will have been rewarded.

W.J.B.Crotch,
General Editor
London, December 1953

---

* We are grateful to the South London Entomological and Natural History Society for prints and permission to use certain photographs on Plates I to V which were used previously in their *Transactions*, 1953. We regret that some of the photographs cannot now be ascribed to individuals, because they were furnished some years ago and are without indication of origin.

EDITORIAL

There has been a huge development in information technology since publication of the third edition of the *Handbook* in 1991. The editors felt that it would serve little purpose to attempt to include an exhaustive list of websites concerned with Coleoptera, equipment and book supply; a 'Google search' using key words will usually produce a wealth of information.

Most of the British-based equipment and book dealers are, amongst entomologists, 'household names' and if an item not stocked by any of these firms is required, a request on 'beetles-britishisles@yahoogroups.com' will usually result in that item being located. Several dealers advertise in the Amateur Entomologist's Society's *Bulletin* and are in attendance at the Annual Exhibition in October.

Reference by name to any particular brand of equipment or entomological supplier in the *Handbook* should not be taken as an endorsement by the Amateur Entomologist's Society of that product or supply business.

J. Cooter
M.V.L. Barclay

INTRODUCTION

by M.V.L. Barclay

When the third edition of the *Coleopterist's Handbook* came out in 1991, I had just finished school. Although I had been fortunate in having an entomologist as a biology teacher, who had introduced me to light trapping and to the AES exhibition (an event more eagerly anticipated than Christmas), I had never actually met a 'real live' coleopterist. I had already decided that the occasional *Nicrophorus* or *Typhaeus* in the light trap was worth any number of moths, so it was clear where my interests would take me, and the little orange book with the gold stag beetle on the cover became a constant companion and source of revelation. I quickly stopped pinning beetles through the thorax, sticking them down with superglue, writing labels in biro and other disgraceful practices. It also opened my eyes to the fact that there was a whole community of people devoted to the study of the British beetle fauna, and that they understood their subject in incredible detail. Reading this distillation of their experiences prompted me to visit the collections of the Natural History Museum, to see what some of the beetles they so enthusiastically described actually looked like, and after 10 years of volunteering there whenever I could, I was given a job, and came to meet and become friends with many of the authors. With this background, I was delighted when Jon asked me to make some small contributions to the fourth edition, and to help him check the text before publication. This edition has been a very unequal partnership, with Jon carrying at least 99% of the editorial workload, and telling me that, in return, I'll have to take the lead in the fifth edition!

Like all specialists, entomologists are concerned about succession, about where the next generation will come from. In my view, the *Coleopterist's Handbooks* are invaluable because they pass basic information and current wisdom from one generation to the next, the kind of information that does not always make it into the mainstream entomological literature, and is usually passed by word of mouth. The enthusiast is strongly recommended to collect all four editions, for they all have their own character, and the earlier ones in particular give a wonderful insight into the collecting techniques and the lives of a past generation who we otherwise know only from papers, collections, and the anecdotes of older colleagues. For example, in the first edition G. B. Walsh suggests collecting large sacks of flood debris from rivers in spate in winter, but strongly urges the reader to seek *'permission from the womenfolk'* before bringing it home to sort, because *'small fry might creep or fly'*! Walsh begins another chapter, on identification, with the sentence; *"When we have experienced the atavistic joy of collecting in the open air and the manipulative pleasure of setting and mounting our chosen beetles there follows the refined intellectual delight of naming them with precision"*. I still think this sentence perfectly encapsulates the enduring appeal of natural history collecting. There is little in life as exciting as a good day in the field, with good company, followed by the thrill of turning out your tubes at home to see exactly what you've got. This especially true towards the beginning of your career, when even the commoner species are new for you. I think you never forget your first *Pyrochroa*, *Rhagium* or *Lucanus*, your first clerid, anthribid or melandryid, and later your first new county record or 'New to Britain'.

I have been asked what my advice would be to the coleopterist who was just starting out, a position I was in myself not long ago. I would suggest that they read the *Coleopterist's Handbooks*, take out a subscription to *The Coleopterist* journal, join the British Entomological and Natural History Society and go on as many of their field days as possible, visit the 'Coleopterist' website (http://www.coleopterist.org.uk/), a treasure house of good things, including a photo gallery of British beetles and Michael Darby's excellent 'Biographical Dictionary of British Coleopterists', and an associated e-group, and finally, but possibly most importantly, become a regular visitor to the beetle collections of a museum, preferably the biggest and busiest one within reach. Museum collections have much to offer coleopterists of all levels of experience, and there is no substitute when it comes to learning a group or a fauna quickly, for simply 'reading through' a well laid out and reasonably complete collection from beginning to end a few times, familiarising yourself with each genus or species in turn. Staff in big museums may also offer help with identification, or provide advice on literature, direct you to helpful specialists, or on obtaining items of equipment.

All the suggestions above will help coleopterists to develop their knowledge and collection quickly and efficiently, to correspond with and meet like-minded people, and to gain useful hints for collecting, maintaining a collection, and identification. It is also useful and pleasant to attend social events like the annual 'Verrall Supper' (usually combined with a day in the collections at The Natural History Museum), and the big entomological fairs in Prague and Germany. Although entomology can be a rather solitary hobby, most of us can identify with the words of a young coleopterist, Charles Darwin, who wrote to his cousin, *'I am dying by inches from not having anybody to talk to about insects'*.

We are very fortunate in this country to have one of the best-studied beetle faunas in the world. Although there is much left to discover, we have a good understanding of the distribution, status, habitat requirements and often the life history of many of the 4000 or so species that occur on these islands; workers on British beetles are adding a valuable layer of knowledge to a long-established tradition. A good knowledge of the British fauna also provides a convenient springboard for those who wish to broaden their horizons by collecting overseas. Most of the major families are present in Britain, and the

British coleopterist will have a good idea in which area he might want to specialise, for some degree of specialisation is generally necessary when confronted with the vast biodiversity of the tropics. This edition of the *Handbook* contains a section on collecting abroad for the first time. Careful collecting abroad will certainly reveal new species to science, and there are opportunities to make great contributions to beetle taxonomy. Britain was formerly a world-leader in this field, and we are well overdue for another crop of international-standard beetle taxonomists. Many important taxonomic collections are housed in The Natural History Museum and other leading national museums, and are available for study by beginners or experts.

Whatever aspect of the study of Coleoptera you concern yourself with, it is impossible to be bored with such a vast and rewarding field of study. To this you can add the satisfaction that your work might one day be cited by entomologists not yet born, and that your collection may one day be a great treasure of a world-famous museum. It is to be hoped that this Fourth Edition of *A Coleopterist's Handbook*, will, like its predecessors, be a true guide for the beginner and a useful reference for the established worker in the field of Coleoptera.

The Beetle Families

INTRODUCTION

by J.Cooter

This section of the *Handbook* gives specific information about the members of the bulk of the beetle families represented in Britain. It is meant to supplement the chapters on general collecting and preparation methods. In several instances acknowledged authorities have very kindly contributed information about their particular speciality; the rest, I have tried to cover as best as I can. Alas, some families receive no mention here. Generally species belonging to the omitted families will turn up by adopting a range of the more general collecting methods (pages 218-239) and will need, more often than not, the standard preparation.

Our knowledge of beetles is increasing daily and the student should always glean what information is available by attending meetings, visiting specialist libraries, subscribing to *The Coleopterist, Entomologist's Monthly Magazine, British Journal of Entomology and Natural History, Entomologist's Gazette* and *Entomologist's Record and Journal of Variation* (or certainly two from that list) and in general conversation with colleagues. New keys to species groups, genera and, less often, whole families are published from time to time in a wide variety of places. Foreign journals and entomological literature are often very useful and sometimes our only source of up-to-date information on a particular group of beetles.

This edition of the *Handbook* largely follows Lawrence and Newton's (1995) classification of Coleoptera *'Families and subfamilies of Coleoptera (with selected genera, notes, references and data on family-group names)'*. This is a bench mark study, and along with the useful *Beetle Families of the World* CD-Rom (Lawrence *et al*, 1999) and Hansen's (1997) *'Phylogeny and classification of staphyliniform beetle families'* is essential reading for all coleopterists. The Curculionoidea follow the classification proposed by Alonso-Zarazaga and Lyal (1999, with amendments 2002) and adopted by Morris (2003). The status of 'Histeroidea' has been updated according to Hansen (1997) while the treatments of Kateretidae and Bolboceratidae have been updated according to ICZN (1999) and Scholtz & Browne (1996) respectively.

Please note, although following modern taxonomic classification, the final order of things is far from settled with only broad agreement having been reached, some families have here been 'combined' into single chapters, for example this affects the 'water beetles' and most, but not all of the Tenebrionoidea, others that have now been sunk as sub-families have here been appended to the parent family for the simple and non-taxonomic reason of having different authors involved. This might be regarded as somewhat eccentric by some, but from the point of view of using the book in a practical way it is hoped the reader will have no trouble 'navigating' the text. In addition, writing to this eclectic format was thought to be more logical as a 'half way house' to diverting the reader from the 'traditional' to the 'modern'. To those offering robust criticism I point out that a revised fifth edition of the *Handbook* will be required in about ten years, so why not volunteer your services?

The sequence to family level proposed by Lawrence and Newton (1995), with minor adjustments as discussed above, but excluding Hansen's (1997) classification of the Staphyliniformia, see later, is as follows (non-British taxa omitted):

Order COLEOPTERA
Suborder MYXOPHAGA Crowson, 1955
      SPHAERIUSIDAE Erichson, 1845
Suborder ADEPHAGA Schellenberg, 1806
      GYRINIDAE Latreille, 1810
      HALIPLIDAE Aubé, 1836
      NOTERIDAE C.G.Thomson, 1860
      HYGROBIIDAE Régimbart, 1878
      DYTISCIDAE Leach, 1815
      CARABIDAE Latreille, 1802
Suborder POLYPHAGA Emery, 1886
    Series STAPHYLINIFORMIA Lameere, 1900
      Superfamily HYDROPHILOIDEA Latreille, 1802
        HYDROPHILIDAE Latreille, 1802
      Superfamily HISTEROIDEA Gyllenhal, 1808
        SPHAERITIDAE Shuckard, 1839
        HISTERIDAE Gyllenhal, 1808
      Superfamily STAPHYLINOIDEA Latreille, 1802
        HYDRAENIDAE Mulsant, 1844
        PTILIIDAE Erichson, 1845 / Motschulsky, 1845
        LEIODIDAE Fleming, 1821
        SCYDMAENIDAE Leach, 1815
        SILPHIDAE Latreille, 1807
        STAPHYLINIDAE Latreille, 1802

Series SCARABAEIFORMIA Crowson, 1960
    Superfamily SCARABAEOIDEA Latreille, 1802
        LUCANIDAE Latreille, 1804
        TROGIDAE MacLeay, 1819
        BOLBOCERATIDAE Laporte de Castelnau, 1840
        GEOTRUPIDAE Latreille, 1802
        SCARABAEIDAE Latreille, 1802
Series ELATERIFORMIA Crowson, 1960
    Superfamily SCIRTOIDEA Fleming, 1821
        EUCINETIDAE Lacordaire, 1857
        CLAMBIDAE, Fischer, 1821
        SCIRTIDAE Fleming, 1821
    Superfamily DASCILLOIDEA Guérin-Méneville, 1843
        DASCILLIDAE Guérin-Méneville, 1843
    Superfamily BUPRESTOIDEA Leach, 1815
        BUPRESTIDAE Leach, 1815
    Superfamily BYRRHOIDEA Latreille, 1804
        BYRRHIDAE Latreille, 1804
        ELMIDAE Curtis, 1830
        DRYOPIDAE Billberg, 1820
        LIMNICHIDAE Erichson, 1845
        HETEROCERIDAE MacLeay, 1825
        PSEPHENIDAE Lacordaire, 1854
        PTILODACTYLIDAE Laporte, 1836
    Superfamily ELATEROIDEA Leach, 1815
        EUCNEMIDAE Eschscholtz, 1829
        THROSCIDAE Laporte, 1840
        ELATERIDAE Leach, 1815
        DRILIDAE Blanchard, 1845
        LYCIDAE Laporte, 1836
        LAMPYRIDAE Latreille, 1817
        CANTHARIDAE Imhoff, 1856
Series BOSTRICHIFORMIA LeConte, 1861
    Superfamily DERODONTOIDEA LeConte, 1861
        DERODONTIDAE LeConte, 1861

Superfamily BOSTRICHOIDEA Latreille, 1802
    DERMESTIDAE Latreille, 1804
    BOSTRICHIDAE Latreille, 1802
    ANOBIIDAE Fleming, 1821
Series CUCUJIFORMIA Lameere, 1938
    Superfamily LYMEXYLOIDEA Fleming, 1821
        LYMEXYLIDAE Fleming, 1821
    Superfamily CLEROIDEA Latreille, 1802
        PHLOIOPHILIDAE Kiesenwetter, 1863
        TROGOSSITIDAE Latreille, 1802
        CLERIDAE Latreille, 1802
        MELYRIDAE Leach, 1815
    Superfamily CUCUJOIDEA Latreille, 1802
        SPHINDIDAE Jacquelin du Val. 1860
        KATERETIDAE Erichson *in* Agassiz, 1846
        NITIDULIDAE Latreille, 1802
        MONOTOMIDAE Laporte, 1840
        SILVANIDAE Kirby, 1837
        CUCUJIDAE Latreille, 1802
        LAEMOPHLOEIDAE Ganglbauer, 1899
        PHALACRIDAE Leach, 1815
        CRYPTOPHAGIDAE Kirby, 1837
        EROTYLIDAE Latreille, 1802
        BYTURIDAE Jacquelin du Val, 1858
        BIPHYLLIDAE LeConte, 1861
        BOTHRIDERIDAE Erichson, 1845
        CERYLONIDAE Billberg, 1820
        ALEXIIDAE Imhoff, 1856
        ENDOMYCHIDAE Leach, 1815
        COCCINELLIDAE Latreille, 1807
        CORYLOPHIDAE LeConte, 1852
        LATRIDIIDAE Erichson, 1842
    Superfamily TENEBRIONOIDEA Latreille, 1802
        MYCETOPHAGIDAE Leach, 1815

CIIDAE Leach *in* Samouelle, 1819
TETRATOMIDAE Billberg, 1820
MELANDRYIDAE Leach, 1815
MORDELLIDAE Latreille, 1802
RHIPIPHORIDAE Gemminger & Harold, 1870
COLYDIIDAE Erichson, 1842
TENEBRIONIDAE Latreille, 1802
OEDEMERIDAE Latreille, 1810
MELOIDAE Gyllenhal, 1810
MYCTERIDAE Blanchard, 1845
PYTHIDAE Solier, 1834
PYROCHROIDAE Latreille, 1807
SALPINGIDAE Leach, 1815
ANTHICIDAE Latreille, 1819
ADERIDAE Winkler, 1927
SCRAPTIIDAE Mulsant, 1856 / Gistel, 1856

Superfamily CHRYSOMELOIDEA Latreille, 1802
CERAMBYCIDAE Latreille, 1802
MEGALOPODIDAE Latreille, 1802
ORSODACNIDAE C.G.Thomson, 1859
CHRYSOMELIDAE Latreille, 1802

Superfamily CURCULIONOIDEA Latreille, 1802
NEMONYCHIDAE Bedel, 1882
ANTHRIBIDAE Billberg, 1820
RHYNCHITIDAE Gistel, 1848
ATTELABIDAE Billberg, 1820
APIONIDAE, Schoenherr, 1823
NANOPHYIDAE Gistel, 1848
DRYOPHTHORIDAE Schoenherr, 1825
ERIRHINIDAE Schoenherr, 1825
RAYMONDIONYMIDAE Reitter, 1912
CURCULIONIDAE Latreille, 1802
PLATYPODIDAE Shuckard, 1840

Hansen's (1997) classification of the Staphyliniformia to family level, non-British taxa omitted, is as follows:

Order COLEOPTERA
    Series STAPHYLINIFORMIA Lameere, 1900
        Superfamily HYDROPHILOIDEA Latreille, 1802
            HELOPHORIDAE Leach, 1815
            GEORISSIDAE Laporte, 1840
            HYDROCHIDAE Thomson, 1859
            SPERCHEIDAE Erichson, 1837
            HYDROPHILIDAE Latreille, 1802
        Superfamily HISTEROIDEA Latreille, 1802
            SPHAERITIDAE Shuckard, 1839
            HISTERIDAE Gyllenhal, 1808
        Superfamily STAPHYLINOIDEA Latreille, 1802
            LEIODIDAE Fleming, 1821
            HYDRAENIDAE Mulsant, 1844
            PTILIIDAE Erichson, 1845 / Motschulsky, 1845
            SCYDMAENIDAE Leach, 1815
            SCAPHIDIIDAE Latreille, 1807
            STAPHYLINIDAE Latreille, 1802
            SILPHIDAE Latreille, 1807

EDITOR'S NOTE

The Palaearctic beetle fauna is currently being catalogued with one new volume expected to appear every 18 months. To date two volumes have appeared:

Löbl, I. and Smetana, A. (editors), 2003. *Catalogue of Palaearctic Coleoptera, 1 Archostemata, Myxophaga, Adephaga.* 819pp. Apollo Books, Stenstrup. ISBN 87-88757-73-0.

Löbl, I. and Smetana, A. (editors), 2003. *Catalogue of Palaearctic Coleoptera, 2 Hydrophiloidea, Histeroidea, Staphylinoidea.* 942pp. Apollo Books, Stenstrup. ISBN 87-88757-74-9.

While both volumes contain separate indices to families, sub-families, tribes and sub-tribes plus genera and sub-genera (volume 1) and family and tribe plus genera and sub-genera (volume 2) the works are enhanced by a free downloadable species index of 587 pages in MS Word format (volume 1) and of 704 pages in MS Word and PDF formats (volume 2).

References

Alonso-Zarazaga, M.A. & Lyal, C.H.C., 1999. *A world catalogue of families and genera of Curculionoidea (excepting Scolytidae and Platypodidae).* Entomopraxis, Barcelona.

Alonso-Zarazaga, M.A. & Lyal, C.H.C., 2000. Addenda and corrigenda to 'A world catalogue of families and genera of Curculionoidea (Insects: Coleoptera)'. *Zootaxa*, **63**: 1-37.

Hansen, M., 1997. Phylogeny and classification of the staphyliniform beetle families (Coleoptera). *Royal Danish Academy of Science and Letters, Biologiske Skrifter*, **48**: 339pp.

International Commission on Zoological Nomenclature, 1999. Opinion 1916: Brachypterinae Zwick, 1973 (Insecta, Plecoptera): spelling emended to Brachypterainae, so removing the homonymy with Brachypterinae Erichson, [1845] (Insecta, Coleoptera); Kateretidae Erichson in Agassiz, [1846]: given precedence over Brachypterinae Erichson. *Bulletin of Zoological Nomenclature*, **56**(1): 82-86.

Lawrence, J. F. & Newton, A. F. 1995. Families and subfamilies of Coleoptera (with selected genera, notes, references and data on family-group names). pp.779-1006. *in* Pakaluk, J. & Ślipiński, S. A. (eds.) *Biology, phylogeny, and classification of Coleoptera: Papers celebrating the 80th Birthday of Roy A. Crowson.* Muzeum i Instytut Zoologii PAN, Warszawa.

Lawrence, J. F., Hastings, A. M., Dallwitz, M. J., Paine, T. A. and Zurcher, E. J., 1999. *Beetles of the World: A Key and Information System for Families and Subfamilies.* CD-ROM, Version 1.0 for MS-Windows. Melbourne: CSIRO Publishing.

Morris, M.G., 2003. An annotated check list of British Curculionoidea (Col.,). *Entomologist's Monthly Magazine*, **193**: 193-225.

Scholtz, C.H. & D.J. Browne, 1996. Polyphyly in the Geotrupidae (Scarabaeoidea: Coleoptera): A case for a new family. *Journal of Natural History* **30**: 597-614.

SPHAERIUSIDAE*

by J.Cooter

At approximately 0.6mm, *Sphaerius acaroides* Waltl is one of our smallest beetles. Generally regarded as excessively scarce in Britain, it may be under recorded due to its small size (Hyman, 1994:95).

In the field *acaroides* can easily be mistaken for a plant mite and is the only occasion I can recall when I have had to use a hand lens in order to establish the creature before me in the field is actually a beetle. The specimens in my collection were found amongst thick wet moss at the edge of a drying out seasonal pool on slumped cliff on the Dorset coast. The beetle was revealed by carefully 'grubbing' through the moss to a depth of about 5-6cm, separating the individual haulms and examining minute black specks with the lens. It was not noted amongst moss growing around dried out pools; the same habitat produced *Thinobius brevipennis* Kiesenwetter (Staphylinidae: Oxytelinae). However, Garth Foster has found *Sphaerius* in France and Spain on bare mud close to, but not amongst vegetation. It is likely the small size of the beetle and secretive habits add to its apparent rarity.

* 'Sphaeriusidae' might be an unfamiliar name to many, the reason for its adoption follows – ICZN Opinion 1957. *Sphaerius* Waltl, 1838 (Insecta,

Coleoptera): conserved; and Sphaeriidae Erichson, 1845 (Coleoptera): spelling emended to Sphaeriusidae, so removing the homonymy with Sphaeriidae Deshayes, 1854 (1820) (Mollusca, Bivalvia) (ICZN, 2000).

References

ICZN, 2000. Opinion 1957. Bulletin of Zoological Nomenclature **57** (3) 182-184.
Hyman, P.S., 1994. *A review of the scarce and threatened Coleoptera of Great Britain (part 2)*. Joint Nature Conservation Committee, Peterborough.

WATER BEETLES
(Plates 1 - 3, 32, 89)

by G.N. Foster

About 360 species in Britain may be conveniently classed as 'water beetles'. These divide into the Myxophaga, Adephaga (or Hydradephaga), five families of truly aquatic, swimming beetles, and a roughly equal number of aquatic species from several families of the Polyphaga. Within the Polyphaga the Hydrophiloidea once included the Hydraenidae, which have now been transferred to the Staphylinoidea, but these two families can still be collectively referred to as 'palpicorns'. A more natural combination occurs in the Dryopoidea, a superfamily embracing the Elmidae, Dryopidae and three very small families, each with only one genus in Britain.

Some species of ground beetle (Carabidae), such as *Oodes helopioides* (F.) and *Odacantha melanura* (L.), and a range of other beetles, particularly rove beetles (Staphylinidae), are exclusively associated with water but this account concerns only the families dominated by aquatic members.

Water beetles have more than their fair share of identification problems. This partly owes to their adaptation to life under water, the streamlined body form not lending itself to accurate description. Another reason is that coleopterists regularly divide into those who are terrestrial and those who are not. Water beetling has become a specialist activity and those who only dabble experience difficulty in familiarising themselves with the field characters of the commoner species. In many genera identification can be made certain only by dissection of the male genitalia (pp. 353-361) and these are signified in the following account. Anyone interested in British water beetles should take advantage of the intense interest in water beetles in Europe as a whole. This has resulted in many identification manuals, of which those not written in English still have invaluable illustrations. Tachet *et al.* (2000) provide a key in pictorial form to the French genera and this is applicable to much of the British fauna. Coleopterists would be well advised to master some elementary German to take advantage of the rich Central European literature, in particular the publications in the series, *Die Käfer Mitteleuropas* (Freude, Harde & Lohse 1971, and its revisions) and *Süßwasserfauna von Mitteleuropa* (Angus 1992, which is in English; Hebauer & Klausnitzer 1998; van Vondel 1997). British species are a blend of species from northern, central and south-western Europe so one must be aware of the potential for omissions from some of these treatments. In particular, keys intended for use in Scandinavia in the *Fauna Entomologica Scandinavica* series will nevertheless prove invaluable (Hansen 1987; Holmen 1987; Nilsson & Holmen 1995). Balfour-Browne's works (1940, 1950, 1958) still provide good guidance on most species, largely because the discursive treatment of each species permits a check on the immediate findings of a key. Friday's treatment (1988), in lacking this discursive element, can lead one astray because the simple keys favoured in the AIDGAP series do not take into account the difficulties associated with identifying beetles, particularly those of a streamlined body form. Nevertheless Laurie Friday's work is invaluable in covering most aquatic families and in using a relatively modern checklist. Would-be water beetlers are also well advised to get used to change. 'Globalisation' was invented by Linnaeus, in creating an internationally agreed set of names. The price to pay for this common language is a sustained rain of changes as our knowledge of the status of type specimens changes, and now as we can take advantage of DNA studies to help in understanding interrelationships. Thus the most recent change has been to transfer several *Agabus* species to *Ilybius*, with more changes of this kind expected. No doubt even more confusion would reign if we adopted common names for water beetles. Very few exist, and those that have been artificially created simply do not stick.

The whirligig beetles (GYRINIDAE) are of characteristic form and habit. They are carnivores, feeding on items falling on the water surface. The larvae are also predatory and are characterised by breathing through filamentous gills rather than by rising to the water's surface to renew their oxygen supply, a characteristic of the larger Dytiscidae. This feature presumably explains why whirligigs are largely confined to deep water bodies. The 'common' species live in flotillas in full view and much time can be wasted in attempting to capture whirligigs staying just beyond the reach of the net. In fact the rarer species, if they are really rarer, live among dense emergent vegetation. The hairy whirligig, *Orectochilus villosus* (Müller), lives in

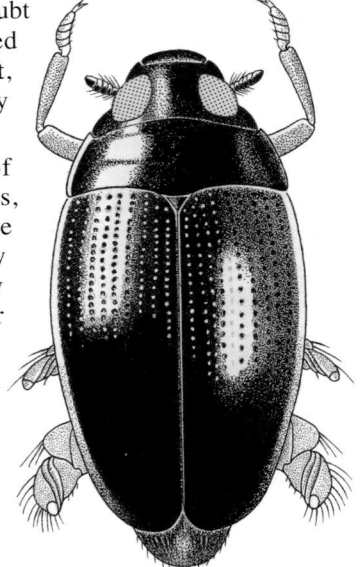

Fig1. *Gyrinus natator* (L.)

running water and in northern lakes; it gyrates on the water's surface at night and spends the day hiding at the water's edge, often clinging under rocks or bridge culverts. Most of the *Gyrinus* are found on still water. Individual species can be difficult to identify with several species-pairs – *aeratus* Stephens and *marinus* Gyllenhal, *paykulli* Ochs and *caspius* Ménétriés, *substriatus* Stephens and *suffriani* Scriba – necessitating examination of the male genitalia. A species that appears to have become extinct in Britain, though it is still present in Ireland, is *natator* (L.) and it can be most easily be distinguished from the commonest British species, *substriatus* by examination of, most unusually, the *female* genitalia.

Beetles of the family HALIPLIDAE are all small, between 3 and 5mm long, with a yellowish ground colour and large flat hind coxal plates. In America they are known as 'crawling water beetles', a reference to their method of swimming with the legs operating alternately, but they have not attracted a common name in Britain. As with most aquatic adult beetles, they retain an air pocket beneath the elytra, but they extend this bubble to the coxal plates so that the hind legs, in making their swimming movement, are constantly agitating and refreshing the water/air interface. Most haliplids are associated with slow-running water or with lakes though at least one species, *Haliplus heydeni* Wehncke, is mainly found in dense vegetation in small water bodies. The larvae feed on algae and duckweeds. At least two species, *H. confinis* Stephens and *H. obliquus* (F.), are dependent upon stoneworts found only in hard water. A common mistake is to misconstrue the dense punctuation of these two species for the finer punctuation found on the otherwise shiny surface of females of the *ruficollis* group (also known as subgenus *Haliplinus*). The adults are less restricted in their diet and are partly carnivorous, *H. lineolatus* Mannerheim, for example, feeding on Hydrozoa, the freshwater relatives of jellyfish. Haliplids are largely confined to lowland habitats and one species, *H. apicalis* Thomson, is rarely found away from brackish water. Members of the *Haliplus ruficollis* group are best identified by examination of the male genitalia, and the male fore and mid tarsi.

The NOTERIDAE have distinctive adults and larvae, another unusual feature being that pupation takes place in cocoons attached to the roots of aquatic plants. The two species of *Noterus* are found in stagnant water, usually in association with dense vegetation including floating rafts of species such as bogbean (*Menyanthes trifoliata*). One species, *N. clavicornis* (De Geer), continues to be known as 'The Larger Noterus' to get over a major naming problem of the past.

HYGROBIIDAE are represented by a single species, the 'squeak' or 'screech' beetle (*Hygrobia hermanni* (F.)), a distinctively convex, large (9mm), yellow and black insect. This species stridulates loudly when disturbed. Both the larva and adult are predatory, grubbing around in soft mud in pools and ditches.

The diving beetles, DYTISCIDAE, are the most attractive species of water beetle to the young coleopterist. Both adults and larvae are carnivorous, though it is certain that many adults are vultures rather than birds of prey. They divide into five subfamilies. The Laccophilinae, with three species of *Laccophilus*, are easily recognised in the field by their jumping ability. The larvae, and to some extent the adults, have a greenish coloration that disappears with death. *Copelatus haemorrhoidalis* (F.) is our sole representative of the largely tropical Copelatinae. The other three subfamilies, Hydroporinae, Colymbetinae and Dytiscinae, roughly conform to small-, medium- and large-sized diving beetles. They tend to be coloured black, yellow and reddish brown, with black stripes on a yellow background being a feature of species associated with sand or gravel. The small diving beetles include the notorious genus *Hydroporus*, with 28 British species found mainly in shallow bog and fen habitats. This genus poses the greatest identification difficulty to those without access to a wide range of material; dissection of the male genitalia is desirable and is best achieved by extrusion from the freshly killed insect, all the structures being depicted in Friday (1988). Many of the other members of the tribe are found over gravel, sand or mud in running water or in ponds. *Hygrotus* species in the subgenus *Coelambus*, with the exception of the northern lake species *novemlineatus* (Stephens), are pioneers, being the first colonists of new ponds. The medium-sized diving beetles first demand attention from would-be coleopterists. The commonest species in Britain is *Agabus bipustulatus* (L.), remarkable for its ability to colonise all forms of aquatic habitat. If anything it most closely resembles in form *Ilybius* species, in which careful attention to size and the extent of metallic reflection is essential to avoid identification errors. Other bronze species, *chalconatus* (Panzer) and *montanus* (Stephens), have recently been transferred to *Ilybius* and can only be distinguished with safety by examination of the tips of the parameres, an exercise not necessarily involving full dissection. Most species are found in the edges of pools or among vegetation at the edges of rivers. *A. guttatus* (Paykull) is a common species of small streams, *biguttatus* (Olivier) being associated with subterranean water including intermittent chalk streams. *A. brunneus* (F.) is a rare species of gravel beds in temporary streams in the New Forest and Cornwall.

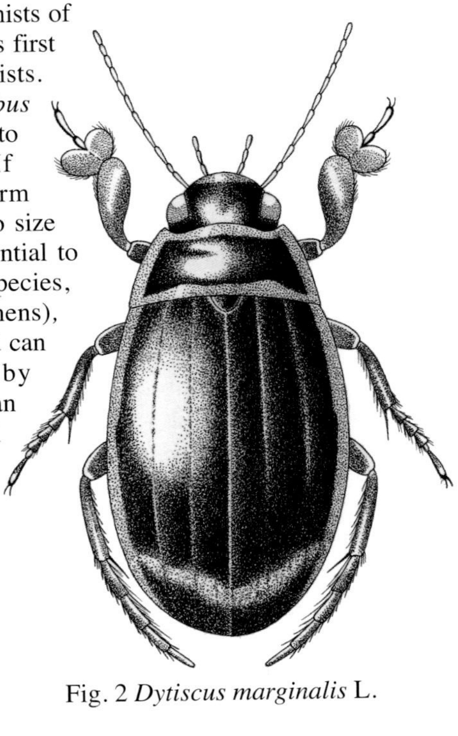

Fig. 2 *Dytiscus marginalis* L.

The subterranean habitat, by virtue of its mystery, has a special appeal to coleopterists. Most species might better be described as 'interstitial', moving down amongst gravel and sand as the water level falls. Only one species, *Hydroporus ferrugineus* Stephens, appears to be fully subterranean throughout its range whereas others, such as *H. obsoletus* Aubé, occur away from springs south of Britain. Nevertheless all such species are best located in Britain by working the net in backwaters or in the outflows of springs after heavy rain; some occur in flood refuse.

The Dytiscinae, the larger diving beetles, are characterised by a strong difference between the sexes, the males having the fore tarsi modified into circular sucker pads and the females often having 'sulcate' or 'fluted' elytra, being provided with hairy grooves. Contrary to general belief a recent analysis of variation indicates that the flutes, far from assisting in copulation, are there to resist it, and the suction cups are the male's answer to this resistance! The voracious larvae of this tribe are largely of the free-swimming type (except *Hydaticus,* whose larvae feed on the bottom like most Colymbetinae). They are the most obvious larvae to be found in the net and can easily be reared through to adults by supplying live food and a pupation site, such as a piece of bark pressed to the side of the aquarium.

There is no shame in experiencing difficulty in identifying the larger water beetles, errors being all too common with the great diving beetles (*Dytiscus*). The rarest species belong to the genus *Graphoderus* and it was only in 1974 that it was recognised that three species, not one, were represented in British collections. The male mid-tarsi provide important characters in this genus and should on no account be glued.

Abroad it is common practice to operate underwater traps to capture the larger dytiscines. The basic principle is that of the minnow trap, with an inverted funnel or funnels leading into a larger chamber baited with rotting meat. Low water temperatures in Britain demand that these traps are operated for weeks rather than days to have any success, one reason for their lack of popularity. If provided with floats, such as blocks of polystyrene foam, they attract undesirable attention but this is the only way to keep specimens alive. This is not only necessary for conservation of unwanted individuals but also because drowned specimens rapidly disintegrate under water. The paramount rule of trapping is never to lose a trap: it is quite conceivable that a trap left too long in a pond with little other food will trap out the whole adult populations.

The aquatic HYDROPHILOIDEA includes four small families. These are the SPERCHEIDAE, GEORISSIDAE, HYDROCHIDAE and HELOPHORIDAE, each represented in Britain by one genus. The remaining HYDROPHILIDAE (in the strict sense) then further divide into the entirely aquatic subfamily Hydrophilinae and the partly terrestrial Sphaeridiinae. Most hydrophilines are able to trap a large bubble of air on their undersides and then perform what to us seems the strange feat of being able to walk, upside down, immediately below the water's surface. They can also float the right way up and in this position they use their clubbed antennae to penetrate the water's surface to renew their air supply, in distinct contrast to the Adephaga, all of which use the bubble trapped under the elytra, the air supply of which is renewed by protruding the tip of the abdomen through to the atmosphere.

Some Hydrophiloidea scattered through several genera have swimming hairs on the mid and hind tibiae and/or tarsi and these species can dive below the water's surface, e.g. the *Berosus* species, performing as agilely as the diving beetles. 'Floaters' would be a useful collective name for the group were it not for such species and it would emphasise the ease with which most hydrophilids can be caught by disturbing the mud or vegetation at the extreme edge of the water and then skimming off the beetles.

Hydrophiloidea are most common in shallow, undisturbed water in pools and ditches, with *Helophorus* species characteristic of impermanent grassy pools and, at the other extreme, *Enochrus* characteristic of permanent water with mosses. All spin silken cases for their eggs, often with ribbon-like masts. *Helochares* species carry their egg cases on the underside of the body. Hydrophiloid larvae are predatory but most adults appear to be either plant-feeders or scavengers.

Although the Sphaeridiinae cannot be classed as water beetles, because *Sphaeridium* and most *Cercyon* are associated with dung, it would be misleading to exclude this tribe from consideration. *Coelostoma orbiculare* (F.) and several of the more local *Cercyon* are exclusively aquatic. Some *Cercyon* are associated with specific habitats such as rotting seaweed (*C. littoralis* (Gyllenhal), *C. depressus* Stephens) or are perhaps nidicolous (*C. laminatus* Sharp). Those with microreticulate elytra (*C. convexiusculus* Stephens, *C. granarius* Erichson, *C. sternalis* Sharp, and *C. tristis* (Illiger)) are found in wet debris. At the other extreme, a few species of *Helophorus* (Helophoridae) are associated with dry ground. A particularly interesting species is *H. tuberculatus* Gyllenhal, associated with burnt heather, the fragments of which it resembles. When identifying hydrophilids one must constantly be on guard for discrepancies between the published descriptions and the beetles under examination. Balfour-Browne's review (1958) is greatly out-of-date in its recognition of species status. If one realises that it is even possible to misidentify the great silver water beetles, represented in Britain solely by *Hydrophilus piceus* (L.), the importance of maintaining a critical stance is appreciated. Dissection of the male genitalia is highly desirable for *Hydrochus, Helophorus, Sphaeridium, Cercyon, Laccobius, Helochares* and the smaller *Enochrus*. It is also important to mount *Anacaena* species upside down or sideways so as to permit viewing of the pubescence on the hind femora, quite free of adhesive.

The family SPERCHEIDAE, represented by *Spercheus emarginatus* (Schaller), is probably extinct in Britain. The GEORISSIDAE are represented by *Georissus crenulatus* (Rossi), a local species of drying mud and silt by river banks and coastal exposures. Detection of this species is not helped by it habit of wearing a cape of mud! Similar habitats are occupied by *Sphaerius acaroides* Waltl, our sole representative of the suborder Myxophaga. *Sphaerius* is so small that it is more often detected by those using a microscope to search through peat and debris for 'subfossils' than it is alive (see above under Sphaeriusidae, by J. Cooter).

The British checklist of HYDRAENIDAE has proved much more resilient in terms of name changes and additions than the Hydrophiloidea, with only one species, *Limnebius crinifer* Rey, added recently. Hydraenids are all small, between 1 and 3mm long, the largest being the male of *L. truncatellus* (Thunberg). They are frequently overlooked because of their small size and sluggishness. *Limnebius* and *Ochthebius* are associated particularly with mud, most of the latter genus living in brackish water. *O. auriculatus* Rey is most often encountered in the drier parts of saltmarshes by sieving grass roots. *O. poweri* Rye occurs on wet sandstone cliff-faces by the sea and *O. lejolisii* Mulsant & Rey can be abundant in rockpools just above the high tide mark. *O. exsculptus* (Germar) lives in fast-running water, a habitat that it shares with several *Hydraena* species. Most of the more interesting *Hydraena* occur in the *edges* of streams, rather than, as is generally perceived, among stones in midstream. Thus 'kick samples', such as are employed in biological surveys by water authorities, when the net is held downstream of the collector's shuffling boots, rarely catch anything other than *H. gracilis* Germar and the elmids. It is better to search for clean backwaters and then to wash down the silt and wait for the hydraenids to float to the surface. It is often claimed that bags of debris and moss will yield good species if taken home and submerged completely so that the beetles are forced to the surface; heat extraction of the dried material is more effective and even then it is rare to find species that could not have been detected in the field.

Assuming that the hydraenid list remains unchanged it is feasible to identify all species without examination of the genitalia, but this is probably *why* the list has stayed the same for so long! Apart from the realistic possibility of new species, the extraordinary structure of *Limnebius* and *Hydraena* male genitalia must be seen to be believed, if not understood.

The ELMIDAE have been named 'riffle beetles' as most are found amongst gravel and stones in the swifter parts of rivers and streams, or on the wave-washed shores of lakes. Two species that had been overlooked, *Oulimnius rivularis* (Rosenhauer), once thought extinct, and *O. major* (Rey), relatively new to the British list, are most definitely not species of riffles, being mainly associated with slower stretches of rivers and with stone-lined fenland drains. *Stenelmis canaliculata* (Gyllenhal) occurs in deep water in rivers and

Macronychus quadrituberculatus Müller appears to be most easily found on submerged logs. The two British species of *Riolus* are found in highly calcareous water and other elmids are commonest in harder waters.

Elmid adults are partly covered with a hydrofugic (water-repellent) pile that functions as a plastron, enabling the beetle to take oxygen from the water without the need to rise to the surface. They will nevertheless float to the surface if disturbed because they are buoyant. As with the Dryopidae, care should be taken when mounting elmids to ensure that they are thoroughly dry before using glue, otherwise it can spread over and disfigure the entire insect.

DRYOPIDAE: The wireworm-like larvae of *Dryops* live in wet soil or sodden wood but the adults are so hydrofugic that some coleopterists will have difficulty in accepting that they are water beetles at all. The commonest British species, *D. luridus* (Erichson), occurs on wet mud by open water whereas the next commonest, *D. ernesti* des Gozis, may often be found in muddy fields well away from truly aquatic habitats. The somewhat stylised illustrations of the male genitalia provided by Olmi (1979) are needed to ensure correct identification of the rarer species associated with fenland. Only *Pomatinus substriatus* (Müller) lives below water, usually in deep streams.

HETEROCERIDAE: *Heterocerus* species are similar to *Dryops* in general appearance but with large mandibles, often with bright elytral markings and with a tendency to escape by flight. The larvae feed under algal mats on mud and most species are restricted to saltmarshes. For their identification see Clarke (1973).

LIMNICHIDAE: Represented in Britain only by *Limnichus pygmaeus* (Sturm), are not aquatic but, living on rocks in splash zones by running water, are most likely to be encountered by water beetlers. Externally very similar to a pill beetle (Byrrhidae) and as with the byrrhids, *Limnichus* draws its appendages and head tightly in against the ventral surface; this, plus its small size (ca 1.75mm) and ovoid form, make it one of the most frustrating beetles to mount with all appendages displayed.

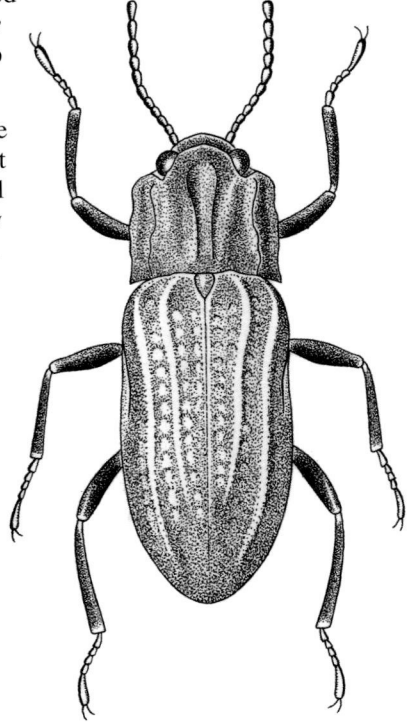

Fig.3 *Stenelmis canaliculata* (Gyllenhal)

PSEPHENIDAE: Our only psephenid, *Eubria palustris* Germar, lives in the same habitat as *Limnichus* but is present so fleetingly as an adult that it is much more likely to be detected as its larva, known as a 'water-penny', with the distinctive form of a flattened woodlouse.

SCIRTIDAE: The Scirtidae (previously known as Helodidae) also have aquatic larvae and soft-bodied adults. To ensure perfect specimens in the cabinet, they are best kept individually in tubes, killed quickly and mounted at once, after dissecting the genitalia, using very dilute glue. Despite the fact that *Elodes* larvae are amongst the commonest water beetles, they are rarely found as adults and fully identified. *Prionocyphon serricornis* (Müller) is found mainly as larvae in rot-holes, particularly in beech. Its larvae are an important feature of this remarkable ecosystem, being the main processors of leaf debris. *Hydrocyphon deflexicollis* (Müller) is associated with running water and its adult can be found on river shingle. *Scirtes* species, found on stream- or lake-side vegetation, have enlarged hind femora and are capable of jumping and might be unwittingly mistaken for flea beetle chrysomelids. *Cyphon* adults are often swept, sometimes in the pond net. Adults of *Elodes, Microcara* and *Cyphon* can also be beaten from hawthorn blossom. The male genitalia of *Cyphon* are flimsy but diagnostic if properly mounted; the most extraordinary feature of this genus is a structure known graphically as 'the crusher' found far up the female genital tract and utterly diagnostic. For *Cyphon*, one must consult Kevan (1962) and Skidmore (1985), or Nyholm (1972). Lohse (1979) provides a well-illustrated key to the whole family, a recent welcome change being replacement of the tongue-twister *Cyphon phragmiteticola* Nyholm with *C. laevipennis* Tournier.

Several leaf beetles (Chrysomelidae) are associated with aquatic plants but only one tribe, the Donaciinae, is highly adapted to the aquatic life, with larvae that breathe air from spaces in the roots of aquatic plants, pupae being attached in the same way. The highly attractive adults can be elusive and prove difficult to identify, particularly the commoner *Plateumaris*. Adults rest on floating and emergent leaves of aquatic plants and appear to feed mainly on pollen. Two species of *Macroplea* occur in Britain, living underwater both as adults and larvae, one species on milfoil (*Myriophyllum*) and *Potamogeton pectinatus* in lakes and canals, the other on eelgrass (*Zostera*) in coastal pools; despite this substantial difference in ecology it is advisable to dissect males. The most recent treatments are by Menzies and Cox (1996) and Bratton and Greenwood (1997).

Many weevils (Curculionidae) feed on aquatic plants and some have plastrons to facilitate respiration whilst crawling on plants below the surface. *Bagous* are perhaps the most elusive of all water beetles, being mainly associated with plants in very slow moving water. Extreme patience is required to detect them as they take so long to move when vegetation is dumped on a plastic sheet to dry. Several species are associated with milfoil in lakes. Many weevils are nocturnal but it is difficult to see how this could be turned to advantage in speeding up their capture. For plant associations see pp. 270-310.

References

Angus, R.B., 1992. *Insecta: Coleoptera: Hydrophilidae: Helophorinae*. Stuttgart, Gustav Fischer. (Süßwasserfauna von Mitteleuropa, No. **20** (10-2)).

Balfour-Browne, F., 1940, 1950, 1958. *British water beetles*. Volumes **1-3**. London, Ray Society.

Bratton, J, & Greenwood, M., 1997. An identification aid to British reed beetles, *Plateumaris* and *Donacia* (Chrysomelidae), using pronotum surface characters. *Latissimus*, **8**: 11-15.

Clarke, R.O.S., 1973. Coleoptera, Heteroceridae. *Handbooks for the Identification of British Insects*, **5**(2c): 14pp. Royal Entomological. Society. London.

Freude, H., Harde, K.W., & Lohse, G.A., 1971. *Die Käfer mitteleuropas*, **3**. Krefeld, Goecke & Evers.

Friday, L.E., 1988. A key to the adults of British water beetles. *Field Studies* **7**: 1-151. Published separately as AIDGAP Book No. **189**, Taunton, Field Studies Council.

Hansen, M., 1987. *The Hydrophiloidea (Coleoptera) of Fennoscandia and Denmark*. Leiden, E.J. Brill. (Fauna Entomologica Scandinavica, No. **18**.)

Hebauer, F., & Klausnitzer, B., 1998. *Insecta: Coleoptera: Hydrophiloidea (exkl*. Helophorus*)*. Stuttgart, Gustav Fischer. (Süßwasserfauna von Mitteleuropa, No. **20** (7, 8, 9, 10-1) part 2)).

Holmen, M., 1987. *The aquatic Adephaga (Coleoptera) of Fennoscandia and Denmark*. I. *Gyrinidae, Haliplidae, Hygrobiidae and Noteridae*. Leiden & Copenhagen, E.J. Brill/Scandinavian Science Press. (Fauna Entomologica Scandinavica, No. **20**.)

Kevan, D.K.. 1962. The British Species of the Genus *Cyphon* Paykull (Col., Helodidae), Including Three New to the British List. *Entomologist's Monthly Magazine*, **97**: 114-121.

Klausnitzer, B., 1971a. Zur Kenntnis der Gattung *Helodes* Lart., *Ent Nachr*. **14**: l77pp

Klausnitzer, B., 1971b. Beiträge zur Insektfauna der DDR: Col., Helodidae. *Beiträge Entomologie*, 21: 477pp.

Menzies, I.S., & Cox, M.L. 1996. Notes on the natural history, distribution and identification of British reed beetles. *British Journal of Entomology and Natural History*, **9**: 137-162.

Nilsson, A.N. (*ed*.) 1996. *Aquatic insects of north Europe - a taxonomic handbook*. Volume **1**: Ephemeroptera, Plecoptera, Heteroptera, Megaloptera, Neuroptera, Coleoptera, Trichoptera and Lepidoptera. 173-194. Apollo Books, Stenstrup.

Nilsson, A.N., & Holmen, M., 1995. *The aquatic Adephaga (Coleoptera) of Fennoscandia and Denmark*. II. *Dytiscidae*, Leiden, E.J. Brill. (Fauna Entomologica Scandinavica, No. **32**.).

Nyholm, T., 1972. Die nordeuropaischen Arten der Gattung *Cyphon* Paykull (Coleoptera). *Entomologica Scandinavica*, Suppl. 3: pp.100.

Olmi, M., 1976. Coleoptera Dryopidae-Elminthidae. *Fauna d'Italia* **12**. Calderini Bologna.

Skidmore, P., 1985. *Cyphon kongsbergensis* Munster (Col., Scirtidae) in Scotland. *Entomologist's Monthly Magazine*, **121**: 249-252.

Tachet, H., Richoux, P., Bournaud, M. & Usseglio-Polatera, P., 2000. *Invertebrés d'eau douce. Systematique, biologie, écologie*. CNRS Editions. [Available from CNRS Editions, La Librairie, 151, bis rue Saint-Jacques, 75005 Paris; e-mail libraire@cnrseditions.fr].

Vondel, B.J. van., 1997. *Insecta: Coleoptera: Haliplidae*. Gustav Fischer, Stuttgart. (Süßwasserfauna von Mitteleuropa, No. **20** (2)).

CARABIDAE
(Plates 4 – 31)

by M.L. Luff

The Carabidae or 'Ground Beetles', with about 350 British species, are one of the most popular beetle families with collectors. They are readily recognisable as a family, and many species are common and easily identifiable while some, especially of *Carabus* itself, are large, conspicuously coloured and sculptured. As now recognised, the family includes the Cicindelinae or Tiger Beetles (five British species), which are treated as a separate family in earlier works. Most ground beetles are in the large subfamilies Carabinae (four tribes), Trechinae (five tribes) and Harpalinae (17 tribes). There are six other small subfamilies: the Elaphrinae, Scaritinae and Broscinae each have a single tribe and few species; the Bombardier Beetles or Brachininae, Omophroninae and Loricerinae each have only one established species.

The Cicindelinae are diurnal predators, active in bright, sunny conditions in dry, open situations where they run and (except for *Cicindela germanica* L.) fly readily. They are commonest in spring and early summer, spending the rest of the year as larvae in vertical burrows in the soil. The adults may be caught by hand or net, or even by pitfall trapping (q.v.). Our single Bombardier beetle *Brachinus creptians* (Brachininae) occurs locally in chalky and other dry districts of southern Britain under stones, often in aggregated groups. Little is known of its biology, but the larvae are probably ectoparasitic on pupae of other beetles, including Hydrophilidae and Staphylinidae.

Fig. 4 *Cicindela hybrida* L.

Our only Omophronine, *Omophron limbatum* (F.), is a semi-aquatic species which burrows in sand at the margins of flooded sand-pits, etc. in the extreme south of England. The single Loricerine, *Loricera pilicornis* (F.), recognisable by the long setae on its antennae, is a widespread species in damp habitats.

The Carabinae include some of the largest species such as the violet ground beetles, *Carabus violaceus* L. and *C. problematicus* Herbst, as well as the very common *Nebria brevicollis* (F.) and the diurnal *Notiophilus* species. The Elaphrinae are also diurnal species, with large eyes and strongly punctured elytra, found near water. The Broscinae and Scaritinae are both modified for burrowing in the soil, with enlarged front legs, and an exceptionally mobile articulation between the pro- and meso-thorax. The Trechinae are mostly small, predatory species, including the largest British genus *Bembidion*, some of which commonly run in bright sunshine. The majority of the remaining ground beetles, in the Harpalinae, are ground living predators or scavengers, many being mainly nocturnal but some *Amara* species are diurnal, and some Dromiini are arboreal, foraging on tree trunks and branches.

Nocturnal species are best sought in their shelters, such as under stones, logs, etc. Isolated stones tend to be more productive than heaps of boulders, but even small pebbles can sometimes cover a specimen hiding beneath. By rivers or lakes many species, especially of the large genus *Bembidion*, occur in shingle, or at the roots of waterside vegetation on sandy, sloping margins. They run rapidly, so the collector must be ready with tube or pooter as the area is searched. In marshy habitats, sieving or searching among litter on the soil surface will yield species of *Agonum*, *Acupalpus* and *Stenolophus*, some of the smaller *Pterostichus*, further *Bembidion* or even *Chlaenius* species, *Oodes helopioides* (F.) or *Badister unipustulatus* Bonelli. Searching in gardens or rough grass will produce the commoner large species of *Carabus, Nebria, Pterostichus, Calathus* and some *Harpalus*. Drier habitats, especially if exposed to the sun, may yield the better *Amara*, *Harpalus* and *Ophonus* species as well as *Notiophilus, Badister bullatus* (Schrank) and possibly *Licinus* species. In woodland one may find *Abax parallelepipedus* (Piller & Mitterpacher), *Leistus* species, *Platynus assimilis* (Paykull), *Calathus rotundicollis* Dejean (=*C. piceus* sensu (Marsham) and the characteristic snail-feeder, *Cychrus caraboides* (L.), which can stridulate loudly if handled. Many species are confined to upland habitats and stone turning in moorland can be very rewarding, as will searching at the roots of heather, which will yield *Trichocellus, Bradycellus, Metabletus* and *Notiophilus* species. Some widespread upland species, such as *Nebria rufescens* (Ström), also occur at lower altitudes, but in more restricted habitats, in this case on stony river banks.

Coastal habitats are particularly rich in Carabidae. Some such as *Dicheirotrichus*, *Pogonus* and *Dyschirius* species are exclusive inhabitants of salt marshes; the latter genus burrows in sandy or clayey banks where they are predatory, usually on the staphylinid genus *Bledius*. Other carabids, such as *Broscus cephalotes* (L.) and *Nebria complanata* (L.) are found under litter at the top of sandy beaches, while several *Bembidion* occur in salt marsh litter. Sand dunes support *Calathus mollis* (Marsham) as well as local *Amara* and *Harpalus* species. The two small species of *Aepus* actually live between the tides, either under stones on sand, or in sand-filled crevices in rocks which are covered by the tides. They can be collected, together with the staphylinid *Micralymma marinum* (Ström), by splitting open such crevices with a hammer and cold chisel. In addition to these specialist coastal species, many carabids which are widely distributed in the south become partly or wholly coastal in their distribution as one moves north in Britain.

Although stone-turning or hand-searching will discover all these beetles in their resting places, they can be caught during their normal activity by pitfall trapping. The simplest traps suitable for carabids consist of plastic beakers or cups set into the soil with their rim flush (or slightly below) the soil surface. If visited frequently the traps can be left dry, but with some risk of the smaller specimens being eaten by the larger ones. A bait can be used, but is not essential. For longer periods of unattended trapping, the most readily available preservatives are saturated salt solution, or commercial motorcar antifreeze (preferably the blue variety, not the red/orange). Salt solution is recommended on grounds of cheapness and non-toxicity, but is less effective as a preservative. About 1cm depth in the bottom of the trap will maintain specimens in identifiable condition for up to two weeks (salt solution) or longer (antifreeze), even if the trap is subsequently filled with rain water.

Most ground beetles are annual, breeding either in spring and early summer (larvae in summer, overwintering as adults) or autumn (larval overwintering). A few large, mainly autumn-breeding, species of *Carabus*, *Pterostichus* and *Harpalus* live for more than one breeding season, so that both larvae and adults overwinter. Overwintering adult carabids may be found, sometimes in large numbers, in habitats such as grass tussocks, which can be dug up and shaken out over a sheet. As well as common *Pterostichus*, *Trechus*, *Notiophilus*, *Amara* and *Bradycellus* one may come across *Lebia chlorocephala* (Hoffmannsegg), whose larva is parasitic on Chrysomelidae. In marshy areas, large numbers of *Carabus*, *Pterostichus* and *Agonum* can be collected in winter sheltering above water level under bark of old logs and tree stumps as well as in moss on trees and walls. Flood refuse is also very productive at this time.

A few British carabids are found more commonly on plants than on the ground. These include *Curtonotus aulicus* (Panzer), which can be swept from rough herbage where it feeds on seeds of Compositae, and the rare *Zabrus tenebrioides* (Goeze) which feeds on cereals and is actually a crop pest in central Europe. Also on plants one finds *Demetrias* and the smaller *Dromius* species, which feed on other insects on the foliage. Arboreal *Dromius* are found under bark of trees in the day, and on the foliage at night, when they can be collected by beating. Also above ground, but mainly on walls, occurs *Ocys quinquestriatus* (Gyllenhal), as well as rare or local species of Tachyini and *Trechus* in crumbling sand and brickwork. In old cellars or outbuildings one may be lucky enough to find the very large and nowadays rare *Sphodrus leucophthalmus* (L.), which is believed to prey on the larvae of the cellar beetle *Blaps* (Tenebrionidae). Also associated with buildings are *Laemostenus* species, one of which (*L. terricola* (Herbst)) also occurs in and near small mammal burrows and runs.

Larvae of most Carabidae occur in the soil or other substrate where the adults are also found, although some species (some *Amara*, *Chlaenius*) migrate by flight between breeding and overwintering sites. The only larvae commonly found active on the surface are those of *Nebria, Notiophilus*, some *Carabus* and *Pterostichus* (particularly those of *P. madidus* (F.)). The larval period of spring-breeding species is quite short, but autumn breeders may spend almost a year as larvae. The pupal stage is short lived in all species. Most larvae are predatory, but those of some *Harpalus* and possibly *Amara* are known to be seed feeders, while *Zabrus* larvae eat leaves and shoots of grasses and cereals.

Following killing by ethyl acetate vapour, most adult Carabidae are robust enough to be relaxed by floating in water if this is needed. Although some collectors prefer to pin the larger species (especially *Carabus*), the slender antennae and legs of such specimens are easily broken: there is in fact no reason why all British Carabidae should not be card mounted. Underside characters (which are seldom needed) can usually be seen if the specimen is viewed from the side, due to the convex shape of the underside of most species. The body should be well glued down, especially if larger specimens are to be sent by post. Antennae and legs can also be well glued, but only on the side nearest to the card; it is difficult or impossible to see essential pubescence or setae on the segments if the appendages are completely covered in glue. The maxillary and labial palpi should if possible be displayed when mounting the specimen. Most species can be sexed by the basal segments of the front tarsi, which are expanded in the male. Some *Pterostichus* have a tooth or ridge on the apical ventral segments of the abdomen, also in the male. The aedeagus (including its internal structures) is a useful confirmatory feature in many species, especially of *Bembidion* and *Ophonus*, but is seldom an essential feature for ground beetle identification. It can readily be extracted from freshly-killed and relaxed specimens by pressure on the abdomen to extend the pygidium, together with prising open the apical abdominal segments, which can be pushed back under the elytra when the specimen is mounted. The most useful orientation of the aedeagus varies according to the species, but it is best mounted dorso-laterally, glued onto the card in a drop of DMHF. The parameres are markedly asymmetrical and should be pulled away from the median lobe so as not to obscure it. The female spermatheca and external genitalia may sometimes show useful identification features, although these are seldom used in the British species as yet.

The most recent check list of the British ground beetles is by Luff & Duff., at www.coleopterist.org.uk. The distribution of the British species was mapped by Luff (1998). The ground beetle mapping scheme is still running, and records should be sent to Dr Mark Telfer care of the RSPB, The Lodge, Sandy, Bedfordshire, SG19 2DL.

For identification, a magnification of up to x50 is needed to see features such as the microsculpture on small species of *Bembidion, Acupalpus* and *Agonum*. Bright but diffuse light is often essential in order to see small pores or setae on the elytra or pronotum. The keys in the R.E.S. *Handbook* (Lindroth, 1974) are the standard means of identification of adult carabids. The main key to genera in this work is, however, difficult to use by inexperienced collectors as it makes no use of subfamilies or tribes. There is also a serious error in the key between couplets 45 and 55: corrections were given in the later reprinted edition of the *Handbook*, and in *Antenna*, 1977, **1**(1):25. The keys to species in Lindroth are generally reliable, although some further notes on particular species have appeared in the *Coleopterist's Newsletter* (May 1981, *Carabus violaceus / problematicus, Nebria brevicollis / salina:* November 1981, *Bembidion lampros / properans, Notiophilus biguttatus / quadripunctatus).*

An alternative, also 'non-systematic' key to all British Carabidae is *Ground Beetles* by Forsythe (2000). The key to genera in Joy (1932) gives a much better idea of the taxonomic 'structure' of the family, but the subfamilies, tribes and ordering of many genera are not as generally accepted nowadays. Systematic keys to, and species descriptions of, British Carabidae are currently being written by the present author as a new, revised edition of the R.E.S. Handbook. There are also good keys in English, that include most British species, in the faunas of Scandinavia by Lindroth (1985-86), Hůrka (1996) Czech and Slovak Republics. The keys (also in English) of Trautner and Geigenmüller (1987) are misleading; the key to tribes and genera is useful, but the specific keys include only selected genera and parts of Europe.

Comprehensive keys to specifically British carabid larvae do not exist, but there are keys to genera of the world (van Emden, 1942) and Europe (Hůrka, 1978). Keys to species that include most known British larvae are in Luff (1993) and Arndt (1991). Keys to British larvae of particular tribes were produced by the present author in *Entomologist* and *Entomologist's Gazette*: tribes up to and including the Trechini and Pogonini have been published.

References

Arndt E., 1991. Familie Carabidae. *In* Klausnitzer, *Die Larven der Käfer Mitteleuropas,* **1**, Goecke & Evers, Krefeld.

Emden, F.I., van, 1942. A key to the genera of larval Carabidae (Col.). *Transactions of the Royal Entomological Society of London,* **92**: 1-99.

Forsythe, T.G., 2000. *Ground Beetles.* Naturalists' Handbook Series no.8 (2nd edn.). 96pp. Richmond Publishing Co. , Slough.

Hůrka, K., 1978. *in* Klausnitzer, *Ordnung Coleoptera, Larvern,* 51-69. Junk, The Hague.

Hůrka, K., 1996. *Carabidae of the Czech and Slovak Republics.* 565pp. Kabourek, Zlín.

Joy, N.H., 1932. *A practical handbook of British beetles.* 2vols. Witherby, London.

Lindroth C.H., 1974. Carabidae. *Handbooks For the Identification of British Insects,* **4**(1) 148pp. Royal Entomological Society, London.

Lindroth, C.H., 1986-6. *The Carabidae (Coleoptera) of Fennoscandia and Denmark* Fauna Entomologica Scandinavica **15** (2 vols.) 497pp. Scandinavian Science Press, Leiden.

Luff, M.L., 1993. *The Carabidae (Coleoptera) larvae of Fennoscandia and Denmark.* Fauna Entomological Scandinavica, **27**. Scandinavian Science Press, Leiden.

Luff M.L., 1998. *Provisional atlas of the ground beetles (Coleoptera, Carabidae) of Britain.* 194pp. Biological Records Centre, Huntingdon.

Trautner, J. & Geigenmüller, K., 1987. *Tiger beetles Ground beetles. Illustrated key to the Cicindelidae and Carabidae of Europe.* 488pp. Margraf, Aichtal.

HISTEROIDEA
(Following Hansen (1997))
(Plate 33)
by Michael Darby

The superfamily Histeroidea includes three families of which two occur in Britain: the SPHAERITIDAE, which is represented by a single species *Sphaerites glabratus* (F.), found rarely in the north of England and Scotland in fungus and in soil under carrion, and the HISTERIDAE. The latter includes forty-nine species of small to medium-sized beetles easily distinguished by their heavily chitinised exoskeletons, their geniculate antennae with solid club and their ability when disturbed to withdraw their heads and retract their legs as if feigning death.

The Histeroidea are the subject of an RES Handbook by Halstead (1963). Since its publication there has been the addition of *Plegaderus vulneratus* Panzer, which is found under bark of pine, mainly in the south, and *Epierus comptus* (Erichson), found firstly near Salisbury in 1982 and more recently in the New Forest, under beech bark.

The HISTERIDAE are divided into six subfamilies as set out below. Underside characters, particularly the presence or absence of a forward pointing projection, called the gular lobe, and the position of the cavities housing the antennae when retracted, are important in distinguishing the major groupings. It should be pointed out that the current British list does not take account of recent research by Continental and other scholars, and numerous changes at subfamily and generic level need to be made. Here Pope (1977) is followed.

TERETIINAE. Includes only one species *Teretrius fabricii* Mazur, which has not been taken in this country since 1907 and may be extinct. It should be looked for in association with *Lyctus* spp. (Bostrichidae) on which it is a predator.

ABRAEINAE. This subfamily includes eight species in five genera, all of which are very small, the largest measuring only 1.5mm. Several occur in interesting habitats: *Halacritus punctum* Aubé, known only from north Devon in recent years, is found exclusively on sand and under seaweed just above high water level; *Aeletes atomarius* (Aubé) in the burrows of the lesser stag beetle

Dorcus parallelipipedus (L); and *Acritus homoeopathicus* Wollaston on burnt ground associated with the fungus *Pyronema confluens*. The commonest members of the subfamily are probably *Abraeus perpusillus* (Marsham) (=*globosus* (Hoffmann)) and *Acritus nigricornis* (Hoffmann) found in rotten wood and rotting vegetation respectively. The three remaining species are all found under bark. *Abraeus granulum* Erichson, is similar to *A.globosus* but may be separated by its smaller size (not exceeding 1.3mm). *Plegaderus vulneratus* and *P. dissectus* Erichson possess a strongly sculpted thorax incorporating a transverse furrow which will quickly separate them from all other Histeridae. The former is associated with conifers and the latter with elm and beech.

SAPRININAE. This subfamily also includes four genera. *Myrmetes*, are represented by single species, is to be found with the ant *Formica rufa* L. Most of the species of *Hypocaccus* and the closely related genus *Saprinus* are also found in dung and carrion, the former being confined to coastal sandhills. Of the five species of *Saprinus* recorded as British, *subnitescens* Bickhardt is now thought to be extinct and *virescens* (Paykull) has been recorded from Cardiganshire only in recent years. The commonest member of the genus is *semistriatus* (Scriba) found on carrion.

DENDROPHILINAE. This subfamily includes only seven species divided between four genera. *Kissister* and *Carcinops*, which are closely related but may be separated by the number of dorsal striae on the elytra, include only one species each, the former being found rarely at the roots of grass in sandy places. *Dendrophilus pygmaeus* (L.) and *D. punctatus* (Herbst) are found in nests of the ant *Formica rufa* L., and species of *Paromalus* (formerly *Microlomalus*) under bark. *P. flavicornis* (Herbst) appears to be becoming more common in the south, whilst *P. parallelepipedus* (Herbst) has not been recorded since 1952 and is now possibly extinct in the UK.

TRIBALINAE. Represented by the single genus *Onthophilus* which may be quickly separated from all other British histerids by the presence of longitudinal keels on the elytra. *O. striatus* (Forster) is the commoner of our two species being found in dung, rotting vegetation, etc. whilst *O. punctatus* (Müller) has only been recorded from three counties in southern Britain in recent years and is associated with underground habitats such as rabbit burrows and mole nests.

HISTERINAE. The species of the five genera represented in this subfamily have at one time or another all been placed in *Hister* Linnaeus. All are of distinctive appearance being rounded, black and shining (shared with many members of the Saprininae) and are found in dung, carrion, rotting vegetation, etc. Some, such as *Hister unicolor* L., *H. striola* Sahlberg and *Paralister carbonarius* (Hoffmann) are three of our commonest Histerids, but other members of the subfamily are rare. *Hister quadrinotatus* Scriba, *H. illigeri* Duftschmid and *Paralister obscurus* (Kugelann) are all thought to be extinct in

Britain, and *H. quadrimaculatus* L., although not uncommon on the Continent, is now rare in this country, being known from a handful of localities in the south only. One record, however, is as recent as 2001, suggesting the possibility of a return to the insect's former status through global warming. *Margarinotus marginatus* (Erichson), is the only British Histerid which is confined exclusively to mole nests.

HETAERIINAE. Represented by the single species *Hetaerius ferrugineus* (Olivier), a rare, small, reddish-coloured beetle found in the nests of the ants *Formica rufa* L. and *F. sanguinea* Lat. mainly in the south.

References

Halstead, D.G.H., 1963. Coleoptera. Histeroidea. *Handbooks for the Identification of British Insects,* **4**(10) 16 pp. Royal Entomological Society London.

Pope, R. D. (Ed.), 1977. A check list of British Insects. Second Edition (completely revised). Part 3: Coleoptera and Strepsiptera. *Handbooks for the Identification of British Insects* **11**(3):i-xiv + 1-105.

PTILIIDAE
by Michael Darby

A large family which includes some of the smallest known insects. In Britain it is represented by seventy-five species, none of which measures much more than 1mm. Adults and larvae feed on fungal hyphae and spores and are consequently found in a wide range of habitats including rotting vegetation, under bark, in rotten wood, dung, leaf litter, and fungi.

Because of their minute size special techniques of collecting Ptiliidae are needed. The author's methods, involving a kitchen sieve and micro-pooter, are described in detail in *Coleopterists Newsletter* No. 13, (1983). The use of traps, particularly grass piles and baited pitfall traps, has proved very effective in catching many species.

The identification of many Ptiliidae, particularly those belonging to the genus *Acrotrichis,* often requires study of the genitalia under high magnification. Slides are best prepared using a mountant such as Euparal on small pieces of 'acetate' pinned with the specimen, as described on pp. 356-361. The most up to date keys to the British species are those in by Besuchet (1971) and in Lohse & Lucht (1989).

In the latest check list (on the Coleopterist website) the Ptiliidae are separated into two subfamilies the Ptiliinae and the Acrotrichinae, which are further divided into six tribes.

NOSSIDIINI. Represented in this country by one species *Nossidium pilosellum* (Marsham) found particularly in rotting elm stumps. This is the only British Ptiliid in which the male aedeagus possesses parameres.

PTENIDIINI. Includes eleven species in one genus, *Ptenidium*, which may be quickly recognised on external characters allied to knowledge of habitat. *P. pusillum* (Gyllenhal) and the much smaller *P. nitidum* (Heer) are the two commonest species, being found in open situations in grass piles, etc. often in large numbers. *P. fuscicorne* Erichson, *P. intermedium* Wankowicz and *P. laevigatum* Erichson are found in wetland habitats the latter often turning up in moles' nests. *P. longicorne* Fuss crawls on mud at the edge of rivers, and the rare *P. brenskei* Flach is found in river shingle. *P. gressneri* Erichson and *P. turgidum* Thomson, are also rare species associated with older woodlands, the first being found in rot holes and the latter in rotten wood. *P. formicetorum* Kraatz is also a woodland species not always found in ants' nests as its name implies, while *P. punctatum* (Gyllenhal) is mainly confined to coastal situations amongst decaying seaweed, although there are some inland records from old dung heaps.

PTILIINI includes the following eight genera. *Euryptilium*, to the single species of this genus listed by Pope, *saxonicum* (Gillmeister), should be added *gillmeisteri* Flach a recently arrived immigrant in the north of England. Both species may be quickly separated from other British Ptiliids by the overall form and presence of elytral humeri. *E. saxonicum*, the commoner of the two, is found more often in the north than the south, in grass heaps particularly.

Genus *Ptiliola* is the name now given to the genus listed in Pope (1977) as *Nanoptilium*. *P. kunzei* (Heer) is the commoner of the two British species, being found in dryish dung, grass heaps, etc.

Genus *Ptiliolum* represented in this country by six species, the females of which may be readily separated on spermathecal characters. The rarest are *P. caledonicum* (Sharp) and *P. sahlbergi* Flach, which are known from a few specimens only, and the commonest is *P. fuscum* (Erichson) found in a wide range of habitats.

The genera *Actidium* and *Oligella* include five species, three have not been taken in this country for many years, and one, *O. intermedia* Besuchet is known from one specimen only in this century taken in Yorkshire in 1988. *O. foveolata* (Allibert), the commonest species, is found in various situations particularly compost- and old dungheaps.

The remaining genera are *Ptilium, Micridium* and *Millidium* all the species of these genera may be quickly distinguished by the presence of three impressed lines on the thorax. Males of one of the commonest species, *P. horioni* Rosskothen, may be quickly separated from the similar but rarer *P. exaratum*

(Allibert) by the presence of a fringe of long hairs between the hind coxae in the former, and a long process, swollen apically, in the latter. *P. caesum* Erichson has been added to our list since Pope (1977) although there are no recent British records. Like *affine* Erichson it is a fenland species. *P. myrmecophilum* (Allibert) is usually found, as its name suggests, in ants' nests.

The only representative of *Micridium* is the very rare *halidaii* (Matthews) found only in old oaks in relict woodland; and the only representative of *Millidium* is the very local and scarce *minutissimum* (Ljungh) found in old dungheaps.

PTINELLINI includes three genera. *Microptilium* – *M. palustre* Kuntzen should be added to the list given by Pope (1977). Both it and *M. pulchellum* (Allibert) are rare marshland species being known from a few localities only.

Ptinella and *Pteryx*. Most of the species are yellowish in colour and live under bark and in rotten wood. Populations are frequently polymorphic, specimens occurring with or without eyes and wings. Some species, e.g. *P. errabunda* Johnson, are parthenogenetic, and appear to be immigrants from Australasia.

NEPHANINI includes three genera. *Baeocara* is immediately recognisable by the presence of the large punctures on the thorax. The single British species *variolosa* (Mulsant and Rey) is found in mouldy horse dung in woods, and is particularly common in the New Forest. *Smicrus* and *Nephanes* are two closely related genera each represented by one species. *S. filicornis* (Fairmaire and Laboulbène) is found very rarely in sedge litter at the edge of ponds, and the much smaller *N. titan* (Newman) is common in old dung heaps.

ACROTRICHINI Includes two genera. *Actinopteryx*, included in the latest check list with a single species, *fucicola* Matthews, a species found in coastal detritus but now thought to be extinct in Britain.

Acrotrichis is the largest Ptiliid genus in the UK consisting of twenty six species found in leaf litter, compost, dung, fungi and similar habitats. They are immediately distinguishable from other Ptiliid genera on external characters, but many require study of the genitalia for individual determination. All of those listed by Pope (1977) as 'incertae sedis' have now been worked out by Johnson and found to be synonymous with existing species. Similarly *fraterna* Johnson has been shown to be synonymous with *rosskotheni* Sundt. The rare *chevrolati*, the smallest species, is now known as *cephalotes* (Allibert).

Two additions have been made to the list given by Pope (1977) *A. sanctaehelenae* Johnson, which may be quickly distinguished from all other species except *A. grandicollis* (Mannerheim), by the presence of long protruding setae, particularly noticeable on the thorax and elytra. *A. josephi* (Matthews), a common N. American species, is similar to *cognata* (Matthews) but lacks the pruinose sheen to the elytra.

The student will find the keys and illustrations given by Besuchet (1971) and Loshe & Lucht (1989), useful for determining individual species.

References

Besuchet, C., 1971. Family Ptiliidae *in* Freude, H., Harde, K.W. and Lohse, G.A., 1971. *Die Käfer Mitteleuropas*, **3**:311-342. Goeke & Evers, Krefeld.

Lohse, G.A. and Lucht, W.H., 1989. *Die Käfer Mitteleuropas*, **12** (supplementband mit katalogteil). Goeke & Evers, Krefeld.

Pope, R. D. (Ed.), 1977. A check list of British Insects. Second Edition (completely revised). Part 3: Coleoptera and Strepsiptera. *Handbooks for the Identification of British Insects* **11**(3):i-xiv + 1-105.

LEIODIDAE

by J. Cooter

The family Leiodidae is divided into six sub-families, four of which, Leiodinae, Coloninae, Cholevinae (= Catopinae) and Platypsyllinae (= Leptininae), occur in Britain. The phylogeny and taxonomic position of the family was for a long time uncertain, for example both Fowler (1889) and Joy (1932) consign the genera to various positions in that taxonomic rag-bag the 'clavicornia.' The biology and ecology of the Leiodidae is poorly known and it is possible for a student to make significant contributions in these fields; for example the larva of the genus *Colon* is not known.

The individual sub-families, and indeed most genera, are easily distinguished; with a little experience this can be achieved in the field without a hand lens. However, identification to species level can be difficult, species of *Leiodes* and *Colon*, especially female specimens, are generally regarded as amongst the 'hardest' British beetles to identify. Their uniform dark coloration, similar shape and small size, combined with difficulties in obtaining sufficient material have all helped to make the Leiodidae unpopular with the amateur.

LEIODINAE: In Britain this sub-family consists 50 species divided unequally between eleven genera in five tribes. The beetles are mostly mycophagous and the biology of most is quite unknown. Identification keys, ecological and some distributional information is given in Cooter (1996).

It is likely that the majority of species of tribes Sogdini and Leiodini inhabit hypogeous fungi (those fungi with underground fruiting bodies including *Tuber* spp. (= truffles)). For a long time it has been known that some species of *Leiodes* are associated with truffles, *Leiodes cinnamomea* (Panzer) having been found, sometimes in numbers, adults and larvae together, in at least four species of truffle (*Tuber* spp.). *Leiodes ciliaris* (Schmidt), *furva* (Erichson), *picea* (Panzer) and the non-British *dubia* (Kugelann) have all been recorded from hypogeal fungi (not truffles), and it is more than likely that *Leiodes oblonga*

(Erichson), *rufipennis* (Paykull) and *calcarata* (Erichson) are similarly associated, but this needs confirmation. Fleischer used truffle as a bait, but attracted only *Leiodes cinnamomea* despite several other *Leiodes* species being on the wing in the vicinity at the time; he also noted as many as ten species of *Leiodes*, adults and larvae, on or near moulds growing on grass roots in a meadow.

Some *Leiodes* and *Cyrtusa* (sensu auct.) have been found on epigeal fungi, but none is known to breed there, though perhaps this might prove likely with *Cyrtusa* (sensu auct.). *Colenis* and *Agaricophagus* (tribe Pseudoliodini) have been found on unidentified (probably *Tuber* sp.) truffles but are also associated with epigeal fungi. *Colenis immunda* has occasionally been noted on Agaricales and *Polyporus squamosus*. The enigmatic *Aglyptinus agathidioides* Blair (tribe Scotocryptini) is known only known from the type series of two specimens found in a moorhen nest in Hertfordshire; no further examples have been found anywhere and the genus is otherwise confined to the Neotropical region and Mexico, with one species recorded from eastern Canada and Louisiana. It is highly likely the Hertfordshire captures represent an accidental importation and the species should not be regarded as British.

Leiodes furva (Erichson) and *Hydnobius punctatus* (Sturm) are typically found on coastal dunes; *L. calcarata* is the most widespread and frequently encountered member of the genus. Some species are excessively scarce, for example *L. silesiaca* (Kraatz) is confined to Scotland and has been taken on three occasions in the Highlands. *Leiodes flavescens* (Schmidt) is known in Britain from a small number of specimens only. *Leiodes picea* (Panzer) is another northern species, but its range extends as far south as Lancashire.

Amongst the Agathidiini all European species of *Anisotoma* are associated with myxomycete sporophores (not plasmodia), this being so for all recorded instances of breeding, though adults have been found in Basidiomycetes. *Agathidium* species might be primarily associated with myxomycetes, but the evidence is less clear. Adults have been found in sporophores of myxomycetes and as often on sporophores of diverse Basidiomycetes especially Polyporaceae. *Amphicyllis* has been found several times on myxomycetes.

Agathidium arcticum Thomson is, in Britain, almost confined to Scotland where it can be found in Caledonian pine forest remnants under fungoid bark of fallen pine trees, especially those attacked by myxomycetes. *Agathidium seminulum* (L.), *rotundatum* (Gyllenhal) and *nigripenne* (F.) likewise occur under bark, but are more widespread and frequent deciduous trees as well. *Agathidium atrum* (Paykull), *convexum* Sharp and *laevigatum* Erichson are often captured by pit-fall trapping or by sifting moss, tussocks and forest-floor litter. *Anisotoma glabra* (F.) and *castanea* (Herbst) are northern species often found together, sometimes with *Agathidium* species under fungoid bark. *Anisotoma humeralis* (F.) is widespread and often found in numbers, *Anisotoma orbicularis* (Herbst) slightly less abundant, in and about myxomycete sporophores.

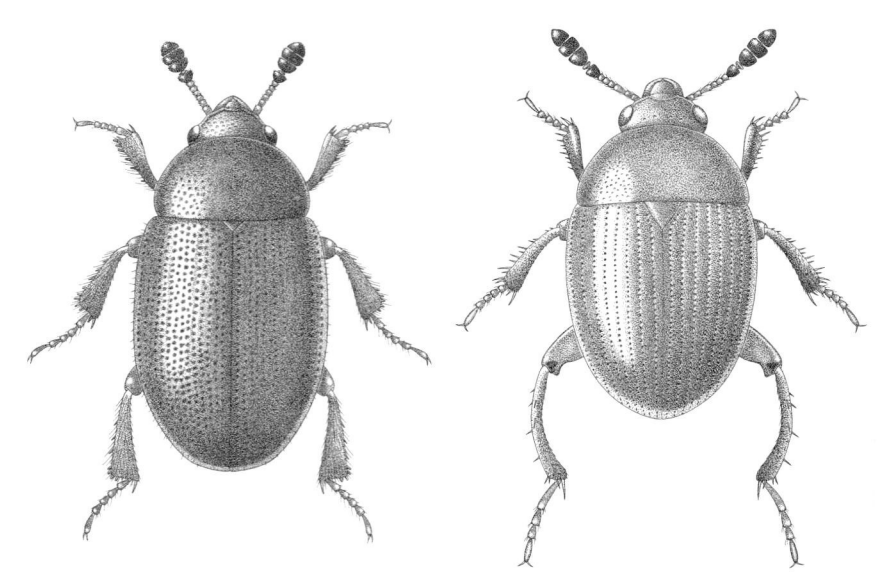

Fig.5 *Sogda suturalis* (Zetterstedt)   Fig.6 *Leiodes litura* Stephens

Occasional specimens of Leiodini will be turned up by sifting grass-tussocks and flood refuse, or even simply by turning stones or examining sand-dunes, but better results will be had by evening sweeping. Pit-fall trapping is often very productive (see Cooter, 1989) but by far the best method of collecting is flight interception trapping (see p. 208). Agathidiini can be found, sometimes in numbers, under fungoid bark, especially that attacked by myxomycetes as well as in myxomycete sporophores. Flight interception trapping is also a reliable method for obtaining Agathidiini.

For information regarding immature stages of Agathidiini the student is directed to a number of papers. The mature larva and pupa of *Agathidium varians* Beck have been described by Angelini and DeMarzo (1984). Růžička (1996) presents a key to the larvae of the European species of *Anisotoma* and descriptions of the third instar larvae of [non-British] *Anisotoma axillaris* Gyllenhal and *A.glabra*. Kilian's (1998) paper is a bench-mark for larval morphology and phylogeny of Agathidiini larval stages.

In the Leiodini, the genitalia of the males exhibit reliable and often marked characters useful in the separation of species and it is highly advisable to extract the aedeagus from every male *Leiodes* specimen as a matter of course prior to mounting. This organ should be prepared as a clove-oil mount (see p. 359). Whereas all female Agathidiini have a sclerotised spermatheca of diagnostic value, amongst the Leiodini, only in genus *Liocyrtusa* is a sclerotised spermatheca to be found. This must be dissected very carefully as it is bulb-like and often possesses a long flagellum; it is best prepared for examination by transmitted light as a clove-oil mount. The aedeagi of male Agathidiini also exhibit specific characters, but it is not necessary to prepare these via clove-oil, they can be glued to the card next to the beetle or kept in glycerol in a van Doesburg tube attached to the insect pin; this enables dorsal, ventral and lateral aspects of the aedeagus to be examined (p. 366). The ovipositor of *Hydnobius* species exhibits specific characters and is sometimes a useful aid to separating females of *punctatus* (Sturm) and *spinipes* (Gyllenhal). It should be prepared as a clove oil mount. Both our Pseudoloidini, *Agaricophagus* and *Colenis* have sclerotised aedeagus and spermatheca, but the species are so distinctive it is not necessary to dissect these organs for identification purposes.

Male Agathidiini have dilated anterior tarsi and/or 5-5-4 tarsal formula; females have rather linear anterior tarsi and 5-4-4 or 4-4-4 tarsal formula. Sexes of the Leiodini are often difficult to assess externally, especially with smaller examples of any particular species; if in doubt dissect. Typically males have broader anterior tarsi than the females and the posterior femora furnished dorsally and ventrally with a plate-like extension which may be rounded, angled or developed into one or more tooth-like projections. Generally the ventral plate is more strongly developed than the dorsal.

The antennae often exhibit important diagnostic characters and great care must be taken when setting these. An absolute minimum of gum should be added to the lower surface of the terminal segment only. All segments, especially those forming the club, must be set flat on the card and checked to ensure they are in the same plane. Being oval in cross-section, a slight twisting will give a marked distortion in relative widths. With the terminal and penultimate segments this is critically important. It ought to go without saying that glue must not reach the dorsal surface.

Some species of *Agathidium* have the ability to roll into a ball with head pressed against the venter. They can be 'unrolled' with little difficulty if properly relaxed, using fine-pointed forceps and a pin. The ventral surface of all Leiodinae must be set flat on the card, care being taken to ensure the head in particular is flat. A slight departure can obscure detail and distort other features.

CHOLEVINAE: These distinctive beetles inhabit various types of organic matter; 32 species occur in Britain.

Species of *Ptomophagus, Nargus, Catops* and *Sciodrepoides* occur in a variety of decaying matter and leaf-litter; *Nemadus colonoides* (Kraatz) frequents the rot-holes in trees that have been used by birds for nesting, both the nest material and associated debris should be searched for this diminutive species. Most species of *Choleva* inhabit mammal nests, (see p. 222) and *Catopidius depressus* (Murray) frequents rabbit burrows, from which it has been taken in large numbers by the use of baited pit-fall traps (see Welch, 1964). *Parabathyscia wollastoni* (Janson) is a subterranean beetle and can be found in autumn when lifting one's potato crop – it occurs in the rotting remains of the seed potato planted in the spring – and other situations where rotting vegetable matter is buried (under domestic compost heaps and doubtless in 'traps' of vegetable peelings buried and left for several months to rot).

Cholevids can be taken occasionally during the course of general collecting, especially evening sweeping, sifting tussocks, in litter and at carrion. They occur in numbers in baited pit-fall traps (see p. 207). If bait is used, the trap must be inspected frequently and a mesh screen or perforated lid placed securely over the trap will prevent admission of large and potentially destructive beetles such as *Nicrophorus* spp. Material thus collected is often quite dirty and will need careful cleaning with alcohol (to disperse fatty deposits). Once again, flight interception trapping is another very productive method of collecting.

Male Cholevinae are easily recognised by their strongly dilated anterior tarsi. Their aedeagi are distinctive and provide reliable characters to assist identification. The females have rather linear anterior tarsi and in only a few genera possess a sclerotised spermatheca of diagnostic value; even these are not widely used as an identification aid. In *Choleva,* the males have characteristically-shaped posterior trochanters and it is often worthwhileremoving one posterior leg whilst mounting the beetle in order to display this feature. Female *Choleva* possess characteristic well chitinised genital sclerites, which in repose are withdrawn into the abdomen and need to be dissected out and glued to the mounting card along-side the beetle or kept in a van Doesburg tube attached to the pin carrying the mounting card (p. 366). Once again it is of prime importance to keep the antennal segments flat and in the same plane and to use an absolute minimum of glue on the terminal segment only. Cholevinae are quite pubescent beetles and care must be taken to keep this clean, especially free from glue. Matting of the pubescence obscures the underlying surface ornament. For degreasing see pp. 335-336.

COLONINAE: Represented in Britain by one genus, *Colon* Illiger with nine British species. They are very poorly known biologically and as yet their larvae are unknown. It is likely and widely assumed that they are associated with hypogeal fungi or moulds, but confirmation is needed.

Odd specimens can be found by general sifting of flood refuse, leaf-litter, moss at the base of trees and in grass-tussocks; they occasionally turn up in pit-fall traps. Evening sweeping is more productive and good results have been obtained by the use of flight interception traps (see p. 208). *Colon* species have been found in rather barren looking areas, a point made by Fowler (1889 page 66) long ago and apparently overlooked subsequently. Joy (1933) describes a novel method of observing *Colon* species; the trick is to find the correct habitat favoured by the beetles (advice that doubtless applies to every beetle on our List).

Colon brunneum (Latreille) is without doubt the most frequently encountered species of the genus, with perhaps *deutipes* (Sahlberg) next. Many collections made by competent coleopterists contain only these two, or even no *Colon* species, such is their scarcity.

In the sub-genus *Eurycolon,* the anterior tarsi of the males are dilated; males of many species have the posterior femora armed with a spine or tooth, but the size and shape of this does vary within species limits. Fortunately *Colon* species are easily sexed externally as the males exhibit five or six abdominal sternites, the females only four. The aedeagi are specifically characteristic and often afford the most reliable characters to facilitate determination. They should be prepared by the van Doesburg method (p. 366) or mounted in Euparal. Preserved by either method the setae and other features are not distorted by drying or obscured by glue. The antennae should be very carefully mounted with all segments in the same plane flat on the card and a minimum of glue on the terminal segment only (ideally no glue if the antennae can be set straight and flat). The relative widths of the terminal and penultimate segments and the shape of the club are important. Gum must be kept off the finely pubescent dorsal surface otherwise valuable diagnostic features will be concealed. When mounting *Colon* species it is good practice to pull the head forward clear of the pronotum exposing the 'neck', thus the dividing line between the punctured area of the head and the unpunctured area of the 'neck' is displayed; the head should be glued flat to the card.

PLATYPSYLLINAE: There is only one member of this sub-family on the British List – *Leptinus testaceus* Müller. It has been recorded from nests of small rodents, birds and bumble-bees; Fowler (1899) in addition records it, rarely, from *Formica fuliginosa* nests.

References and further reading

Allen, A.A., 1966. Annotated Corrections to the List of British species of *Leiodes* Latr. (Col.. Leiodidae). *Entomologist's Monthly Magazine,* **101**(1965):178-184.

Allen, A.A., 1968. *Leiodes clavicornis* Rye (Col., Leiodidae) new to England; with diagnostic notes. *Entomologist's Monthly Magazine,* **103**(1967):262-263.

Angelini, F., & DeMarzo, L., 1984. Morfologia della larva matura e della pupa in *Agathidium varians* Beck (Col., Leiodidae, Anisotomini), *Entomologica Bari* **19**:51-60.

Besuchet, C., 1980. Reviosion des *Leptinus* paléarctiques (Coleoptera: Leptinidae). *Revue Suisse de Zoologie,* **87**(1):131-142.

Cooter, J., 1978. The British species of *Agathidium* Panzer (Col., Leiodidae). *Entomologist's Monthly Magazine*, **113** (1977): pp. 125-135.

Cooter, J., 1989. Some notes on the British *Leiodes* Latreille (Col., Leiodidae). *Entomologist's Gazette* **40**:329-335.

Cooter, J., 1996. Annotated keys to the British Leiodinae (Col., Leioididae). *Entomologist's Monthly Magazine*, **132**:205-272.

Daffner, H., 1983. Revision der paläarktischen Arten der Tribus Leiodini Leach (Col., Leiodidae). *Folia Entomologica Hungarica*, **44**(2):9-163.

Fowler, W.W., 1889. *Coleoptera of the British Islands* 3:66. L. Reeve & Co., London.

Joy, N.H., 1911. A Revision of the British Species of *Leiodes* Latreille (*Anisotoma* Brit. Cat.). *Entomologist's Monthly Magazine*, **47**:166-179.

Joy, N.H., 1932. *A practical handbook of British beetles*. 2vols. Witherby, London.

Joy, N.H., 1933. *British beetles their homes and habits*. F. Warne & Co.

Kevan, D.K., 1945. The aedeagi of the British species of the genus *Catops* Pk., (Cot. Cholevidae) *Entomologist's Monthly Magazine*, **81**:69-72.

Kevan, D.K., 1945. The aedeagi of the British species of the genera *Ptomophagus* Ill., *Nemadus* Th., *Nargus* Th., and *Bathyscia* Sch. (Col., Cholevidae). *Entomologist's Monthly Magazine*, **81**:121-125.

Kevan. D.K., 1946. The sexual characters of the British species of the genus *Choleva* Latr., including *C. cisteloides* (Frölich) new to the British List (Col., Cholevidae). *Entomologist's Monthly Magazine*, **82**:122-130.

Kevan, D.K., 1946. *Catops nigriclavis* Gerh. (Col., Cholevidae) new to the British List. *Entomologist's Monthly Magazine*, **82**:155-157.

Kevan, D.K., 1946. The aedeagus of *Catopidius depressus* (Murray) (Col., Cholevidae). *Entomologist's Monthly Magazine*, **82**:308-309.

Kevan, D.K., 1947. A revision of the British species of the genus *Colon* Hbst. (Col., Cholevidae). *Entomologist's Monthly Magazine*, **83**:249-267.

Kevan, D.K., 1964. The spermatheca of the British species of *Ptomophagus* Ill., and *Parabathyscia wollastoni* (Jansen) (Col., Catopidae). *Entomologist's Monthly Magazine*, **99**:216.

Kilian, A., 1998. Morphology and phylogeny of the larval stages of the tribe Agathidiini (Coleoptera: Leiodidae: Leiodinae). *Annales Zoologici (Warszawa)* **48**(3/4):125-220.

Růžička, J., 1996. Description of the third instar larvae of *Anisotoma axillaris* and *A.glabra* (Coleoptera: Leiodidae), with a key to larvae of European *Anisotoma* species. *Acta Societatis Zoologicae Bohemicae*, **60**:455-468.

Szymczakowski, W., 1969. Die mitteleuropäischen Arten der Gattung *Colon* Herbst (Col., Colonidae). *Entomologische Abhandlungen* **36**(8):303-339.

Welch, R.C., 1964. A simple method of collecting insects from rabbit burrows. *Entomologist's Monthly Magazine*, **100**:99-100.

Welch, R.C., 1964. *Catopidius depressus* (Murray) (Col., Anisotomidae) in large numbers in a rabbit warren in Berkshire. *Entomologist's Monthly Magazine*, **100**:101-104.

SCYDMAENIDAE
by Michael Darby

The Scydmaenidae is a moderate-sized family of small Coleoptera represented in this country by thirty species divided between four tribes. It has often been linked with the Pselaphidae, but although there is no dispute that both are correctly placed in the Staphylinoidea, the similarity of the adults is now thought to be the result of parallel adaptation, and the Scydmaenidae are considered to be more primitive than the Pselaphidae. Henry Denny's *Monographia Pselaphidarum et Scydmaenidarum Britanniae,* which was published in Norwich in 1825, still remains the only monographic treatment although the British fauna has been included in two major publications on the European species, by Croissandeau (1891-1900), and more recently by Franz and Besuchet (1971 & 1989). The latter provides the basis for the listing in Pope (1977), and up-dates the keys and other information given by Joy (1932), and by Allen (1969). Serious students will also find Brown and Crowson's (1979) paper useful.

Brown and Crowson's research confirmed that the Scydmaenidae are probably placed correctly as near relatives of the Leiodidae, and suggested that two genera *Eutheia* and *Cephennium* deserve subfamily ranking. These groups are treated as tribes by Pope along with the Stechniini, with by far the largest number of species, and the Scydmaenini. More information about individual genera and species is set out below. Scydmaenidae may be quickly separated from Pselaphidae by the form of the elytra which more completely cover the abdomen.

EUTHEIINI. Includes the genera *Euthiconus* and *Eutheia*. The first is represented by a single species, *conicicollis* (Fairmaire and Laboulbène), added to the British list as recently as 1999 on the basis of a specimen taken in a flight intercept trap at Silwood Park, near Windsor. Like many other Eutheiini it is small (1-1.2mm) and associated with old woodland. *Eutheia* is represented by five species which may be quickly distinguished from all other British Scydmaenids by their flattish, more parallel appearance and truncated elytra. Three are very rare: *formicetorum* Reitter, *linearis* Mulsant and *plicata* (Gyllenhal) known from a handful of recent records only, and two: *schaumi* Kiesenwetter and *scydmaenoides* Stephens, which are found in cut grass, manure, compost and other rotting vegetation, more common though still Notable species.

CEPHENNIINI. Some dozen or so species of the genus *Cephennium*, which is the only representative of this tribe, are found on the continent, but only one in Britain, *gallicum* Ganglbauer, one of our commonest scydmaenids, occurring in litter, compost, etc.

STENICHNINI. This tribe includes twenty-five species in five genera most of which are rare. In Joy (1932) they are divided between *Neuraphes*, *Euconnus* and *Stenichnus*, but more recently *Microscydmus* has been shown to be the correct genus for the minute *nanus* (Schaum). This genus is further represented by *minimus* (Chaudoir), which is also found locally in rotten wood and is the rarer of the two species; *helvolus* (Schaum) and the larger and lighter *sparshalli* (Denny), formerly assigned to *Neuraphes*, are now placed in *Scydmoraphes*. The species in this tribe most likely to be met with is *Stenichnus collaris* (Müller & Kunze) found in leaf litter and other vegetable refuse.

SCYDMAENINI. *Scydmaenus rufus* Müller & Kunze, and *S. tarsatus* Müller & Kunze are the only species in this tribe. Of the two, *rufus* is only known from four counties since 1969 the larger *tarsatus* is the more common, being found in vegetable refuse. The two species may be separated by the presence of fovea at the base of the thorax in *tarsatus* which are absent in *rufus*.

References

Allen, A.A., 1969. Notes on some British Scydmaenidae with corrections to the British list. *Entomologist's Record and Journal of Variation,* **81**:239-246.

Brown, C and Crowson, R.A., 1979. Observations on scydmaenid larvae with a tentative key to the main British genera. *Entomologist's Monthly Magazine,* **115**:49-59.

Croissandeau, J., 1891-1900. Scydmaenidae européens et circa-mediterraneens, *Annales Societé Entomologique de France,* **62**:199-238; 409-42; **63**: 351-400; **66**: 402-30; **67**: 105-67; **69**:116-60.

Denny, H., 1825. *Monographia Pselaphidarum et Scydmaenidarum Britanniae.* vi + 74pp and 14 hand coloured plates. S.Wilkin, Norwich.

SILPHIDAE
(Plates 34 – 36)

by J. Cooter

Members of this distinctive family are commonly known as the 'carrion beetles'; in Britain it is represented by 21 species plus another, *Nicrophorus germanicus* (L.), presumed extinct in Britain (if indeed it ever was native). The genus *Nicrophorus* is familiar to the general naturalist as the Burying or Sexton Beetles.

The largest species, *Nicrophorus humator* (Gleditsch) is, except for its bright orange antennal club, wholly black; other members of the genus have distinctive orange bands on their elytra.

Necrodes littoralis (L.) bears a superficial resemblance to *Nicrophorus humator* but differs most markedly in having matt, strongly striate elytra, a much less abrupt antennal club, and in the male, thickened posterior femora. It is more frequently noted on the coast, but many inland instances have been recorded.

Thanatophilus species are associated with carrion, as is *Oiceoptoma thoracium* (L.) which is also to be found in the fruiting bodies of the Stinkhorn fungus and also decaying vegetation, fungi and carrion. Both species of *Aclypea* are vegetarian and have been reported as pests of beet and turnip crops. *Dendroxena quadrimaculata* (Scopoli) is a nocturnal predator of certain lepidopterous larvae in trees and bushes. During the day it can be found at the roots of grass and in litter at the base of trees. Some species of *Silpha* are predatory on snails, but frequent carrion as well. *Silpha atrata* L. can be found, sometimes in small numbers, under the bark of trees, particularly *Salix* species, during the winter months.

Nicrophorus species are large enough to be direct pinned (p. 321), but most people opt to card them as this gives added protection from accidental damage. Identification poses little problem and with experience the majority will be recognised in the field or simply run down with the aid of a hand lens and key.

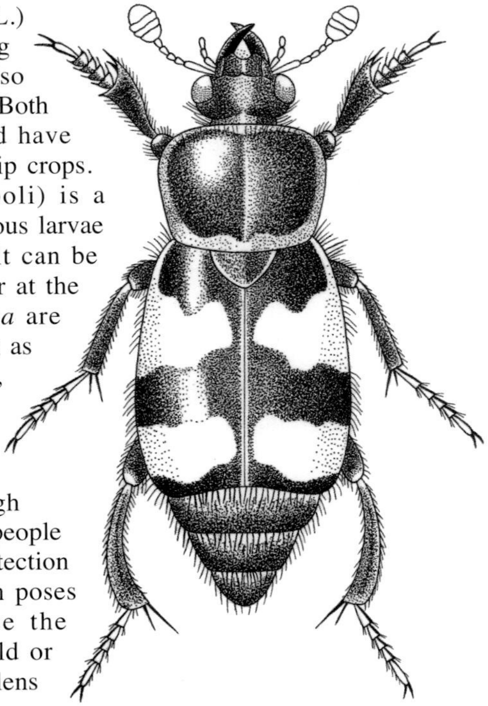

Fig. 7 *Nicrophorus vespillio* (L.)

STAPHYLINIDAE: PSELAPHINAE
by Michael Darby

The Pselaphinae, like the Scydmaenidae, are small beetles (0.7-2.5mm with the exception of *Trichonyx* 2.8-3.5mm in the UK though larger species are found abroad especially in the tropics) which feed on mites. Unlike the Scydmaenidae, the Pselaphinae have long been popular with entomologists. On the continent Reichenbach published an important monograph on the family as early as 1816, which was quickly followed by another by de Leach in the following year. Henry Denny's *Monographia Pselaphidarum et Scydmaenidarum Britanniae,* published in Norwich in 1825, has already been mentioned. The males and females of many Pselaphinae exhibit sexual dimorphism which caused early authors to attribute them to different species. In spite of this, six of the new species described by Denny still survive in the British list, which now numbers fifty two species.

Students of the British fauna will find the Reverend Pearce's R.E.S. Handbook (1957) an invaluable guide to species and their identification. Many useful tips are provided about collecting and habitats, but it is important that the *Handbook* be read in conjunction with the same author's 1974 paper which brings the *Handbook* up to date in the light of research by Dr. Claude Besuchet, the leading continental authority. Besuchet's (1974 and 1989) work includes all the British species, is essential reading for the serious student not least because it includes drawings of many aedeagi not illustrated by Pearce, which are essential for the accurate determination of some species.

Pearce divided the Pselaphinae into two sub families, the Clavigerinae and the Pselaphinae and the latter into two tribes. Besuchet, however, favoured a division of the family into eight tribes (as set out below), and this is the classification used by Pope (1977) and repeated in the latest British Checklist, with the addition of the Trichonychini, (as part of Staphylinidae on the '*Coleopterist*' web site) based on Newton and Thayer (1992 and 1995). The following more specific notes will aid in the capture and identification of specimens; the '*Coleopterist*' web site Pselaphinae check list contains useful annotation (www.coleopterist.org.uk/staphylinidae-list.htm).

EUPLECTINI. This is the largest tribe with five genera including twenty six species. Morphologically it is distinct, all representatives having the hind coxae adjacent or contiguous, and the hind body more linear and parallel sided. *Trimium* is represented a single species which is rare, being recorded only from N.E.Yorkshire since 1969. Similarly both species of *Plectophloeus* are also rare. The same is again true of no less than ten representatives of the two largest genera *Euplectus,* (fifteen species), and *Bibloplectus,* (six species), which have a RDB classification. Several other species, however, such as *Euplectus karstenii* (Reichenbach), *E. piceus* Motschulsky and *E. sanguineus* Denny are more common, being found in a wide range of habitats, but the species of both genera are notoriously difficult to separate and dissection of the aedeagi is essential for most males. Some females, not associated with males, cannot be identified with certainty.

BATRISINI. Contains two genera *Batrisus*, represented by the extinct *formicarius* Aubé, and *Batrisodes* with three species. All are normally found associated with the ant *Lasius brunneus* Latreille, although *B. venustus* (Reichenbach.) is sometimes found in other habitats, particularly under bark and in rotten wood of oaks. Both *adnexus* (Hampe) and *delaporti* (Aubé) are classed as RDB1 being known from only one or two localities in the UK.

BYTHININI. Includes three genera of which the rare *Tychobythinus glabratus* (Rye) is also myrmecophilous, being found with the ant *Ponera coarctata* (Latrielle) in the south and south east. The species of *Bythinus* and *Bryaxis* exhibit sexual dimorphism, the males having the first two segments of the antennae enlarged. This is particularly noticeable in *Bryaxis bulbifer* (Reichenbach), which is one of the commonest British Pselaphids. *Bryaxis clavicornis* (Panzer) has recently been deleted from the British list.

TRICHONYCHINI. Includes two rare species only *Trichonyx sulcicollis* (Reichenbach) and *Amauronyx maerkelii* (Aubé) both found in association with various ants. The range of *sulcicollis* used to extend to Yorkshire but it is now known from only three southern counties in England, whereas the range of *maerkelii,* which is likewise known from three counties only, extends further north to Denbighshire and Worcestershire. At 2.8 to 3.5 mm *Trichonyx sulcicollis* is the largest British pselaphid.

TYCHINI. Includes two species in the genus *Tychus,* of which *T. niger* (Paykull) is the commoner. The males of this species may be quickly distinguished from the very rare *T. striola* Guillebeau (found only once in UK, Seaton, S.Devon, 1957) by the presence of the enlarged fifth segment of the antennae. *Tychus* species generally may be distinguished by the second segment of the maxillary palpus, which is almost as long as the third, apparently last, segment.

BRACHYGLUTINI. Includes four genera of which two, *Reichenbachia* and *Trissemus,* are represented by one species each. Both are found in association with water. This is also the habitat favoured by the species of the genera *Rybaxis* and *Brachygluta, B. helferi* (Schmidt-Göebel) and *B. simplex* (Waterhouse) being confined to salt marshes where they may be found crawling on the mud amongst vegetation.

PSELAPHINI. Includes two species formerly ascribed to the genus *Pselaphus,* but one of which, *dresdensis* Herbst, is now placed in *Pselaphaulax*. Both are found in moss and tussocks, especially in bogs.

CLAVIGERINI. The two species in this tribe, both in the genus *Claviger,* may be quickly distinguished from all other British Pselaphinae by the absence of eyes and the truncated 6-segmented antennae. They are both myrmecophilous, being found in the nests of ants of the genus *Lasius*, although *longicornis* Müller has not been seen in the UK since 1945.

References

Besuchet, C., 1974, Familie Pselaphidae, *in* Freude, H, Harde, K.W. and Lohse, G.A., *Die Käfer Mitteleuropas,* **5**:305-362. Goeke & Evers, Krefeld.

Besuchet, C., 1989. Familie Pselaphidae, *in* Lohse, G.A. and Lucht, W.H., *Die Käfer Mitteleuropas,* **12**:240-243. Goeke & Evers, Krefeld.

Denny, H., 1825. *Monographia Pselaphidarum et Scydmaenidarum Britanniae.* vi + 74pp and 14 hand coloured plates. S.Wilkin, Norwich.

Newton, A.F. Jnr. and Thayer, M.K., 1992. Current classification and family-group names in Staphyliniformia (Coleoptera). *Fieldiana: Zoology* n.s.**67**:1-92.

Newton, A.F. Jnr. and Thayer, M.K., 1995. Protopselaphinae new subfamily for *Protopselaphus* new genus from Malaysia, with a phylogenetic analysis and review of the Omaliine Group of Staphylinidae including Pselaphidae (Coleoptera). *in* Pakaluk, J. & Ślipiński, S.A. (eds.): *Biology, Phylogeny and Classification of Coleoptera: Papers celebrating the 80th Birthday of Roy A. Crowson,* **1**:219-320. Muzeum i Instytut Zoologii PAN, Warsaw.

Pearce, E.J., 1957. Coleoptera: Pselaphidae. *Handbooks for the Identification of British Insects,* **4**(9):32pp. Royal Entomological Society, London.

Pearce, E.J., 1974. A revised annotated list of the British Pselaphidae. *Entomologist's Monthly Magazine,* **110**:13-26.

STAPHYLINIDAE: SCAPHIDIINAE

by J. Cooter

As a result of studying larval morphology, Kasule (1966) showed the Scaphidiidae should be regarded a sub-family of the Staphylinidae, an opinion subsequently supported by Lawrence and Newton (1982) though not by Hansen (1997).

The sub-family is represented in Britain by only five species, one of which, *Scaphium immaculatum* (Olivier), is known from Britain from only a few specimens captured between 1918 and 1936 in East Kent and is now regarded as extinct in Britain. Another, *Scaphisoma assimile* Erichson, has only been captured in Britain in 1864 (1specimen) (see Rye, 1865) and 1974 (two specimens) (see Philp, 1990). The strikingly marked and unmistakeable *Scaphidium quadrimaculatum* Olivier is not uncommon under bark especially so during winter. The species of *Scaphisoma* should be looked for in moss, rotten wood and decaying fungi and leaf litter, especially in rotten leaves under logs. The three British species resemble each other quite closely and should be set to show the antennae clearly, as the relative lengths of the antennal segments are important, as too are the characters displayed at the base of the elytra. The male genitalia are characteristic and should be dissected, ideally, before mounting. These are figured by Freude (1971) and by Löbl (1970) who, in addition, figures the antennae of each species. S. *agaricinum* (L.) is the most frequently met with and widespread, S. *boleti* (Panzer) being more scarce and localised.

References

Freude. H., 1971. Scaphidiidae in Freude, H.. Harde, K.W., and Lohse, GA., *Die Käfer Mitteleoropas*, **3**:343-347.

Hansen, M., 1997. Phylogeny and classification of the staphyliniform beetle families (Coleoptera). *Royal Danish Academy of Science and Letters, Biologiske Skrifter*, **48**:339pp.

Kasule, F.K., 1966. The subfamilies of the larvae of Staphylinidae (Coleoptera) with keys to the larvae of British genera of Steninae and Proteininae. *Transactions of the Royal Entomological Society of London*, **118**: 261-283.

Lawrence, J.F. & Newton, A.F., 1982. Evolution and classification of beetles. *Annual Review of Ecology and Systematics*, **13**:261-290.

Löbl. I., 1970. *Klucze do Oznaczania Owadow Poiski* (Keys for the Identification of Polish Insects) Coleoptera, zeszyt **23** Scaphidiidae: l6pp.

Philp. E.G., 1990. *Scaphisoma assimile* Erichson (Col., Scaphidiidae) in Kent. *Entomologist's Record and Journal of Variation*, **102**:116.

Rye, E.C., 1865. Occurrence of a species of *Scaphisoma* new to Britain. *Entomologist's Monthly Magazine*, **2**:139-141.

STAPHYLINIDAE

(Plates 37 – 40)

by S.A. Williams

In Britain the family Staphylinidae is divided into 18 subfamilies which are, following the classification of Lawrence and Newton (1995), Omaliinae, Proteininae, Micropeplinae, Pselaphinae, Pholoeocharinae, Tachyporinae, Trichophyinae, Habrocerinae, Aleocharinae, Scaphidiinae, Piestinae, Oxytelinae, Oxyporinae, Steninae, Euaesthetinae, Pseudopsinae, Paederinae and Staphylininae.

If one accepts that beetles are found in all habitats throughout the British Isles, then it is almost true to say that Staphylinidae are too. There are only a few phytophagus species but sweeping and beating will produce Staphylinidae, many by chance.

The family to too large to cover in detail in a book of this type which is a pity as the Staphylinidae are amongst the most interesting of beetles. Unfortunately they can be difficult to identify and the student will have to acquire a range of non-British literature to achieve this. Many species are small and somewhat delicate, their integument being comparatively weakly sclerotised. They should thus be treated carefully and smaller species, the aleocharines in particular, are best killed prior to preparation rather than in the field. The student should as routine carefully dissect the male and female genitalia from aleocharines and gain familiarity with the secondary sexual characters possessed by many genera, invariably the tergites of the distal abdominal segments; such characters are extremely useful in identification of such genera as *Gyrophaena* and *Tachinus*.

Hodge & Jones (1995) will be found very useful when working the Staphylinidae. Excellent up to date information about the British Staphylinidae is to be found on the '*Coleopterist*' website (http://www.coleopterist.org.uk) including an invaluable annotated checklist.

The sub-families are summarised below.

OMALIINAE (78 species). The Omaliines are distinguished by having a pair of ocelli on the head level with the hind margin of the eyes. They are mostly small inconspicuous beetles, many are flattened particularly those that are found under bark such as certain *Phyllodrepa*. Many, for instance *Omalium* species are found in rotting vegetation, dung and carrion, whilst the genus *Eusphalerum* can be swept from flowers, sometimes in numbers and the lucky collector may find the uncommon *E. primulae* (Stephens) on primroses in the spring. Certain species are found in birds nests and piles of droppings, for example *Phyllodrepa* which is associated with feral pigeons, whilst others, for example the rare *Eudectus whitei* Sharp, are found in barren montane regions above 600m.

PROTEININAE (11 species). Small broad insects comprising three genera: *Metopsia* and *Megarthrus* both have the hind edge of the pronotum strongly excised whilst *Metopsia* also has a single ocellus in the centre of the head. The third genus *Proteinus* has the hind angle entire. All three genera are found in rotting fungi and vegetable matter and in addition *Proteinus* is often present in carrion, sometimes in large numbers.

The identity of *Proteinus* species is best confirmed by reference to the aedeagus of the male. These are illustrated in Tottenham (1954) and Lohse (1964).

MICROPEPLINAE (five species). Species of *Micropeplus* are easily recognised by the strongly ribbed pronotum, elytra and abdomen. They occur in decaying vegetation; patience is needed when searching for them as they move very slowly and are often covered with detritus making them hard to see.

PSELAPHINAE (51 species) (by M. Darby – see earlier, p. 38)

PHLOEOCHARINAE (one species) *Phloeocharis subtillissima* Mannerheim is a small beetle not exceeding 2mm. Rather broad and somewhat flattened, covered in greyish pubescence; easily be mistaken for an Omaliine but has no ocelli on the head. It is found under loose bark and moss on trees.

TACHYPORINAE (66 species) Mostly medium sized beetles, fusiform with the abdomen strongly tapering to the apex. They are attractive insects of characteristic appearance often shining and brightly coloured. Nine genera, with *Tachyporus* and *Tachinus* the most commonly met. Indeed *Tachyporus hypnorum* (F.) is one of the most common British beetles and it is unusual to sieve moss or tufts of grass without finding it, along with other members of the genus. *Tachyporus* are not always easy to identify and earlier keys based on colour differences are not entirely reliable, it is advisable to use the recent key based on elytral setae (Booth, 1984). Good diffuse lighting is needed to see the setae (or if abraded, the point of insertion) and although the key may seem difficult at first, it is worth persevering. On certain occasions it can be helpful to remove an elytron from the beetle and view on a glass slide in clove oil, as this enables the setae and seta-bearing punctures to be more clearly observed. It is a good idea to save up *Tachyporus* until a number are available for identification on the sound principle that it is often easier to identify several than the odd one. It is all a matter of gaining familiarity.

Two further genera that can be troublesome are *Sepedophilus* and *Mycetoporus*, on account of further species being added to the British List since the publication of Joy (1932); the males are best dissected and identified using Hammond (1972) for Sepedophilus and Lohse (1964 and 1989) for *Mycetoporus*.

Tachyporines occur in moss, grass tussocks and leaf litter. Clumps of 'toadstool' fungi growing on cut stumps etc are likely to produce species of *Lordithon* with some *Tachinus* also in fungi, carrion or dung (especially badger 'latrines'); some species of *Sepedophilus* occur in rotting wood, as does *Tachinus bipunctatus* (Gravenhorst). A number of *Mycetoporus* species are restricted to high ground and mountains where they can be found by 'grubbing' and sieving moss, litter etc.

TRICHOPHYINAE (one species) *Trichophya pilicornis* (Gyllenhal) is a rare species associated with wood chippings and sawdust in woodland, particularly pine. It flies readily and has been taken in autocatchers and flight interception traps. The antennae are very similar to those of *Habrocerus* but more slender, it is a delicate beetle with the body roughly sculptured and might be mistaken in the field for an Aleocharine.

HABROCERINAE (one species) *Habrocerus capillaricornis* (Gyllenhal) similar in general form to *Tachyporus* spp. but with very slender antennae with long hairs ringed around the apex of most segments. The body is shining, with long bristles along the sides. It is found commonly in vegetable refuse.

ALEOCHARINAE (340 species) The Aleocharinae are by far the largest subfamily in the Staphylinidae and pose the greatest number of problems relating to identification and classification. They are characterised by having the antennae inserted on the upper surface of the head at or near the level of the front of the eyes and not under the front or on the side as in the other subfamilies. Aleocharines are a diverse and a most interesting group of small beetles, often rather soft bodied and delicate requiring careful handling if they are to stay in one piece (for example *Myllaena* species; these should be handled with great care and killed immediately prior to mounting. A minimum of very dilute glue should be used as the appendages will easily be broken off and antennal segments separated).

Aleocharines are to be found in most habitats; rotting vegetation probably produces the greatest number of individuals and species closely followed by dung, carrion and fungi particularly when decaying, birds and mammals nests, banks of rivers and lakes, the seashore, high ground and mountains all can produce characteristic Aleocharines. Many are rare or very local whilst others are amongst the commonest of beetles. They are usually easier to name if the genitalia is dissected out and kept with the specimen either on the card mount or underneath on acrylic sheet (pp. 359-361) Most can be named by using Joy (1932) supplemented by Freude, Harde, & Lohse (1974 and 1989). The latter is particularly useful as it contains figures of the male and female genitalia of the Athetini, a difficult tribe that can often be hard to identify as the species closely resemble one another and the most recent key in English, that of Joy (1932) is out of date and difficult to use unless you have reliably named Athetini for comparison. Here the student will invariably need the assistance of a coleopterist with specialist knowledge in this difficult tribe. Such help is usually given if material is sent in small quantities (with return postage) and due notice is given.

However, with some experience and perseverance these problems can be overcome and the Athetini will be a source of considerable interest and pleasure.

SCAPHIDIINAE (five species) (by J.Cooter – see earlier, p. 40)

PIESTINAE (one species) *Siagonium quadricorne* Kirby is a flat, shiny insect that should be looked for under bark. The male has long sharp horns on the mandibles making it an attractive beetle of medium size and cannot be confused with anything else. It is widespread, often common where it occurs.

OXYTELINAE (90 species) A subfamily of 15 genera mostly occurring in marshy places or near water, an exception is genus *Anotylus* of which the majority are to be found in dung and rotting vegetation including seaweed. The large genus *Bledius* are to be found near water where they burrow in the sandy or clayey margins of rivers, ponds and sea cliffs and can be detected by the little mounds that they throw up. They tend to congregate in colonies neglecting areas that, to the human eye, seem to be just as suitable. They come to light and are occasionally caught in moth traps. *Carpelimus* are to be found in similar situations, often in numbers and are best captured by splashing the banks with water which brings them out from the mud or sand where they make shallow galleries. They are also found in litter and grass roots in marshy places. *Thinobius* are very small beetles seldom exceeding 2mm that may be sieved from sand or fine shingle on the banks of rivers, particularly in Scotland. Patience is needed when collecting such small beetles and a tray or camping table on which to carry out the sieving is often preferable to crouching over a sheet. The group can be readily named with Tottenham (1954) with allowances for additional species and name changes (for details of which see Hodge and Jones (*op. cit.*) and the check list posted on the '*Coleopterist*' web site).

OXYPORINAE (one species) *Oxyporus rufus* (L.) is a robust shining beetle of moderate size (11mm) bright orange-red with black head, elytra and apex of the body. The large mandibles are black and at the base of each elytron is a yellow patch; it is a very handsome distinctive insect unlikely to be confused with any other; it can be found, not uncommonly, in the gills of fungi.

STENINAE (74 species) *Stenus* are insects of distinctive appearance; broad head with bulging eyes, barrel shaped pronotum and abdomen tapering to the apex. Easily recognised and most are readily identified using Tottenham (1954), the exception being certain small species in the sub genus *Nestus*, which can be troublesome unless they are dissected males. Mostly found in damp places in moss and grass roots at the edges of rivers and ponds.

EUAESTHETINAE (three species) Small beetles not exceeding 2mm, of robust build with the foreparts strongly punctured. Reddish brown in colour, usually found in damp marshy places at the roots of grass and rush (*Scirpus*); the three species are local in their distribution but usually common where they occur.

PSEUDOPSINAE (one species) *Pseudopsis sulcata* Newman is a curious uncommon species that is found in the bottom of piles of hay* or straw with a mixture of dung. It is readily recognised by the presence of strong ridges on the pronotum and elytra. Easily distinguished from *Micropeplus* (which has similar ridges) by its larger size, narrower build and absence of an antennal club. [*See comments in 'Collecting – Sample Habitats' 'Winter – Damp refuse', page 220].

PAEDERINAE (60 species) Paederines are small to medium sized beetles characterised by having the last segment of the maxillary palpi extremely small and the penultimate segment widened apically; the body parallel sided. Most are found in wet places; some are not uncommon whilst certain *Medon* and *Scopaeus* are very rare and seldom encountered. Other species are to be looked for in grass tufts, decaying hay bales and general sweeping. The species of *Rugilus* could be confused with large ants at first glance, whilst the genus

Paederus are very attractive beetles; shining orange-red with the head and apex of the abdomen black and the elytra metallic blue. The group is dealt with very well by Joy (1932) and no great problems should occur during identification.

STAPHYLININAE (180 species) This subfamily contains the largest staphylinids with several species including *Ocypus olens* (Müller) the familiar 'Devils Coach Horse' exceeding 25mm in length and many *Philonthus* and *Quedius* reach 12mm. They are found in rotting compost, fungi, dung, under bark and in carrion, whilst the genus *Cafius* should be looked for in rotting seaweed; *Emus hirtus* (L.) on the Kentish marshes in fresh dung; the rare *Velleius dilatatus* (F.) occurs in hornets nests in the New Forest and Windsor Forest and has been captured at the sap of 'Goat moth' (*Cossus*) trees. *Bisnius subuliformis* (Gravenhorst) and *Quedius brevicornis* (Thomson) occur in birds nests in rotten trees.

The Staphylininae are a fairly characteristic group and often among the first Staphylinids that the coleopterist recognises; most can be identified using Joy (1932). The males of most species, especially the genera *Philonthus*, *Gabrius*, *Heterothops*, *Quedius*, etc. should be dissected as many of the aedeagi are figured by Freude, Harde & Lohse, (1964 and 1989) and Szujecki (1980). The single paramere should be carefully dissected from the median lobe and mounted to show its inner face. In general, males have the anterior tarsi more dilated than the females; females of this tribe do not possess a sclerotised spermatheca of use in identification.

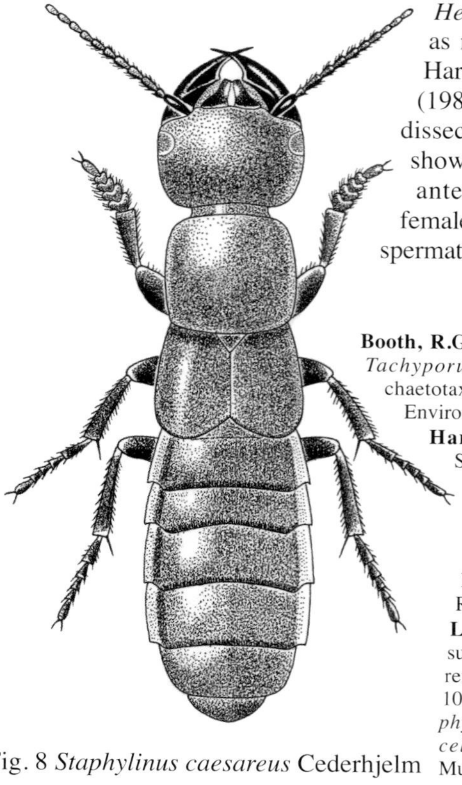

References

Booth, R.G., 1984. A provisional key to the British species of *Tachyporus* (Coleoptera) Staphylinidae) based on elytral chaetotaxy. *Circaea*,**2**:15-19 [Bulletin of the Association for Environmental Archaeology, York].

Hammond, P.M., 1972. Notes on the British Staphylinidae 3. The British species of *Sepedophilus* Gistel (*Conosomus* auctt.). *Entomologist's Monthly Magazine*, **108**:130-165.

Hodge, P.J. & Jones, R.A., 1995. *New British beetles: species not in Joy's practical handbook*. British Entomological and Natural History Society, Reading. xvi+175pp.

Lawrence, J. F. & Newton, A. F. 1995. Families and subfamilies of Coleoptera (with selected genera, notes, references and data on family-group names). pp.779-1006. in Pakaluk, J. & Ślipiński, S. A. (eds.) *Biology, phylogeny, and classification of Coleoptera: Papers celebrating the 80th Birthday of Roy A. Crowson*. Muzeum i Instytyt Zoologii PAN, Warszawa.

Fig. 8 *Staphylinus caesareus* Cederhjelm

Lohse, G.A., 1964. *in* Freude, H., Harde, K.W. & Lohse, G.A., Die *Käfer Mitteleuropas*, **4** (Micropeplinae bis Tachyporinae), 264pp. Goecke & Evers, Krefeld.
Lohse, G.A. and Lucht, W.H., 1989. *Die Käfer Mitteleuropas*, **12** (1 supplementband mit katalogteil), 346pp. Goecke & Evers, Krefeld.
Szujecki, A., 1980. *Klucze do Oznaczania Owadow Polski* (Keys for the Identification of Polish Insects) Coleoptera, zeszyt **24e** Staphylinidae (Staphylininae): 164pp. Polish Entomological Society, Warsaw.
Tottenham, C.E., 1954. Staphylinidae. *Handbooks For the Identification of British. Insects,* **4**(8a). Royal Entomological Society, London, 79pp.

SCARABAEOIDEA
(Plates 41 – 48)

by Darren J. Mann

The Scarabaeoidea are a well-defined group of beetle families, which share a number of characteristics, the most obvious is their antennae, which possesses a 3-8 segmented strongly asymmetrical club that is often lamellate. This gave rise to the old group name Lamellicornia. Worldwide, the Scarabaeoidea are one of the largest groups of the Coleoptera, containing some 31,000 species distributed among 14 families, with the family Scarabaeidae accounting for approximately 90% of all species. The group occurs throughout the world, though it is most abundant in the tropics. Scarabaeoids are found in most habitats and have a wide range of feeding habits, with coprophagous, fungivorous, herbivorous, necrophorous, saprophagous and even carnivorous species. The group also includes a number of mymecophilous, termitophilous, and even ectoparasitic species. Scarabaeoids have an impact on humans through being agricultural pests, and in their use in removal of dung, and hence indirect biological control of dung inhabiting flies (Diptera).

The scarabaeoids contain a number of species that possess conspicuous ornamentation (e.g. horns), and/or beautiful colours and patterning, this along with their large size, interesting biologies and ecologies, have made them a popular group of insects. Scarabs played an important role in early religious beliefs, such as the shamanic Indian tribes (e.g. the Chaco of South America) and the Taoists (of Asian origin), both of which pre-date the well-known use of the 'sacred' scarab in Ancient Egypt, where it was associated with the sun god Khepri. This symbolism came from the association of the beetle's dung ball with the sun, and gave rise to the belief that a giant scarab rolled the sun through the sky making the sunrise and set. The first scarab worshipped was probably *Kheper aegyptorum* (Latreille). However, later, figures of *Scarabaeus sacer* L. became more common, possibly indicating some climatic change affecting the distribution of the two species in the region. Some Egyptian scholars also suggest that mummification was first derived through observations of the life history of the scarab, and that the mummy mimics the pupae of the scarab, through which the resurrection of the person (metamorphosis of the scarab) is ensured (e.g. see Camberfort, 1994).

Dung beetles (*Scarabaeus* and *Copris*) were among the favourite insects of the famous French entomologist Jean Henri Fabre, who in his publication *Souvenirs Entomologiques* (especially volume 5, 1897) wrote extensively on their biology and behavior, thus popularising the science of insect behaviour. The dung beetles (i.e. coprophagic members of the Geotrupidae and Scarabaeidae, especially the subfamilies: Aphodiinae, Coprinae, Scarabaeinae) have been extensively studied by ecologists as they exploit a so-called 'minor habitat' and can be used in the understanding of ecological patterns and processes, such as modeling population dynamics and resource competition (e.g. Hanksi & Camberfort, 1991 and references therein). Other scarabaeoids, especially the Lucanidae, and the scarabaeid subfamilies Cetoniinae, Dynastinae and certain Rutelinae, are also very popular with collectors, and high prices are often attached to the biggest and rarest specimens.

The classification of this group is still not fully resolved, with the monophyly of some groups still under question. This article follows the checklist of Mann (2002a), which follows that of Lawrence and Newton (1995) at the family level, but differs in its treatment of some subfamilies and tribes.

LUCANIDAE (stag beetles): There are about 800 species of Lucanidae in the world, with Asia being the most species-rich region. They range in size from 10-90mm, and are associated with woodland, where their larvae feed on decaying wood.

TROGIDAE (hide beetles): The 300 or so species of this family widespread family feed upon the dry remains of dead animals and the debris in the nest of birds and mammals. The three genera world wide (*Trox*, *Omorgus* and *Afromorgus*) range in size from 2-20mm and are often attracted to light.

GEOTRUPIDAE (dor beetles; earth-boring dung beetles): The geotrupids are distributed worldwide, though they are mostly restricted to the holarctic, the 300 or so species range in size from 10-45mm. The adults construct tunnels in the soil and provision them with dung, humus or leaf litter for their larvae.

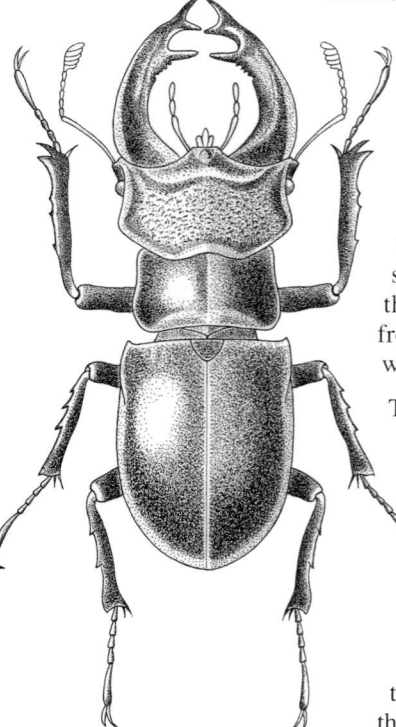

Fig. 9 *Lucanus cervus* (L.) The Stag Beetle (male)

BOLBOCERATIDAE: This family was formerly part of the Geotrupidae, and is still considered as such by some authors. The 400 species are cosmopolitan in their distribution, though they are most species-rich in the southern continents, with Australia having the richest fauna. The adults provision their burrows with humus and leaf detritus for the larvae. Represented in Britain by the genus *Odonteus* (formerly *Odontaeus*).

SCARABAEIDAE (scarab beetles, chafers, dung beetles etc.): The Scarabaeidae is the largest family within the Scarabaeoidea, with 27,000 species distributed throughout the world. The family ranges in size from 2-60mm, and includes some of the largest insects, such as the Goliath and Hercules beetles. The subfamily classification of this group is yet to be fully resolved, with many subfamilies and tribes being ranked at different levels by different authors. The following subfamilies are given by Lawrence & Newton (1995): Aclopinae; Allidiostomatinae; Aphodiinae; Cetoniinae; Dynamopodinae; Dynastinae; Euchirinae; Melolonthinae; Orphninae; Pachypodinae; Phaenomeridinae; Rutelinae; Scarabaeinae. However, several authors adopt the use of other subfamilies listed as tribes by Lawrence & Newton (1995), for example the Coprinae, which is a tribe of Scarabaeinae *sensu* Lawrence & Newton, 1995) and the Aegialiianae, Corythoderinae, Eupariinae, Psammodiinae which are all in the Aphodiinae *sensu* Lawrence & Newton (1995). The life histories of the group are very diverse and the adults and larvae of the scarabs feed on a wide range of foods from dung, fungi and carrion to pollen, fruits and plant roots.

In Britain, we have just 102 species of Scarabaeoidea in five families: Lucanidae (four spp.), Trogidae (three spp.), Bolboceratidae (one sp.), Geotrupidae (seven spp.) and Scarabaeidae (87 spp.), although many of the subfamilies and tribes within the latter family are elevated to higher ranks by some authors. The Aphodiinae and Coprinae (along with the non-British Scarabaeinae) are mainly dung feeding beetles, and number about 68 species in Britain, two-thirds of which belong to one of the largest of insect genera, *Aphodius*. The six subfamilies Cetoniinae, Trichiinae, Hopliinae, Rutelinae, Sericinae and Melolonthinae comprise the chafers, and are either detritivorous or phytophagous. The pest status of some chafers has ensured the financing of much research into their biology and control in recent years. The remaining

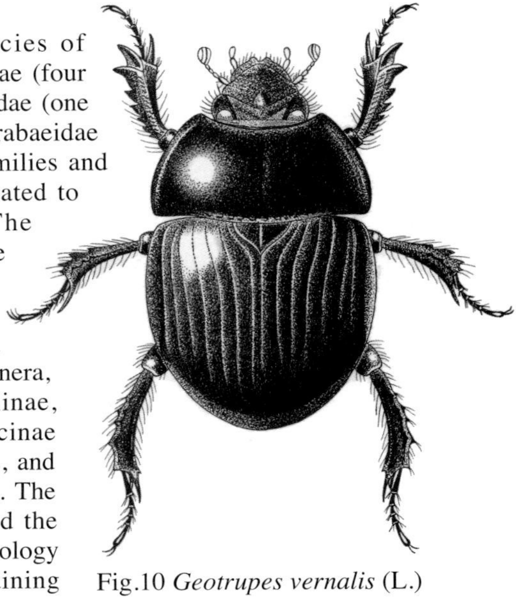

Fig.10 *Geotrupes vernalis* (L.)

three small subfamilies, Psammodiinae (six spp.), Aegialiinae (three spp.) and Eupariinae (two spp.), are often treated within the Aphodiinae. They are mostly detritivorous, many in coastal habitats, and the Eupariinae associated with dead wood.

Most adult scarabs are heavily-built, their flattened anterior tibiae and powerful spiny posterior tibiae being adaptations for burrowing. They tend to walk slowly and clumsily, but most are capable of strong flight. Sexual dimorphism is often strongly marked, with the males having greatly enlarged mandibles (e.g. stag beetles), or horn-like projections on the head and pronotum (e.g. Bolboceratidae, Geotrupidae, Coprinae and Scarabaeinae, and many of the tropical groups). The whitish, fleshy larvae have a characteristically C-shaped or 'scarabaeiform' appearance. The larval life may be unusually long in some species – 3.5 years or more in the stag beetle - in contrast to the few weeks of active life in the adult. However, the adults of more species remain inert in the pupal cell from late summer until the following spring or summer before emerging. The habitats in which the larvae and adults live are diverse: the following accounts are arranged according to the larval food source.

1. DEAD WOOD

The Lucanidae and most Cetoniinae breed in decaying wood, and excavation in suitable wood will often reveal both adults and larvae of the commoner species such as *Dorcus parallelipipedus* (L.) [the lesser stag] and *Sinodendron cylindricum* (L.), and sometimes *Cetonia aurata* (L.) [rose chafer]. Others are more restricted, in their distribution such as *Lucanus cervus* (L.) [stag beetle] in the southeast, *Trichius fasciatus* (L.) [bee beetle] in birch in Wales and Scotland, and the rare *Gnorimus nobilis* (L.) in dead wood, especially that oak, willow and of old orchard trees in the fruit-growing counties. *G. variabilis* (L.) once more widespread in the London area, is confined in Britain to ancient oaks in Windsor Forest and Great Park. [The presecnce of *Gnorimus* larvae in trees is often ascertained by finding their characteristic hard faecal pellets (frass) in cavities or spilling from the hollow interior. Larvae of *Gnorimus* (plate 46) do not 'burrow' and make tunnels like lucanid larvae, but eat the rotten fungoid wood mould at the edges of pre-existing cavities. The dry hard frass (plate 48) thus accumulates within the hollow tree. JC]. The small eupariines, *Saprosites mendax* (Blackburn) and *S. natalensis* (Peringuey) are thought to have originated in Australia, and are now established in Arundel Park and the London area respectively, where they have been found in the borings of *Dorcus* and *Sinodendron*, under the bark of dead trees and on the undersides of cut logs. The larvae of these species are relatively easily reared, in suitable containers with a quantity of the substrate. Adult lucanids may be taken in flight at dusk in June and July, whereas most of the cetoniine chafers (which are very active fliers) may be taken on flowers in sunshine. The adults of both *Saprosites* species are found in similar habits to the larvae.

2. DECAYING VEGETATION

Most species of the Aegialiinae and Psammodiinae feed on decaying plant debris (humus/detritus) in sandy soils, a habit sometimes regarded as ancestral to feeding on dung or on plant roots. The members of these groups usually occur among coastal sand dunes, an exception being *Psammoporus sabuleti* (Panzer), which is often found in shingle and in damp sandy soils on riverbanks. The rare *Diasticus vulneratus* (Sturm), found only on the sandy heaths of the Breckland, also belongs to this category: its suggested association with rabbit dung and burrows is unconfirmed. *Aphodius niger* (Panzer) feeds on decaying matter in damp soil, and in Britain is apparently confined to the edges of a number of ponds and streams in the New Forest. *Oxyomus sylvestris* (Scopoli) and *A. granarius* (L.) have catholic tastes, and are often found among decaying vegetable and compost matter, as are occasional individuals of other *Aphodius* species more often associated with dung. Although the adults are most often found in dung, the larvae of species in the subgenera *Melinopterus* (i.e. *consputus* Creutzer, *prodromus* (Brahm), *punctatosulcatus* Sturm, *sphacelatus* (Panzer) and *Nimbus* (i.e. *contaminatus* (Herbst), *obliteratus* Panzer) are thought to be general detritivores in the soil. *Oxyomus sylvestris* is recorded from compost heaps, manure heaps and rabbit dung middens, and is often quite numerous in old 'trodden-in dung and soil mix' in shaded areas, especially where deer and sheep use trees/bushes for shelter.

The chafer *Protaetia cuprea* (F.) also belongs in this guild. Its larvae should be sought in the nest mounds of wood ants, particularly those of *Formica aquilonia* Yarrow and *F. lugubris* Zetterstedt (Hymenoptera: Formicidae) in the old Caledonian pine forest of Scotland, and are quite easy to rear. Its relative *Cetonia aurata* has sometimes been found in similar situations in the south.

3. PLANT ROOTS

These constitute the larval food of chafers of the subfamilies Melolonthinae, Sericinae and Rutelinae, most of which appear to be less common than formerly. In some years, however, *Melolontha melolontha* (L.) [cockchafer or maybug], *Amphimallon solstitiale* (L.) [summer chafer] and *Phyllopertha horticola* (L.) [garden chafer] can be locally abundant, and may become serious pests of turf and occasionally of cereal crops. The adults feed on leaves. Two species are usually coastal, *Anomala dubia* (Scopoli) among sand dunes, and *Amphimallon fallenii* (gyllenhal) (=*ochraceum* (Knoch)) on grassy cliff tops in a few areas on the western and southern coasts. *Melolontha hippocastani* F. is restricted to the northwest of Britain, where it is found in woodland, and has declined dramatically. Many of the chafers may be found flying about actively by day, whereas the dusk fliers (*Serica brunnea* (L.), *M. melolontha* and *hippocastani* and *A. solstitiale*) may be attracted to light in large numbers.

4. FUNGI

The single member of the Bolboceratidae, *Odonteus armiger* (Scopoli), is said to breed in hypogeal fungi such as truffles, apparently one of the most primitive habits seen in the scarabaeoids. It has perhaps always been rare, with most records coming from the chalky and sandy areas of the south and southeast. It flies readily to light at dusk and the majority of recent records are from this source. *Aphodius plagiatus* (L.) does not feed on dung, unlike most its relatives, but on small fungi growing in damp hollows among sand dunes, etc.

5. ANIMAL REMAINS

Members of the Trogidae specialise in decaying animal material, and are usually found among the dry skins of old carcasses and the nest debris and pellets of hole-nesting birds such as owls and jackdaws. *Trox sabulosus* (L.) is a very local day-flier confined to sandy areas, whereas the widespread *T. scaber* (L.) flies at night and may be taken at light. The third British species; *perlatus* was last recorded in the 1930s on the Dorset cliffs. Occasionally, the adults of *Aphodius, Onthophagus* and geotrupid species are found in carcasses.

6. DUNG

The geotrupids (i.e. *Typhaeus, Anoplotrupes, Geotrupes* and *Trypocopris*), together with the Coprinae (i.e. *Copris* and *Onthophagus),* excavate burrows beneath or just to the side of the dung. The burrows are 1-1.5m deep in *Typhaeus* (the minotaur), 40-60cm in *Anoplotrupes, Geotrupes* and *Trypocopris,* 10-20cm in *Copris,* and 5-20cm in *Onthophagus*. Pairs cooperate in excavating the burrows, which have one or more brood chambers, and in furnishing them with supplies of dung for the larvae. Parental care is taken still further in *Copris lunaris* (L.) [horned dung beetle], whose female remains in the brood chamber until the young adults emerge. The larvae of such burrowers may be reared, with a greater chance of success where a brood chamber has been removed intact. Sometimes the burrows of geotrupids are found to contain *Aphodius* specimens that appear to behave as 'cuckoo parasites' (kleptoparasites), such as *A. porcus* (F.); others may simply use the burrows as over-wintering sites. Geotrupids fly at dusk and may come to light, which has also been known to attract *Copris*.

Most of the Aphodiinae, in contrast, remain in the surface dung throughout their larval life and pupate nearby. Though, *A. erraticus* (L.) constructs a brood mass of dung in the soil beneath the dung. *Aphodius* species hibernate as eggs; third instar larvae, pre-pupae or adults according to species. The numerous species lend themselves to rewarding studies in succession, competition, habitat and seasonal distribution. There has been some success in rearing their larvae, but surface dung is very liable to become moldy (see rearing techniques, p. 367). Only three or four species fly at night and are attracted to light, most notably *Aphodius rufipes* (L.) and *A. rufus* (Moll). Aphodiines can sometimes be found in large numbers among the debris left by winter and spring flooding.

Dung beetles are usually obtained by searching the dung pads, with the aid of a trowel, stick or spatula. Digging out and sifting the soil beneath the dung is also very productive, especially for *Onthophagus* species and some species of *Aphodius*, such as *erraticus* and *granarius*. Sieving dung into a tray, can be profitable, and is especially useful for searching for smaller aphodiines such as *Aphodius merdarius* and *Oxyomus*.

Dung baited pitfall traps are an effective method of collecting coprophagic scarabs (see Doube & Giller, 1990; Aveiga *et al.*, 1989), especially since they can be left on-site for a period of time, thus collecting the species attracted to the various successional stages of the dung. There are various designs, but a basic trap can be made easily by using a mesh cover (*ca.* 10mm gage) over a large standard pitfall, on which a dung pad is left. A second method, which has been employed with great success in Africa, is the 'tripod trap'; a net bag filled with fresh dung is suspended (using a tripod as support) over the pitfall trap mouth. If large quantities of dung are to be examined, it may be collected in buckets or plastic bags and later dispersed in water, in the floatation technique or merely washed through a suitable mesh sized sieve.

Specimens are often covered in dung, so using tissue as a filler in the collecting vessel enables the beetles to move freely and thus clean themselves, and also to excrete digested dung making them less likely to smell after mounting. Sometimes, they may require a final wash between killing and mounting. Boiling water is a very good method of killing coprophagic scarabs, as it not only helps clean the specimen, but also kills all microbial activity in the gut.

In the past much has been made of the mammalian source of the dung preferred by different scarabs, and in general, sheep dung is the most productive (Finn & Giller, 2002). However, the picture is not that simple. The factors considered to be much more important than the species of dung are:

- Habitat preference: Most species will utilise *any* dung within a restricted habitat, e.g. *Aphodius zenkeri* Germar (under dense shade in deciduous woodland and parkland) and *A. nemoralis* Erichson (shady woodland in the north & northwest).
- The unit size of the dung. The large *A. fossor* (L.), for instance, has a strong preference for cow dung.
- The stage in the aging process of the dung. *Aphodius merdarius* (F.) is one of the first beetle visitors to fresh dung, while *A. fimetarius* (L.) is found in most of the successional stages of dung.
- Soil type and site aspect.

It is evident from the above that one or two general habitats are particularly worth searching for scarabs. Many of the local species are confined to sandy ground, especially among coastal sand dunes or on inland heaths. For several of these species the requirement for light, freely drained soil is also satisfied on chalk downland and other calcareous soils. Generally, the southern coastal areas are the most species rich. On upland moors, in contrast, fewer species are present but they can occur in very large numbers.

ACTIVITY PERIOD

The adults of many species are found in May and June, but several species must be sought earlier or later than this. Some species, such as *Euheptaulacus villosus* (Gyllenhal), are only active for a very short period and are easily missed. Those reaching a peak in March and April include *A. constans* Duftschmid, *A. nemoralis* Erichson and *Heptaulacus testudinarius* (F.). Spring species such as *Typhaeus typhoeus* (L.) and *A. sphacelatus* (Panzer) reappear in smaller numbers in the autumn. Species confined to late summer and autumn (July to October) include *Geotrupes spiniger* Marsham, *A. foetens* (F.), *A. ictericus* (Laicharting), *A. rufipes* (L.) *A. rufus* (Moll), *A. zenkeri* Germar, and from September *A. contaminatus* (Herbst), *A. obliteratus* Panzer and *A. porcus* (F.). Some species, such as *A. consputus* Creutzer, *A. obliteratus*, *A. distinctus* (Müller), *A. paykulli* Bedel, *A. sphacelatus* and *A. prodromus* are found throughout the winter months, especially during mild periods. There may be slight variation in time and length of the flight period with differing latitudes and altitudes e.g. *A. contaminatus* is often found as early as August in the Black Mountains in Wales.

MOUNTING/PRESERVATION

The traditional method of card mounting specimens less than about 10mm in length is adequate for the British species, though at least a few specimens should be side mounted to allow examination of underside characters, which can be beneficial in difficult species pairs. Outside of Britain, many workers use card points attached to the underside of the specimen, as the method of mounting the smaller scarabs, this is beneficial for displaying underside characters, but can leave the specimen vulnerable to damage. Larger species are generally robust and may be pinned through the right elytron. Care should be taken to display the antennae, mouthparts and legs; the hind tarsi in particular are frequently used in identification. Dissection of the aedeagus and endophallus is not always necessary, at least for the British fauna, but can be helpful with such critical pairs as *Aphodius fimetarius* / *pedellus*, *Aphodius sphacelatus* / *punctatosulcatus*, *Amphimallon solstitiale* / *fallenii*, *Melolontha melolontha* / *hippocastani*, *Onthophagus fracticornis* / *similis*, and can aid the identification of some of the more tricky specimens of *Aphodius* and *Onthophagus*. The diagnostic features of the female genitalia have not been used extensively in the Scarabaeoidea, though they have been found to include a number of valuable species separation characteristics (e.g. in *Onthophagus*).

In some species, and especially if one works on faunas further afield, the dissection and examination of the epipharynx is a valuable diagnostic character. This best done by removing then macerating the head in potassium hydroxide (20%) for 5-10mins depending on the age and size of the specimen (i.e. old specimens take longer than fresh). Then placing the entire head in acid alcohol (to neutralize the KOH), the head is then washed in 75% IMS. The mouthparts (mentum, lacinia, palps) can then be teased away and the epipharynx gently removed. The epipharynx and mouthparts should then be mounted on the specimen mount; DMHF is the best mountant for the epipharynx.

Larvae are best fixed before they are preserved; this fixation prevents further deterioration and darkening of the specimens due to microbial activity. Fixatives, such as Carnoy's fluid or larval preserving fluid [57 parts distilled water; 28 parts IMS (95%); 11 parts Formaldehyde (35%); 4 parts glacial acetic acid] (see larval preservation techniques, p. 196) are suitable for scarabaeoid larvae. Long-term storage is best achieved in 75-80% IMS, adding a few drops of glycerol to each specimen tube will reduce evaporation, loss of alcohol and enable the specimens to be 're-wetted' should the alcohol dry out.

IDENTIFICATION

Fortunately, a reasonably up-to-date key is available (Jessop, 1986) and at time of this *Handbook* just three species were absent from this text (*Aphodius pedellus, A. punctatosulcatus* and *Saprosites natalensis*), although literature is available on their identification (e.g. Wilson, 2001; Mann, 2002b; Angus *et al.* 2003). An up-to-date checklist of the British species is available on the website of *The Coleopterist* (Mann, 2002a). The works of Baraud (1992), Krell & Ferry (1992) and Bunalski (1999) illustrate the male genitalia for all our species, and the colour plates of habitus figures in Bunalski (1999) are a useful identification guide.

The members of the genus *Aphodius* can be very variable, and some collectors have, in the past, been very keen on hunting for colour varieties. This is not so fashionable nowadays, although Dellacassa (1983) illustrates 65 varieties of *A. distinctus,* divided among 13 named forms. This exercise may seem a little pointless to most, however, it is often these extremes forms that can be the most difficult to identify so it is worth checking publications such as this if a specimen does not fit within the normal colour patterns.

Jessop's book will also aid the identification of larvae as far as genus, and van Emden (1941) includes keys to species in a number of genera. However, much more work is still needed on the immature stages of this group. Of allied interest is P. Skidmore's *Insects of the cowdung community,* (Skidmore, 1991).

THE STATUS OF SCARABS IN BRITAIN

Accounts of species with a conservation status (marked thus*) are given in Hyman, 1992; 1994 or Shirt, 1987. Most species become scarcer in the north. Unfortunately, as with most of the British coleopterous fauna many of the scarabaeoids have declined, mostly through habitat loss. The dung feeding species have suffered most dramatically, for the most part through agricultural intensification. The use of insecticidal treatments (e.g. avermectins) of livestock for the control of parasites has also had an affect on dung beetle populations, as the livestock excrete these chemicals in their dung, which have a residual detrimental affect on the dung beetles (e.g. see Strong, 1992; Strong & Wall, 1994).

1. NOT RECORDED SINCE BEFORE 1900: *Platycerus caraboides* (L.), *Aphodius obscurus* (F.), *A. satellitius* (Herbst), *A. scrofa* (F.), *A sturmi* Harold, *A. varians* Duftschmidt, *Rhyssemus germanus* (L.), *Pleurophorus caesus* (Creutzer *in* Panzer), *Onthophagus taurus* (Schreber) and *Polyphylla fullo* (L.).

2. NOT RECORDED POST 1970: *Trox perlatus* Goeze*, *Onthophagus nutans* (F.)*, *Aphodius punctatosulcatus*, *A. subterraneus* (L.)*

3. VERY RARE (THREATENED): *Aegialia rufa* (F.)*, *Aphodius brevis* Erichson*, *A. lividus* (Olivier)*, *A. quadrimaculatus* (L.)*, *Brindalus porcicollis* (Illiger)*, *Diastictus vulneratus* (Sturm)*, *Copris lunaris**, *Gnorimus variabilis** and *Onthophagus fracticornis*.*

4. RARE: *Odonteus armiger**, *A niger**, *Euheptaulacus sus**, *Heptaulacus testudinarius**, *Melolontha hippocastani** and *Gnorimus nobilis**.

5. VERY SCARCE TO LOCAL: *Trox sabulosus**, *Geotrupes mutator* (Marsham)*, *Trypocopris pyrenaeus* (Charpentier)*, *Aphodius coenosus* (Panzer)*, *A. conspurcatus* (L.)*, *A. consputus* Creutzer*, *A. constans* Duftschmidt, *A. distinctus**, *A. nemoralis**, *A. paykulli**, *A. plagiatus**, *A. porcus*, *A. putridus* (Fourcroy)*, *A. sordidus* (F.), *Euheptaulacus villosus**, *Psammodius asper* (F.), *Onthophagus nuchicornis* (L.)*, *O. vacca* (L.)*, *Omaloplia ruricola* (F.)* and *Amphimallon fallenii* (Gyllenhal)*.

6. LOCAL TO COMMON IN N + W, SCARCE TO ABSENT IN SE: *Anoplotrupes stercorosus* (Scriba), *Trypocopris vernalis* (L.)*, *Psammoporus sabuleti* (Panzer)*, *Aphodius depressus* (Kugelann), *A. fasciatus* (Olivier)*, *A. lapponum* Gyllenhal, *Protaetia cuprea** and *Trichius fasciatus**.

7. LOCAL TO COMMON IN S, SCARCE TO ABSENT IN N: *Lucanus cervus**, *Dorcus parallelipipedus*, *Trox scaber*, *Typhaeus typhoeus*, *Aphodius borealis* Gyllenhal, *A. foetidus* (Herbst), *A. granarius*, *A. haemorrhoidalis* (L.), *A. zenkeri**, *Oxyomus sylvestris*, *Onthophagus coenobita* (Herbst), *O. joannae* Goljan, *O. similis* (Scriba), *Hoplia philanthus* (Fuessly), *Amphimallon solstitiale*, *Anomala dubia* and *Cetonia aurata*.

8. GENERALLY DISTRIBUTED, LOCAL TO ABUNDANT: *Sinodendron cylindricum, Geotrupes spiniger* (Marsham), *G. stercorarius* (L.), *Aegialia arenaria* (F.), *Aphodius ater* (De Geer), *A. contaminatus* (Herbst), *A. erraticus* (Panzer), *A. fimetarius* (L.), *A. foetens* (F.), *A. fossor, A. ictericus, A. luridus* (F.), *A. merdarius* (F.), *A. obliteratus* Panzer, *A. pedellus, A. prodromus* (Brahm), *A. pusillus* (Herbst), *A. rufipes, A. rufus, A. sphacelatus, A. sticticus* (Panzer), *Serica brunnea* (L.), *Melolontha melolontha* and *Phyllopertha horticola* (L.).

9. ESTABLISHED ALIENS: *Saprosites mendax* (Sussex), *S. natalensis* (London area), *Tesarius caelatus* (Lancashire & Glamorgan).

10. VAGRANTS: *Oxythyrea funesta* (Poda) and *Trichius zonatus* Germar.

References

Angus, R, B, CJ Wilson, JF Maté, PM Hammond, and DJ Mann. 2003. *Saprosites mendax* (Blackburn) and *S. natalensis* (Peringuey) (Scarabaeoidea, Aphodiidae), two species introduced into Britian. Proceedings of the Second Pan-European Conference on Saproxylic Beetles, **18**:1-4.

Aveiga, C. M., Lobo, J. M. & Martin, I. F., 1989. The Dung-Baited Pitfall Traps Possibilities in the Study of Dung Beetles Scarabaeoidea Coleoptera I. Analysis of Effectiveness. *Revue d'Ecologie et de Biologie du Sol*, **26**(1): 91-110.

Baraud, J., 1992. Coleopteres Scarabaeoidea d'Europe. *Faune de France*, **78**: i-ix, 1-856.

Bunalski, M., 1999. *Die Blatthornkäfer Mitteleuropas: Coleoptera, Scarabaeoidea. Bestimmung - Verbreitung - Ökologie.* Frantisek Slamka, Bratislava. 1-80.

Camberfort, Y. 1994. Beetles as Religious Symbols. *Cultural Entomology*, **2**: 15-23.

Dellacasa, G., 1983. Sistematica e nomenclatura degli Aphodiini italiani (Coleoptera Scarabaeidae: Aphodiinae). *Museo Regionale di Scienze Naturali Monografie*, **1**: 1-465.

Dellacasa, M., 1988. Contribution to a world-wide catalogue of Aegialiidae, Aphodiidae, Aulonocnemidae, Termitotrogidae (Coleoptera, Scarabaeoidea). *Memorie della Societa Entomologica Italiana*, **66**: 1-455.

Dellacasa, M., 1988. Contribution a world-wide catalogue of Aegialiidae, Aphodiidae, Aulonocnemidae, Termitotrogidae (Coleoptera, Scarabaeoidea). Part 2. *Memorie della Societa Entomologica Italiana*, **67**(1): 1-229.

Dellacasa, G., Bordat, P. & Dellacasa, M., 2001. A revisional essay of world genus-group taxa of Aphodiinae (Coleoptera Aphodiidae). *Memorie della Societa Entomologica Italiana*, **79**: 1-482.

Doube, B M. & Giller, P. S., 1990. A Comparison of Two Types of Trap for Sampling Dung Beetle Populations Coleoptera Scarabaeidae. *Bulletin of Entomological Research*, **80**(3): 259-264.

Finn, J.A. & Giller, P.S., 2000. Patch size and colonisation patterns: An experimental analysis using north temperate coprophagous dung beetles. *Ecography*, **23**: 315-327.

Finn, J.A. & Giller, P.S., 2002. Experimental investigations of colonization of north temperate dung beetles of different types of domestic herbivore dung. *Applied Soil Ecology*, **20**(1): 1-13.

Finn, J.A., Gittings, T. & Giller, P.S., 1998. *Aphodius* dung beetle assemblage stability at different spatial and temporal scales. *Applied Soil Ecology*, **10**: 27-36.

Gittings, T. & Giller, P.S., 1997. Life history traits and resource utilisation in an assemblage of north temperate *Aphodius* dung beetles (Coleoptera: Scarabaeidae). *Ecography*, **20**: 55-66.

Holter, P., 1982. Resource utilization and local coexistence in a guild of scarabaeid dung beetles (*Aphodius* spp.). *Oikos*, **39**: 213-227.

Jessop, L., 1986. Dung beetles and chafers. Coleoptera: Scarabaeoidea. *Handbooks for the Identification of British Insects*, **5**(11):1-53. Royal Entomological Society, London.

Krell, F.T. & Fery, H., 1992. Trogidae, Geotrupidae, Scarabaeidae, Lucanidae. *in* Lohse, G.A. & Lucht, W.H. [Eds]. *Die Käfer Mitteleuropas. 2 Supplementband mit Katalogteil.* (Volume 13). Goecke & Evers. Krefeld. 1-375. Chapter pagination: 201-253.

Mann, D.J., 2002a. *Checklist of Beetles of the British Isles. Families: Lucanidae; Trogidae; Bolboceratidae; Geotrupidae; Scarabaeidae.* http://www.coleopterist.org.uk/checklist.htm

Mann, D.J., 2002b. Changes to the British Coleoptera List published in 2000 and 2001. *The Coleopterist*, **11**(2): 52-63.

Hanski, I. & Cambefort, Y., [Eds]. 1991. *Dung beetle ecology.* Princeton University Press, Princeton, New Jersey. i-xii, 1-481.

Skidmore, P., 1991. *Insects of the British cow-dung community.* Field Studies Council Occasional Publication, **21**: 1-166.

Strong, L., 1992. Avermectins: a review of their impact on insects of cattle dung. *Bulletin of Entomological Research*, **82**: 265-274.

Strong, L. & Wall, R., 1994. Effects of ivermectin and moxidectin on the insects of cattle dung. *Bulletin of Entomological Research*, **84**(3): 403-409.

van Emden, F.I., 1941. Larvae of British Beetles. II. A key to the British Lamellicornia Larvae. *Entomologists Monthly Magazine*, **77**: 117-127, 181-192.

Wilson, C.J., 2001. *Aphodius pedellus* (De Geer), a species distinct from *A. fimetarius* (L.) (Coleoptera: Aphodiidae). *Tijdschrift voor Entomologie*, **144**(1): 137-143.

CLAMBIDAE

by J. Cooter

Another small family, both numerically and in the size of the beetles. In Britain there are ten species in two genera, nine of these can be identified by using the RES Handbook (Johnson 1966) and details of their distribution and habits see Johnson 1992. For the tenth species, *Clambus simsoni* Blackburn, see Johnson (1997). One name change has been necessary since the species that will key to *radula* Endrödi-Younga in Johnson is in fact *gibbulus* (Leconte).

References

Johnson, C., 1966. Coleoptera, Clambidae. *Handbooks for the Identification of British Insects*, **4**(6(a)): 13pp. Royal Entomological Society, London.

Johnson, C., 1992. A bionomic review of the British Clambidae (Coleoptera). *Entomologist's Gazette*, **43**(1):67-71.

Johnson, C., 1997. *Clambus simsoni* Blackburn (Col., Clambidae) new to Britain, with notes on its wider distribution. *Entomologist's Monthly Magazine*, **133**:161-164.

BUPRESTIDAE

(Plate 49)

by Howard Mendel, updated by Keith Alexander

The attractive coloration, unusual form and rarity of most of the British Buprestidae contribute towards the special fascination the group has always held for coleopterists. It is unfortunate therefore that only 14 of the approximately 200 European species have been found in the wild in Britain. Most of these are restricted to the south and none is found in Ireland.

The *Agrilus* species are the most likely to be encountered and may be taken by beating the appropriate host trees in the summer months or occasionally by general sweeping. The larvae develop in, or under, bark on dying trees and shrubs and are characterised by the greatly enlarged pro-thoracic segment of their bodies (Fig. 34). Adults may be found across much of the summer. *A. laticornis* (Illiger) is the most widespread and is mostly associated with oak, while *A. angustulus* (Illiger) is associated more with hazel and especially where growing in open sunny situations such as in active coppice woodland. The characteristic emergence holes of *A. biguttatus* (Fabricius) (previously known as *A. pannonicus* (Piller & Mitterpacher) in the UK), *A. sinuatus* (Olivier) and *A. viridis* (L.) are found more frequently than the adults, in oak, hawthorn and willow respectively. The sinuate, subcortical larval burrows are also distinctive. *A. biguttatus* was formerly a great rarity but has become much more widespread across the south-east in recent decades, probably due to climate change. Another oak species, *A. sulcicollis* Lacordaire has turned up locally in parts of the south-east in recent years and appears to be a recent colonist from the near continent (James, 1994). Species with larvae that live sub-cortically or within the bark are not difficult to rear. The larval stage lasts between one and three years. In contrast, *Buprestis aurulenta* L., a large American species occasionally imported into Britain with timber, has a larval stage which may last over thirty years, and adults often emerge long after the wood supporting them has been processed.

Melanophila acuminata (De Geer) adults are attracted to fire-damaged trees on lowland heaths in July and, when at rest, resemble small pieces of charred bark. *Anthaxia nitidula* (L.) is probably our rarest species and has not been reported for many years; most of the specimens taken have been beaten from hawthorn blossom, *Crataegus* spp., or found on the flowers of Lesser Celandine, *Ranunculus ficaria* L., although the larvae develop under the bark of dying rosaceous trees and shrubs. *A. quadripunctata* (L.) develops under bark on dying pine and spruce, and has been reported from England on one occasion (Cooter, 1982).

The smaller species need specially targeted searching, focusing on the larval habits.

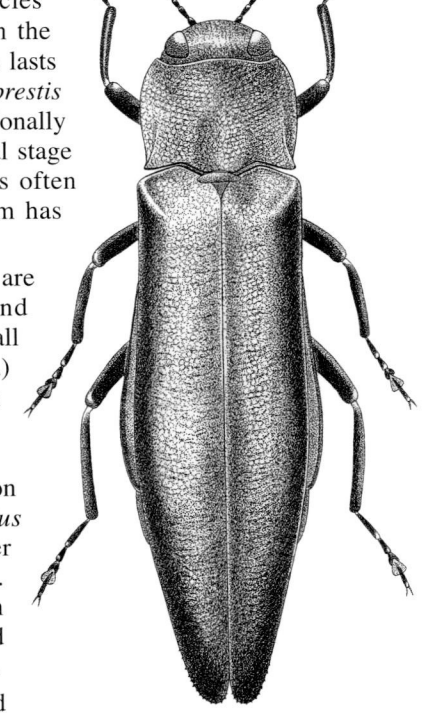

Fig. 11 *Agrilus viridis* (L.)

The larvae of the two British species of *Aphanisticus* develop in the stems of rushes, *Juncus* spp. and sedges, *Carex* spp.. *A. pusillus* (Olivier) is most readily found on calcareous grassland hillsides where there are large stands of glaucous sedge *Carex flacca* (Alexander, 2000). The larvae of *Trachys* species are leaf miners; *T. minutus* (L.) is usually associated with Willows or Hazel, *Corylus avellana* L.; *T. scrobiculatus* Kiesenwetter with Ground Ivy, *Glechoma hederacea* L., and *T. troglodytes* Gyllenhal with Devil's-bit Scabious, *Succisa pratensis* Moench. Sites with these species may be most easily recognised by looking for the characteristic larval leaf mines or shiny black eggs on the foodplant (Alexander, 1989 & 1990) or the adult feeding damage in the form of notched leaf margins (Porter, 1988). *Aphanisticus* and *Trachys* adults are long-lived and over-winter, and may be found by sieving litter or moss beneath the larval food-plants. In suitable areas they are occasionally found in flood refuse, but the use of modern portable suction samplers is much the easiest way of finding them (see pp. 215-217).

All of the British species can be identified from the dorsal surface without resort to the genitalia. However, examination of the prosternal process helps with the separation of *Agrilus angustulus* and *A. laticornis* females and *Trachys scrobiculatus* and *T. troglodytes*. A small race of *T. troglodytes* from the Breckland of East Anglia may cause confusion (Allen, 1968). Provisional distribution maps are available (Alexander, 2003).

References and further reading

Alexander, K.N.A., 1989. *Trachys troglodytes* Gyllenhal (Col., Buprestidae) widespread in the Cotswold Limestone grasslands of Gloucestershire. *British Journal of Entomology and Natural History* , **2**: 91-2

Alexander, K.N.A., 1990. *Agrilus sinuatus* (Olivier) (Col., Buprestidae) widespread in Gloucestershire and at a Herefordshire locality. *British Journal of Entomology and Natural History*, **3**: 31-2.

Alexander, K.N.A., 2000. *Aphanisticus pusillus* (Olivier) (Buprestidae) new to Gloucestershire. *Coleopterist* **9**: 40.

Alexander, K.N.A., 2003. Provisional atlas *of the Cantharoidea and Buprestoidea (Coleoptera) of Britain and Ireland*. Biological Records Centre, CEH/JNCC, Peterborough.

Allen, A.A., 1968. Notes on some British serricorn Coleoptera, with adjustments to the list. 1. – Sternoxia. *Entomologist's Monthly Magazine*, **104**: pp. 208-216

Bílý, S., 1982. *The Buprestidae (Coleoptera) of Fennoscandia and Denmark*. Kiempenborg: Scandinavian Science Press Ltd. [written in English and includes all of the British species except *Agritus sinuatus*].

Cooter, J., 1992. *Anthaxia quadripunctata* (Linnaeus, 1758) (Coleoptera: Buprestidae) in England: an enigma. *Entomologist's Gazette*, **43**: 75.

James, T., 1994. *Agrilus sulcicollis* Lacordaire (Buprestidae): a jewel beetle new to Britain. *Coleopterist* **3**: 33-35.

Levey, B., 1977. Coleoptera Buprestidae, *Handbooks for the Identification of British Insects*, **5**(1b): 8pp. Royal Entomological Society, London.

Porter, K., 1988, *Dyfed Invertebrate Group Newsletter*, no. 11.

BYRRHIDAE
(Plate 50)

by J. Cooter

The thirteen British species of Byrrhidae are to be found under stones, by grubbing and by sifting litter and moss on moorland and heathland, especially in areas of established heather which is not managed for grouse and other game by cyclic burning. River banks in such places are favoured by Byrrhids as are sandy/chalky areas and vegetated shingle at the coast. In early spring some species are to be found walking in the open on moorland and generally, as the year progresses, the Byrrhids become abraded.

The Byrrhids are commonly known as the 'Pill Beetles', a name derived from their ovoid form which is enhanced when the beetles are disturbed or in death by tightly drawing their appendages against the ventral surface and flexing the head ventrally. Some, for example *Byrrhus* species, have tibiae furnished externally with a sulcus for reception of the tarsi. Such habit and structure makes it necessary to carefully tease out the appendages prior to mounting. This is best achieved by placing the beetle on its dorsal surface and holding it gently with a pair of forceps and arranging the appendages with the aid of a pin and/or fine forceps; the head should also be gently pulled forward. Care must be taken at this stage not to abrade the setae and scales of the dorsum. Dissection of male *Byrrhus* species is advisable as the external characters are quite variable but the aedeagus offers reliable features.

With the passage of time, Joy (1932) has become out of date, but will enable identification to genus, bearing in mind *Syncalypta* has been divided into the two genera *Chaetophora* and *Curimopsis,* see Johnson (1978). For *Byrrhus* Johnson 1965 is recommended. Alternatively Paulus (1979) or Mroczkowski (1958) present keys illustrated covering all the British species.

The most widespread and abundant byrrhid is *Simplocaria semistriata* (F.), the other member of this genus, *maculosa* Er., is excessively scarce with only three British specimens known. *Morychus aeneus* (F.) is a northern species known as far south as Lancashire, but the majority of specimens in collections originate from the Scottish Highlands. *Cytilus sericeus* (Forster) occurs in a wide range of habitats including woodlands and moors throughout Britain. *Byrrhus* species are typical of moorland and heaths with *arietinus* Steffahny, the most uncommon member of the genus, having a decidedly northern distribution. *Porcinolus murinus* (F.) is a rare heathland insect known from the Breck, New Forest and Surrey heaths; there appear to be no recent records. *Curimopsis nigrita* (Palm), recently added to our List, is known only from Thorne Waste, Yorkshire. *Curimopsis maritima* (Marsh.), as its name implies, is to be found near the coast in dry places such as vegetated shingle and sandy/chalky areas. It occurs in England and is not known from the southwest and west coasts, but there are in addition a few records from inland counties. *Curimopsis setigera* (Ill.) on the other hand is a western species found in dry sandy places from Dumfries to Hampshire and with one inland county record.

References

Johnson, C., 1965. The British species of the genus *Byrrhus* L., including *B. arietinus* Steffahny (Col., Byrrhidae) new to the British List. *Entomologist's Monthly Magazine*, **101** (1966):111-115.

Johnson, C., 1978. Notes on Byrrhidae (Col.); with Special Reference to, and a Species New to the British Fauna. *Entomologist's Record and Journal of Variation*, **90**:141-147.

Mroczkowski, M., 1958. Byrrhidae – Nosodendridae: Coleoptera, zeszyt **50-51** 30pp. Klueze do Oznaczania Owadow Poiski, Warsaw.

Paulus, H.F., 1979. *in* Freude, H., Harde, K.W., & Lohse, G.A., *Die Käfer Mitteleuropas*, **6**:47. Fam. Byrrhidae: pp. 328-350. Goeke & Evers, Krefeld.

PTILODACTYLIDAE
by D.J. Mann

The family Ptilodactylidae, at the time of writing (January, 2005), has not yet been brought forward as British, see Mann, *in press*. The single species, *Ptilodactyla exotica* Chapin, 1927, known to occur in Britain has been recorded from glass houses at Kew and Cambridge University Botanic Gardens where it has been found in numbers; the first British capture is dated 1907.

Reference

Mann, D.J., *in press*. *Ptilodactyla exotica* Chapin, 1927 (Coleoptera: Ptilodactylidae) established breeding under glass in Britain, with a brief discussion on the family Ptilodactylidae.

EUCNEMIDAE
by Howard Mendel

Of the six species on the British list only three were known prior to the 1950s, and it is quite possible that others await discovery. Eucnemids are secretive woodland species seldom found by the casual collector, but they may occasionally be seen in numbers at breeding sites. The larvae live in dead wood but only those of *Melasis buprestoides* (L.) and *Epiphanis cornutus* Eschscholtz are likely to be found.

M. buprestoides is the largest (up to 10mm) and most common species of the family in Britain. It is found as far north as Yorkshire and it is the only eucnemid known from Ireland. Its larvae are not of the typical eucnemid type but somewhat resemble those of Buprestids, having an enlarged pro-thoracic segment. They are most frequently found in dead beech, less commonly in oak, hornbeam or other dead wood; often with dead adults. Their u-shaped tunnels, across the grain of the wood, are characteristic. *Microrhagus pygmaeus* (F.) is also widespread, though very local, and found from southern England to the north of Scotland. Most specimens are swept in June or July beneath the ancient oaks in which the species breeds. *Eucnemis capucina* Ahrens, an ancient forest relict, is a great rarity. The larvae develop in dead wood, usually beech, and in warm days in June and July adults may be found crawling on the surface of the dead trunks.

The two species of *Hylis* and *Epiphanis cornutus* are relatively recent discoveries in Britain. *H. olexai* (Palm) is found sparingly in the south-east, associated with a range of tree species, including conifers. *H. cariniceps* (Reitter) is known only from a very few sites in southern Britain and has not yet been associated with any particular type of timber. It is not known for certain whether or not our two species of *Hylis* are long over looked forest relicts or recent arrivals. The latter seems more likely and both species appear to be expanding their range in Britain. *Epiphanis cornutus* was most likely introduced from North America and is also extending its range in Britain; in recent years there have been records in Scotland, well away from the core distribution in the south, perhaps the result of a separate introduction. It is associated equally with hardwoods and softwoods.

All of the British eucnemids can be identified by external characters of the dorsal surface. Allen (1968) provides a very useful key to the British species. It also includes *Cerophytum elateroides* Latreille (Cerophytidae), at one time thought to be a native British species. Muona (1993) is a valuable source of information about species found in Britain.

References

Allen, A. A., 1954. *Hypocoelus procerulus* Mannh. (Col., Eucnemidae, Anelastini) in Kent and Surrey: a tribe, genus and species new to Britain. *Entomologist's Monthly Magazine,* **90**:228-230.

Allen, A. A., 1969. Notes on some British serricorn Coleoptera, with adjustments to the list. 1. Sternoxia. *Entomologist's Monthly Magazine,* **104** (1968):208-216.

Mendel, H. & Clarke, R. E., 1996. *Provisional atlas of the click beetles (Coleoptera: Elateroidea) of Britain and Ireland.* Ipswich: Ipswich Borough Council.

Muona, J., 1993. Review of the phylogeny, classification and biology of the family Eucnemidae. *Entomologica scandinavica,* supplement no. 44.

Skidmore, P., 1966. *Epiphanis cornutus* Eschsch. (Col.. Eucnemidae) new to the British list. *Entomologist,* **99**:137-139.

THROSCIDAE
by Howard Mendel

The small family Throscidae (some 200 species worldwide) is represented in the British Isles by two genera, *Trixagus* (*Throscus* in older works) with four species, and a single *Aulonothroscus,* which together comprise most of the European fauna. The Throscidae are closely related to the true click-beetles and are able to spring in the same way, although this ability is less well developed and seldom observed. The British species are superficially very similar; each less than 3.5mm in length, darker or lighter brown, elongate-oval with pointed pronotal hind angles and clubbed antennae.

When disturbed, throscids tuck their antennae and legs into grooves on the underside of their bodies and remain motionless assuming a seed-like appearance. They react in this way in the killing bottle and are notoriously difficult to set. However, a little time spent extending the antennae and legs can save considerably more time when it comes to identification. Excess glue on the mounting card may also make identification difficult.

The male genitalia are quite distinct in each species but it will seldom be necessary to resort to dissection. The British species can all be reliably separated on external characters and, with a little practice, careful examination of the head alone is all that is required for identification. The most useful characters are the extent of division of the eyes and the presence/absence and type of carinae between the eyes. Size, pronotal shape, and puncturation of the pronotum and elytra provide additional characters for identification. There are also good characters on the underside which can be useful when identifying spirit preserved material. All of the British species show sexual dimorphism and this can provide a short-cut to identification. For example, only the males of *T. gracilis* (= *elateroides* auctt. Brit.) have a distinct tooth near the base of the middle tibiae, and the dense comb of pubescence on the elytral margin of male *T. obtusus* (Curtis) is similarly diagnostic. Males of *T. dermestoides* (L.) and *T. carinifrons* (de Bonvouloir) have a distinctly expanded antennal club compared with the females.

Except for *Aulonothroscus brevicollis* (de Bonvouloir) , discovered 'new to Britain' by Ashe (1942), Fowler or Joy will satisfactorily identify the British species. This addition to the British fauna was recognised as an *Aulonothroscus* rather than a *Trixagus* by Burakowski (1975); it is the only British throscid without distinctly emarginate eyes. It is a species of quality pasture woodland sites (Mendel, 1985) and is being found in an increasing number of localities. Where it occurs, beetles may be beaten from old trees (usually oaks), or found in wood mould or under bark.

General collecting methods such as sweeping and beating (particularly in the late afternoon or evening) will produce throscids in suitable localities. They fly on warm still evenings and are often attracted to light. Our commonest species, *T. dermestoides*, is the only species recorded from either Scotland or Ireland. *T. gracilis* and *T. obtusus* are most likely to be found in saltmarsh localities in S.E. England and East Anglia. They may be swept in the summer or collected from grass tussocks or flood refuse in the winter months.

Burakowski (1975) provides a detailed description of the larva and pupa of *T. dermestoides* and an excellent account of its life history. The larvae live in the soil and show little activity, feeding on ectotrophic mycorrhizae of various trees. The life cycle is usually completed in two years, the newly eclosed adults over-wintering within the pupal cell. Little is known about the life histories of the other British species.

References

Ashe, G. H., 1942. *Trixagus* (= *Throscus*) *brevicollis* Bonv. (Col., Trixagidae), a species new to Britain. *Entomologist's Monthly Magazine*, **78**:287.

Burakowski, B., 1975. Development, distribution and habits of *Trixagus dermestoides* (L.) with notes on the Throscidae and Lissomidae (Coleoptera, Elateroidea). *Annales Zoologici*, **32**:376-405.

Leseigneur, L., 1997. Réhabilitation de *Trixagus gracilis* Wollaston (Coleoptera, Throscidae). *Bulletin de la Société entomologique de France*, **102**:137-142.

Mendel. H., 1985. *Trixagus brevicollis* (de Bonvouloir) (Col. Throscidae) in Britain. *Entomologist's Monthly Magazine*, **121**:58.

Mendel, H. & Clarke, R. E., 1996. *Provisional atlas of the click beetles (Coleoptera: Elateroidea) of Britain and Ireland.* Ipswich: Ipswich Borough Council.

ELATERIDAE
(Plates 51 – 53)

by Howard Mendel

These elongate beetles with characteristic pointed hind-angles to the pronotum are immediately recognised by their ability to 'spring' into the air when disturbed; often with an audible 'click'. They are only likely to be confused with beetles belonging to the related families Throscidae and Eucnemidae.

The larvae of click beetles are collectively known as 'wireworms', and are almost as well known as the adults. The majority are elongate with a tough cuticle so that the term 'wireworm' is entirely appropriate (fig. 35 and plate 52). However, larvae of *Lacon querceus* (Herbst) and *Agrypnus murinus* (L.) (Agrypninae) are more fleshy and grub-like, whilst those of *Cardiophorus* spp. and *Dicronychus equisetioides* Lohse (Cardiophorinae) are incredibly long and thread-like with pseudo-segmentation and an expandable, membranous cuticle. A few species can cause serious crop damage and because of this have been well studied.

Emden (1945) wrote a useful key to larvae of British species known at that time, amended by Cooper (1945). There is also a detailed account of the larvae of *Ampedus* (Emden, 1956). Larvae of most species are now generally regarded as omnivorous, relying to a greater or lesser extent on other insect larvae, worms and various other invertebrates, as well as resorting to cannibalism. Superficially similar larvae of *Dryops* spp. and some of the Tenebrionidae lack the characteristic pygopodium of the 10th. abdominal segment of the Elateridae. The length of larval life in Elateridae is not fixed, but rather determined by the conditions and the availability of suitable food. Species that may normally develop in two or three years may take five or six years if food is in short supply.

Although a few click-beetles are extremely common and widespread, and will be found in the course of general collecting (under stones, by beating, sweeping or in pit-fall traps), a large proportion are genuinely scarce and have precise habitat requirements. Such species are seldom encountered accidentally and to find them a knowledge of their life-histories and distributions (see Mendel & Clarke, 1996) is most helpful. This work also includes the most up-to-date published check list of species found in the British Isles. *Selatosomus cruciatus* (L.), *Ampedus sanguineus* (L.), *Cardiophorus gramineus* (Scopoli), *C. ruficollis* (L.) and *Zorochros meridionalis* (Castelnau) have not been recorded for over a century and should be regarded as extinct in Britain.

The British species divide nearly equally between those with larvae that inhabit the soil and those associated with dead wood. It is amongst those inhabiting the soil that the pest species are found and the genus *Agriotes* is particularly well known in this respect. However, some species of *Athous*, *Ctenicera* and *Aplotarsus* may also be serious pests. Although there is overlap, the various species tend to inhabit different soil types, so that heavy lowland soils, thin peaty upland soils, heathlands, woodlands etc. will all have their own particular suite of species. Whether or not and how often the ground is flooded is also an important environmental factor determining where elaterid larvae live.

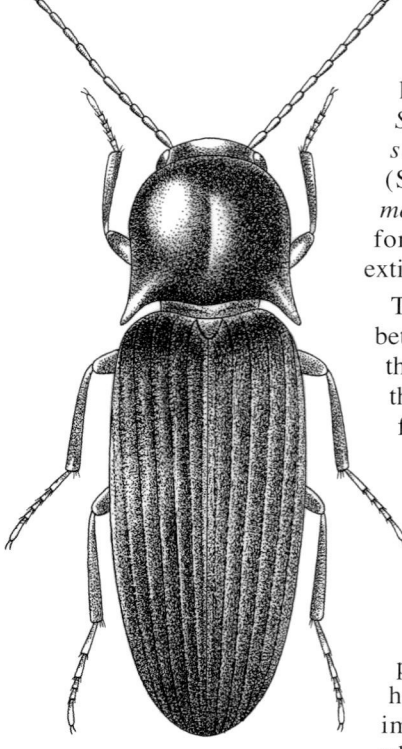

Fig. 12 *Agriotes* sp.

Two particular soil habitats are of special note. *Cardiophorus asellus* Erichson, *Dicronychus, Melanotus punctolineatus* (Pelerin) and *Cidnopus aeruginosus* (Olivier) are associated with sandy soils and dune systems. Searching suitable areas on fine days in spring when the beetles may be found struggling over areas of bare sand or flying in the sunshine, can be very productive. *Negastrius* spp., *Zorochros minimus* (Boisduval & Lacordaire) and *Fleutiauxellus maritimus* (Curtis) are found amongst sand and shingle on river banks. These beetles are also best searched for in the spring or early summer and each species seems to have a preferred micro-habitat. *F. maritimus*, for example, likes areas of 'clean' (unvegetated and with minimal silt) shingle, *Zorochros* can tolerate shingle with sand and a certain amount of mud, whereas the species of *Negastrius* prefer well-graded coarse sand or grit with sparse vegetation, often occupying a distinct zone some distance from the water.

The deadwood species have amongst their ranks some of the rarest of British beetles; relicts of the wildwood, now confined to a very few localities. *Lacon, Ampedus nigerrimus* (Lacordaire) and *Brachygonus* are known today only from Windsor Forest. *Limoniscus violaceus* (Müller) ('violet click beetle'), also known from Windsor, has been added to Schedule 5 of the Wildlife and Countryside Act 1981, and should not be collected. Many of the deadwood species are associated with rot-holes in the living trees, developing in the red-rot (*Lacon, Ampedus cardinalis* (Schiødte)) or black-rot (*Ischnodes, Elater*). Other species are commonly found under bark (*Melanotus villosus* (Geoffroy *in* Fourcroy), *Stenagostus, Denticollis*) or in the wood itself (*Ampedus* spp., *Procraerus, Megapenthes*). Larvae are often found in 'colonies' of other wood-boring beetles (Lucanidae, Cerambycidae and Curculionidae) on which they prey. A strong knife, or even better a small pick, is needed to work suitable dead wood, but care must be exercised as a great deal of damage can be done in a short time by an eager collector. *Procraerus*, or even *Megapenthes*, may be found by the fortunate collector by beating the blossom of hawthorn in ancient woodland areas.

Two basic types of life-histories are found in the Elateridae. Some species eclose in July and August but remain in the pupal cell until the following spring. Searching for fully mature adults in their pupal cells, in the winter months, can be very rewarding. *Procraerus* and the species of *Ampedus* are most easily found in this way. In other species, adults emerge from the pupae in the summer and leave the pupal cell as soon as the cuticle has hardened. It is surprising just how infrequently many of the species in this latter group are found as adults. The large black larvae of *Stenagostus*, for example, are not uncommon under loose bark in suitable localities across much of southern Britain, but adults in the open are rarely seen except perhaps at the lepidopterist's lamp. Cryptic behaviour and a short adult life combine to make some species elusive rather than genuinely rare.

The larvae of many species are more frequently encountered than the adults, so rearing can be a very productive way of obtaining specimens as well as a good way of learning more about the life-histories of Elateridae. Deadwood species, in particular, can easily be reared by placing sufficient pabulum in a jar and occasionally adding a few drops of water. The contents of the jar must not be allowed to become either too dry or sodden. Insect larvae may also have to be added as food, from time to time, but cheese or other high protein prey substitutes have been used successfully. Patience will often be required.

Considering that click-beetles form a well defined group which is generally popular with collectors, it is surprising how difficult some of the species can be to identify. The species of *Ampedus* are notorious. As a general guide, in the majority of species the males have longer and differently formed antennae. Although the genitalia may be of value in separating certain species, the structure is of little use overall for identification. Unfortunately, comparative differences in shape and in punctuation still have to be relied upon.

Fowler (vol. 4) and Joy are still the standard works, although Platia (1994) and Laibner (2000) have keys in English to most of the species found in the British Isles. Leseigneur (1972) has numerous figures and interesting notes on biology; it will prove useful to coleopterists who can cope with a French text. In the following bibliography, the number of references to additions and adjustments to the British list, since Joy (1932), are a good indication of the inadequacy of easily available keys.

References

Allen. A. A., 1936. *Adelocera quercea* Herbst (Col., Elateridae) established as British. *Entomologist's Monthly Magazine,* **72**:267-269.

Allen, A. A., 1937. *Limoniscus violaceus* Müll. (Elateridae), a genus and species of Coleoptera new to Britain. *Entomologist's Record and Journal of Variation,* **49**:110-111.

Allen, A. A., 1938. *Elater ruficeps* Muls.; a beetle new to Britain. *Entomologist's Monthly Magazine,* **74**:172.

Allen, A. A., 1966. The rarer Sternoxia (Col.) of Windsor Forest. *Entomologist's Record and Journal of Variation,* **78**:14-23.

Allen, A. A., 1969. Notes on some British serricorn Coleoptera with adjustments to the list. 1. – Sternoxia. *Entomologist's Monthly Magazine,* **104** (1968):208-216.

Allen, A. A., 1990. Note on, and key to, the often confused British species of *Ampedus* Germ. (Col.: Elateridae) with corrections to some erroneous records. *Entomologist's Record and Journal of Variation,* **102**:121-127.

Cooper, B. A., 1945. Notes on certain elaterid (Col.) larvae. *Entomologist's Monthly Magazine,***81**:128-130.

Cooter, J., 1983. *Zorochros flavipes* (Aubé) (Col., Elateridae) new to Britain. *Entomologist's Monthly Magazine,* **119**:233-236.

Emden, F. I. van, 1945. Larvae of British beetles. V. Elateridae. *Entomologist's Monthly Magazine,* **81**:31-37.

Emden, H. F. van, 1956. Morphology and identification of the British larvae of the genus *Elater* (Col., Elateridae). *Entomologist's Monthly Magazine,* **92**:167-188.

Hignett, J., 1940. *Corymbites angustulus* Kies.; an Elaterid new to the list of British Coleoptera. *Entomologist's Monthly Magazine,* **76**:14.

Laibner, S., 2000. *Elateridae of the Czech and Slovak Republics.* Zlín: Kabourek.
Leseigneur, L., 1972. Coléoptères Elateridae de la faune de France continentale et de Corse. *Bulletin Mensuel de la Société Linnéenne de Lyon,* **41** (suppl.).
Mendel, H., 1990. *Zorochros meridionalis* (Lap.) (Coleoptera: Elateridae) a British species? *Coleopterist's Newsletter,* **39**:2-3.
Mendel, H., 1990. The status of *Ampedus pomone* (Stephens) *A. praeustus* (F.) and *A. quercicola* (du Buysson) (Coleoptera, Elateridae) in the British Isles. *Entomologist's Gazette,* **41**:23-30.
Mendel, H., 2002. Notes on British Elateridae: *Dicronychus equisetioides* Lohse, 1976 and *Negastrius arenicola* (Boheman, 1853) recorded from Britain. *Coleoptertist,* **11**:77-80.
Mendel, H. & Clarke, R. E., 1996. *Provisional atlas of the click beetles (Coleoptera: Elateroidea) of Britain and Ireland.* Ipswich: Ipswich Borough Council.
Owen, J. A., Allen, A. A., Carter, 1. S. & Hayek, C. M. F. von, 1985. *Panspoeus guttatus* Sharp (Col., Elateridae) new to Britain. *Entomologist's Monthly Magazine,* **121**:91-95.
Platia, G., 1994. Coleoptera Elateridae. *Fauna d'Italia,* **33**. Edizioni Calderini, Bologna.
Speight, M.C.D., 1986. *Asaphidion curtum, Dorylomorpha maculata, Selatosomus melancholicus* and *Syntormon miki:* insects new to Ireland. *Irish Naturalists Journal,* **22**:20-23.

LYCIDAE
by Keith Alexander

The Lycidae are amongst the most attractive of Britain's beetles, being relatively large and of a distinctive matt crimson colour. They all develop in dead wood that is being decayed by white-rot fungi and are primarily associated with relict old forest areas, ancient woodland and wood pastures. They are therefore all rare. *Dictyoptera aurora* (Herbst) and *Pyropterus nigroruber* (De Geer) are found in the Scottish Highlands in conifer and broadleaves, respectively, while the latter also may be found in the NE Midlands and the Killarney area of SW Ireland. In contrast, the other two British species are more south-eastern in distribution, with *Platycis minutus* (F.) relatively widespread, but *Platycis cosnardi* (Chevrolat) has rarely been seen at all. Joy keys out the first three only, *P. cosnardi* having been added to the British List subsequently (Airy-Shaw, 1944). Hodge & Jones (1995) give a brief description of how to distinguish this species from *P. minutus*.

These species are best sought by eye, scanning any decaying wood, but sweep-netting can also be productive. They are very seasonal – *P. cosnardi* the earliest in May, followed by *D. aurora* in May and June, *P. nigroruber* in July and August, and finally *P. minutus* in August and September.

References
Airy-Shaw, H.K., 1944. *Dictyopterus (Platycis) cosnardi* Chevr. (Col., Cantharidae, Lycidae) new to Britain. *Enomologist's Monthly Magazine,* **80:**204-205.
Hodge, P., & Jones, R.A., 1995. *New British beetles: species not in Joy's practical handbook.* 175pp. British Entomological and Natural History Society, Reading.

LAMPYRIDAE
(Plate 54)

by J. Cooter, updated by Keith Alexander

This family contains the 'fire-flies' and 'glowworms' which are represented in the tropics by a large number of species.

In Britain only one species, *Lampyris noctiluca* L., is likely to be encountered. The larviform female is strongly bioluminescent, the male only slightly less so; this ability to emit light has made the 'glow-worm' one of our more familiar beetles, at least in name. It occurs from the south coast as far as Argyllshire in the west and Northumberland in the east. The female is totally devoid of wings and elytra, the male possesses functional wings and full elytra. The life-history of the 'glow-worm' has been outlined by Wootton (1976) and further elucidated by Tyler (2002), the main food being snails.

Phosphaenus hemipterus (Goeze) has been found occasionally at various sites across south-east England, but no established population is known, and it appears most likely to be a casual introduction. Sites tend to be in towns, gardens, churchyards, or associated with building rubble. The larviform female is weakly luminescent, the male possesses abbreviated strongly dehiscent elytra leaving the bulk of the abdomen exposed.

A third species, *Lamprohiza splendidula* (L.) is known in Britain from two male specimens taken in Kent during 1884 (see Allen, 1989); their origin can only be speculated. The male bears a superficial resemblance to the widespread *Lampyris noctiluca*.

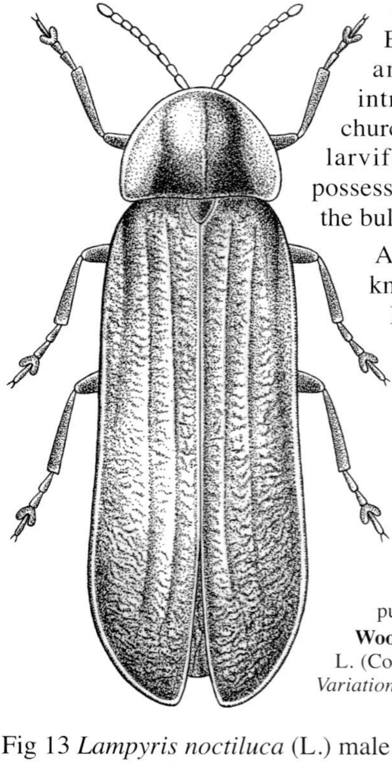

References

Allen, A.A., 1989. *Lamprohiza splendidula* (L.) (Col., Lampyridae) taken in Kent in 1884. *Entomologist's Monthly Magazine,* **125:**182.

Tyler, J., 2002. *The Glow-worm.* Sevenoaks: privately published.

Wootton, A., 1976. Rearing the Glow-worm, *Lampyris noctiluca* L. (Coleoptera: Lampyridae). *Entomologist's Record and Journal of Variation,* **88**(3):64-67.

Fig 13 *Lampyris noctiluca* (L.) male. 'the glow-worm'

CANTHARIDAE
(Plates 55 – 57)

by Keith Alexander

The Cantharidae are readily split in two. The larger, brightly coloured species form the Cantharinae, of which Britain has 25 species currently recognised. The sub-family is the Malthininae, which comprises 15 smaller dark species, generally having yellowish tips to their elytra. Although the bright colours and conspicuous behaviour of the Cantharinae make them an attractive group, their study is much neglected. Identification is not always easy as many species can be extremely variable in colour, and there may be considerable size variation. Malthininae also have their problems. In particular, female *Malthodes* can often only be keyed down to two possibilities. The males of this genus are however very straightforward, with their characteristic terminal abdominal segments. Joy (1932) is still the most readily accessible identification guide, although there have been many changes in nomenclature – check Kloet and Hincks, one species split – *Cantharis cryptica* Ashe from *C. pallida* Goeze (Ashe, 1946 & 1947), and one synonymised – *C. darwiniana* (Sharp) is now regarded as an extreme form of *C. rufa* L. Mike Fitton's (1973) Ph.D. thesis *Studies in the Biology and Ecology of Cantharidae (Coleoptera)* (Imperial College, 1973) includes a more up to day key with better detail and illustrations – copies of this key appear to be in wide circulation. A key to larval genera is available (Fitton, 1975).

The bright colours of the Cantharinae have attracted sufficient attention for them to be given English names: 'soldier beetles' for the reddish and yellowish species, and 'sailor beetles' for the black and bluish ones. However, the variation in colour possible in some species makes the distinction unhelpful, and all Cantharidae are generally known simply as 'soldier beetles'. The adult beetles are generally found amongst vegetation or on flower heads, and are most readily captured by sweeping or beating. Tall, well-developed vegetation and trees and shrubs are the most productive sources, with the majority of species living in deciduous woodlands, hedgerows and scrub, or in marshland and wet meadows. Fewer species occur in dry grasslands or on heath and moor - although these have their specialists and should not be neglected. The velvety larvae are found in similar situations and are also present in the surface layers of the soil and in leaf litter. Both adults and larvae are mainly carnivorous, feeding on dead or injured invertebrates and also smaller and slower healthy ones. They are fluid feeders, and will also feed from plant material. The very abundant *Rhagonycha fulva* (Scopoli) reddish-yellow with black elytral tips, so characteristic of flower heads in July and August, presumably feeds extensively on nectar.

Larvae can be reared in captivity by feeding with inactive insect larvae, pupae, small worms, injured larvae and woodlice, or cut pieces of vegetables such as carrot or potato. Wheat grains will also be fed upon. A soft substrate of some kind should be provided as a pupation site – moist sterile sand will do; natural sites include dead wood and fungi.

Adult Malthininae are also found by sweeping and beating, although mainly in and around woodland or old trees. Their larvae are mainly found under bark and within decaying timber. The feeding habits are most probably very similar to Cantharinae.

Deciduous woodlands, parks and hedgerows are the most productive sites, producing *Podabrus alpinus* (Paykull), *Cantharis cryptica* Ashe, *C. decipiens* Baudi , *C. livida* L., *C. nigricans* (Müller), *C. pellucida* F., *C. rufa* L., *Rhagonycha limbata* Thomson, *R. lignosa* (Müller) and *Malthodes marginatus* (Latreille) from about mid-May until early July. The very rare *M. crassicornis* (Mäklin) is also an early species. Other species occur from about mid-June to mid-August: *Malthinus balteatus* Suffrian, *M. flaveolus* (Herbst), *M. frontalis* (Marsham), *M. seriepunctatus* Kiesenwetter, and *Rhagonycha translucida* (Krynicki). Damper sites are preferred by *Malthodes dispar* (Germar) and *Rhagonycha testacea* (L.). Other species are most readily found in the deciduous woodlands of the hill country of northern and western Britain: *Ancistronycha abdominalis* (F.), *Cantharis obscura* L., *Malthodes flavoguttatus* Kiesenwetter, *M. fuscus* (Waltl), *M. guttifer* Kiesenwetter and *M. mysticus* Kiesenwetter. *Malthodes fibulatus* Kiesenwetter is most often found in the woods of chalk and limestone country, and *Rhagonycha elongata* (Fallén) is associated with the old pine forests of Scotland.

Marshes, wet meadows and similar places are the habitat of *Cantharis figurata* Mannerheim, *C. lateralis* L., *C. nigra* (De Geer), *C. pallida* Goeze, *C. thoracica* (Olivier) and *Silis ruficollis* (F.) - the latter mainly in East Anglia and sparingly from Kent to S. Wales. June and July are the best months. *Cantharis paludosa* Fallén is numerous on the peat moors and mosses of the North and West. *C. rustica* Fallén is a common species of dry grasslands in May and June, but also occurs widely in other situations. *Rhagonycha fulva* is ubiquitous in open flowery situations in July and August.

Some species are commoner in southern Britain than in the north - *Cantharis lateralis, C. rustica, Rhagonycha lutea, Malthinus balteatus, M. seriepunctatus,* and *Malthodes minimus*. For the known distribution of British Cautharidae see Alexander, 2003.

Care needs to be taken when mounting Malthininae to ensure that the male terminal abdominal segments remain free of gum. Pinning is a good alternative for the family since abdominal characters can be important in determination of the species. Antennae also may need to be examined and the minimum of gum should be used in their setting.

Distinguishing the sexes is not always straightforward. Male *Malthodes* are readily distinguished by the complex development of the terminal abdominal segments already referred to. In this and in other genera, the head of the male is often larger, especially broader, with the eyes more prominent. The antennal segments are often longer, and the middle segments have a fine groove or impressed line along their length in several *Cantharis* (*livida, pellucida,* and *nigricans*). The claws of *Ancistronycha abdominalis* are toothed in both sexes, but in the female the tooth is much longer and spine-like.

References

Alexander, K.N.A., 2003. *Provisional Atlas of the Cantharoidea and Buprestoidea (Coleoptera) of Britain and Ireland.* 81pp. Biological Records Centere, Huntingdon.

Ashe, G.H., 1946. A new British *Cantharis* confused with *C.pallida* Goeze (= *bicolor* Br. Cat.) (Col., Cantharidae). *Entomologist's Monthly Magazine,* **82**:138-9

Ashe, G.H., 1947. *Cantharis cryptica* sp.n. (Col., Cantharidae), a British species new to science. *Entomologist's Monthly Magazine,* **83**:59.

Fitton, M.G., 1973. *Studies in the biology and ecology of Cantharidae (Coleoptera).* Unpublished Ph.D. thesis, Imperial College, University of London.

Fitton, M.G., 1975. The larvae of the British genera of Cantharidae (Coleoptera). *Journal of Entomology,* (B), **44**(3): 243-245.

Joy, N.H., 1932. *A practical handbook of British beetles.* 2vols. Witherby, London.

DERODONTIDAE
by J.Cooter

Our single representative of this family, *Laricobius erichsoni* Rosehauer was first noticed in Britain during 1971 (see Hammond & Barham, 1982 and also Peacock, 1993). Its arrival here was not unexpected as it had been increasing its range north and westwards across Europe for some years helped by the increase in planting of conifer trees during the past 100 years.

Since the first finding in Suffolk, *erichsoni* has spread widely in England and Scotland and continues to increase its range. Associated with conifer trees it feeds on aphids and adelgids both as adult and larva.

Reference

Hammond, P.M. & Barham, C.S., 1982. *Laricobius erichsoni* Rosehauer (Coleoptera: Derodontidae), a species and superfamily new to Briatin. *Entomologist's Gazette,* **33**:35-40.

Peacock, E.A., 1993. Adults and larvae of hide, larder and carpet beetles and their relatives (Coleoptera: Dermestidae) and of Derodontid beetles (Coleoptera: Derodontidae). *Handbooks for the Identification of British Insects.,* **5**(3):144pages. Royal Entomological Society, London.

DERMESTIDAE
(Plates 58 and 114a)

by J. Cooter

The dermestid beetles are familiar as pests of stored products (see p. 258) and as household and museum pests, but a few are to be found out-of-doors. A Royal Entomological Society Handbook for the Dermestidae was published after the last edition of the *Coleopterist's Handbook* appeared (Peacock, 1993) and covers the family in Britain most thoroughly.

Many of the dermestids on the British List are cosmopolitan having been spread by commerce. The methods of transporting, packaging and storing raw materials and food stuffs is constantly changing, as are the methods and processes involved in producing the finished product. Health & Safety, Food Hygiene and other legislation have all conspired to restrict the occurrence of 'pests' in the industrial setting. Furthermore these laws conspire against a coleopterist gaining access to industrial premises. It is probably true to say that the bulk of stored product dermestids and other beetles being added to collections in the past 10-20 years have come from cultured stock rather than 'wild' (if that can be the appropriate term) collected.

Upon removal from the killing tube, it will be found that many, if not all, dermestids have their legs and antennae drawn tightly to the ventral surface. Careful teasing out of the appeadages is necessary and while doing this great care must be taken not to abrade the scales and setae of the dorsal surface (the task invariably being performed with the beetle 'on its back') particularly so with species of *Anthrenus*. Likewise the dorsal surface and hence the scales should be kept free of glue during mounting.

Reference to Peacock (1993) will familiarise the student with the anatomy, ecology and habits of the Dermestidae. The book is doubly useful as it contains keys to larvae (which can be recognised with the naked eye after a little practice and are relatively easy to breed).

Reference

Peacock, E.A., 1993. Adults and larvae of hide, larder and carpet beetles and their relatives (Coleoptera: Dermestidae) and of Derodontid beetles (Coleoptera: Derodontidae). *Handbooks For the Identification of British. Insects.*, **5**(3):144pages. Royal Entomological Society, London.

ANOBIIDAE
(Plates 59, 60)

by J. Cooter

The Anobiidae, following Lawrence & Newton (1995) in Britain comprises the following the sub-families: Eucradinae, Ptininae, Dryophilinae, Ernobiinae, Anobiinae, Ptilininae, Xyletininae and Dorcatominae.

Several of the British species of Anobiidae are synanthropic and some are of considerable economic importance; for example, *Ptinus tectus* Boieldieu, *P. fur* (L.) and *Niptus hololeucus* (Faldermann) ('spider beetles') *Anobium punctatum* (De Geer) (woodworm or furniture beetle), *Xestobium rufovillosum* (De Geer) (the 'death watch beetle') and *Stegobium paniceum* (L.) (the 'biscuit beetle' or 'drug-store beetle'). The latter, *Lasioderma serricorne* (F.) (the 'cigarette beetle' ot 'tobacco beetle') and several Ptininae are pests of a range of stored products (see p. 258) while many anobiids are associated with dead and dying timber including that used in buildings.

EUCRADINAE: The sole representative in Britain is the prettily marked *Ptinomorphus imperialis* (L.). It can be found by beating, and possibly prefers dead twigs and thin branches.

PTININAE: Many species of this sub-family are synanthropic and have been spread world wide by commerce. Of the other species *Ptinus lichenum* Marsham, *palliatus* Perris *pilosus* Müller and *subpilosus* Sturm are associated with ancient timber. *Ptinus fur* also inhabits dead trees exploiting a range of organic matter from fungi to scraps in bird nests.

The only recent records of *Gibbium aequinoctiale* Boieldieu are from deep coal mines, the beetles being found in 'latrines' used by the miners.

DRYOPHILINAE: *Dryophilus anobioides* Chevrolat frequents broom in the Breckland of East Anglia; *D. pusillus* (Gyllenhal), a naturalised species, is associated with conifer trees, *Pinus, Larix*. *Grynobius planus* (F.) breeds in a variety of broad-leaved trees, it is widespread but local.

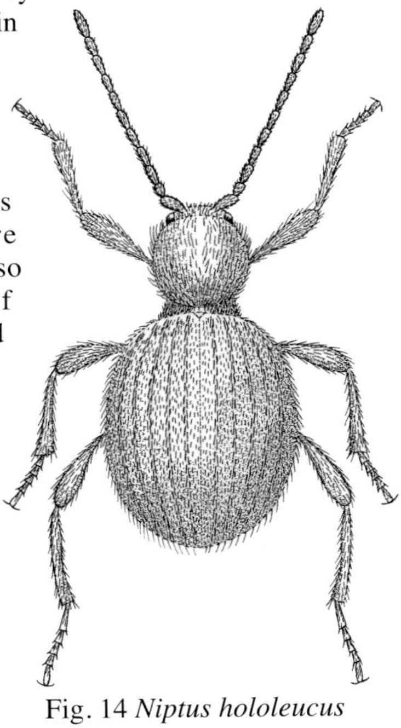

Fig. 14 *Niptus hololeucus* (Faldermann) Ptininae

ERNOBIINAE: *Ochina ptinoides* (Marsham) can be beaten from ivy; it has a scattered and disjointed distribution.

Ernobius species may be beaten from pines. *E. abietis* (F.) and *angusticollis* (Ratzeburg) are vagrant species unlikely to be encountered; *E. gigas* (Mulsant & Rey) and *pini* (Sturm) are naturalised species frequenting southern England. *Ernobius mollis* (L.) is the most widespread species of the genus. *E. nigrinus* (Sturm), a Scottish species spread by forestry into England, develops in the thin branches of pine and spruce which have been attacked by *Magdalis* and *Hylastes* species (Curculionidae).

Xestobium rufovillosum (De Geer) the notorius pest 'death watch beetle' can be found very rarely in the wild particularly in ancient dead oak trees where fungal decay and sufficient moisture exist. According to Alexander (2002) its preferred host in southern England is willow and in the Midlands, oak.

ANOBIINAE: *Gastrallus immarginatus* (Müller) breeds in ancient field maple, the adult beetle is crepuscular in habits but infested trees show groups of small exit holes. It is very localised in Britain and for many years known only from Windsor Forest and Great Park but recently has been found in a few localities on the Cotswolds.

Hemicoelus nitidus (Herbst) is another rarity first found in Britain in 1980 (Mendel, 1982) on grey poplar and subsequently in Windsor Great Park on field maple.

Adults of *Hemicoelus fulvicornis* (Herbst), *Anobium punctatum* and *Hadrobregmus denticollis* (Creutzer) can be obtained by beating hardwood trees, particularly mature/over mature oak with dead and dying branches. *Anobium inexpectatum* Lohse breeds in thick stems of ivy.

Priobium carpini (Herbst) is a recent addition to the British List, though, to date, all captures have been made indoors. Elsewhere in its range it occurs in ancient woodland, developing in dry dead wood. Barclay (2005) in adding the species to our List gives a detailed account of the species' history in Britain, its taxonomic placement and its identification.

Stegobium paniceum commonly known as 'the biscuit beetle' and 'the drug store beetle' is a notorious pest of a very wide range of dried vegetable matter and other materials.

PTILININAE: *Ptilinus pectinocornis* (L.) breeds in exposed dry heart-wood of various broad-leaved trees, preferring upright timber. The male has greatly enlarged pectinate antennae and is more often encountered than the female.

XYLETININAE: The late A.M. Massee once found *Xyletinus longitarsis* Jansson breeding in numbers in powdery dry broom growing at Dungeness. Unfortunately contemporary and later collectors came to regard broom as its preferred host, if anything, its occurrence in broom was a marked exception. I have found this scarce beetle frequently in my home county, Herefordshire on dead oak trunks and by beating the thin crown branches and twigs of dead or fallen trees (see Cooter, 1992).

Lasioderma serricorne commonly known as 'the cigarette beetle' or 'the tobacco beetle' is a pest of dried vegetable matters and animal protein. It is occasionally imported into Britain and thrives in a warm environment.

DORCATOMINAE: *Caenocara affinis* (Sturm) occurs in *Lycoperdon perlatum* puff-ball fungi and is known only from the Breck of Suffolk; there are no post 1970 records. *C. bovistae* (Hoffmann) is more widespread occurring in *Lycoperdon bovistae* puff-balls. *Anitys rubens* (Hoffmann) is found in ancient woodland where it breeds in red-rotten oak; it is often found in company with *Mycetophagus piceus* (F.).

Our five species of *Dorcatoma* breed in wood-rotting fungi and with ancient timber attacked by fungus. *D. flavicornis* (F.) and *chrysomelina* Sturm in (mainly) oak and beech attacked by *Laetiporus sulphureus*. *D. dresdensis* Herbst in hard bracket fungi – *Ganoderma* and *Fomes fomentarius* on beech and oak (mainly). *D. serra* Panzer in *Inonotus dryadeus* on ancient broad leaf trees. *D. amberjoerni* Baranowski (see Mendel and Owen, 1991) has been bred from brackets of *Inonotus cuticularis* on ancient beech and also in the red-rotten heartwood.

Joy (1932) will be found quite adequate for identification purposes of most common species. A more modern reference to Joy can be found in Freude, Harde and Lohse (1969) and Lohse and Lucht (1992). *Ernobius*, the British species are treated by Johnson 1966; *Gastrallus immarginatus* (Müller) see Donisthorpe 1936; *Hemicoelus nitidus* (Herbst) see Mendel, 1982; *Anobium inexpectatum* Lohse see Allen (1977); A good, well-illustrated key to *Dorcatoma* which includes figures of the male genitalia of all the British species is given by Baranowski 1985. Joy's keys to the 'Ptinidae' are very out of date and omit many species; the worker is referred to Hodge and Jones (1995) and references therein.

References

Alexander, K.N.A., 2002. *The invertebrates of living & decaying timber in Britain and Ireland – a provisional annotated checklist*. English Nature Research Reports No., 467:142pp. English Nature, Peterborough.

Allen, A. A., 1977. Notes on some British serricorn Coleoptera with adjustments to the list. 3. – A new British *Anobium*, with notes on three others of the family. *Entomologist's Monthly Magazine*, **112**: 51-54.

Baranowski. R., 1985. Central and Northern European *Dorcatoma* (Coleoptera: Anobiidae), with a key and description of a New Species. *Entomologica Scandinavica*, **16**:203-207.

Barclay, M.V.L., 2005. *Priobium carpini* (Herbst, 1793) (Coleoptera: Anobiidae) a European woodworm established in Britain. *Entomologist's Monthly Magazine*, **141**:43-47.

Cooter, J., 1992. *Xyletinus longitarsis* Jansson (Col., Anobiidae) in Herefordshire. *Entomologist's Monthly Magazine*, **128**:183.

Donisthorpe. H.St.J., 1936. *Gastrallus laevigatus* Ol.. (Col., Anobiidae) A genus and Species of Coleoptera New to Britain. *Entomologist's Monthly Magazine*. **72**:200.
Freude, H., Harde, K.W. and Lohse, G.A., 1969. *Die Käfer Mitteleuropas* **8**. 388 pages. Goeke & Evers, Krefeld.
Hodge, P.J. & Jones, R.A., 1995. *New British Beetles – Species not in Joy's Practical Handbook*. British Entomological and Natural History Society, Reading.
Johnson, C., 1966. The Fennoscandian, Danish and British Species of the Genus *Ernobius* Thomson (Cot., Anobiidae). *Opuscula Entomologica*, **31**:81-92.
Lohse, G.A. and Lucht, W.H., 1992. *Die Käfer Mitteleuropas* **13** (supplementband 2 mit katalogteil). 375pp. Goeke & Evers, Krefeld.
Mendel. H., 1982. *Hemicoelus nitidus* (Hbst.) (Col., Anobiidae) New to Britain. *Entomologist's Monthly Magazine*.. **118**:253-254.
Mendel, H. and Owen, J.A., 1991 . *Dorcatoma ambjoerni* Baranowski (Col., Anobiidae), another Windsor specialty? *Coleopterist's Newsletter*. **43**: 12-13.

CLEROIDEA
(Plates 61 – 62)

BY J. COOTER

This superfamily comprises, in Britain, the families (and subfamilies) Phloiophilidae, Trogossitidae (Peltinae, Lophocaterinae, Trogossitinae), Cleridae (Thaneroclerinae, Tillinae, Clerinae, Tarsosteninae, Korynetinae) and Melyridae (Melyrinae, Rhadalinae, Dasytinae and Malachiinae).

With a few exceptions their identification is straightforward and no special methods need be adopted in their preparation.

PHLOIOPHILIDAE: The sole British species, *Phloiophilus edwardsi* Stephens can be found during the autumn and winter; it is associated with the fungus *Peniophora quercina* (K.N.A. Alexander, *pers. comm.*)which grows on dead wood. Its ecology is the subject of an interesting paper by Crowson (1964).

TROGOSSITIDAE: *Lophocateres pusillus* (Klug) (Lophocaterinae) occurs in Britain only under artificial conditions and is a stored product pest. It is a cosmopolitan species spread by commerce and for years was omitted from the British List even in the 'introduced' category though arguably has a stronger case for inclusion than some other species, *Thaneroclerus buqueti* for example (see below). Our two British representatives of subfamily Trogossitinae have widely different habits. *Nemozoma elongatum* (L.), a great rarity, is associated with the bark beetle *Acrantus vittatus* (F.) upon which it preys. *Tenebriodes mauritanicus* (L.), the 'Cadelle', is a pest of grain, flour, etc. in mills, warehouses and bakeries (see p. 258). Of the two British representatives of subfamily Peltinae, *Thymalus limbatus* (F.) is the most widespread species and occurs under bark of dead and dying conifers and hardwood trees. It bears a strong superficial resemblance to a 'tortoise beetle' *(Cassida* spp.).

Ostoma ferrugineum (L.) is a very scarce species, confined to ancient Caledonian forest remnants where it lives under the bark of fallen pine trees with the fungus *Phaeolus schweinitzii*; it makes characteristic semi-circular emergence holes in the bark.

CLERIDAE: Of the 14 species on the British List, we are likely to encounter only eight. Of the other six species, four, *Tilloidea unifasciatus* (F.), *Trichodes alvearius* (F.), *T. apiarus* (L.) and *Tarsostenus univittatus* (Rossi) have not been recorded in Britain since the 19th century. The remaining two, *Thaneroclerus buqueti* (Lefebvre) and *Paratillus carus* (Newman) are considered accidentally introduced species of doubtful status, *T. buqueti*, found once in imported root ginger, as mentioned above, having a very dubious claim for inclusion on the British List.

The three *Necrobia* species are associated with dry carcases and bones in old nests where carrion has been the food, around bone mills and in dried meat products in pet shops. *Korynetes caeruleus* (De Geer) is associated with timber and is a predator of anobiid and scolytine beetles. It is also well known from certain animal products such as dried meat and leather. *Tillus elongatus* (L.) is a predator of various anobiid beetles, particularly *Ptilinus pectinicornis* (L.) and *Anobium punctatum* (De Geer); its larvae hunt nocturnally under bark and on the outside of the tree, the adult beetles are to be found under loose bark, particularly of beech. The female, with red pronotum and elytra expanded posteriorly, is more often noted than the wholly black, more parallel-sided male. *Thanasimus*

Fig. 15 *Hypebaeus flavipes* (F.) male

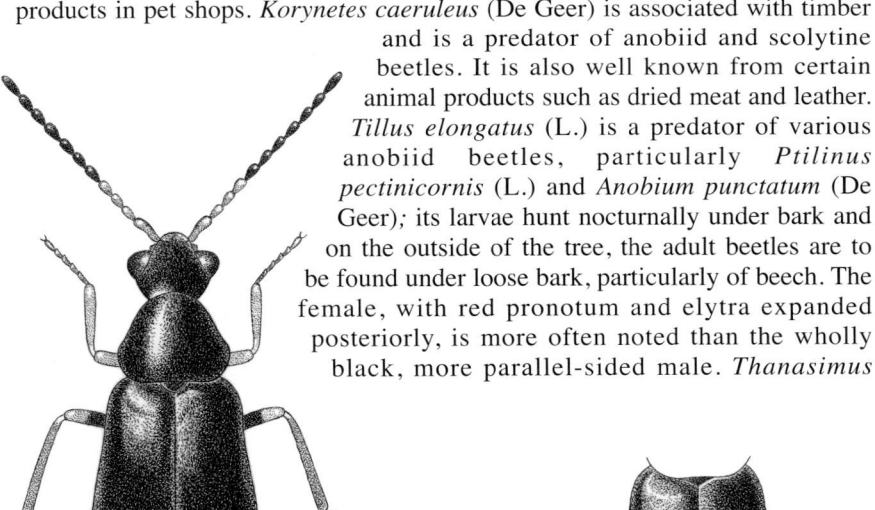

Fig.15a *Hypebaeus flavipes* (F.) female, elytra

formicarius (L.) is a widespread woodland species while its scarce relative, *T. femoralis* (Zetterstedt), is confined to N.E. Scotland and in particular the Caledonian Pine Forest remnants, but has also been recorded from plantations. Both are mimics of mutilid wasps. *Opilo mollis* (L.) is a scare nocturnal predator of a variety of anobiid beetles, for example *Xestobium*, *Hadrobregmus* and *Xyletinus*; it is occasionally met with in daytime by beating oak.

MELYRIDAE: Included here are a number of species found by sweeping flower-rich meadows during early summer (*Dasytes*, and *Malachius*) or by beating trees and sweeping thereunder (*Hypebaeus* and *Aplocnemus*), the latter also occurring under bark. Several are predominantly coastal – *Psilothrix* can be found in numbers by sweeping dunes and grasslands in southern coastal areas of England and Wales; *Dolichosoma lineare* (Rossi) inhabits the East Anglian and Thames saltmarshes; *Dasytes puncticollis* Reitter is particularly fond of cliff grassland, *Malachius barnevillei* Puton, *M.marginellus* Olivier and *M.vulneratus* Abeille are recorded from coastal/saltmarsh localities in East Anglia and south-east England. Wetland species include *Cerapheles terminatus* (Ménétriés) and *Anthocomus rufus* (Herbst).

In the Malachiinae, the sexes often differ markedly, the males exhibiting a complex process at the elytral apices, or enlarged antennal segments; these parts, in the female, are simple.

For identification, Joy (1932) will be found adequate, provided the user is aware of the fact that eleven species are not included and the key (but see Hodge and Jones, 1995) for *Dasytes* species does not work (it separates, *Dasytes niger* and *aeratus* only; the key in Freude, Harde & Lohse (1979) is essential for the separation *of puncticollis* and *plumbeus*).

References

Crowson, R.A., 1964. Habitat and Life-cycle of *Phloiophilus edwardsi* Stph. (Col., Phloiophilidae). *Proc. Royal Entomological Soc. Lond.,* **39**(a):15 1-152.

Freude, H., Harde, K.W., & Lohse, G.A., 1979. *Die Käfer Mitteleuropas,* **6**. Goeke & Evers, Krefeld.

Hodge, P.J. & Jones, R.A., 1995. *New British Beetles – Species not in Joy's Practical Handbook*. British Entomological and Natural History Society, Reading.

Pope, R. D., 1977. Kloet and Hincks a check list of British insects (2nd ed completely revised) pt 3 Coleoptera and Strepsiptera. *Handbooks for the Identification of British Insects,* **11**(3):105pp. Royal Entomological Society, London.

KATERETIDAE
by A.H. Kirk-Spriggs

The Kateretidae is a very small family which was formerly the subfamily Cateretinae of the Nitidulidae (Kloet & Hincks, 1977: p56). It has, however, been split off from the Nitidulidae based on male genital studies and is regarded as a distinct family (Audisio 1984 and Kirejtshuk, 1986).

The Kateretidae are small (1.5mm – 4.2mm), dull-coloured beetles with clubbed antennae. The British fauna comprises only nine species in three genera: *Brachypterus* (2 spp.); *Kateretes* (3 spp.); and *Brachypterolus* (4 spp.). All have adults and larvae that develop on flowers, feeding on buds and pollen.

The British species are readily identified using Kirk-Spriggs 1996.

Brachypterus glaber (Stephens) and *B. urticae* (F.) are both very common beetles which have species of nettles *(Urtica* spp.) as their host-plants. They can be easily found on nettles in woodland, roadsides, field margins and waste ground etc., being commonest at the flowering period June to September. Both species are often found together on the same plant, with *B. urticae* being the more abundant.

The genus *Kateretes* is split into two subgenera, *Pulion* Des Gozis with only one British species *K. (Pulion) rufilabris* (Latreille) and *Kateretes (s.str.),* which contains the other two species *K. (s.str.) pedicularius* (L.) and *K. (s.str.) pusillus* Thunberg (= *K. bipustulatus* (Paykull)). They are rather flat beetles having characteristically developed basal antennal segments in males. All three species are to be found in wetland areas, particularly marshes and bogs, where adults and larvae feed on rushes and sedges, *K. pedicularius* and *K. pusillus* on *Carex* spp. and *K. rufilabris* on *Carex* and *Juncus* spp. *Carex* species flower from May to August and *Juncus* from May to September. By carefully examining flowering heads of these plants many examples can be collected. It does not appear to be known how many species of *Carex* these beetles utilise as hosts and this can only be ascertained by careful rearing on the individual sedge species.

Of the four species of *Brachypterolus* which now occur in the British Isles two are native species: *B. linariae* (Stephens) and *B. pulicarius* (Linnaeus). These two species were formerly both regarded as *B. pulicarius,* but a key for their separation has been published (Johnson, 1967:143). Both these species are to be found feeding on flowers of *Linaria* species on roadsides, disused railway lines, woodlands, etc. The remaining two species *B. vestitus* (Kies.), and *B. antirrhini* (Murray) (= *B. villiger* (Reitter)), are introduced, *Brachypterolus vestitus* first being taken in Britain by Fryer (1929) and *B. villiger* by Williams (1926). They are to be found as pests on cultivated *Antirrhinum* flowers in parks and gardens. Many accounts of the damage they cause have been published, notably those of Jarvis (1944), Henderson, (1946), and Gimingham & Perkins (1944). The larvae and damage caused have been described in detail by Tempère (1926).

All these genera can be collected by sweeping and by searching host-plants. *Kateretes* spp. have also been collected using pitfall traps in wetland areas.

Males of genus *Brachypterolus* should have their aedeagi dissected and the median lobe and tegmen carefully separated. These can be mounted, dorsal side uppermost, with a minimum of gum beside the beetle or kept in a small vial of glycerol attached to the mounting pin, see p. 366.

References

Audisio, P., 1984. Necessità di ridefinizione della Sottofamiglie nei Nitidulidae e nuove prospeltive per Ia ricostruzione filogenetica del gruppo (Coleoptera). *Bollettino Zoologia.,* **54** (suppl.):1-5.
Fryer, J.C.F., 1929. *Brachypterolus (Heterostomus) vestitus* Kiesenwetter, in Britain. *Entomologist's Monthly Magazine,* **65**:101-102.
Gimingham, C.T. & Perkins, J.F., 1944. *Brachypterolus (Heterostomus) vestitus* Kies. (Col., Nitidulidae) in Hertfordshire. *Entomologist's Monthly Magazine,* **80**:290.
Henderson, J.L., 1947. 10th July 1946, Exhibits. *Proceedings of the South London Entomological and Natural History Society,* 1946-47: 18.
Jarvis, C.M., 1944. *Brachypterolus (Heterostomus) vestitus* Kies. (Col., Nitidulidae). *Entomologist's Monthly Magazine,* **80**:237.
Johnson, C., 1967. The identity of *Brachypterolus linariae* (Stephens) (Col., Nitidulidae), with notes on its occurrence in Britain. *The Entomologist,* **100**:142-144.
Kirejtshuk, A.G., 1986. [An analysis of the genitalia morphology and its use in reconstructing the phylogeny and basis of the system of Nitidulidae (Coleoptera)]. *Trudy Vsesoiuznogo Entomologischeskoe Obshchostva,* **68**:22-28. [in Russian].
Kirk-Spriggs, A.H., 1996. Pollen beetles. Coleoptera Kateretidae and Nitidulidae: Meligethinae. *Handbooks for the Identification of British Insects,* **5**(6a):157 pages. Royal Entomological Society, London.
Tempère, G., 1926. Un Coléoptère nitidulide du muflier des jardins. *Revue de Zoologie Agricole et Appliquée.* Bordeaux, **25**:155-158.
Williams, B.S., 1926. *Brachypterolus (Heterostomus) villiger* Reitt., a clavicorn beetle new to Britain. *Entomologist's Monthly Magazine,* **62**:262-263.

NITIDULIDAE

by A.H. Kirk-Spriggs updated by J. Cooter

The classification of the family Nitidulidae has changed considerably since the publication of the check-list of British Insects (Kloet & Hincks, (1977: pp. 56-57). The nomenclature and higher classification used here are according to Audisio (1984) and Kirejtshuk (1986). The major change has been the removal of the subfamily Cateretinae from the Nitidulidae and its elevation to family status, the Kateretidae (above). This new family comprising the genera *Brachypterus, Brachypterolus* and *Kateretes* is dealt with above.

The Nitidulidae are small- to medium-sized, obovate to oblong beetles, having 11-segmented antennae with a compact 3-segmented club, the elytra are foreshortened and one to three abdominal tergites are exposed, the tarsi are 5-segmented with the fourth segment always shorter than the others.

The British nitidulids are divided into five subfamilies, the Meligethinae (*Meligethes, Pria*); the Carpophilinae, with two tribes, the Carpophilini (*Carpophilus*) and Epuraeini (*Epuraea*) the Nitidulinae (*Nitidula, Omosita, Soronia, Amphotis, Pocadius, Thalycra, Cychramus*), the Cryptarchinae (*Cryptarcha, Pityophagus, Glischrochilus*) and the Cybocephalinae (*Cybocephalus*).

The genus *Meligethes* is the largest of our British genera, with thirty-six species. In Britain they are to be found feeding as larvae and adults on the unopened buds and flowers of the families Cistaceae, Rosaceae, Campanulaceae, Cruciferae, Labiatae, Papilionaceae and Boraginaceae. Each species is specific to an individual plant species, on which the larvae develop. In the case of our two commonest species *Meligethes aeneus* (Fabricius) and *M. viridescens* (Fabricius) the true British host plant, *Sinapis arvensis* L. (Cruciferae), is not the only larval host plant utilised. Unlike other British species they are both capable of completing there larval development on other species of yellow Cruciferae, notably oilseed rape, swede, turnip, cabbage and black mustard, as a result of which they have become very serious pests of these crops. Two other species of the genus are minor pests: *M. flavimanus* Stephens of cultivated roses and *M. nigrescens* Stephens of sweet pea flowers.

Although species are restricted to particular plants they are very commonly collected from a wide range of flowers, particularly before and after the flowering period of the larval host-plants. This has led to a great deal of confusion in the past, as to their true associations, for example the first edition of this *Handbook* has an extensive list of plants with associated beetles but less than a quarter of these entries are correctly associated for this genus and should be ignored.

The adult beetles emerge from hibernation in the spring, and on sunny days fly to their host-plants, or feed on pollen of other flowers prior to locating their flowering host-plants. Eggs are laid on developing buds, on which the larvae feed, larval development is usually very short, in the case of *M. aeneus* taking between 9 and 13 days under laboratory conditions (Osbourne, 1965:748). After this period the mature larvae drop to the soil and bury themselves, forming an earthen cell in which to pupate. After a few weeks the adults emerge and feed on flowers, before seeking winter quarters, usually in the soil, in which to hibernate. There is usually only one generation per year. The larvae of *M. aeneus* and *M. viridescens* have been described by Osbourne (1965); *M. nanus* Erichson and *M. ruficornis* (Marsham) (as *M. flavipes*) by Perris (1873); and *M. difficilis* (Heer) by Rey (1866:174-175). Parasites have been discussed by Easton (1962); Askew (1979) and Osbourne (1955 & 1960). Recent papers on the control of *M. aeneus* include Hokkanen (1989); Hokkanen *et. al*. (1986 & 1988).

The genus *Pria* is superficially similar to *Meligethes,* differing by the lack of arched impressions on the last abdominal sternite. Species of *Pria* also exhibit a high degree of sexual dimorphism, particularly in antennal structure, where the male has a four-segmented club and the female a three-segmented club. The genus has its centre of distribution in Africa and is represented by only one European species, *Pria dulcamarae* (Scopoli), which also occurs in the Yemen and East Africa. Seventy three species have been described so far and an excellent key is given by Cooper (1982).

Very little is known of the biology of *Pria*. The African species do not appear to be as host specific as *Meligethes,* but *P. dulcamarae* seems to be restricted to *Solanum dulcamara* L. and to a lesser extent *S. nigrum* L. (Solanaceae) in Britain. The life cycle has been described by Perris, (1875) and Norgaard, (1919). Eggs are laid amongst the stamens, on which the larvae develop; pupation takes place within the soil. The duration of the life cycle is not known, but Perris notes that adults are always found within flowers from May to September.

All the British Meligethinae can be identified using Kirk-Spriggs (1996), in addition this book contains a wealth of information about bionomics and distribution.

The genus *Carpophilus* (Carpophilini) includes generally cosmopolitan species which have been introduced from their tropical centres of distribution, being commonly associated with man as pests of dried fruit, particularly currants, raisins and figs. They only occur in large numbers as storage pests when the fruit has become slightly mouldy. Those species which have managed to establish themselves in non-synanthropic situations are encountered on fungi and mouldy fruits. The larvae are campodeiform with short feeble legs, the abdomen bears a pair of horns at the tip, with a pair of smaller horns just above, they are whitish to pale yellow in colour.

A figure of the rear end of a *Carpophilus* larva is given by Munro (1966:101); a description of larvae is given by Wickham, (1894). Nikitskij (1980:45) notes the genus as being predatory on Scolytidae; nematode parasites have been discussed by Remillett & Waerebeke (1975).

Ten species of *Carpophilus* are recorded from the British Isles. According to Kloet & Hincks (1977: 56), three species, *C. dimidiatus* (Fabricius); *C. freemani* Dobson and *C. ligneus* Murray, occurring in Britain only under artificial circumstances, while four species, *C. humeralis* (Fabricius) (as *Urophorus); C. flavipes* Murray, and *C. maculatus* Murray are of doubtful occurrence, this leaves three species as occurring in natural situations in Britain.

A key to the species of economic importance as stored products pests is given by Mound (1989:22-23).

The second largest genus of nitidulids in Britain is the taxonomically difficult *Epuraea,* with twenty-two species. The larvae and in many cases adult biology of the majority of species can only be guessed at or is completely unknown.

Larvae are generally found in the galleries of boring beetles (Scolytinae), on flowing tree sap and in fungus, with adults also occurring in flowers. Spornraft (1967:51) gives details of some species, upon which these notes are based. *Epuraea rufomarginata* (Stephens) has been recorded from the galleries of wood boring beetles, under the bark of spruce and on flowing sap of deciduous trees, particularly birch. It has also been recorded from the fungus *Daldinia concentrica*. Twenty-one *Epuraea* species have been recorded in association with fungi in Europe, two tentatively associated, *E. silacea* (Herbst) and *E. limbata* (Fab.) and two have, with reservation, been associated, *E. deleta* Sturm and *E. unicolor* (Olivier). Spornraft (1967: 51) has suggested that *E. distincta* (Grimmer), may also be associated with fungi.

Saalas (1951) deals with eleven *Epuraea* species associated with spruce, galleries of boring beetles under spruce bark being typical but not the only biotype utilised, these species being *E. boreella* (Zetterstedt), *E. angustula* (Sturm), *E. pygmaea* (Gyllenhal), *E. thoracica* Tournier, *E. deubeli* Reitter and *E. abietina* J.Sahlberg. Three further species are common according to Saalas on other coniferous trees, *E. pusilla* Illiger, *E. oblonga* (Herbst) and *E. laeviuscula* (Gyllenhal). Not all of these species are British. *Epuraea laeviuscula*, a non-British species is known to be associated with the boring beetle *Xyloterus lineatus* (Olivier) (Scolytinae).

Epuraea aestiva (L.), *E. melina* Erichson, *E. longula* Erichson and *E. melanocephala* (Marsham) are all to be found on flowers. In the case of *E. aestiva* it is known with certainty that it develops in subterranean nests, especially those of bumblebees, and the morphology of the immature stages has been described by Scott (1920). Those British species not mentioned above are in many cases associated with tree sap, but the biology of these is almost completely unknown.

Larvae and adults of *Epuraea* species associated with wood-boring beetles have been seen to devour their hosts. In some cases these have been dead and mouldy boring beetle larva. The true associations remain a mystery. Further notes on bark beetle associations are given by Nikitskij (1980 pages 47 & 108); larval morphology is described by Pototskaya (1978).

Within the subfamily Nitidulinae are two genera found in association with bones, dried carrion and similar substances; the genus *Nitidula* with four British species, and the genus *Omosita* with three British species. Some of these species, particularly *N.bipunctata* (L.) and *O.discoidea* (F.), are occasionally encountered in larders and pantries, where meat or other substances have been allowed to dry-out in airy conditions. Some have been taken in birds' nests, particularly those of birds of prey, where animal remains are present in the nests. *Nitidula* species are rather convex, dark beetles often with small spots or other markings on the elytra, *Omosita,* on the other hand, are markedly flattened and have the elytra and pronotum with blotchy markings more extensively developed. The larva and pupa of *Omosita colon* L. has been described by Eichelbaum (1903).

The genus *Amphotis* is represented in Britain by only one species, *A. marginata* (F.). This species is myrmecophilous, living in the nests of the ant *Lasius fuliginosus*. Donisthorpe (1927) gives some very interesting notes on this species (pp. 25-27). He states that *A. marginata* is only very rarely collected away from the nests of this ant, and that the ants feed the beetles with honey as well as the beetles consuming flies and other prey given by their hosts. He gives an illustration of an ant feeding an adult beetle on page 27.

The genus *Soronia* has two British species, *S. grisea* (L.) and *S. punctatissima* (Illiger), which are to be found on tree sap, fermenting vegetable material, and especially in the tunnels and galleries of *Cossus* spp. (Lepidoptera:Cossidae). The remaining nitiduline genera, *Pocadius; Thalycra* and *Cychramus*, are all to be found in association with one sort of fungus or another. *Pocadius ferrugineus* (F.) is in powdery fungi, particularly puff balls, *Thalycra fervida* (Olivier) in hypogeous fungi such as truffles and puff balls, and *Cychramus luteus* (F.) is associated with fungi and also found in flowers, hawthorn in spring being particularly productive. Very little appears to be known of the biology of any of these species.

The fourth subfamily, the Cryptarchinae, comprises three British genera. *Cryptarcha* has two species, *C. strigata* (F.) and *C. undata* (Olivier), which are for the most part on running sap of deciduous trees. *Pityophagus ferrugineus* (L.) is frequently found under the bark of coniferous trees. In Britain the genus *Glischrochilus*, has three species, *G. hortensis* (Fourcroy); *G. quadriguttatus* (F.) and *G. quadripunctatus* (L.), which can be collected on flowing tree sap, in the galleries of boring beetles and occasionally on rotting fruit, etc. in autumn.

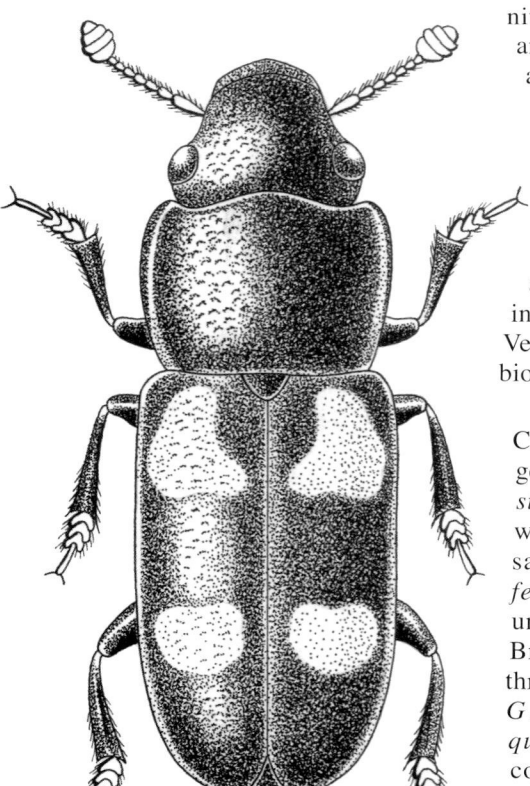

Fig. 16 *Glischrochilus quadripunctatus* (L.)

The works which includes the entire British nitidulid fauna currently available are Spornraft (1967 and 1992 (partly revised and up-dated)) written in German, and Audisio (1993) written in Italian, but with keys in English.

Mounting: In the genera *Carpophilus, Meligethes, Epuraea, Nitidula, Soronia* and *Glischrochilus,* reference to the genitalia of the male or ovipositor of the female is often desirable. Because of its poorly sclerotised and tubular form, the ovipositor is best made up as a clove-oil preparation or kept in glycerol (see p. 366). Dissection proceeds in the normal way (see p. 356) but with *Meligethes* it is important to remove only the apical three segments of the abdomen otherwise the caudal marginal line of the hind coxae which exhibits, in some cases, specific characters may be destroyed.

The aedeagus is dissected in the normal way, but after the careful removal of excess tissues, the median lobe and tegmen should be carefully separated and mounted dorsally separately. They can be mounted beside the beetle with a minimum of glue.

In order to appreciate certain diagnostic characters, *Meligethes* species require a special mounting technique. They can be card-pointed but specimens so treated are vulnerable to accidental damage and can easily be dislodged from the card-point unless adequate protection is afforded by the data labels. The preferred method is to mount the beetles on their right-hand side with the appendages of the left teased out and the right anterior leg removed and mounted dorsally alongside the beetle. In this way the teeth of the anterior tibiae on both sides of the specimen are visible and characters of the ventral and dorsal surfaces are easily observed.

Useful characters are displayed on the intermediate tibiae of male *Epuraea* species. The male genitalia in this genus require examination from above and in profile, for this reason they are best stored in glycerol vials pinned beneath the specimen (see p. 366) or glued upright, being attached to the card by their base.

The species of *Glischrochilus* are also best mounted on their sides as they exhibit diagnostic ventral characters; there is, however, no necessity to remove the right anterior leg.

The single species of Cybocephalinae occurring in Britain is the minute (1.1-1.4mm) *Cybocephalus fodori* Endrödy-Younga, 1965. It has been found low down on birch trees on Putney Heath, south London (2000-2002); it feeds on scale insects (Coccoidea, Diaspididae). See Prance, 2001 and Endrödy-Younga, 1967.

References

Askew, R.R., 1979. The biology and larval morphology of *Chrysolampus thenae* (Walker) (Hym., Pteromalidae). *Entomologist's Monthly Magazine,* **115**:155-159.

Audisio, P., 1984. Necessità di. ridelinizione delle Sottofamiglie nei Nitidulidae e nuove prospeltive per la ricostruzione filogenetica del gruppo (Coleoptera). *Bollettino Zoologia,* 54(suppl.):1-5.

Audisio, P., 1993. Nitidulidae-Kateretidae. *Fauna d'Italia vol.23 Coleoptera.* Xvi +971pp. Edizione Calderini, Bologna.

Cooper, M.C., 1982. The species of the genus *Pria* Stephens (Coleoptera: Nitidulidae). *Zoological Journal of the Linnaean Society,* **75**(4):327-390.

Donisthorpe, H. St.J. K., 1927. *The Guests of British Ants, their habits and life-histories.* George Routledge and Son Limited, London xiii + 244pp.

Easton, A.M., 1962. Mites parasitising *Meligethes* spp., (Col. Nitidulidae). *Entomologist's Monthly Magazine,* **98**:41.

Endrödy-Younga, S., 1967. 51. Familie: Cybocephalidae, *in* Freude, H., Harde, K.W. and Lohse, G.A., *Die Käfer Mitteleuropas, Band 7. Clavicornia.* [Cybocephalidae: 77-79]. Goecke & Evers, Krefeld, 310pp.

Eichelbaum, F., 1903. Larve und Puppe von *Omosita colon* L. *Allgemeine Zeitschrift Fuer Entomologie,* **8**(5):81-87.

Hokkanen, H.M.T., 1989. Biological and agrotechnical control of the rape blossom beetle *Meligethes aeneus* (Coleoptera, Nitidulidae). *Acta Entomologica Fennica,* **53**:25-29.

Hokkanen, H., Granlund, H., Husberg, G-B., & Markkula, M., 1986. Trap crops used to control *Meligethes aeneus* (Col., Nitidulidac), the rape blossom beetle. *Annales Entomological Fennici,* **52**:115-120.

Hokkanen, H., Husberg, G-B., & Soderblom, M., 1988. Natural enemy conservation for the integrated control of the rape blossom beetle *Meligethes aeneus* F. *Annales Agriculturae Fenniae,* **27**:281-294.

Kirejtshuk, A.G., 1986. [An analysis of the genitalia morphology and its use in reconstructing the phylogeny and basis of the system of Nitidulidae (Coleoptera)]. 97 *Trudy Vsesoiuznogo Entomologischeskoe Obshchostva,* **68**:22-28 [in Russian].

Kirk-Spriggs, A.H., 1996. Pollen beetles. Coleoptera Kateretidae and Nitidulidae: Meligethinae. *Handbooks for the Identification of British Insects,* **5**(6a):157 pages. Royal Entomological Society, London.

Mound, L., (ed.) 1989. *Common insect pests of stored food products a guide to their identification, Seventh Edition.* British Museum (Natural History), ix + 68pp.

Munro, J.W., 1966. *Pests of Stored Products.* The Rentokil Library. Hutchinson, London, 234pp.

Nikitskij, N.B., 1980. *[Insect predators of bark beetles and their ecology].* Nauka, Moscow, 237pp. [in Russian].

Norgaard, A., 1919. Om *Pria dulcamarae* Scopoli og dens lavevis. *Entomologiske Meddelelser,* **12**:128-136.

Osbourne, P., 1955. The occurrence of five hymenopterous parasites of *Meligethes aeneus* F. and *M. viridescens* F. (Col., Nitidulidae). *Entomologist's Monthly Magazine,* **91**:47.

Osbourne, P., 1960. Observations on the natural enemies of *Meligethes aeneus* (F.) and *M. viridescens* (F.) [Coleoptera:Nitidulidae]. *Parasitology,* **50**:91-110.

Osbourne, P., 1965. Morphology of the immature stages of *Meligethes aeneus* (F.) and *M. viridescens* (F.) (Coleoptera:Nitidulidae). *Bulletin of Entomological Research,* **55**(4):747-759.

Perris, E., 1873. Résultats de quelques promenades entomologiques. *Ann. Soc. Ent. Fr.,* **5**(3):61-98 and 249-252.

Perris, E., 1875. Larves de Coléoptères. *Annales de la Société Linnéenne de Lyon,* **22**:289.

Potoskaya, V.A., 1978. Larval morphology and ecology of beetles of the genus *Epuraea* (Coleoptera:Nitidulidae), *Enomologicheskoe Obozrenie,* **57**(3):570-577. [translated *Entomological Review,* **57**(3) 1978(1979):391-296].

Prance, D.A., 2001. *Cybocephalus fodori* Endrödy-Younga, (Coleoptera: Cybocephalidae) new to Britain. *Entomologist's Gazette,* **52**(2):125-127.

Remillet, M., & Waerebeke, van D., 1975. Description et cycle biologique de *Howardula madescassa* n. sp. et *Howardula truncati* n. sp. (Nematoda:Sphaerulariidae) parasites de *Carpophilus* (Coleoptera: Nitidulidae). *Nematologica,* **21**(2):192-206.

Rey, C., 1866. Larve de Coléoptères. *Annales de la Société Linnéenne de Lyon*, **33**:131254.
Saalas, U., 1951. Einiges über Charakterarten der Käferbestände an Fichten von verschiedener Beschaffenheit. *Z. angew. Ent. Berlin*, **33**:12-18.
Scott, H., 1920. Notes on biology of some inquilines and parasites in nest of *Bombus derhamellus*. *Transactions of the Royal Entomological Society of London* (1920), pp105-124.
Spornraft, K. von., 1967. 50. Familie: Nitidulidae, in Freude, H., Harde, K.W. and Lohse, G.A., *Die Käfer Mitteleuropas, Band 7. Clavicornia*. [Nitidulidae: 20-76]. Goecke & Evers, Krefeld, 310pp.
Spornraft, K. von., 1992. Familie Nitidulidae. (in) *Die Käfer Mitteleuropas, 2 supplementband mit katalogteil*. Goecke & Evers, Krefeld, 375pp.
Walsh, G.B., & Dibb, J.R., 1954. *A Coleopterist's Handbook*, Amateur Entomologist's Society, 120pp.
Wickham, H.F., 1894. Description of the larvae of *Tritoma, Carpophilus* and *Cyllodes*. *Entomological News*, **5**: pp. 260-263.

MONOTOMIDAE
by J. Cooter

This small *family* includes twenty-two British species, divided between the subfamilies Rhizophaginae (thirteen species) and Monotominae (nine species). Members of both divisions are of characteristic appearance and easily recognised in the field. The majority of the Rhizophaginae are to be found under sappy bark of dead and dying trees, sometimes in numbers and occasionally two or three species together. *Rhizophagus parallelocollis* (the 'coffin beetle') and *perforatus* are subterranean and *cribratus* often occurs in vegetable refuse, fungi and the like. *Cyanostolus aeneus* occurs under bark of trees near to or partially submerged in water. *Rhizophagus grandis* Gyllenhal has recently been introduced to Britain in an effort to control the scolytid beetle *Dendroctonus micans;* it is not included in the *R.E.S. Handbook* (Peacock, 1977).

Species of the genus *Monotoma* are very slow moving beetles which inhabit rotting vegetable matter, such as manure and compost heaps. Two species *Monotoma conicicollis* and *angusticollis* are myrmecophilus with *Formica* species. *Monotoma* species, because of their habit of keeping still for a long time and subsequent sluggish movement, are easily passed over whilst examining siftings on a tray in the field or in an extractor.

The family was previously known as the Rhizophagidae, but Monotomidae has priority.

Reference
Peacock, E.R., 1977. Coleoptera, Rhizophagidae. *Handbooks for the Identification of British Insects*, **5**(5a): 23pp. Royal Entomological Society. London.

PHALACRIDAE
by J. Cooter

There are fifteen British species belonging to this distinctive family – *Phalacrus* five spp., *Olibrus* seven spp., and *Stilbus* three spp.

They are commonly collected by sweeping, but sometimes it is necessary to search plants. *Phalacrus* species are associated with smutted grasses and sedges, *Olibrus* with Compositae and *Stilbus* with *Typha* and can be sifted from hay and other vegetable debris.

The beetles are of characteristic appearance, though separation to species level requires high magnification and dissection. There is an *R.E.S. Handbook* (Thompson, 1958) available for this family; it is well-illustrated and straightforward to use. When preparing Phalacrids it is of prime importance to keep the dorsal surface clean as the microsculpture is easily obscured by glue and grease. The anterior tibiae of *Olibrus* and *Phalacrus* possess a varying number of spurs which must be kept free of glue. The ovipositor of the female often exhibits reliable specific characters and must be dissected with care and mounted in Euparal (after first passing through alcohol and xylene) on a celluloid or acetate strip. The male genitalia should also be dissected and the median lobe separated from the tegmen; these parts can be glued to the card behind the beetle, but as it is sometimes necessary to view the tegmen from more than one angle, storage in a van Doesburg tube (see p. 366) is preferable; the ovipositor can be likewise treated.

Reference
Thompson, R.T., 1958. Coleoptera, Phalacridae. *Handbooks for the Idenification of British Insects*, 5(5a):16pp. Royal Entomological Society. London.

CRYPTOPHAGIDAE
by R.J. Marsh

This large family is represented in the British Isles by about 100 species, Hypocoprinae with one species, and five species in Telmatophilinae, with the remainder about equally divided between Atomariinae and Cryptophaginae.

HYPOCOPRINAE comprises a single genus with one somewhat enigmatic species, *Hypocoprus latridioides* Motschulsky. Unrecorded in Britain since the early 20th century, it is associated with animal dung.

TELMATOPHILINAE comprises the single genus *Telmatophilus* which may be found by sweeping marsh plants such as *Typha* or *Carex* or in the ground litter beneath. *T. typhae* (Fallén) being the commonest is most easily located between the leaf layers of *Typha* stems.

CRYPTOPHAGINAE comprises the genera *Paramecosoma, Henoticus, Cryptophagus, Micrambe* and *Antherophagus*. *Paramecosoma melanocephalum* (Herbst) may turn up in flood refuse but has been found in rotting bracket fungi on trees. *Henoticus* may be searched for in various bracket fungi or beaten from dead branches where it probably feeds on fungi under loose bark. The large genus *Cryptophagus* occurs in a wide variety of habitats, particularly stored products and other synanthropic situations (e.g. *cellaris* (Scopoli), *saginatus* Sturm and *scutellatus* Newman). Decaying and putrefying substances such as compost and manure heaps, rotting straw and piles of cut grass may support some common species (e.g. *acutangulus* Gyllenhal, *pilosus* Gyllenhal and *pseudodentatus* Bruce). Some rarer species (e.g. *labilis* Erichson, *confusus* Bruce and *acuminatus* Coombs & Woodroffe) are associated with fungal growths in old timber such as oak, beech and elm. *C. badius* Sturm, *fallax* Balfour-Browne and *lapponicus* Gyllenhal have been recorded from nests of birds and mammals, whilst the nests of bees and wasps have yielded *setulosus* Sturm and *populi* Paykull. *C. micaceus* Rey has been found in the nests of tree-dwelling Hymenoptera including the Hornet *Vespa crabro* L.. Rotting fungi may be searched for *lycoperdi* (Scopoli), *ruficornis* Stephens and *dentatus* (Herbst).

Micrambe vini (Panzer) and *villosus* (Heer) may be beaten from gorse *Ulex*, whilst the rare *lindbergorum* (Bruce) has been found by shaking dead stems of thistle *Cirsium* spp..

Antherophagus species are associated with the nests of bumblebees *Bombus*.

Caenoscelis species have been recorded from light traps and by general sweeping along river banks (*ferruginea* (Sahlberg, C.R.) and *subdeplanata* Brisout). The single British specimen of *C. sibirica* Reitter was found in leaf litter in woodland.

ATOMARIINAE comprises the genera *Atomaria, Ootypus* and *Ephistemus*.

Of Atomaria, some of the more common and widespread species (e.g. *nitidula* (Marsham), *testacea* Stephens and *lewisi* Reitter) may be found by sieving decaying, but not too wet, plant material, compost and manure heaps or by general sweeping of low vegetation during warm weather. Some much less common species appear to be specialists, in ground litter in saltmarshes (*rhenana* Kraatz), under bark of fallen conifers (*lohsei* Johnson & Strand), or in the fungus *Coprinus comatus* (*fimetarii* (F.)). Evening sweeping in marshy places may yield *mesomela* (Herbst), and the more rare species *atra* (Herbst), *gutta* Newman and *barani* Brisout. Sawn ends and sap runs in coniferous timber may produce *pulchra* Erichson and *strandi* Johnson.

Ootypus globosus (Waltl) is found mainly in the dung of large herbivorous mammals (mainly horse and cow), whilst the extremely common and widespread *Ephistemus globulus* (Paykull) is a grassland species of rotting and decaying habitats and occurs in large numbers in piles of grass cuttings and compost heaps.

Joy (1932), although extremely out of date, is still useful in identifying to genera. Identification to species level is more problematic especially with *Atomaria* and C*ryptophagus*. This difficulty is compounded by the absence of an up-to-date, comprehensive and authoritative checklist, the most recent of which is Kloet & Hincks (1977), which does not include the more recent additions to our fauna, for which the reader is referred to Hodge and Jones (1995). Identification of *Cryptophagus* is problematic with many of the species and is best attempted when large numbers of specimens have been accumulated, as there is considerable variation within species limits. The most useful key for determining *Cryptophagus* is undoubtedly that of Coombs & Woodroffe (1955), who revised the British species. Some species such as *scutellatus* Newman and *ruficornis* Stephens are usually distinguishable fairly readily on colour, build or size, but in many areas such as the *dentatus*-group (*dentatus* (Herbst), *pseudodentatus* Bruce, *acuminatus* Coombs and Woodroffe and *intermedius* Bruce) dissection of males is recommended as a matter of course, as the genitalia offer reliable characters. Unassociated females of this group are very difficult to identify with certainty. Sexing of specimens of *Cryptophagus* and *Micrambe* (except *abietis* (Paykull)) is straightforward in that the male hind tarsi possess four segments, whereas in females this number is five. If dissection is proposed Coombs and Woodroffe (1955) give excellent figures of the genitalia. The median lobe should be separated from the parameres and these parts displayed on separate transparent mounts in Euparal below and slightly in front of the whole insect, which should be set with antennae flat on the card mount, all segments set in the same plane ideally with an absolute minimum of gum applied only to the terminal segment. Alternatively the genitalia may be mounted behind the insect on the same piece of card. Glue should be kept from the dorsal surface, as pubescence and sculpture are important diagnostic features. Other papers are necessary reading such as Johnson (1988).

The limitations of Joy (1932) above also apply to *Atomaria*, as fourteen or so species have been added to our list since that date. Johnson's (1993) atlas gives a revised checklist of all our Atomariinae as well as habitat information for each species with a national distribution map; this atlas was updated in Johnson (2002). Again, Hodge & Jones (1995) give details of *Atomaria* species that have appeared since Joy's handbook. Lohse (1967) gives keys (in German) for the mid-European fauna, and Johnson (1992a) revises this key to include all the British species. As with *Cryptophagus*, specimens should be mounted dorsal surface uppermost with the antennae displayed normally and flat, keeping gum from contaminating the sculpture and pubescence. The papers by Johnson 1992b and 1986 are also very useful.

References

Coombs, C.W. & Woodroffe G.E., 1955. A revision of the British species of *Cryptophagus* (Herbst) (Coleoptera: Cryptophagidae). *Transactions of the Royal Entomological Society of London,* **106**:237-282.

Hodge, P.J. & Jones, R.A., 1995. *New British Beetles – Species not in Joy's Practical Handbook.* British Entomological and Natural History Society, Reading.

Johnson, C., 1986. New synonymy and changes in the nomenclature of European Cryptophagidae (Coleoptera). *Entomologist's Gazette,* **37**:129-132.

Johnson, C., 1988. Notes on some British *Cryptophagus* Herbst (Coleoptera: Cryptophagidae), including *confusus* Bruce new to Britain. *Entomologist's Gazette,* **39**:329-335.

Johnson, C., 1992a. 55. Familie: Cryptophagidae. *In* Lohse, G.A. & Lucht, W.H. *Die Käfer mitteleuropas,* **13**:114-134. Goecke & Evers, Krefeld.

Johnson, C., 1992b. Further changes in the nomenclature of European *Atomaria* Stephens (Coleoptera: Cryptophagidae). *Entomologist's Gazette,* **43**:145-146.

Johnson, C., 1993. Provisional *Atlas of the Cryptophagidae-Atomariinae (Coleoptera) of Britain and Ireland.* Institute of Terrestrial Ecology, Monks Wood.

Johnson, C., 2002. Provisional atlas of the Cryptophagidae: Atomariinae (Coleoptera) of Britain and Ireland: supplement to the 1993 atlas and *Atomaria turgida* Erichson newly recorded for the fauna. *Entomologist's Gazette,* **53**:183-189.

Joy, N.H., 1932. A Practical Handbook of British Beetles (2 vols.) Witherby, London. (Reprint by E.W.Classey, Faringdon, 1976).

Kloet, G.S. & Hincks, W.D., 1977. A check list of British insects, Coleoptera and Strepsiptera. 2nd edition, completely revised. *Handbooks for the Identification of British Insects,* **11**(3). Royal Entomological Society, London.

Lohse, G.A., 1967. 55. Familie: Cryptophagidae. *in* Freude, H., Harde, K.W. and Lohse, G.A., *Die Käfer Mitteleuropas,* **7**:110-158. Goecke & Evers, Krefeld.

EROTYLIDAE
(Plate 64)

by Z. Simmons

The family Erotylidae is mainly evident in tropical and sub-tropical regions but can be found worldwide. There is within the region of ca 2500 known species though much of the classification for this family is still under revision and due to the difficulty associated with collecting this family the number of worldwide species may well be appreciably higher. Early taxonomic work on this family involved the use of colour and pattern as determining characteristics and many modern systems still rely at least in part, on these characteristics, despite the large amount of variation represented within individual species. The three genera represented in the British fauna are widespread across the Palæarctic, Nearctic and extend into the Asian/Oriental region and display some of the most common erotylid traits.

British erotylid species are small (in the region of 3-6 mm) and lack the conspicuous bright colouration and distinctive patterning of their foreign relatives which can reach sizes upwards of 30 mm. In Britain the usual colour combination is black and red/orange, though *Triplax* species have a metallic blue sheen to the elytra that is especially noticeable on *aenea*. The British fauna comprises seven species, which ought not present problems in identification for which Joy (1932) can be used, though for the genus *Dacne* (two species) reference should be made to Shaw (1951) to ensure correct identification.

Erotylidae have a 5-5-5 tarsal formula with the third segment being strongly bilobed and the fourth greatly reduced (pseudotetramerous) in all of the British species except *Dacne* which has equally proportioned tarsi. All of the species have well developed three-segmented antennal clubs and securiform (triangular/hatchet shaped) palps. *Triplax scutellaris* differs in that it apparently has an antennal club composed of four segments (the seventh segment is more pronounced than in other *Triplax* species). *Dacne* species have a more rounded and obviously flattened antennal club than the other erotylid species.

Erotylids are mycophagous and species of the three British genera; *Triplax* (plate 64), *Tritoma* and *Dacne* are to be found on certain tree fungi and under fungoid bark sometimes in numbers. The widespread *Triplax aenea* and the scarcer species *lacordairii* and *scutellaris* have been recorded from *Pleurotus*; and *russica* in the western part of the country from *Inonotus hispidus* and from *Fomes* in the north. *Dacne bipustulatus* has been reared from *Laetiporus sulphureus* and *Piptoporus betulinus*; *Dacne* and *Tritoma* seem to prefer the brackets of softer polypore fungi, but can also be found under fungoid bark. Polypore fungi either grow continually or re-appear at the same site each year so a detailed record of the location will allow populations to be monitored. Breeding and larval development is thought to occur on the fungus or surrounding bark and over-wintering sites may well be on the fungal host. Fungal preferences as either food or breeding sites may exist, possibly helping to explain distribution patterns for this family. It is common for species within this family to be nocturnal and thus it may require a good eye to spot these beetles unless they are found in high density. It is also worthwhile to carefully check the underside of fungus and search for these beetles at night.

The British erotylids can be mounted ventrally on cards; there is no need to dissect their genitalia.

References

Joy, N.H., 1932. *A practical handbook of British beetles*. 2vols. Witherby, London.

Shaw, S., 1951. The British species of the genus *Dacne* Latreille (Col., Erotylidae). *Journal of the Society for British Entomology*, **3**(6): 276-278.

BYTURIDAE, BIPHYLLIDAE, BOTHRIDERIDAE, CERYLONIDAE and ALEXIIDAE

by J.Cooter

These five families are grouped together here for convenience, each is poorly represented in Britain and they follow each other taxonomically.

There are two species of family BYTURIDAE on the British List, *Byturus tomentosus* (De Geer) a sometimes garden/horticultural pest known commonly as 'the raspberry beetle' and its less common, slightly larger relative *B. ochraceus* (Scriba). In isolation the two can be a little tricky to identify, but the males have distinctive genitalia making separation easy, see Vogt, 1967.

Our two representatives of family BIPHYLLIDAE are woodland species. *Biphyllus lunatus* (F.) develops in the fruiting body of the fungus *Daldinia concentrica* on ash and occasionally other trees. Adults of *Diplocoelus fagi* Guérin-Méneville are primarily found under the loose outer bark of dead beech but in more recent years have been found associated with sooty bark disease on sycamore.

There are five British species of the family BOTHRIDERIDAE. The species of sub-family Teredinae, *Teredus cylindricus* (Olivier), *Oxylaemus cylindricus* (Panzer) and *O. variolosus* (Dufour) are associated with ancient forest habitats, and are seldom encountered; *O. cylindricus* is most likely extinct in Britain, there having been no records of capture since the 19th century. *Teredus* is assumed to be predatory and is often found in ancient oak trees in association with the ant *Lasius brunneus* or anobiid beetles; most records of this species are from Windsor and Sherwood Forests. *Oxylaemus variolosus* has been sieved from litter in ancient hollow oaks and in the fungus *Collybia fuscipes* at the base of red-rotten oak. It is likely that *Oxylaemus* species are subterranean; they have been captured using subterranean pit-fall traps.

Our two species of sub-family Anommatinae, *Anommatus duodecimstriatus* (Müller) and *A. diecki* Reitter (see Eccles and Bowestead, 1987) are subterranean. *A. duodecimstriatus* can be found in decaying seed potatoes (ie the decaying remains of the seed potato planted in spring, which are often present when the crop is lifted in the autumn) and buried vegetable matter. *A. diecki* has been found on the fungoid underside of rotten wood laying on the ground.

In CERYLONIDAE, subfamily Ceryloninae comprises three species in of the genus *Cerylon* in Britain, *C. fagi* Brisout, *C. ferrugineum* Stephens and *C. histeroides* (F.), all of which can be found under bark and in decaying wood of ancient hardwood trees, especially beech and oak. Sub-family Murmidiinae has two British species, *Murmidius segregatus* Waterhouse is generally regarded as a stored product introduction, the other, *M. ovalis* (Beck) has not been recorded from Britain since 1831; it used to be found in cellars, hay-stack bottoms (when indeed hay-stacks were a common rural feature) and a range of stored products.

Our sole representative of the ALEXIIDAE, the diminutive (ca 1.3mm) *Sphaerosoma piliferum* (Müller) has been found by sieving fungoid litter and decaying vegetation.

The species belonging to all these families can be identified from Joy (1932) and Eccles and Bowestead (1987), but the reader will not be surprised to learn their taxonomic placement in Joy is rather eclectic even by his standards. No special treatment is required during their preparation.

References

Eccles, T.M. and Bowstead, S. 1987. *Anommatus diecki* Reitter (Coleoptera: Cerylonidae) new to Britain. *Entomologist's Gazette*, **38**: 225-227.

Joy, N.H., 1932. *A practical handbook of British beetles*. 2vols. Witherby, London.

Vogt, H., 1969. 49 Familie: Byturidae. In Freude, H., Harde, K.W. and Lohse, (*eds*.) G.A., *Die Käfer Mitteleuropas, Band 7. Clavicornia*. Goecke & Evers, Krefeld, 310pp.

ENDOMYCHIDAE
(Plate 65)

by Z. Simmons

Endomychids are similar in habit to erotylids and are also mainly evident in tropical and sub-tropical regions. There are approximately 1300 described species worldwide though this number is still increasing as more effort is made to locate new species in warmer climes. There has been much work done in recent years on the taxonomy and phylogeny of this family and the higher classification is still under debate with many changes taking place in the last 50 years. The taxonomy of the British species however, now appears to be stable.

This family ranges in size from the minute (1-2 mm) to approximately 25mm long. Larger tropical endomycids tend to be bright in colour and have spotted patterns similar to some erotylid species and may also exhibit strong morphological adaptations in the form of spines and flattened edges to the elytra. Species that are small are generally light to dark brown and may be densely pubescent. In Britain, five of our eight species fall into the latter category and require a good hand-lens or microscope to identify. The remaining three species are larger, being in the region of 4-6 mm long, having little to no pubescence and brighter, more striking colouration. The best example of this is *Endomychus coccineus* (L.) (plate 65), which superficially resembles an *Adalia* (Coccinellidae: Coccinellinae) species.

Endomychidae are typically mycophagous, with different sub-families feeding on spores and hyphae of microfungi (Holoparamecinae) or larger Basidiomycetes (*Endomychus, Lycoperdina*). They have also been recorded as inhabiting leaf litter though this is probably related to the fungal content.

Endomychidae possess a 4-4-4 tarsal formula and lack subcoxal lines on abdominal sternite 1. The three larger British species have two characteristic grooves at the basal third of the pronotum extending anteriorly. They also have relatively long antennae in proportion to their body (in the region of one third of their total length) and distinct antennal insertions that can be seen clearly from above. The smaller British species have rounded bodies, discreet antennal insertions and dense pubescence that conceals any grooves that are present on the thorax. All endomychids posses a three segmented antennal club but the strength of definition depends on the species.

The eight species represented in Britain are from four sub-families. Endomychidae: Endomychinae: The striking *Endomychus coccineus* (L.) (plate 65) can be found under fungoid bark, especially beech bark, sometimes in numbers and with its larvae. Lycoperdininae: The British species of *Lycoperdina* are associated with puff-ball fungi and are listed as nationally rare, *L. succincta* (L.) being confined to the Breck of East Anglia (see Collecting: Habitats, p 237). *L. bovistae* (F.) has recently been found in the 'earth star' fungus *Geastrum fimbriatum* (see Green, 1997). Anamorphinae: The small pubescent species of Mycetaeinae *Mycetaea subterranea* (F.) and *Symbiotes latus* Redtenbacher can be found by sieving fungoid wood and litter. Holoparamecinae species are to be found by sieving litter.

The British species are easily identified using Joy (1932), only the genus *Holoparamecus* is likely to present problems (see Peez, 1967). It is worth pointing out Joy (1932) places *Symbiotes latus* in the Mycetophagidae and *Mycetaea subterranea* in the Latridiidae. *Sphaerosoma piliferum* is no longer included in the Endomychidae since its sub-family has been raised to family status, the Alexiidae (above).

The British endomychids can be mounted ventrally on cards, there is no need to dissect their genitalia.

References

Green, D.M., 1997. New beetle national rarity on a different fungus than usual: *Lycoperdina bovistae* (Endomychidae). *Worcestershire Record*, **1**(3) (November 1997). (Worcestershire biological Records Centre).

Joy, N.H., 1932. *A practical handbook of British beetles*. 2vols. Witherby, London.

Peez, Alexander, von, 1967. 55: Familie Lathridiidae. In Freude, H., Harde, K.W. and Lohse, G.A. (*eds*) *Die Käfer Mitteleuropas*, **7**:168-190. Goecke & Evers, Krefeld.

COCCINELLIDAE
(Plates 66 – 68)

by J. Muggleton

This is a family of world-wide distribution, numbering in excess of 3,500 species, most of which are instantly recognisable as ladybirds. Five subfamilies and ten tribes are represented in the British fauna. The tribes and the genera within them are the Chilocorini (*Chilocorus* and *Exochomus*), Platynaspini (*Platynaspis*), Scymnini (*Scymnus*, *Nephus* and *Clitostethus*), Stethorini (*Stethorus*), Hyperaspini (*Hyperaspis*), Coccinellini (*Adalia*, *Anatis*, *Anisosticta*, *Aphidecta*, *Calvia*, *Coccinella*, *Harmonia*, *Hippodamia*, *Myrrha*, and *Myzia*), Tytthaspini (*Tytthaspis*), Psylloborini (*Psyllobora* and *Halyzia*), Epilachnini (*Epilachna* and *Subcoccinella*) and Coccidulini (*Coccidula* and *Rhyzobius*). Currently 43 species are considered to be breeding in the British Isles. Another species, *Hippodamia 13-punctata* (L.), has not been reliably recorded since the 1950's and may only become temporarily established as a result of migration. A further four species, *Exochomus nigromaculatus* (Goeze), *Coccinula 14-pustulata* (L.), *Vibidia 12-guttata* (Poda) and *Nephus bisignatus* Boheman were recorded in the nineteenth century but no longer appear to breed here, although single specimens of the first two were found in Yorkshire in 1967 and 1993 respectively (Muggleton, 1999). Of the 43 breeding species two, *Epilachna argus* (Geoffroy) and *Rhyzobius chrysomeloides* (Herbst), were first recorded from the British Isles in the late 1990s (Hawkins, 2000). The former is slowly spreading, the latter may have been present, but overlooked, for some time. In addition several exotic species, introduced into glasshouses as biological control agents, escape into the wild from time to time.

Because of their distinctive coloration and shape, coccinellids must be among the most easily recognised of the Coleoptera. A number of the smaller species in the subfamily Scymninae are less than

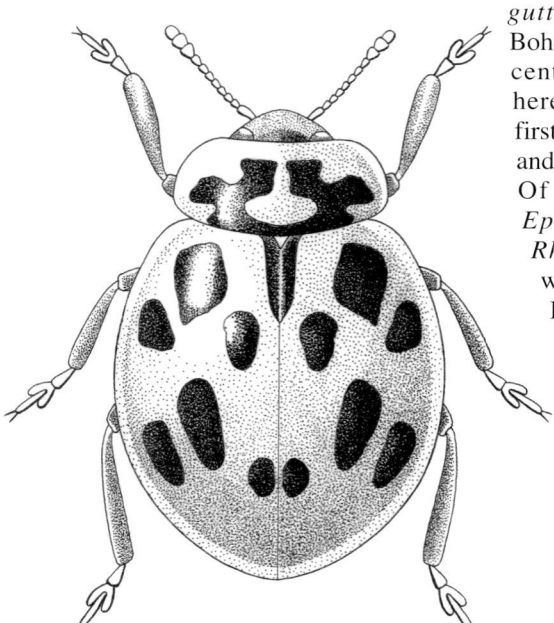

Fig 17 *Propylea quattuordecimpunctata* (L.)
'The 14-spot ladybird'

3mm or even 2mm long and may escape attention because of their size, but when viewed through a microscope they are clearly coccinellids. Of the British species only four might confuse the inexperienced, the two species of *Rhyzobius* because of their brown coloration and the two species of *Coccidula* which are brown and elongate in shape. Although coccinellids are easily recognisable, identification to species level can be tricky because, in a number of species, their most obvious character, the dorsal colour pattern, is subject to extreme variation. One of our commonest species, *Adalia 2-punctata* (L.), may vary from having all red elytra to all black elytra, and from having elytra with twelve red spots on a black background to having elytra with fourteen black spots on a red background. The colour of the pronotum may also vary. This species and its close relative, *A. 10-punctata* (L.), are our most variable species, but other species may vary according to the number of spots, whether the spots are confluent or by the presence of all black forms. Even more confusing is the fact that several species may share the same or similar colour patterns. The variation in colour pattern is genetically determined and is an example of a colour polymorphism: a phenomenon where two or more colour forms are present in the population at above the frequency at which mutations occur. The proportion of one colour form to another is likely to be controlled by environmental factors.

It follows that caution must be exercised when using identification keys that rely solely on dorsal colour patterns. There are two keys to the British species that largely ignore colour patterns, one by Pope (1953), which is long out of print and difficult to obtain, the other by Majerus & Kearns (1989). Both of these keys use anatomical features, many of which are on the underside and are rather obscure and difficult to use for the novice or anyone without a microscope. Majerus & Kearns (1989) also give an easier to use field key to the larger species. The tiny species belonging to the genera *Nephus* and *Scymnus* were covered in considerable detail by Pope (1973), although this does not illustrate the female genitalia of *Nephus* and *Scymnus*. Illustrations of the female genitalia are, however, to be found in Volume 7 of *Die Käfer Mitteleuropas* (Fürsch, 1967) which also provides keys (in German) that will allow the identification of the two species recently added to the British list as well as of those no longer found here. Keys (in French) to all the European members of the subfamily Coccinellinae have been provided by Iablokoff-Khnzorian (1982). Majerus & Kearns (1989) provide a key to the final instar larvae of our Coccinellinae, but a more comprehensive key is given by Hodek (1973).

Because most of the species are small, coccinellids are best mounted on card. This should be done after the species has been identified as identification may depend on ventral characters or require the dissection of the genitalia. Perhaps the most frequently encountered problem with mounting coccinellids is getting their appendages, especially the antennae, displayed. This can be achieved by placing the beetle on its dorsal surface onto a small square made up of three or four layers of damp kitchen towel. It can then be held in place with a light downward pressure from fine-pointed forceps and the appendages carefully teased out with the aid of a fine pin. For the best results the specimens need to be completely relaxed and for this reason coccinellids are best killed by putting them in a deep freeze and leaving them there until they need to be mounted.

Coccinellids are regarded as beneficial insects because many of the species feed on aphids. However, aphids are not the only plant pests that are eaten and not all coccinellids are predators. Aphids are the main food of all but two of the British members of the Coccinellini, the exceptions being *Aphidecta obliterata* (L.) which feeds on adelgids and *Coccinella hieroglyphica* L. for which both aphids and chrysomelid larvae are quoted as prey. In fact, many of the aphid feeding species will also take other prey such as adelgids, coccids and mites. The British Coccidulini also feed on aphids. The prey of the Scymnini are not so well documented but it seems that *Scymnus* spp. generally feed on aphids, *Nephus* spp. on coccids and *Clitostethus arcuatus* (Rossi) feeds on whitefly. The Chilocorini feed on coccids. Of our remaining predatory species, *Hyperaspis pseudopustulata* Mulsant feeds on coccids, *Platynaspis luteorubra* (Goeze) feeds on aphids and *Stethorus punctillum* (Weise) feeds on mites. Three of our species, *Halyzia 16-guttata* (L.), *Psyllobora 22-punctata* (L.) and *Tytthaspis 16-punctata* (L.) feed on mildew (Erysiphaceae), although the latter species may also take thrips and mites. Finally we have two herbivorous species, *Epilachna argus* which eats Cucurbitaceae, including White Bryony (*Bryonia dioica*) in the wild and various cucumbers, melons etc. in gardens, and *Subcoccinella 24-punctata* (L.) which normally feeds on campions and trefoils but which has recently been recorded (Hawkins, 2000) feeding on false oat grass (*Arrhenatherum elatius*).

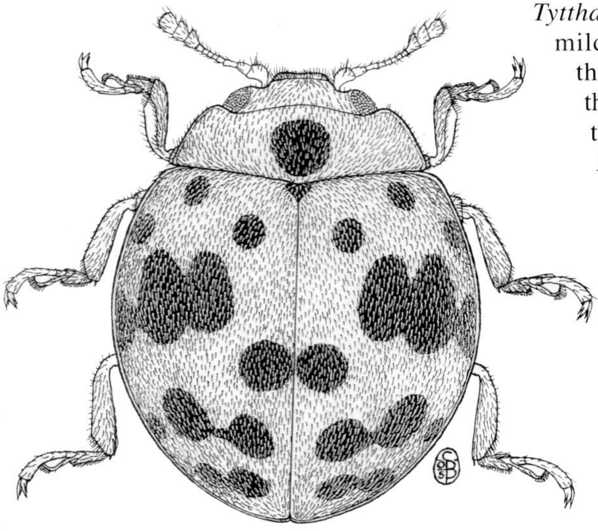

Except at times of hibernation or migration, the easiest way to find

Fig. 18 *Subcoccinella vigintiquatuorpunctata* (L.)
'The 24-spot ladybird'

coccinellids is to first locate their prey or foodplants. Once this has been done the larger species can usually be found by searching, but beating and sweeping will give the best results, especially for the smaller species. A number of species hibernate in large groups of adults and can be found during the winter in cracks and crevices in trees, posts and walls and in window frames. These aggregations are often found at prominent high spots such as hill tops or telephone posts in a flat landscape. In spring the adults, newly emerged from hibernation, will often be found sunning themselves on evergreen shrubs and trees. Migration appears to occur at times of food shortage and may only involve the aphid feeding species. When food is short these coccinellids move in search of prey and eventually reach the coast where on-shore winds force them down in the sea or on the beach (Muggleton, 1977). Such migratory swarms are made up of those species found in the coastal hinterlands and do not originate from continental Europe.

The distribution of coccinellids in the British Isles has been mapped on a 10km square basis (Majerus, Majerus, Bertrand & Walker, 1997). Many of our Coccinellidae are widespread but a small number are decidedly scarce and others are restricted to certain areas of the British Isles. One species, *Clitostethus arcuatus*, is classified as endangered (RDB1) and seems to be associated with ivy and *Viburnum*. It has only been found at a few sites in south and central England and even then, there are very few records. *Nephus 4-maculatus* (Herbst) is classified as vulnerable (RDB2) having been found only in Kent and Suffolk but it has recently been found in Surrey, where it appears to be spreading, and in Essex and Middlesex. It also appears to be associated with ivy. *Coccinella 5-punctata* L. is associated with unstable river shingles where it is most frequently encountered during the spring, roaming the shingle banks; during the summer it may be taken by sweeping riverside vegetation. It is now only known from the Spey valley and west Wales and there are old records from south Devon. This species has RDB3 (Scarce) status. Another species with an interesting distribution is *Coccinella magnifica* (Redtenbacher) which, in Britain, is only found in association with the wood ant (*Formica rufa* L.). One species regarded as uncommon is *Hippodamia variegata* (Goeze). This species seems to specialise in brownfield sites, a habitat often ignored by the coleopterist and hence its undeserved notable status. The distribution of our coccinellids is by no means fixed. *Harmonia 4-punctata* (Pontoppidan), first recorded from Suffolk in 1937 has now spread to south-west England, Wales and Scotland. *Halyzia 16-guttata* (L.), known less than twenty years ago from the Scottish Highlands and a few scattered sites in southern England, has undergone an explosive spread in the southern half of England and in Wales in the last decade. The arrival of two new breeding species from continental Europe has already been mentioned. There are several more species in continental Europe that would seem to be likely invaders, especially *Calvia 10-guttata* (L.) and *Oenopia conglobata* (L.), so a watch should be kept for these and other species.

We are fortunate in having our coccinellids well documented and, to date, the only group of British beetles that warrant a volume in the *New Naturalist* series (Majerus 1994). While a considerable amount of information has been obtained concerning the biology, ecology and distribution of British Coccinellini, there are still many gaps in our knowledge and the biology and ecology of the other coccinellids is poorly understood. A good idea of topics that would warrant further investigation can be found in Hawkins (2000). This, despite its parochial title, is one of two essential books for the student of British coccinellids; the other is Majerus & Kearns (1989).

[NB: I would be failing in my duty as Editor if I did not bring to the reader's attention the fact that the new edition of the ICZN code specifically rules against the use of the hyphenated shorthand for numbers in scientific names, ie the abbreviated name with number is an invalid name. It is, however, incredibly convenient of course, compared to the unspellable latin numbers for example *vigintiquattuorpunctata = 24-punctata* and *quattuordecimpunctata = 14-punctata* etc - JC].

[Subsequent to the submission of this chapter, in September 2004, *Harmonia axyridis* (Pallas) was recorded in Britain. There have since been numerous new UK records of this Asian species, which has colonised much of North America and Europe, and has been described by Majerus (in website below) as 'the most invasive ladybird in Earth' and 'a threat to our native aphid-feeding ladybirds'. This very variable species ranges from black with red spots to almost plain orange, and it is frequently referred to as the 'Harlequin Ladybird'.The websites http://www.ladybird-survey.pwp.blueyonder.co.uk/H_axyridis.htm and http://www.harlequin-survey.org/ provide further information – MVLB].

References

Fürsch, H. von., 1967. In Freude, H., Harde, K.M. & Lohse, G.A., (eds.) *Die Käfer Mitteleuropas*, **7**, 310pp. Goeke & Evers, Krefeld.

Hawkins, R.D., 2000. *Ladybirds of Surrey*, 136pp. Surrey Wildlife Trust, Woking.

Hodek, I., 1973. *Biology of Coccinellidae*, 260 pp. W. Junk, The Hague.

Iablokoff-Khnzorian, S.M., 1982. *Les Coccinelles*, 568pp. Boubée, Paris.

Majerus, M.E.N., 1994. *Ladybirds* (The New Naturalist series), 366pp. Harper Collins, London.

Majerus, M.E.N. & Kearns, P., 1989. *Ladybirds* (Naturalists' Handbooks 10) 103pp. Richmond Publishing, Slough.

Majerus, M.E.N., Majerus, T.M.O., Bertrand, D. & Walker, L.E., 1997. The geographic distribution of ladybirds (Coleoptera:Coccinellidae) in Britain (1984-1994). *Entomologist's Monthly Magazine*, **133**:181-203.

Muggleton, J., 1977. Ladybirds on the move. *Country Life*, **161**: 601.

Muggleton, J., 1999. *Coccinula 14-pustulata* (L.) and *Exochomus nigromaculatus* (Goeze) (Col., Coccinellidae) in Britain. *Entomologist's Monthly Magazine*, **135**:169-172.

Pope, R.D., 1953. Coccinellidae and Sphindidae. *Handbooks for the Identification of British Insects* **5**(7), 12pp. Royal Entomological Society, London.

Pope, R.D., 1973. The species of *Scymnus* (s.str.), *Scymnus* (Pullus) and *Nephus (Col., Coccinellidae)* occurring in the British Isles. *Entomologist's Monthly Magazine*, **109**: 3-39.

CORYLOPHIDAE
by S. Bowestead

This family of minute, rather globular beetles (average size 1mm. or less) is represented in the British fauna by only nine species in four genera: *Sericoderus*, one sp.; *Corylophus*, two sp.; *Orthoperus*, five sp.; *Rypobius*, one sp.

Corylophids are mould feeders and are most easily found in mould ridden situations under bark or in rotten wood or grass heaps. Sweeping under trees or in wet areas in the evening, especially after hot weather can be productive as this is the time when the smallest beetles fly. Debris should be sifted over a tray and examined (I use a x2 headband magnifier at this stage). Remove the specimen from there to a small tube containing 70% alcohol + 5% acetic acid which helps to prevent hardening. At home, sample specimens should have their abdomen removed on a watchglass with a little water, before setting in a tiny blob of gum tragacanth with the abdomen set behind the specimen on the card. To examine the genitalia of these minute beetles requires very fine micropins attached to the end of a matchstick or something similar. I use one straight pin and one with a very fine right-angle at the apex. The abdomen of a fresh specimen can be placed in a cavity slide with a drop or two of water and the aedeagus carefully teased out from between the sclerites. At this stage the tegmen can be slid off the median lobe.

The median lobe can then be conveniently examined in glycerine under a compound microscope at a magnification between x200 – x400. The same general procedure applies to the examination of the spermatheca. Sexing requires some practice but in general the females have evenly curved apical sternites, in males these are slightly truncate. In *Orthoperus* the males have a small keel on the median line of the metasternum. Orientation of these genitalia preparations is very important for accurate determination. For this reason, the glycerine provides a convenient means for manipulation of the specimen so that orientation can be checked. After examination the genitalia can be conveniently stored in a very small blob of glycerine inside a microtube whose

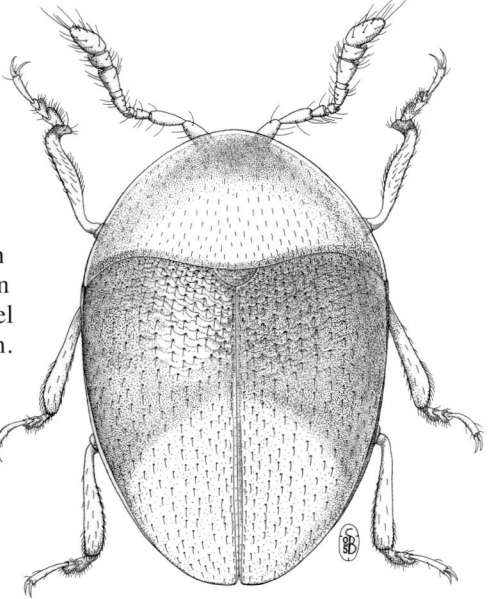

Fig.19 *Corylophus cassidoides* (Marsham).

cork is impaled beneath the specimen on the pin (see p. 366). Alternatively the genitalia can be gradually dehydrated in increasing concentrations of alcohol and then transferred to a blob of clove oil and from there to an acetate strip and covered with a drop of Euparal, the strip then being pinned beneath the card on which the specimen is mounted. Make every effort to keep gum away from the dorsal surface of the beetle at this stage as the microsculpture of these beetles is a most important character for their reliable determination.

My revision of the West Palaearctic Corylophidae (Bowestead, 1999) has brought about several changes to the British catalogue (revised Kloet and Hinks, 1977). Full synonymy and diagnostic information, instructions for dissection and preparation of genitalia, supported by figures of: habitus outline; microsculpture; aedeagus with tegmen removed and spermatheca for all species is provided in (Bowestead, 1999) and a complete habitus of each of the genera in (Bowestead & Leschen, 2002).

There are plans to publish distribution maps of the British species but until then the following general distribution and habitat preference of the British species is as follows:

Sericoderus lateralis (Gyllenhal) – Dry, warm and sun exposed areas in cut grass and compost heaps. Southern to mid-England.

Corylophus cassidoides (Marsham) – Fenland. Southern and eastern England.

Corylophus sublaevipennis Jacquelin du Val – Dry areas on south facing cliffs or downland, at roots of plants. Southern to mid-England.

Orthoperus aequalis Sharp – Old forest and parkland, in rotten wood and under bark of oak and beech. Southern to northwest and mid-England.

Orthoperus atomus (Gyllenhal) – In pine debris and under bark. Throughout Britain.

Orthoperus brunnipes (Gyllenhal) – Fenland, by sweeping, in cut debris and at the base of stacked vegetation in wet places. Southern and eastern England.

Orthoperus corticalis (Redtenbacher) – Old forest and parkland, under bark and on fungi of oak and beech. Southern to mid-England.

Orthoperus nigrescens Stephens – Old forest and parkland, by sweeping under oak and on fungi on various hardwoods.

Rypobius praetermissus Bowestead – In plant debris or under stones near plant roots, in Europe this species has usually been found near salt marshes and at the edge of inland waters, but I have also taken it in numbers at the roots of plants in stony fields. South coast of England.

The synanthropic species *Orthoperus atomarius* (Heer) has not been recorded here for more than 70 years but may be rediscovered in an old wine cellar, as it has been imported into many European cities in the past on mouldy wine barrels or the packing that is around them.

References

Allen, A. A., 1970. Revisional Notes on the British species of *Orthoperus* Steph. (Col., Corylophidae). *Entomologist's Record and Journal of Variation.*, **82**: 112-120.

Bowestead, S., 1999. A Revision of the Corylophidae (Coleoptera) of the West Palaearctic Region. *Instrumenta Biodiversitatis* 3. 203 pages (426 figs). Muséum d'histoire naturelle, Geneva.

Bowestead, S. & Leschen, R., 2002. *American Beetles*, Vol. 2: 390 – 394, Chapter 94. Corylophidae CRC Press, Boca Raton, Florida, USA. pp. 1: 443; 2: 861.

Hammond, P. M., 1971. *Rypobius ruficollis* Jacqu. (Col., Corylophidae) A Genus and Species New to Britain. *Entomologist's Gazette,* 22: 241-243.

Matthews, A., 1899. *A Monograph of the Coleopterous Families Corylophidae and Sphaeriidae.* 220 pages +8 plates. Janson & Son, London.

LATRIDIIDAE
by C. Johnson

Fifty-four British species, with at least one other soon to be added, are roughly evenly divided between the mostly glabrous Latridiinae (including 'Plaster Beetles') and the pubescent Corticariinae.

The beetles occur in a wide range of habitats, although dung is not favoured. Since most of the beetles are detritus feeders, they are to be found in humid, mouldy and dark places. Typical haunts such as heaped garden and farm refuse and haystack bottoms will yield *Cartodere (Aridius) nodifer* (Westwood) (common plaster beetle) – probably our most ubiquitous species, closely followed by *C. (A.) bifasciata* (Reitter), *Enicmus histrio* Joy & Tomlin, *Latridius minutus*-group members, the minute *Dienerella ruficollis* (Marsham) and *Corticaria elongata* (Gyllenhal). Cut ends of logs, as well as fungoid twigs and mouldy fungi, are always worth checking for other *L. minutus*-group as well as *Dienerella* (*clathrata* (Mannerheim) and *elongata* (Curtis)). Ripe (powdery) slimefungi (Myxomycetes) are the breeding site for three species of *Enicmus* (*fungicola* Thomson, *rugosus* (Herbst) and *testaceus* (Stephens)). Under tight fungoid bark occurs *Enicmus brevicornis* (Mannerheim), rarer *Corticaria* (*alleni* Johnson on broadleafed, *linearis* (Paykull) on conifers). Beating conifers produces others (*ferruginea* Marsham and *Corticarina similata* (Gyllenhal)). Flowers, foliage and plant litter yields *Stephostethus lardarius* (De Geer), *Cortinicara gibbosa* (Herbst) and some *Corticarina* (*fuscula* (Gyllenhal) and *fulvipes* (Comolli)). Litter is home to other species of *Corticaria* (*impressa* (Olivier)), especially by saltmarshes (*crenulata* (Gyllenhal)) and rivers (*punctulata* Marsham). Leaf litter from broadleafed trees is the characteristic habitat of *E. transversus* (Olivier). In old and dark buildings (especially museums and toilets in pubs), mouldy patches, old wall paper, old wood (in wood stores), soiled tiles and cupboard interiors produce several species which only live indoors (being hardly ever found outside), *Lithostygnus serripennis* Broun, rarer *Dienerella* (*filiformis* (Gyllenhal) and *filum* (Aubé)), *Cartodere constricta* (Gyllenhal), *Thes bergrothi* (Reitter), *Adistemia watsoni* (Wollaston) and some *Corticaria* (*fulva* (Comolli) and *inconspicua* Wollaston). Despite their small sizes, many of these beetles can be easily seen in the morning when they have fallen onto any pale surface – especially window ledges, sinks and crockery. Some of these species are also associated with stored products (see Part III) and were monographed by Hinton (1941).

When preparing latridiids, as with all small beetles, it is very important that glue is kept off the dorsal surface. Antennae should be carefully displayed with all segments in the same plane, flat on a card. It is necessary to dissect and sex specimens, since most males have important sexual characters (secondary) on the legs, and in the male genitalia. The male genital tergite is also important as a character in the *L. minutus*-group.

Within the family there are several difficult groups and as Joy (1932) is not adequate (Hodge & Jones, 1995) for determination pupases, it is recommended that Peez (1967) together with Rücker's (1992) important update is used. The nomenclature in the family has changed much in recent years – even the family spelling, as the world fauna has become better known. Several useful papers on the British fauna have also appeared in the entomological press, and these are listed below in the references.

References

Allen, A. A., 1951. *Lathridius bifasciatus* Reitt. (Col., Lathridiidae): an Australian beetle in Britain. *Entomologist's Monthly Magazine*, **87**:114-115.

Allen, A. A., 1952. *Lathridius norvegicus* A.Strand (Col., Lathridiidae) rediscovered: an addition to the British List. *Entomologist's Monthly Magazine*, **88**:282-283.

Allen, A. A., 1966. A Clarification of the Status of *Cartodere separanda* Reitt. (Col., Lathridiidae); and *C. schueppeli* Reitt. New to Britain. *Entomologist's Monthly Magazine*, **102**:192-198.

Hinton, H. E., 1941. The Lathridiidae of Economic Importance. *Bulletin of Entomological Research*, **32**:191-247, figs. 67.

Hodge, P. J. & Jones, R. A., 1995. *New British Beetles. Species not in Joy's practical handbook*. British Entomological and Natural HistorySociety, Reading.

Johnson, C., 1974. Studies on the genus *Corticaria* Marsham (Col., Lathridiidae). Part 1. *Ann. Entomologica Fennica*, **40**(3):97-107.

Johnson, C., 1986. Notes on some Palaearctic *Melanophthalma* Motschulsky (Coleoptera: Lathridiidae), with special reference to *transversalis* auctt.. *Entomologist's Gazette*, **37**:117-125.

Johnson, C., 1992. Additions and Corrections to the British List of Coleoptera. *Entomologist's Record & Journal of Variation*, **104**:305-310.

Levey, B., 1997. *Stephostethus alternans* (Mannerheim) (Lathridiidae) – a species new to Britain. *Coleopterist* **6**(2):49-51.

Lyszkowski, R. M., Owen, J. A. & Taylor, S., 1992. *Corticaria abietorum* Motschulsky (Col. Lathridiidae) new to Britain. *Entomologist's Record & Journal of Variation,* **104**:67-69.
Peez, A. von., 1967. *in* Freude, H., Harde, K. W. & Lohse, G. A., *Die Käfer Mitteleuropas,* **7**:169-190. 58 Fam. Lathridiidae. Goecke & Evers, Krefeld.
Rücker, W. H., 1992. *in* Lohse, G. A. & Lucht, W. H. *Die Käfer Mitteleuropas, 2. Supplementband mit Katalogteil,* **13**:139-160. 58. Familie Latridiidae. Goecke & Evers, Krefeld.
Tozer, E. R., 1973. On the British species of *Lathridius* Herbst (Col., Lathridiidae). *Entomologist's Monthly Magazine,* **108**(1972):193-200.

CIIDAE
by Glenda M. Orledge

The name of this family is of Greek derivation, and its correct spelling has been a matter of some debate. Cioidae, Cissidae, Cisidae and Ciidae have all been used, but Ciidae is now widely favoured. The nomenclature of British taxa requires some revision at the generic level. Since this work is not yet complete, the names used by Pope (1977) are followed here.

The British Ciidae currently comprise twenty one native species, plus the introduced *Cis bilamellatus* Wood, which appears to have arrived during the nineteenth century in Australian fungus sent to the Royal Botanic Gardens, Kew (Paviour-Smith, 1960). The most recently discovered species is the rare native, *C. dentatus* Mellié (Aubrook, 1970). Ciids occur throughout England, Wales and Scotland, although species richness declines from south to north. However, the progressive northwards loss of *C. coluber* (Abeille), *Sulcacis bicornis* (Mellié) (for which, see map of Orledge & Smith, 1999), *C. micans* (F.), *C. setiger* (Mellié), *C. pygmaeus* (Marsham) and *S. affinis* (Gyllenhal), is partially offset by the gain of *C. dentatus*, *C. jacquemarti* (Mellié) and *Rhopalodontus perforatus* (Gyllenhal) which are known only from Scotland. Of the remaining species, *C. boleti* (Scopoli), *C. festivus* (Panzer), *C. nitidus* (F.) and *Octotemnus glabriculus* (Gyllenhal) are widespread and common, with *C. alni* Gyllenhal and *C. bidentatus* (Olivier) also widespread, but more local in occurrence. *Cis fagi* Waltl, *C. hispidus* (Paykull), *C. vestitus* Mellié and *Ennearthron cornutum* (Gyllenhal) are generally recorded less frequently, particularly in the north, and Scottish records for these species are few. In contrast, records for *C. lineatocribratus* Mellié and *C. punctulatus* Gyllenhal are concentrated in the north, although *C. punctulatus* is known from East Anglia and Kent (Allen, 1935, 1937: Mendel, 1987) and *C. lineatocribratus* occurs in the New Forest. *Cis bilamellatus* is now common throughout much of England, Wales and southern Scotland. Its invasive spread has been studied by Paviour-Smith (1960).

Adult British ciids are small (1mm-4mm long) black or brown beetles with deflexed heads. Their antennae have eight to ten segments, including a three-segmented club. The males of most species have the front of the clypeus, and sometimes also the front edge of the pronotum, either raised to form a sinuous ridge, developed into plates, or with a pair of tubercles or teeth. With the exception of *E. cornutum* and *C. punctulatus*, males also have a setose fovea (sometimes very small) in the middle of the first visible abdominal sternite. These features, together with the shape of the front tibia, also the form of pronotal and elytral bristles or hairs and their associated puncturation, are important characters for separating the species. However, morphological differences between species can be slight, whilst within species there can be variation in size, colour and the degree of male head and pronotal development. The consequent identification uncertainties, and the lack of adequate British keys, have contributed to the ciids' lack of popularity and earned them their reputation as a 'tricky' group. It is hoped that new keys (Orledge, in preparation) will go some way to alleviating this situation, but, for now, the German key of Lohse (1967) is the best available and includes all native British species. The keys of Joy (1932) can be confusing, especially for the beginner, as they rely heavily on comparative characters. Additional reference to Kevan (1967) and Allen (1990) can sometimes be of help for the *C. festivus* group species. The illustrated German key to ciid larvae of Holter, Milewski & Reibnitz (1999) includes all British species except *C. bilamellatus* and *C. coluber*.

Ciids are fungivores. Typically both larvae and adults exploit the fruiting bodies – and in some instances the vegetative mycelium – of poroid, wood rotting basidiomycetes. These are the fungi which produce the familiar 'brackets' and hard encrustations which appear on the trunks and branches of trees and dead wood. It is in these fruiting bodies that ciids are found most readily, although they also occur in fungoid wood and mats of fungal mycelium. Some species can be found by beating dry dead branches, especially of deciduous Oaks, *Quercus* species. Both larvae and adults of at least some ciids may be found throughout the year. Development time from egg to adult varies with temperature, but is around two months in the summer (see, for example, Paviour-Smith, 1968).

Typically ciids live and breed in association with only a subset of known ciid hosts, and some are restricted to a single fungal genus. *Trametes* supports the greatest number of ciid species, although other genera, including *Ganoderma*, *Fomes*, *Piptoporus* and *Stereum*, are also important hosts. *Trametes* brackets will often yield specimens of *C. boleti* and *O. glabriculus*, whilst old *Ganoderma* brackets frequently support *C. nitidus*. *Cis festivus* may be found in association with *Stereum*, and *Piptoporus* brackets are a good place to look for *C. bilamellatus*. Just a small fragment taken from the edge of an old, well-eaten bracket can yield a number of ciid specimens, sometimes of more than one species. Such minimal samples, from different host taxa, are likely to produce three or four species at any one locality, with the possibility of perhaps six or more at some sites. This careful and restrained collecting is not only generally sufficient, but is also essential in order to safeguard the patchy and frequently scarce fungal hosts and their dependent organisms. Certainly, the complete removal of a large bracket, or group of small brackets, also the disturbance of isolated brackets, should be avoided.

Ciids' small size and morphological similarities mean that their identification is best attempted using a microscope with a magnification of at least x60-x80, and an adjustable light source. Whilst it is possible, with experience, to identify live specimens, and specimens preserved in spirit, some characters are undoubtedly easiest to see on mounted specimens. Ciids should be mounted on card, using a minimum of water-soluble glue, and with the head visible, also at least one antenna and front leg extended and fully visible. If a (short) series of specimens of a species is collected, at least one example of each sex should be mounted dorsal side uppermost and a male should be mounted ventral side uppermost. Given only a single specimen, this is best either kept in spirit, or mounted on its side to allow inspection of both dorsal and ventral characters.

Dissection of male genitalia is occasionally necessary. This is most easily achieved on freshly killed and mounted specimens, and by separating the elytra before removing the genitalia through an opening torn in the terminal abdominal segments. The elytra can then be returned to their original position. Ideally, the median lobe and parameres should be separated before mounting in dimethyl hydantoin formaldehyde (DMHF) beside the specimen from which they have been taken.

As already noted, identification requires careful microscopic examination and may not be straightforward. However, with practice the main species groups and the more distinctive species may be readily recognised. A small reference collection comprising two or three labelled examples of each sex of each species is an invaluable aid. Where appropriate, data labels should include information about the fungal host and host tree. Surplus live specimens should be returned to their capture site, or, if this proves impossible, should be preserved as additional reference material. They should never be released elsewhere.

References

Allen, A. A., 1935. Some recent capture of Coleoptera. *Entomologist's Monthly Magazine*, **71**: 64-67.

Allen, A. A., 1937. Two 'Northern' beetles in Kent. *Entomologist's Record and Journal of Variation.*, **49**: 60-61.

Allen, A. A., 1990. Notes on the species-pair *Cis festivus* Panz. and *C. vestitus* Mell. (Col.: Cisidae). *Entomologist's Record and Journal of Variation*, **102**: 177-179.

Aubrook, E. W., 1970. *Cis dentatus* Mell. (Col. Cisidae): an addition to the British list. *The Entomologist*, **103**: 250-251.

Holter, U., Milewski, I. & Reibnitz, J., 1999. Familie: Cisidae. *In* Klausnitzer, B. *Die Larven der Käfer Mitteleuropas*, **5**: pp. 222-236. Goecke & Evers, Krefeld.

Joy, N. H., 1932. *A Practical Handbook of British Beetles.* Volumes 1 (text) and 2 (plates). Witherby Ltd., London. xxvii + 622pp. and 194pp.

Kevan, D. K., 1967. On the apparent conspecificity of *Cis pygmaeus* (Marsh.) and *C. rhododactylus* (Marsh.) and on other closely allied species (Col., Ciidae). *Entomologist's Monthly Magazine,* **102**: 138-144.

Lohse, G. A., 1967. Familie: Cisidae. *In* Freude H., Harde K. W. & Lohse G. A. *Die Käfer Mitteleuropas,* **7**: pp. 280-295. Goecke & Evers, Krefeld.

Mendel, H., 1967. *Cis punctulatus* Gyllenhal (Coleoptera: Cisidae): a northern species established in Suffolk. *Entomologist's Record and Journal of Variation,* **99**: 156.

Orledge, G. M. & Smith P. A., 1999. *Sulcacis bicornis* (Mellié) (Ciidae) new to Wales, with notes on the species in Britain. *The Coleopterist,* **8** 113-115.

Paviour-Smith K., 1960. The invasion of Britain by *Cis bilamellatus* Fowler (Coleoptera: Ciidae). *Proceedings of the Royal Entomological Society of London (A),* **35**: 145-155.

Paviour-Smith, K., 1968. A 'summer' development time for *Cis bilamellatus* Wood (Col., Ciidae). *Entomologist's Monthly Magazine,* **103**: 247-249.

Pope, R. D., 1977. Kloet & Hincks. A Check List of British Insects. Part 3: Coleoptera and Strepsiptera. 2nd. revised edition. *Handbooks for the Identification of British Insects,* **11**: xiv + 105pp.

MORDELLIDAE
(Plate 71)

by B. Levey

The Mordellidae are a taxonomically difficult group. Several species new to Britain have been added in recent years, and two new species have been recently described from British material, and later synonymised. 17 species are known from Britain and it is possible further unrecognised species may occur.

Mordellidae are of characteristic form and appearance, most occur on a range of pollen-rich flowers, especially Umbelliferae and Rosaceae, but *Tomoxia bucephala* Costa is usually taken on dead trunks and large boughs of beech and more rarely oak, its larval hosts. They are especially active around mid-day in warm sunny windless weather from late May to August. Some *Mordellistena* species have been collected by suction sampling.

Larval hosts of British Mordellidae are not well known. Many, if not all of the black *Mordellistena* species probably live in the pithy stems of herbaceous plants. The collection and rearing of adults from the stems of possible hosts would greatly increase our knowledge of these species. Larvae of *T. bucephala*, *Mordellochroa abdominalis* (F.), and probably all the mainly yellow *Mordellistena* species related to *M. humeralis* (L.), live in woody tissues of deciduous trees and shrubs.

When mounting it is important to keep gum off the posterior femora and tibia as well as the upper surface especially in pubescent species. Killed with ethyl acetate, the beetles are transferred from the killing bottle to a sheet of white absorbent paper (kitchen tissue, filter paper etc.) and turned on their dorsal surface. The legs, head, mouth-parts, and antennae are frisked out with the aid of a fine pin, the beetle is held gently with the forceps. Male genitalia are dissected carefully with a fine pin and/or forceps via the terminal abdominal segment. The dissected genitalia are then transferred to a dissecting dish and the median lobe and minute parameres, paired and asymmetrical, are painstakingly separated. They can be lightly glued to, or mounted in a small amount of DMHF resin, or other alcohol or water-soluble mountant on the mounting card alongside the beetle. The appearance of the parameres alters depending on their orientation, so it is essential for comparison that the parameres are always mounted in the same orientation. The beetle is then carefully transferred to the mounting card, upon which waits a small elongate blob of glue. The head, mouth-parts, antennae and two anterior pairs of legs are then suitably arranged and lightly glued to the card. The posterior pair of legs should not be gummed but raised so that the tibia is level with the dorsal surface of the elytra. The form and number of ridges on the hind tibia and tarsi are important in identification, as are the antennae the fore tibia, and maxillary palpi in some species.

For identification, the student is directed to the paper by Batten (1986), Levey (1999; 2002;) and Cooter (1991), both Joy (1932) and Buck (1954) being long out of date. For collecting and preparing, full detail is contained in Batten (1988).

References

Allen, A.A., 1995. On the British *Mordellistena humeralis* (L.) (Col.: Mordellidae) and its allies. *Entomologist's Record and Journal of Variation*, **107**:181-184.

Batten, R., 1986. A Review of the British Mordellidae (Coleoptera), *Entomologist's Gazette*, **37**:225-235.

Batten, R., 1988. Mordellidae (Coleoptera): catching, preparing and mounting. *Nieuwsbrief European Invertebrate Survey – Nederland*, **18**:9-10.

Buck, F.D., 1954. Coleoptera (Largiidae, Alleculidae etc.) *Handbooks for the Identification of British Insects* 5(9). Royal Entomological Society. London.

Cooter, J., 1991. *Mordellistena pygmaeola* Ermisch, 1956 (Coleoptera: Mordellidae) new to Britain. *Entomologist's Gazette*, **42**:97-98.

Levey, B., 1999. *Mordellistena secreta* Horak (Coleoptera: Mordellidae), a species new to Britain. *British Journal of Eentomology and Natural History*, **12** : 227-229.

Levey, B., 2002. A revision of the British species of *Mordellistena* (Coleoptera, Mordellidae) belonging to the *parvula* group and the subgenus *Pseudomordellina* Ermisch. *British Journal of Eentomology and Natural History*, **15** : 83-90.

SCRAPTIIDAE
by B. Levey

In Britain this family (sometimes regarded as a subfamily of the Mordellidae) is confined to two genera – *Scraptia* with three species and *Anaspis* with eleven.

The occurrence in Britain of *Scraptia dubia* (Olivier) is based on old records from Berkshire and Dorset; it is a much larger beetle than either of its two congeners and despite being found not infrequently in France and the rest of the Continent has not been taken in Britain for over a hundred years, and may be extinct. *Scraptia fuscula* Müller has been known from the Windsor area for many years and there are recent records from Richmond Park, Surrey and East Gloucestershire. *S. testacea* Allen is much more widespread and has been recorded from about 11 vice counties, mostly in ancient parklands and pasture woodlands in southern England. They can be beaten from over-mature oaks during June and early July. *S. testacea* sometimes also occurs on other deciduous trees. The larvae, which are distinctive, occur in dead wood, wood mould and beneath the bark of trees (Švácha 1995). Great care must be taken when mounting as the beetles are very delicate and can easily be broken to pieces even with a fine hair brush; a minimum of very dilute glue is recommended.

Anaspis are also delicate and easily damaged. Females are difficult to identify, especially when captured in isolation. Fortunately, the bulk of the species occur in numbers. Males should be mounted ventrally after first removing the abdomen, this should be mounted dorsally alongside the beetle. The shape and form of the abdominal laciniae (or their absence) and the fore tarsi are of great diagnostic value in the males. The antennae and palpi and mid tibia are also significantly different in some species. The antennae, palpi and legs should be set in the normal way.

Most *Anaspis* species are not uncommon and are widely distributed. They can be obtained in numbers by beating blossoms – especially hawthorn and other spring flowering shrubs - and by sweeping umbelliferous plants. *A. lurida* Stephens however is most commonly collected by beating or fogging oak trees. Most species are found in June but some also occur in July and August. A few species are rare or uncommon. *Anaspis bohemica* Schilsky is confined to the Scottish Highlands and is rarely collected. *A. thoracica* (L.), which may be the same species as *A. septentrionalis* Champion (Levey, 2002), is widespread but usually not taken in numbers. *A. costai* Emery is also scarce and is largely confined to southern England (Levey, 2003). The very rare *A. melanostoma* Costa is probably not a British species (Levey, 2003). Larvae of most species probably occur in dead wood of some description but a few species such as *A. pulicaria* Costa, which often occurs away from woodland situations, possibly live in the stems of non woody plants.

For identification, both Joy (1932) and Buck (1954) are now out of date or otherwise unreliable. There are a number of more recent papers which are necessary reading (Allen, 1975; Levey, 1996; 2002; 2003). Batten (1976) has good illustrations but the key is in Dutch. Ermisch (1969) covers all the British species but the illustrations are less good.

References

Allen, A.A., 1975. Two Species of *Anaspis* (Col., Mordellidae) new to Britain, with a Consideration of the status of *A. hudsoni* Donis. etc. *Entomologist's Record and Journal of Variation*, **87**: 269-274.

Batten, R. 1976. De Nederlandse soorten van de kevefamilie Mordellidae. *Zoologische Bijdragen*, **19**: 1-37.

Buck, F.D., 1954. Coleoptera (Largiidae, Alleculidae etc.) *Handbooks for the Identification of British Insects*, **5**(9). Royal Entomological Society. London.

Levey, B., 1996. Anaspis septentrionalis Champion, a senior synonym of A. schilskyana Csiki (Scraptiidae). *Coleopterist* **5(2)** : 58.

Levey, B., 2002. Are Anaspis septentrionalis Champion and A. thoracica (Linnaeus) a single variable species? *Coleopterist* **11 (1)** : 1-5.

Levey, B., 2003. Taxonomic notes on European Anaspis (Nassipa) (Coleoptera: Scraptiidae), with general notes on the British species. *Entomologist's Gazette*, **54**: 197-206.

Švácha, P., 1995. The larva of Scraptia fuscula *(P.W.J. Müller) (Coleoptera: Scraptiidae): autotomy and regeneration of the caudal appendage. In* Pakaluk, J. and Ślipiński, S.A. (eds.) *Biology, Phylogeny, and Classification of Coleoptera: Papers Celebrating the 80th Birthday of Roy A. Crowson*, Vol.1 pp. 473-489. Warsaw: Muzeum i Instytut Zoologii PAN.

TENEBRIONOIDEA
(Plates 63, 69, 70, 72 – 76)

by M.V.L. Barclay

The superfamily Tenebrionoidea is a large and very diverse group of beetles comprising some 30 families (Lawrence & Newton, 1995), of which 17 have been recorded in the British Isles: Mycetophagidae, Ciidae, Tetratomidae, Melandryidae, Mordellidae, Rhipiphoridae, Colydiidae, Tenebrionidae, Oedemeridae, Meloidae, Mycteridae, Pythidae, Pyrochroidae, Salpingidae, Anthicidae, Aderidae and Scraptiidae. Members of these families are generally characterised by a tarsal formula of 5-5-4. The superfamily corresponds closely to the group 'heteromera' of earlier authors, although the 'heteromera' of Joy (1932) did not include Mycetophagidae or Ciidae, and that of Pope (1977) excluded Ciidae.

Exactly two hundred species of tenebrionoid are currently recognised from the British Isles. Only a handful are common and widespread, and more than half are given a conservation status by Hyman (1992). Many taxa are associated with ancient trees and their fungi, but the group also includes pests of stored products, and a number of coastal species. A large proportion of species, especially the forest taxa, have a very short-lived and elusive adult stage. In these species, a more realistic picture of their distribution and abundance could be gained if more attention was paid to the larvae, especially in taxa such as Pyrochroidae and Pythidae, where immature stages are conspicuous and easily recorded. New techniques for sampling adults, such as emergence traps, arboreal flight-interception (canopy) trapping and canopy fogging may also increase our knowledge of these insects, many of which inhabit wood and fungi high up in trees. While some of the species on the British list have not been seen for over a century, and are almost certainly extinct, other species new to Britain almost certainly await discovery. Only 71 species are recorded from Ireland (Anderson *et al.*, 1997).

With the notable exceptions of the difficult genus *Mordellistena* (Mordellidae), some Ciidae and some female Scraptiidae and Oedemeridae, identification of British Tenebrionoidea is quite straightforward, and can be achieved using Buck (1954), Brendell (1975) and a combination of Joy (1932) and papers listed in Hodge & Jones (1995). Almost 50 British species are figured and discussed by Harde (1984), and 16 by Chinery (1986).

MYCETOPHAGIDAE. As the name suggests, these small beetles, represented by 13 species in Britain, are generally associated with fungi. None are particularly host-specific, but *Mycetophagus multipunctatus* F. and *Pseudotriphyllus suturalis* (F.) can be reared in numbers from birch bracket fungi *Piptoporus betulinus*, while *Mycetophagus atomarius* (F.) is more often found in large specimens of *Daldinia concentrica* (Bolt.) ('King Alfred's Cakes' fungus), often together with *Biphyllus lunatus* (F.) (Biphyllidae) and *Platyrhinus resinosus* (Scopoli) (Anthribidae). *Typhaea stercorea* (L.) can be collected from mouldy straw, reed refuse, under dry bark etc. and occasionally occurs in neglected stored products. *Mycetophagus quadripustulatus* (L.) and *Litargus connexus* (Fourcroy) are not uncommon under bark of decaying logs. *Eulagius filicornis* (Reitter) was recently discovered new to Britain, on the fungus *Stereum* in the Reading area (Harrison, 1996).

CIIDAE. see p. 107. Treated in a separate chapter by Glenda Orledge

TETRATOMIDAE. The four British species of this genus are all associated with fungi. The slender *Hallomenus binotatus* (Quensel) was recently transferred from Melandryidae (Nikitsky, 1998). The remaining three species are all in the genus *Tetratoma*. The blue and orange *T. fungorum* F. is not uncommon on bracket fungi, especially *Piptoporus*, from late autumn to early spring, and is also attracted to light. *T. desmaresti* Latreille has been found, very sparingly, by sieving fungoid debris under old oaks in early spring.

MELANDRYIDAE. The 17 British species of melandryid are all associated with ancient trees or their fungi, and generally occur in old woodland. The finding of any melandryid is an event, and all except one species, *Orchesia undulata* Kraatz, are given conservation status (Hyman, 1992). *O. undulata* is occasionally found, generally singly, in the frass under the loose bark of well rotted logs lying on the ground. *Orchesia micans* (Panzer) develops in arboreal bracket fungi, especially *Inonotus hispidus* (Bull. ex Fr.); examination of this fungus in winter will almost always reveal the pale pink larvae, which can be reared along with their braconid parasitoid *Meteorus obfuscatus* (Nees). The adults are extremely elusive, and probably live mainly in the tree canopy, as do the adults of *O. minor* Walker. Species of *Abdera* are usually found on the fungus-encrusted cut surfaces of fallen trees, or can be collected, as can *Anisoxya fuscula* (Illiger), and very occasionally *Hypulus quercinus* (Quensel), by beating dead and moribund deciduous trees. *Zilora ferruginea* (Paykull) and *Xylita laevigata* (Hellenius) are species of the Caledonian Pine Forests. *Melandrya barbata* (F.) is one of our scarcest beetles, and is known from only a handful of chance-met specimens, mostly taken by beating in old oak pasture woodlands in the southeast.

MORDELLIDAE. see p.110. Treated in a separate chapter by Brian Levey.

RHIPIPHORIDAE. The only British rhipiphorid, *Metoecus paradoxus* (L.), is a brood parasite of the Common Wasp *Vespula vulgaris* (L.). Eggs are laid on dry wood, and the triangulin larvae are transported to the wasps' nest by workers collecting wood pulp for nest building. There, they invade a cell and feed on the wasp larva. *Metoecus* is probably quite common and widespread, although rarely seen. Adults are most conspicuous in autumn.

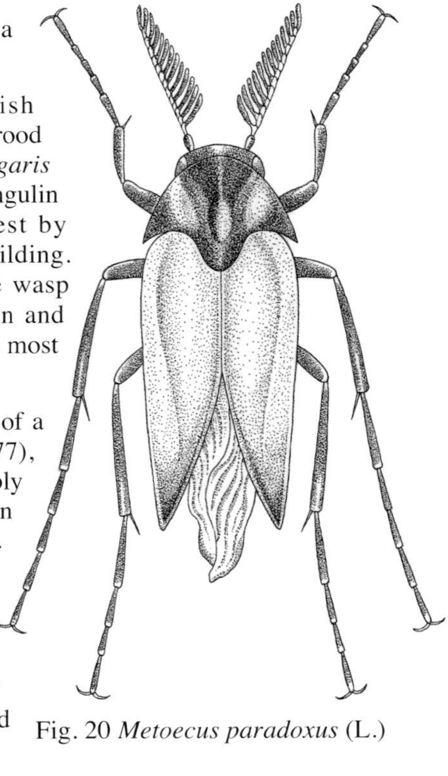

COLYDIIDAE. This family was something of a taxonomic ragbag at the time of Pope (1977), and its composition has changed considerably since. *Teredus* and *Oxylaemus* have been transferred to Bothrideridae (Cucujoidea). *Myrmechixenus* has been moved to Tenebrionidae, and *Aglenus* to Salpingidae. Furthermore, Colydiidae is increasingly treated as a subfamily of the Zopheridae (Lawrence et al, 1999). The most likely species to be encountered is the four-spotted

Fig. 20 *Metoecus paradoxus* (L.)

Bitoma crenata (F.), commonly found under loose bark of deciduous trees. *Colydium elongatum* (F.) has increased since the mid-1990s, probably in response to the rapid increase of the platypodid *Platypus cylindrus* (F.), on which it feeds; adult *Colydium* can often be found in the conspicuous colonies of *Platypus*. *Cicones undatus* Guérin-Méneville and *Synchita separanda* (Reitter), in company with the latridiid *Enicmus brevicornis* (Mannerheim), are often recorded under the bark of sycamore *Acer pseudoplatanus* (L.) infected with sooty bark disease fungus *Cryptostroma corticale* Gregory & Walker. The naturalised *Pycnomerus fuliginosus* Erichson occurs, very locally, under bark in wet woodland. All the remaining species are woodland specialists, except for *Orthocerus clavicornis* (L.), which is found among lichens, moss and roots on sand dunes and sandpits, and *Langelandia anophthalma* Aubé, a subterranean species that has been shown to be quite widespread with the advent, in Britain, of deep pitfall trapping.

TENEBRIONIDAE. This family is the largest in the Tenebrionoidea, with more than 20,000 species described world-wide. 47 are recorded from Britain. Many, including *Blaps*, *Diaperus* and *Bolitophagus* have an extremely pungent smell produced as a chemical defence. Most Tenebrionidae, unlike most other members of the superfamily, are fairly to very long-lived as adults. Exceptions include the Lagriinae and Alleculinae, occasionally given family status, but here treated as subfamilies of Tenebrionidae.

British Tenebrionidae fall into several distinct ecological categories; a number are, to a greater or lesser extent, synanthropic, associated with stored food products or with human and animal housing, including the genera *Tribolium*, *Latheticus*, *Gnatocerus*, *Alphitophagus*, *Alphitobius*, *Palorus*, *Tenebrio* and *Blaps*. Many of these species have declined with the demise of horse-drawn transport and domestic cellars, and with increased domestic and retail hygiene, although *Tribolium* remains an important pest, and *Blaps mucronata* Latreille (cellar or churchyard beetle) still occurs in stables, zoos etc. *Tenebrio molitor* L. is the familiar 'mealworm', sold as pet food. Another group is associated with old woodland, and many such species are distinctly uncommon. *Pentaphyllus testaceus* (Hellwig), not recorded for over 100 years, was recently rediscovered at Windsor (P.M. Hammond pers. comm.) in some numbers. *Diaperus boleti* (L.) is seen only sporadically in *Piptoporus betulinus* (Bull. ex Fr.) on birch. *Corticeus* spp. are associated with the borings of other Coleoptera, including Scolytinae and Lymexylidae, while *Myrmechixenus* spp. are associated with ants. *Nalassus laevioctostriatus* (Goeze) (formerly *Cylindrinotus*) is one of the few truly abundant Tenebrionoidea, and is found in woodlands, on moorlands and seacliffs, and even in graveyards; the larvae feed in the soil, while adults graze at night on algae and lichens. Several species are primarily coastal, including *Xanthomus pallidus* (Curtis), *Crypticus quisquilius* (L.), *Opatrum sabulosum* (L.), *Phylan gibbus* (F.) and *Melanimon tibiale* (F.).

Phaleria cadaverina (F.) can be found among strandline debris. Alleculinae and Lagriinae are more prone to flight than other Tenebrionidae. The abundant *Lagria hirta* (L.) is common in urban settings and flies to house lights. The alleculines *Isomira murina* (L.), *Gonodera luperus* (Herbst) can be beaten from various blossoms or collected by sweeping. *Mycetochara humeralis* (F.), *Pseudocistela ceramboides* (L.) and *Prionychus* spp. are nocturnal, and best searched for with a head torch on dead wood after dark. *Pseudocistela* and *Prionychus* are also attracted to light.

OEDEMERIDAE. This family includes ten species in Britain, including the extremely abundant grassland species *Oedemera nobilis* (Scopoli) and *O. lurida* (Marsham), which develop in herbaceous stems. *Nacerdes melanura* (L.) is a cosmopolitan beetle that develops in waterlogged wood, especially driftwood. It also develops in railway sleepers, and has been found in city streets and on trains in summer. The pale, long limbed *Oncomera femorata* (F.) is the only nocturnal oedemerid in Britain, and is occasionally collected on ivy blossom and at light. The figure in Harde (1984) supposedly representing *Oncomera* is a quite different non-British species *Oedemera femorata* (Scopoli). *Chrysanthia nigricornis* (Westhoff) is a recent discovery (Skidmore, 1973) from the Scottish Highlands. *Ischnomera* develop in dead wood. *I. caerulea sensu auct.* has recently been shown to include three species in Britain, the more widespread *I. cyanea* (F.), and the rarer true *I. caerulea* (L.) and *I. cinerascens* Pandellé. These species should be dissected, and can be identified using Mendel (1990).

MELOIDAE. 11 species have been recorded in Britain, of which none are common. The bright green *Lytta vesicatoria* (L.), an occasional visitor to these shores, is the source of the aphrodisiac cantharidin (Spanish Fly), which the beetle uses as a defensive secretion and an aggregation chemical. This chemical is a powerful vesicant and can cause allergic reactions in those handling the beetles. It is secreted, in varying quantities by many genera of Meloidae, and some Oedemeridae, and can

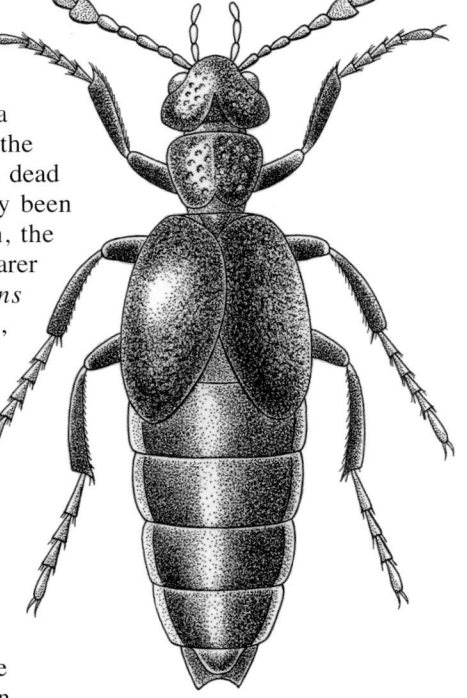

Fig. 21 *Meloe violaceus* Marsham. An oil beetle

be used as an attractant for a range of insects (although it is not always easy to obtain). *Sitaris muralis* (Forster) (formerly *Apalus*) is associated with the nests of solitary bees in old walls, but like many meloids, *Trichodes* spp. (Cleridae) and other thermophilic Coleoptera associated with Hymenoptera, it has apparently disappeared from Britain; there are no reliable recent records. The remaining nine species belong to the genus *Meloe*, and only three, *M. violaceus* Marsham, *M. rugosus* Marsham and *M. proscarabaeus* L. are likely to be encountered by any but the luckiest or most persistent coleopterist. Several species are presumed extinct in Britain, or were vagrants that formed temporary colonies that did not persist. The fascinating behaviour of this group of beetles has been discussed in detail by many authors, including an excellent account by J.H. Fabre (1919). The huge females lay vast numbers of eggs, which hatch into triangulin larvae; these climb onto flowerheads and wait for a pollinating insect, and if that insect is a bee of an appropriate genus, they are carried to the nest where they develop as brood parasites. Thousands perish by attaching themselves to hoverflies, wasps, inappropriate bees etc. It is in the phoretic triangulin phase that these beetles can be carried into Britain from abroad, as the adults are unable to fly. Meloid triangulins should always be looked for in the residue at the bottom of malaise trap bottles, because they can be identified to species using Emden (1943). Recently, a further species *Stenoria analis* (Schaum) has been collected in the Chausey Islands (Livory, 1998) and Guernsey (C. David pers. comm. 2003). This beetle appears to be associated with the autumn-flying bee *Colletes hederae* Schmidt & Westrich, a recent colonist of Britain (Cross, 2002). This beetle, and its triangulin larvae, should be searched for on the British mainland.

MYCTERIDAE. *Mycterus curculioides* (F.) is the only species of this small family recorded from Britain. The beetle is very easily collected by sweeping and beating where it occurs, and it seems unlikely that it can have been overlooked for over 100 years. It is presumed extinct in Britain.

PYTHIDAE. The single British species of this small family, *Pytho depressus* (L.), can be found under the bark of pine in Scotland. The large,

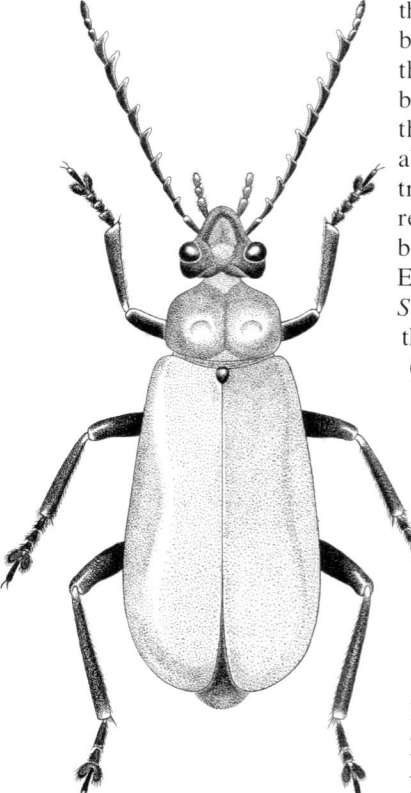

Fig.22 *Pyrochroa coccinea* (L.)

orange, flattened larvae, very similar to those of Pyrochroidae, are much more commonly found than the adults, and can easily be reared in a bag or tin containing sufficient pabulum, kept from becoming too dry or too damp.

PYROCHROIDAE. The 'Cardinal Beetles'. Three species occur in Britain. The large, flattened orange-brown larvae are often very abundant among the frass under loose bark of deciduous trees. The adults, although large and brightly coloured, are short-lived and surprisingly easily overlooked. The red-headed *Pyrochroa serraticornis* (Scopoli) is widespread, while the black-headed *P. coccinea* (L.) has a more southern distribution, and the smaller *Schizotus pectinicornis* (L.) is generally restricted to Scotland and the Welsh borders.

SALPINGIDAE. The 11 species in Britain are mostly associated with dead wood. Many species superficially resemble small weevils (Curculionoidea). *Vincenzellus ruficollis* (Panzer), *Salpingus planirostris* (F.) and *S. ruficollis* (L.) (both formerly *Rhinosimus*) are abundant, and can be found under bark or beaten from dead twigs. *Aglenus brunneus* (Gyllenhal), recently transferred from Colydiidae (see Lawrence & Newton, 1995), is a near-cosmopolitan stored product pest. Species of *Lissodema*, *Sphaeriestes* (*Salpingus* sensu auct.) and *Rabocerus* are generally uncommon, and associated with dead, often burnt, twigs and brushwood.

ANTHICIDAE. These small beetles superficially resemble ants or small Carabidae. Most are associated with decaying vegetable matter, often in salt-marshes (e.g. *Cordicomus instabilis* (Schmidt), *Cyclodinus constrictus* (Curtis)) or in man-made habitat (e.g. *Sticticomus tobias* (Marseul) and *Omonadus formicarius* (Goeze), in compost, chicken litter etc.). 13 species are recorded from Britain. Most are ground living, although some, especially *Notoxus monocerus* (L.) can be taken by sweeping.

ADERIDAE. Three species of these small, delicate, forest beetles occur in Britain. All are uncommon, and *Aderus brevicornis* (Perris) is a great rarity. The beetles can be beaten from old trees in mid to late summer, and occasionally from hawthorn blossom. Their biology is poorly understood, and *A. brevicornis* has once

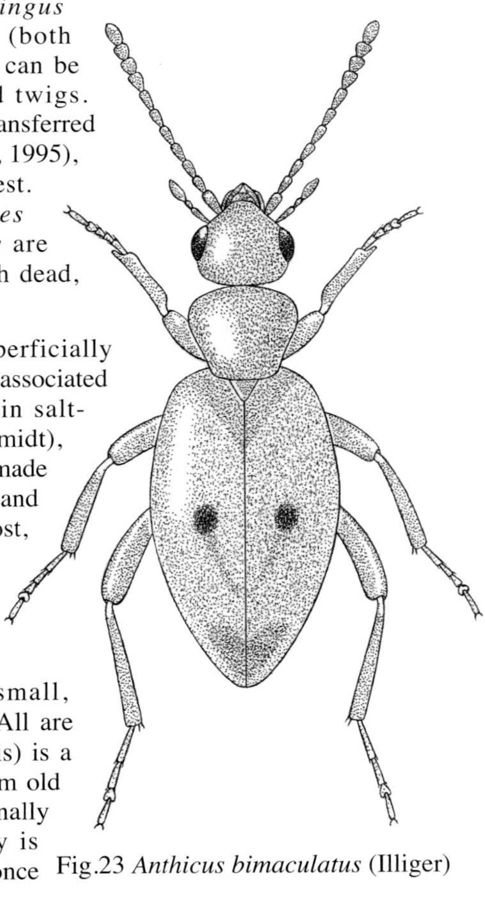

Fig.23 *Anthicus bimaculatus* (Illiger)

been recorded from a very rotten fence-post some distance from woodland (J. Cooter pers. comm.).

SCRAPTIIDAE. see p. 112. Treated in a separate chapter by Brian Levey.

References

Anderson, R., Nash, R. and O' Connor, J.P., 1997. Irish Coleoptera: A revised and annotated list. *Irish Naturalists' Journal*. (special edition) 1-81.

Brendell. M.J.D., 1975. Coleoptera Tenebrionidae. *Handbooks for the Identification of British Insects*, 5(10), 22pp. Royal Entomological Society, London.

Buck, F.D., 1954. Coleoptera (Lagriidae, Alleculidae, Tetratomidae, Melandryidae, Salpingidae, Pythidae, Mycteridae, Oedemeridae, Mordellidae, Scraptiidae, Pyrochroidae, Rhipiphoridae, Anthicidae, Aderidae and Meloidae). *Handbooks for the Identification of British Insects* , 5(9), 30pp. Royal Entomological Society, London.

Chinery, M., 1986. *Collins Guide to the Insects of Britain & Western Europe*. Collins, London.

Cross, I.C., 2002. *Colletes hederae* Schmidt & Westrich (Hym., Apidae) new to Mainland Britain with notes on its ecology in Dorset. *Entomologist's Monthly Magazine*,**138**: 201-203.

Emden, F.I. van, 1943. Larvae of British Beetles. IV. Various Small Families. *Entomologist's Monthly Magazine* **79**: 209-223.

Fabre, J. H., 1919. *The Glow Worm and other beetles* The Works of Fabre, transl. A. Teixeira de Mattos viii, 487 pp. London.

Harde, K. W., 1984. *(Revised and updated by Hammond, P.M.)* A field guide in colour to beetles. Octopus (recently reprinted in 1998 by Blitz Editions. 333pp.

Harrison, T.D., 1996. *Eulagius filicornis* (Reitter) (Mycetophagidae) apparently established in Britain. *Coleopterist* **4**(3): 65-67.

Hodge, P.J. & Jones, R.A., 1995. *New British beetles: species not in Joy's practical handbook*. British Entomological and Natural History Society, Reading. xvi+175pp.

Hyman, P.S., 1992. *A review of the scarce and threatened Coleoptera of Great Britain* (part 1). 484pages. Joint Nature Conservation Committee, Peterborough.

Joy, N.H., 1932. *A practical Handbook of British beetles* (2 vols). Witherby, London.

Lawrence, J. F. & Newton, A. F., 1995. Families and subfamilies of Coleoptera (with selected genera, notes and references, and data on family-group names). pp 779 – 1006, *in* Pakaluk, J. and Ślipiński, S. A. (eds), *Biology, Phylogeny, and Classification of Coleoptera: Papers Celebrating the 80th Birthday of Roy A. Crowson*. Warsaw: Muzeum i Instytut Zoologii PAN.

Lawrence, J. F., Hastings, A. M., Dallwitz, M. J., Paine, T. A. and Zurcher, E. J., 1999. *Beetles of the World: A Key and Information System for Families and Subfamilies*. CD-ROM, Version 1.0 for MS-Windows. Melbourne: CSIRO Publishing.

Livory, A., 1998. Faune chausiaise: une surprise de taille! *Argiope*, **22**, 13-18.

Mendel, H., 1990. The identification of British *Ischnomera* Stephens (Coleoptera: Oedemeridae). *Entomologist's Gazette*, **41**(4): 209-211.

Nikitskii, N. B., 1983 Morphology of the *Myrmechixenus subterraneus* larva and some remarks on systematics of the genus *Myrmechixenus*. *Byulleten'-Moskovskogo-Obshchestva-Ispytatelei-Prirody-Otdel-Biologicheskii*. **88** (2): 59-63.

Nikitsky, N. B., 1998. *Generic classification of the beetle family Tetratomidae (Coleoptera, Tenebrionoidea) of the world, with description of new taxa*. Pensoft Publishers, Faunistica No. 9., Sofia & Moscow. 1-80.

Pope, R. D., 1977. Kloet & Hincks. A Check List of British Insects. Part 3: Coleoptera and Strepsiptera. 2nd. revised edition. *Handbooks for the Identification of British Insects*, **11**: xiv + 105pp.

Skidmore, P., 1973. *Chrysanthia nigricornis* Westh: (Col., Oedemeridae) in Scotland, a genus and species new to the British list. *Entomologist* **106**: 234-237.

CERAMBYCIDAE
(Plates 77 – 88, 114a and 117)
by Martin Rejzek

Introduction

The family Cerambycidae in its broadest sense (also known as longhorn beetles or timbermen) currently comprises about 25,000 described species world-wide and thus represents one of the largest families within the order Coleoptera, even after exclusion of certain small exotic groups (see below). Together with the leaf beetles (Chrysomelidae *sensu lato*, now usually also divided into several families) it forms the superfamily Chrysomeloidea (e.g. Crowson, 1981, Lawrence & Newton, 1995). Alternatively, some authors have classified longhorn beetles as a separate superfamily, Cerambycoidea (e.g. Böving & Craighead, 1931, Švácha & Danilevsky, 1987). Family classification of the cerambyciform line of the Chrysomeloidea (or the Cerambycoidea of some authors) is not settled. The majority of earlier authors treat it as a single family, Cerambycidae (with or without Parandrinae), or also give the family rank to the large subfamilies Prioninae and/or Lamiinae. Recent authors retain all those subfamilies within the Cerambycidae. On the other hand there is an increasing tendency to exclude several small, non-British groups, which have often been regarded as having an uncertain relationship. This includes the former cerambycid subfamilies Disteniinae (Linsley, 1961), Vesperinae, Philinae and Oxypeltinae (Crowson, 1981) and Anoplodermatinae (Švácha & Danilevsky, 1987). All British species belong to the Cerambycidae *sensu stricto*.

The subfamily classification of the family Cerambycidae *sensu stricto* has only recently become relatively stable and the following subfamilies are usually recognised in current publications: Parandrinae, Prioninae, Lepturinae, Necydalinae, Spondylidinae (including the tribe Asemini), Cerambycinae and Lamiinae (e.g. Sama, 2002). The subfamily Apatophyseinae, established by Danilevsky (1979), should also be added to this list, although it is still not recognised by many authors and is usually incorrectly retained in the subfamily Lepturinae. In Britain representatives from the following five subfamilies (listed in taxonomic order) are known: Prioninae, Lepturinae, Spondylidinae, Cerambycinae and Lamiinae.

Cerambycids are usually elongate, subcylindrical to moderately flattened beetles. Frequently their elytra are broader than the thorax, often considerably so. The antennae usually articulate on raised tubercles and are capable of being folded back along the body. The pedicel (the second antennal segment) is usually very short. The antennae can be filiform, serrate, pectinate or flabellate. Their length varies but frequently they are at least as long as the body. The tarsi are usually five-segmented, but the fourth segment is generally very short and inconspicuous and can be obscured by the third, which is almost always strongly bilobed. As a result the tarsi may appear to be only four-segmented (pseudotetramerous type). In some Lamiinae, segments 4 and 5 are entirely fused. All tibiae bear distinct apical spurs. Eyes are large, often emarginate and partially surround the antennal base. In some species they are divided into two portions. Most species possess a mesonotal stridulatory plate (absent in Prioninae, Parandrinae and a few other species).

It can be quite difficult to distinguish cerambycids from the leaf beetles. No characteristics are reliably distinct. This is particularly the case when comparing cerambycids with the families Megalopodidae and Orsodacnidae. The frequently mentioned (e.g. Bense, 1995) distinguishing character of all tibiae bearing distinct apical spurs in cerambycids, may occur in leaf beetles as well (e.g. *Zeugophora*, *Orsodacne* but also in some Chrysomelidae *sensu stricto* like Criocerinae). As in all Phytophaga the Chrysomelidae have the same tarsal formula as the Cerambycidae. Cerambycids can usually be distinguished from the leaf beetles by their more elongate body form and proportionately longer legs and antennae. In Chrysomelidae the body is likely to be oval and the eyes are usually not emarginate.

British Cerambycidae are readily recognisable as a family and most species are easily identifiable as they are frequently large and conspicuously coloured. Even the less spectacular species can usually be identified using any of the following standard works. Duffy (1952) provides a concise key to British Cerambycidae, which also covers some imported species recorded in Britain. This work does not, however, go into great detail and is now slightly outdated. Bense's (1995) bilingual publication (English / German) offers user-friendly keys, distribution maps and biological data for most European cerambycids including all British ones. The recently published first volume of Sama's (2002) *Atlas of the Cerambycidae of Europe and Mediterranean area* includes all known British cerambycids. This work offers original keys to facilitate identification of western Palaearctic species of some genera. One or more photographs are provided for each species: usually a typical male, sometimes both sexes. The work provides detailed information on every species occurring within the study area and represents an indispensable monograph for every Cerambycidae specialist. The nomenclature and higher classification of Cerambycidae used in this text (see also **Table 1**) follow Rejzek (2004).

Status of British Cerambycidae

Of the estimated 550 European cerambycid species, 64 are considered to be either native or naturalised in Britain (Twinn & Harding, 1998). Out of the total number of British species, there are five that are thought to be extinct. These are *Lepturobosca virens* (L.), *Strangalia attenuata* (L.), *Cerambyx scopolii* Füessly, *Obrium cantharinum* (L.) and *Plagionotus arcuatus* (L.) (Twinn & Harding, 1998). This means that there are 59 species currently known to occur in Britain.

Out of these, four species are believed to be introductions (Twinn & Harding, 1998). Furthermore, *Cerambyx cerdo* L. is believed to have been lost to British native fauna before entomological recording began and is, therefore, not included in these numbers.

The British Cerambycidae fauna has been studied intensively in the past and a great number of papers and monographs dealing with morphology (both adult and larval), taxonomy, biology, ecology and distribution of British Cerambycidae have been published. It would be beyond the scope of this chapter to mention all of them, especially when detailed lists of literature exist in comprehensive works such as Duffy (1953), Uhthoff-Kaufmann (1985, 1987, 1988, 1989, 1989a, 1989b, 1990, 1990a, 1990b, 1990c, 1990d, 1991, 1991a, 1991b, 1991c, 1991d, 1992 and 1992a) and Twinn & Harding (1998). The last summarises distribution in the form of very informative distribution maps and also provides concise original information on the biology of every British cerambycid.

Most British cerambycids are associated with trees and shrubs of the original post-glacial forests. Although there are no exact studies known to me, it seems that populations of many British cerambycids are very weak, some almost dwindling on the verge of extinction. This unfortunately includes species that are otherwise quite frequent and locally even abundant in some Continental countries (e.g. *Dinoptera collaris* (L.) or *Oberea oculata* (L.)). It is certainly a very alarming fact that in Britain even species such as *Plagionotus arcuatus*, present and locally common in coastal regions of northern France, Belgium, the Netherlands, Germany, Denmark and even Norway (Bense, 1995), are regarded as extinct.

It seems that the number of species occurring in Britain today does not reflect its relatively large area and variety of habitats. At least five species have apparently been lost during the last two centuries and an unknown number of species became extinct even before entomological recording began. Rather than natural factors, this situation reflects the state of British woodland, which in the past has been reduced in size, over-fragmented and over-exploited. Moreover, its structure and plant species composition have been altered to such an extent that many woodland species and Cerambycidae are no exception, are verging on extinction or have become extinct already.

Ecology and Conservation of Cerambycids

Although species do occur in herbaceous plants, longhorn beetles are primarily forest insects. The ecological role of longhorn beetles is to decompose plant material such as tissues of woody (in conditions ranging from healthy to moribund to dead and decaying) and herbaceous (both living and dead) plants to humus (Linsley, 1959, Booth *et al.*, 1990). Thus they play an important role in nutrient recycling in forest ecosystems. Their specific place in the succession of insects found in a gradually disintegrating tree varies from species to species, depending upon their life history and habits and also from host to host and from region to region. Death and decomposition are vital to the health of the woodland ecosystem, for they enable the recycling of nutrients, which would otherwise be permanently locked up in wood and leaves. The chief agents of tree decay are fungi, often aided by a variety of longhorn beetles. For these reasons Cerambycidae are insects largely beneficial to man.

However, a limited number of Cerambycidae species are associated with cultured plants and in certain cases they might compete with their production or cause damage to products made of plant tissues. They usually show a high degree of adaptability to habitat changes and can occasionally become very abundant. Such species are recognised as pests and their larvae may be detrimental to timber, tree plantations, fruit or decorative trees and a variety of associated products. *Hylotrupes bajulus* (L.), a species detrimental to structural timber, is a well documented example of a Cerambycidae species recognised as a pest. *Tetropium castaneum* (L.) and *Tetropium gabrieli* Weise are frequently underestimated pests of coniferous plantations. Under certain circumstances these beetles can act as physiological pests (attacking seemingly sound trees that are, however, exposed to a range of other stress factors) and substantially contribute to the final death of the host trees. In addition these species very likely serve as vectors of fungi harmful to trees. *Aromia moschata* (L.), *Gracilia minuta* (F.), *Nathrius brevipennis* (Mulsant), *Callidium violaceum* (L.), *Phymatodes testaceus* (L.), *Saperda carcharias* (L.) and *Saperda populnea* (L.) have also sometimes been regarded as pests in the past but their significance has decreased along with the decreased need for the plants or products to which they cause damage.

The majority of longhorn beetles, however, do not show such adaptability to habitat changes and survive primarily in virgin forests or areas, which have escaped major habitat modification. In Britain such areas are mainly represented by the last remnants of ancient woodland (woodland areas that are likely to have been subjected to some sort of management in the past, that, however, have never been completely cleared and have been forested throughout historical time). No species recognised as truly virgin forest relicts are known to occur in Britain.

Cerambycidae associated with ancient woodland are among the most vulnerable organisms native to Britain and, therefore, they and their habitats deserve the highest degree of protection. They comprise *Judolia sexmaculata* (L.), *Pedostrangalia revestita* (L.), *Stictoleptura scutellata* (F.), *Anoplodera sexguttata* (F.) and *Pyrrhidium sanguineum* (L.). Ancient woodland, however, will always harbour populations of a broad variety of other specialists, including other less habitat-quality sensitive cerambycids.

The major group of British species are mainly associated with old trees within old, semi-natural woodland areas, woodland pastures or parks and show at least some ability to penetrate newly created habitats. To name just a few: *Prionus coriarius* (L.), *Rhagium mordax* (De Geer), *Dinoptera collaris*, *Stenocorus meridianus* (L.), *Grammoptera abdominalis* (Stephens), *G. ustulata* (Schaller), *Leptura aurulenta* F., *Poecilium alni* (L.), *Anaglyptus mysticus* (L.), *Mesosa nebulosa* (F.), *Pogonocherus hispidulus* (Piller & Mitterpacher) and *Saperda scalaris* (L.). A substantial group of British Cerambycidae is formed by species showing high adaptability to habitat modification, or species that actually benefit from extractive woodland management but which are not recognised as pests. They are *Grammoptera ruficornis* (F.), *Leptura quadrifasciata* L., *Stictoleptura rubra* (L.), *Rutpela maculata* (Poda), *Asemum striatum* (L.), *Arhopalus rusticus* (L.), *A. ferus* (Mulsant), *Molorchus minor* (L.), *Clytus arietis* (L.), *Pogonocherus hispidus* (L.), *Leiopus nebulosus* (L.), *Oberea oculata* and *Tetrops praeustus* (L.).

Protection of longhorn beetles should primarily concentrate on protection of their natural habitats. Occurrence of a broad variety of longhorn beetle species is a good hallmark of woodland habitat quality. Especially indicative are species classified as ancient woodland dwellers. These should always be given preference in conservation as they indicate sites with a high quality of biodiversity. If such sites are conserved, a wider range of genetic resources in specialist species will be protected (Hambler & Speight, 1995).

It is especially important that Britain's few remnants of ancient woodland are given the highest possible protection. These woodlands frequently harbour the last remnants of species indigenous to the British Isles and should therefore be preserved. Native tree species within ancient woodlands should be excluded from any felling activity, including coppicing, which are detrimental to most woodland insects (Hambler & Speight, 1995). Dead wood should always be left in the woodland and allowed to decompose. Exotic trees and plants within ancient woodland should be replaced by native species, but gradually so, to avoid possible damage caused by creating open areas and edges prone to dramatic changes in microclimate. A variety of tree ages should always be present in an existing woodland. Only a reasonable level of management, such as re-introductions, importation of freshly cut wood and removal of non-indigenous plant species, should be employed. Where possible, non-intervention management should be given a preference at these sites. The ultimate goal should be to regain self-sustainable late-successional high forest supporting a high diversity of woodland species truly indigenous to the British Isles. Moreover, attempts should be undertaken to extend existing ancient woodlands by planting directly adjacent areas. Native tree species typical to the ancient core should be used for this purpose.

In man-made woodland habitats, a reduction in the felling of old and dying trees and a reduction in the removal of dead wood, can substantially increase the chances of survival for beneficial insect populations. Similarly to ancient woodland, gradual replacement of imported and exotic trees and shrubs would contribute to the health of native species' populations. A decrease in habitat fragmentation by creating wildlife corridors composed of native plants (road side plantations, hedges, parks, etc.) is one important measure that should be taken to avoid extinction of even more species native to Britain.

Adult Habits and Collecting

Adult cerambycids are usually short-lived. The sole purpose of the adult stage is to ensure population continuity. As a rule this happens very quickly and after copulation and oviposition is finished the adults die. Cerambycids spend most of their lives in immature stages (egg, larva, pupa). There are exceptions however; in Lamiinae active life may be substantially extended, facilitated mainly by adult feeding (Hanks, 1999). Moreover, *Lamia textor* (L.) and *Pogonocherus fasciculatus* (De Geer) pupate and emerge in late autumn and adults overwinter outside their larval substrates. This naturally prolongs the time they spend in the adult stage to several months.

Most adult cerambycids can be found on the plants or plant tissues in which their larvae develop. Generally, Prioninae, Spondylidinae and Cerambycinae do not seem to require any sort of food (apart perhaps from water) and copulate shortly after they emerge. Many species, however, require nourishment during the adult stage. In Lamiinae this seems to be a rule. Both sexes feed on fresh plant tissues (usually of the same plant serving as larval host) or even on the fruiting bodies of certain fungi, mainly to attain sexual maturity. Their head being characteristically perpendicular to the body axis and bearing mouthparts pointed distinctly (postero)ventrally in fact prevents Lamiinae from feeding on pollen and particularly nectar. Consequently adult Lamiinae generally do not visit flowers. Lepturinae, on the other hand, frequently assemble on flower-heads to feed on pollen. Most cerambycids oviposit in cracks of bark or wood. In all Lamiinae, the egg is inserted in an incision created by the female using its mandibles. This behaviour is virtually unknown in other subfamilies.

In Britain most adult cerambycid appear between May and August, the majority of them, however, can be found in June. Exceptions are species overwintering as adults either outside or in their larval substrates. Such species usually emerge very early (April). *Acanthocinus aedilis* (L.), for instance, appears in April and adults can be found sitting on fresh pine stumps or large diameter lying pine trunks.

Most cerambycids are diurnal and can be conspicuously coloured. Their colouration may closely resemble other organisms, such as wasps, in order to deter predators. Frequently, the colouration of their bodies mimics the environment the adults spend most time in. Several cerambycids are known to have a chemical defence system of their own. *Agapanthia villosoviridescens* (De Geer) and *Aromia moschata* are the best examples. Both species, when disturbed, produce a characteristic scent protecting the beetles from predators. Some cerambycids only appear at dusk or at night. Such species are called crepuscular or nocturnal and their bodies are usually of brown to black colour, some of them are attracted to light. All cerambycids when disturbed quickly fall to the ground and either feign death or quickly disappear to the surrounding vegetation. In good weather diurnal species may fly away.

To look for adult cerambycids it is first necessary to choose an appropriate habitat. Ancient woodland is the ideal case. Many other woodland habitats containing a spectrum of tree ages (from young saplings to mature and moribund trees) with plenty of continuously available dead wood will reveal a range of cerambycids. Woodland containing native trees will support a substantially broader spectrum of species than a plantation of exotic trees. However, even such plantations are better than arable land with no trees at all. Plantations of conifers (*Picea* spp., *Pinus* spp., *Abies* spp., *Larix* spp.) native to Continental Europe or the North American subcontinent, support several species of Cerambycidae like *Rhagium inquisitor* (L.), *R. bifasciatum* F., *Stictoleptura rubra*, *Asemum striatum*, *Tetropium castaneum*, *T. gabrieli*, *Arhopalus rusticus*, *A. ferus*, *Obrium brunneum* (F.), *Molorchus minor*, *Callidium violaceum*, *Pogonocherus fasciculatus* and *Acanthocinus aedilis*. However, a broadleaved woodland, which is native to most of Britain, or a native pine wood in Scotland will always show a broader and more interesting spectrum of cerambycids than a tree plantation.

Timing is also very important. Repeated visits to the same place between May and August will yield different species. It should also be noted that even diurnal species are not equally active during all periods of the day. The most intensive periods of activity (flight, copulation, feeding, oviposition) are usually in the morning (10 am to 12 am) and then again in the afternoon (3 pm to 6 pm). *Oberea oculata* and both British *Tetrops* species exhibit flying activity in the late afternoon (about 5 pm). Dusk is then a critical time, because it reveals species active only at this precise moment. *Prionus coriarius* for example mainly flies at dusk. Nocturnal species can be found later by inspecting their host plants. *Arhopalus rusticus* and *A. ferus* for example can be found at night sitting on large trunks of dead pines. Certain species such as *Prionus coriarius*, *Arhopalus rusticus* and *A. ferus* can be attracted to lights. *Saperda carcharias* has also been reported as being attracted to lights. Night collecting is, however, usually more productive in late summer (July and August) and is usually better in lowlands than in mountainous areas.

Many cerambycids can be found using standard entomological equipment such as a beating tray or a sweeping net. A sweeping net has only limited use for finding British cerambycids. On the other hand a beating tray will prove to be extremely useful. The main target for a beating tray should always be the branches of recently felled or fallen trees, piles of branches, recently dead branches hanging from living trees and shrubs, recently dead or moribund ivy and similar substrates. Among others, beating such places will very likely yield *Poecilium alni* (L.), *Mesosa nebulosa*, *Leiopus nebulosus*, *Pogonocherus hispidus*, *P. hispidulus*. It may also yield *Gracilia minuta* (F.) and *Nathrius brevipennis* when deciduous trees are inspected or *Pogonocherus fasciculatus* when coniferous trees are beaten. Certain cerambycids can be collected by beating from living trees and shrubs. Shrubs and trees growing on woodland edges are especially attractive for these beetles. When their leaves are beaten species including *Stenocorus meridianus*, *Pedostrangalia revestita*, *Saperda populnea* (*Populus tremula*), *Stenostola dubia* (Laicharting) (*Tilia* spp.), *Oberea oculata* (*Salix* spp.), *Tetrops praeustus* (from various Rosaceae) and *Tetrops starkii* Chevrolat (*Fraxinus* spp.) can be found. Especially attractive for longhorn beetles are flowering shrubs and by far the best one is hawthorn. Beating a flowering hawthorn growing on the edge of ancient deciduous woodland, which includes some pine trees, can reveal species such as *Rhagium mordax*, *Dinoptera collaris*, *Grammoptera ustulata*, *G. abdominalis*, *G. ruficornis*, *Cerambyx scopolii*, *Obrium brunneum*, *Molorchus minor*, *Glaphyra umbellatarum* (von Schreber), *Clytus arietis*, *Anaglyptus mysticus* and *Saperda scalaris*.

Many cerambycids can also be collected individually without using any equipment. Recently dead wood is especially attractive for cerambycids. Individual trunks of fallen or freshly felled trees, piles of trunks or cordwood in both shady and sunny locations should always be closely inspected. It is only likely, however, that a few insects will show themselves. Most of them will be hiding in bark crevices or sitting on the lower part of the wood. When inspecting wood of deciduous trees, species like *Rhagium mordax*, *Pyrrhidium sanguineum*, *Phymatodes testaceus*, *Poecilium alni*, *Clytus arietis*, *Plagionotus arcuatus*, *Anaglyptus mysticus*, *Mesosa nebulosa*, *Leiopus nebulosus* or *Saperda scalaris* may be recorded. Wood of coniferous trees on the other hand will yield *Rhagium inquisitor*, *R. bifasciatum*, *Tetropium castaneum*, *T. gabrieli*, *Hylotrupes bajulus* and *Callidium violaceum*. Certain beetles are attracted mainly to the fresh stumps of pine trees. When such places are inspected very early in the spring (April and May) species such as *Acanthocinus aedilis* and *Asemum striatum* can be found.

A variety of cerambycids, especially Lepturinae, assemble on flower-heads to feed on pollen and can easily be found there and collected individually. Among the species, which can be collected in this way are *Dinoptera collaris*, *Grammoptera ruficornis*, *Lepturobosca virens*, *Leptura quadrifasciata*, *L. aurulenta*, *Anastrangalia sanguinolenta* (L.), *Stictoleptura rubra*, *S. scutellata*, *Paracorymbia fulva* (De Geer), *Anoplodera sexguttata*, *Judolia sexmaculata*, *Pachytodes cerambyciformis* (Schrank), *Alosterna tabacicolor* (De Geer), *Pseudovadonia livida* (F.), *Strangalia attenuata*, *Rutpela maculata*, *Stenurella melanura*, *S. nigra* L. but also *Cerambyx scopolii* (*Sambucus nigra* seems to be especially attractive for this species) and *Obrium brunneum*.

Two British cerambycids develop in herbaceous plants. They are *Agapanthia villosoviridescens* mainly associated with *Heracleum sphondylium* (Hogweed) and *Phytoecia cylindrica* (L.) associated with a variety of Apiaceae plants including *Anthriscus sylvestris* (Cow Parsley), *Anthriscus caucalis* (Bur Chervil) or *Daucus carota* (Wild Carrot), but no doubt many others too. Adults of these beetles can be found sitting on their host plants (not in the flower, however) or alternatively can be collected by sweeping convenient vegetation.

The above mentioned methods roughly summarise ways of finding British adult Cerambycidae and after a little practice a spectrum of species will be discovered. Some British cerambycids, however, are very rare and only a focused search will lead to their discovery. This search should always be aided by a detailed knowledge of their biology. Information on host plants, type and quality of larval substrate, time of adult emergence, adult habits and habitat preferences will certainly help to find the desired insect.

For instance a focused beating of the living branches of *Fraxinus excelsior* can yield a very rare species, *Tetrops starkii*. *Fraxinus excelsior* is the species' host and the adults spend most of the day sitting motionlessly on the leaves of this tree on which they probably feed.

Focused beating of *Populus tremula* can, at least in theory, yield *Obrium cantharinum* (L.), a species believed to be extinct in Britain. *Obrium cantharinum* develops in the dead branches of several poplar species but aspen is usually preferred. Adults are in general very difficult to collect as they do not visit flowers.

Aromia moschata can be found on living *Salix* spp. (willows). Despite its large size and metallic green colour, the adult is often difficult to find sitting motionless on twigs and branches in the canopy; this species only occasionally visits flowers so focused searching of willow branches is likely to prove the best method for its discovery.

Saperda carcharias can be found sitting on the leaves of various poplar species. The adults feed on the leaves leaving a very characteristic frass pattern. This can serve as a hallmark of the species' presence. Larvae are impossible to rear because most of their development takes place in living trees and consequently the search must be focused on adults.

Lamia textor undoubtedly belongs to the group of rarest British cerambycids. The species never visits flowers and it effectively mimics the colour of its surroundings (the colour of willow bark). To make things even more difficult larvae feed in the roots and wood of trunks of living *Salix* spp. and consequently are difficult to find. For these reasons the best chance of finding this species is to search for adults. The size of the beetle suggests that it should develop in large mature trees. The opposite is, however, true. The species develops in shrubby willows hardly exceeding two metres in height. Such willows, when growing on river banks, usually have very rich undergrowth, which effectively covers their large, more or less horizontal trunks and extensive, partially exposed, root systems. Very frequently this undergrowth comprises nettles. To look for this beetle, it is necessary to get as close to the trunk and roots as possible so that they can be carefully inspected. The adult beetles usually sit motionlessly on the bark, partially hidden in crevices or in vegetation usually very close to, or directly on the ground. Occasionally adults can be found sitting on the terminal twigs of their hosts, but this only happens late in the afternoon on warm, late autumn days. During winter, hibernating adults can be found in the ground close to willow roots or trunks and last but not least, floods cause the beetles to crawl up the willow branches where they can be seen more easily.

Pedostrangalia revestita is perhaps Britain's most difficult cerambycid to find. The adults are short lived (about 2 weeks at the most) and so precise timing is required. They appear when temperatures first exceed 20°C, which is usually towards the end of May or slightly later. The species is thermophilous and consequently the preferred habitats are oak woods growing on south-facing slopes or sunny woodland edges. Especially attractive to the beetles are old stunted oaks with plenty of wounds and dead branches still attached to the living trees. The adults will sit on oak leaves and in the late afternoon, when the temperature is high enough, they will fly quickly around the tree canopies. Only at this time can they be seen easily and captured on the wing. The adults hardly ever visit flowers.

Immature Stages – The immature stages of cerambycids and in particular larvae, have been studied intensively in the past. Duffy's (1953) 'Monograph of the immature stages of British and imported timber beetles' certainly belongs to the classic works in this field. More recently Švácha & Danilevsky (1987, 1988, 1989) published a monograph on 'Cerambycoid Larvae of Europe and Soviet Union' covering all Cerambycidae subfamilies except Lamiinae. Lamiinae, including all British examples, are elaborated in Švácha (2001). These monographs provide detailed information on the techniques of collection, preservation and study of larvae, larval morphology, keys to larvae going to the specific level, very reliable data on the biology of every species treated and an indispensable list of literature.

When compared to the classical approach of adult collecting, larval collecting has several advantages. The most obvious one is the fact that larvae of the whole spectrum of species present in a particular location can be collected in one day even though adults may emerge at different times. In addition, larval collecting and their subsequent rearing to adults provides substantially more detailed information about the biology of each particular species. The host plant, parasitic insects, time of emergence (when the wood is kept outside), adult habits and much more may be discovered; little of which would be available by finding an adult in the field. However, it should always be borne in mind that information gained from rearing larvae under laboratory conditions may not always reflect behaviour in the wild. This is particularly the case with regard to emergence times, which will be artificially early when larvae are kept indoors.

The main disadvantage of larval collecting is its potentially destructive nature. The method should never be used in nature reserves or other protected areas unless absolutely necessary and only with permission. Living trees and shrubs should never be used as potential substrate targets anywhere. Larval collecting should only be applied where collecting of adults is impossible or very difficult. In general the method should always be applied to a restricted part of the biotope and plenty of potential larval substrate should always be left intact. Always bear in mind that this method is very destructive and could potentially endanger the already very weak populations of British cerambycids, which is certainly not our aim. On the other hand, if stored wood destined for burning or industrial use is attacked by beetles, then no special limitations to larval collecting need be observed.

Another disadvantage of larval collecting is the somewhat fragile nature of the larvae and the need to successfully rear them if adults are desired. Successful rearing can be learned through experience. If the aim of larval collecting is to rear out adults, then the timing is crucial. The best results will be obtained when last larval instars, prepupae, pupae or even adults in pupal cells are collected. In Britain, this means that the season for larval collecting starts in January and lasts until the end of May. There is very little chance to successfully rear larvae collected outside this period, although it is possible with substantially more expertise.

Late larval instars do not feed any more and can, therefore, usually be taken out of their pupal cells and placed into rearing vials.

Typically such a vial is a glass tube equipped with a plastic stopper (see plate 111). A small aperture in the stopper will allow ventilation but also excess moisture to disappear. For most species the tube should be roughly 50 mm long and its internal diameter should be about 9 mm. Larger species can be kept in tubes 50 mm long with the internal diameter of 11 mm. Most importantly the tubes must be equipped with a paper insert ensuring that the liquids produced by the larvae can be collected effectively. The paper inserts for the smaller diameter vials can conveniently be made out of a standard paper cut into rectangular pieces 70 mm long and 47 mm wide. The piece of paper is wrapped around a pen (diameter of about 8 mm) with one end of the paper overlapping the end of the pen. The overlapping end of the paper is then pinched together to form a sealed end to the tube. The paper tube is then inserted into the vial sealed end first. The paper inserts absorb excess moisture, provide the larva with insulation, sometimes even with nutrients, and very importantly support the larvae and pupae during ecdysis.

The larvae must always be placed in vials separately and must never come in contact. The vials containing larvae should be kept in safe boxes (metal or dark plastic) at all times. If possible the vials are kept horizontally in these boxes. The box will protect the glass tubes from braking, excessive light, parasites and predators in the field, when being transported and in the laboratory. The larvae in vials must never be exposed to subzero or elevated temperatures. In the field special care should be paid to avoiding long lasting exposure of especially the metal boxes to direct sunshine, or storage of the boxes in parked cars exposed to sunshine. It is important to keep field data for each vial with the vial and, ideally, a copy in a note book.

In the laboratory the boxes should be kept at room temperature in a dark place. The vials should be checked on a regular basis. If signs of mould or excessive moisture appear on the paper insert surface the insect should be moved into a new vial.

When a late larval instar, prepupa or a pupa is collected in the field and then kept in a rearing vial at room temperature, the adult will usually hatch after several weeks or up to two months, but not longer. If no pupation is observed during this time it indicates that the larva is of a younger instar and it will need to feed for longer in a suitable substrate and to hibernate.

If hibernation is required the larvae must first be allowed to re-enter a substrate as close to the original as possible. The substrate containing the larvae is then exposed to temperatures ideally alternating between –5 to +5 °C for a time period of at least two months. Refrigerators can not be used for this purpose.

Many longhorn beetles can be reared in vials. For example, nearly all British Lepturinae, *Plagionotus arcuatus*, *Anaglyptus mysticus*, *Mesosa nebulosa* and *Agapanthia villosoviridescens* can be reared successfully in this way.

Saperda scalaris is an especially good example for the use of vials. Adults of *Saperda scalaris* spend most of their life high in the tree canopies where they feed on leaves and are difficult to see, they only rarely visit flowers. The only time one is likely to see this spectacular beetle is when the females oviposit. The species is extremely polyphagous but oak is always a welcome substrate. The females oviposit into bark crevices of recently dead, mainly standing trunks of large diameter. The larvae feed first subcortically and before they pupate they enter the sapwood where they construct a shallow pupal cell. The entrance to this cell is then secured by a characteristic wad of fibrous frass. This wad and the larval galleries are clearly visible on the wood surface when bark is removed. The insect can then be carefully removed from the pupal cell using a firm knife or even better, a chisel, before being placed into the above mentioned vial. If bark of dead standing trunks is peeled off, always bear in mind that this is a rare and indispensable micro-habitat so leave most of the bark intact.

Rearing adults of all British Spondylidinae from larvae in vials is generally very difficult. If attempted, it usually leads to malformations in the adults. For this reason the following method of collecting and rearing larvae in their own substrate, should be applied.

In the case of especially sensitive larvae (Spondylidinae) or when early-instar larvae, which are still feeding are collected (the almost transparent cuticle shows dark material inside the larval digestive system), the safest method of rearing them is to collect and keep them in their original substrate. Once certain that the desired larvae are present inside the selected substrate, any exposed galleries or pupal cells containing larvae should be secured using a piece of bark and an adhesive tape (or thread in the case of very fragile substrates). The substrate should then be kept in an airy container at room temperature. Moisture control of the larval substrate is the crucial task with both mould or over-desiccation being deadly conditions for the larvae. Moderate water spraying of the substrate surface can be carried out, but usually after a couple of weeks or so the substrate loses its normal water content due to low humidity inside an artificially heated room. For this reason, on a regular basis (once a month or so), the substrate should be placed into water for about an hour and then allowed to dry (overnight) before being placed back into the container. When the substrate is collected in the main season (January to May) the larvae quickly pupate and adults hatch within a few weeks to two months at the longest. In this way many British Lepturinae, Spondylidinae, Cerambycinae and Lamiinae can be reared.

Recently dead branches of *Hedera helix* (ivy) frequently will yield *Pogonocherus hispidus*, one of the most common species in Britain sometimes with *Leiopus nebulosus*.

Obrium cantharinum is an example of a species that is extremely difficult to obtain in its adult stage. The adults hardly ever visit flowers and remain on their host for most of their short lives. It is, however, very easy to find the larvae. The host of the species is mainly *Populus tremula*. The species develops in recently dead wood (from twigs to branches of a large diameter), which is always quite dry. The phloem must be of a yellow rather than black colour (black signals a fungal attack that is not tolerated by this insect species). Larvae feed subcortically creating a very characteristic pattern of galleries. Shortly before pupation the larvae enter the sapwood and create a pupal cell there. Once found, the dead wood can be taken home and adults will emerge from it.

Larvae of *Tetrops starkii* are also easier to find than the adults. The species is associated with *Fraxinus excelsior*. Larvae feed subcortically in recently dead, quite moist twigs, that usually lay on the ground under older living host trees. The undergrowth cover and contact with soil ensures higher moisture content within the larval substrate. Pupation occurs in a shallow pupal cell built either under the bark or in the sapwood. Once a larva of this tiny and rare species is detected it is advisable to collect most of the branch without further inspecting it (i.e. peeling of the bark). The larvae of this species are gregarious and consequently it is fairly likely that there will be more. Adults will hatch very soon after the branches have been taken to room temperature.

It is obvious that a focused search for larvae always needs to be supported by a detailed knowledge of the species' biology. A great deal of reliable information on the biology of Cerambycidae is summarised in monographs (with the exception of *Paracorymbia fulva* who's biology has not yet been discovered). Demelt (1966) and Horion (1974) deal with Central Europe but cover most British species. The most relevant text to British Cerambycidae is, however, the work by Duffy (1953). Švácha & Danilevsky (1987, 1988, 1989), Švácha (2001) and Sama (2002) give very reliable accounts of the biology of every species including all British ones.

Longhorn beetles are phytophagous insects whose larvae usually develop in living or dead woody plants (xylophagous species), but some develop in living tissues of herbaceous plants (herb feeders). They show varying host specificity (monophagous, oligophagous to broadly polyphagous). Polyphagous species can still be somewhat specialised and feed in deciduous trees (DT), coniferous trees (CT) or herbaceous plants (HP) only. Host plants of British Cerambycidae are summarised in **Table 1**. The list of host plants reflects preferences of British cerambycids and does not necessarily comprise all plants known to host the species in the whole of its distribution range. The arrangement of taxa within **Table 1** follows Rejzek (2004). The larval substrates of cerambycids also vary considerably from living tissues to decaying wood. Nevertheless, certain general patterns can be recognised. **Table 1** summarises the biology of British Cerambycidae using these general patterns; if not stated otherwise, the biology data in **Table 1** reflects the author's opinion. All British Cerambycidae can be categorised by larval host conditions using categories suggested by Hanks (1999). These categories are based on the condition of the larval host at the time of colonisation and they comprise: healthy host (HH), weakened host (WH), stressed host (SH) and dead host (DH). Furthermore, cerambycid larvae differ in the type of host tissues they spend most time feeding in. For British species the following categories (**Table 1**) will suffice: bark (b), root (r), subcortical zone tissues (su), stem (st) and wood in general (w). Wood can further be divided to sapwood (sw) and heartwood (hw). Pupation occurs in the same tissues or in soil. The exact location of pupal cells within the host may depend on bark thickness. Any dead woody tissues utilised by larvae can either be sound (^{s}) or decaying (^{d}). The nutritional quality of larval substrates in dead woody plants may be greatly improved by proliferation of fungal hyphae, which may favour development of wood borers. Certain British cerambycids are only found in such situations, thus *Grammoptera abdominalis* is associated with the fungus *Vuilleminia comedens*. Larvae of *Pseudovadonia livida* live in soil infested with mycelium of the fungus *Marasmius oreades*, which probably constitutes their main food source. The larvae have been shown to also feed on other soil components of organic origin.

Table 1: Biology of British Cerambycidae

Species	H	LS	P	Host Plant and/or Fungus
Prioninae Latreille, 1802				
Prionus coriarius (Linnaeus, 1758)	DH	r^d	soil	polyphagous DT & CT (< *Quercus*)
Lepturinae Latreille, 1802				
Rhagium (s. str.) *inquisitor* (Linnaeus, 1758)	DH	su	su	polyphagous DT << CT
R. (Hagrium) bifasciatum Fabricius, 1775	DH	w^d	w^d	polyphagous DT & CT
R. (Megarhagium) mordax (De Geer, 1775)	DH	b, su	b, su	polyphagous DT > CT (< *Quercus*)
Stenocorus meridianus (Linnaeus, 1758)	DH	r^d	soil	polyphagous DT
Dinoptera collaris (Linnaeus, 1758)	DH	su	soil	polyphagous DT
Grammoptera ustulata (Schaller, 1783)	DH	su	su	polyphagous DT (<< *Quercus*)
G. abdominalis (Stephens, 1831)	DH	su	w^d	*Quercus* + *Vuilleminia comedens*
G. ruficornis (Fabricius, 1781)	DH	su	su	polyphagous DT + fungus
Pedostrangalia revestita (Linnaeus, 1767)	DH+ HH	w^d	w^d	polyphagous DT
Lepturobosca virens (Linnaeus, 1758)	DH	w^d	w^d	*Betula*, *Pinus*, *Picea*
Leptura quadrifasciata Linnaeus, 1758	DH	w^d	w^d	polyphagous DT
L. aurulenta Fabricius, 1792	DH	w^d	w^d	polyphagous DT (< *Fagus*)
Anastrangalia sanguinolenta (Linnaeus, 1761)	DH	w^d	w^d	*Pinus*, *Picea*
Stictoleptura rubra (Linnaeus, 1758)	DH	w^d	w^d	*Pinus*, *Picea*, *Larix*, *Abies*
S. scutellata (Fabricius, 1781)	DH	w^d	w^d	polyphagous DT (< *Fagus*) + white-rot fungi
Paracorymbia fulva (De Geer, 1775)	DH	?	?	unknown
Anoplodera sexguttata (Fabricius, 1775)	DH	hw^d	hw^d	*Quercus*
Judolia sexmaculata (Linnaeus, 1758)	DH	w^d	w^d or soil	polyphagous DT < CT
Pachytodes cerambyciformis (Schrank, 1781)	DH	r	soil	polyphagous DT & CT
Alosterna tabacicolor (De Geer, 1775)	DH	b, su, w^d	b, su, w^d	polyphagous DT & CT + fungus

Pseudovadonia livida (Fabricius, 1776)	DH	soil	soil	*Marasmius oreades*
Strangalia attenuata (Linnaeus, 1758)	DH	w^d	w^d	*Pinus, Quercus, Tilia, Betula*
Rutpela maculata (Poda, 1761)	DH	w^d	w^d	polyphagous DT >> CT + white-rot fungi
Stenurella melanura (Linnaeus, 1758)	DH	w^d	w^d	polyphagous DT & CT
S. nigra (Linnaeus, 1758)	DH	w^d	w^d	polyphagous DT
Spondylidinae Serville, 1832				
Asemum striatum (Linnaeus, 1758)	DH	r^d-w^d	r^d-w^d	*Pinus > Picea, Abies, Larix*
Tetropium castaneum (Linnaeus, 1758)	SH	su	sw	*Picea > Abies, Pinus < Larix*
T. gabrieli Weise, 1905	SH	su	su, sw	*Larix* >>> Duffy 1953: *Pinus*
Arhopalus rusticus (Linnaeus, 1758)	DH	w^d	w^d	*Pinus* >> *Picea, Abies, Larix*
A. ferus (Mulsant, 1839)	DH	w^d	w^d	*Pinus* >> *Picea*
Cerambycinae Latreille, 1802				
Trinophylum cribratum Bates, 1878	DH	*su	?	**polyphagous DT (< *Quercus*) & *Pinus sylvestris* and *Larix*
Cerambyx scopolii Füessly, 1775	DH	su-w	w	polyphagous DT
Gracilia minuta (Fabricius, 1781)	DH	w	w	polyphagous DT (< *Salix*)
Obrium brunneum (Fabricius, 1792)	DH	su	w	polyphagous CT
O. cantharinum (Linnaeus, 1767)	DH	su	sw	polyphagous DT (<<< *Populus*)
Nathrius brevipennis (Mulsant, 1839)	DH	su-w	w	polyphagous DT
Molorchus minor (Linnaeus, 1758)	SH	su	w	polyphagous CT (<<< *Picea*)
Glaphyra umbellatarum (Schreber, 1759)	DH	su	w	polyphagous DT (< Rosaceae)
Aromia moschata (Linnaeus, 1758)	WH	w^s	w^s	*Salix* >>> *Populus*
Hylotrupes bajulus (Linnaeus, 1758)	DH	w	w	polyphagous CT
Callidium violaceum (Linnaeus, 1758)	SH	su-sw	sw	polyphagous CT
Pyrrhidium sanguineum (Linnaeus, 1758)	SH	su	b or sw	polyphagous DT (<< *Quercus*)
Phymatodes testaceus (Linnaeus, 1758)	SH	su	su or sw	polyphagous DT (<< *Quercus*)
Poecilium alni (Linnaeus, 1767)	SH	su	w	polyphagous DT (<< *Quercus*)
Clytus arietis (Linnaeus, 1758)	DH	su-w	w	polyphagous DT
Plagionotus arcuatus (Linnaeus, 1758)	SH	su-sw	sw	polyphagous DT (<< *Quercus*)
Anaglyptus mysticus (Linnaeus, 1758)	DH	sw	sw	polyphagous DT (< hardwoods)
Lamiinae Latreille, 1825				
Mesosa nebulosa (Fabricius, 1781)	DH	su-w^d	w^d	polyphagous DT + fungus
Agapanthia villosoviridescens (De Geer, 1775)	HH	st	st	polyphagous HP (<< *Heracleum sphondylium*)

Lamia textor (Linnaeus, 1758)	WH	r-w	r-w	*Salix* >> *Populus*
Pogonocherus fasciculatus (De Geer, 1775)	SH	su	w	polyphagous CT (<< *Pinus*)
P. hispidus (Linnaeus, 1758)	SH	su	w	polyphagous DT
P. hispidulus (Piller et Mitterpacher, 1783)	SH	su	w	polyphagous DT
Acanthocinus aedilis (Linnaeus, 1758)	SH	su	b, su, sw	polyphagous CT (<<< *Pinus*)
Leiopus nebulosus (Linnaeus, 1758)	SH	su	b, su	polyphagous DT >> CT
Saperda carcharias (Linnaeus, 1758)	HH	su-w	w	*Populus* >> *Salix*
S. scalaris (Linnaeus, 1758)	SH	su	sw	polyphagous DT >> CT (< *Quercus*)
S. populnea (Linnaeus, 1758)	HH	w	w	*Populus* (< *P. tremula*) >>> *Salix*
Stenostola dubia (Laicharting, 1784)	SH	su	w	polyphagous DT (< *Tilia* & *Fraxinus*)
Phytoecia cylindrica (Linnaeus, 1758)	HH	st - r	r	oligophagous on Apiaceae
Oberea (s. str.) *oculata* (Linnaeus, 1758)	HH	w	w	*Salix* spp. (< *Salix caprea*)
Tetrops praeustus (Linnaeus, 1758)	SH	su	w	***oligophagous on Rosaceae
T. starkii Chevrolat, 1859	SH	su	su-w	*Fraxinus excelsior*

Legend: H - Host, **LS** - Larval Substrate, **P**- Pupation, **DT** - deciduous trees, **CT** - coniferous trees, **HP** - herbaceous plants, **HH** - healthy host, **WH** - weakened host, **SH** - stressed host, **DH** - dead host, **b** - bark, **r** - root, **su** - subcortical zone tissues, **st** - stem, **w** - wood in general, **sw** - sapwood, **hw** - heartwood, **s** - sound, **d** - decaying, < - slight, << - strong, <<< - very strong preference.

*Whitehead (pers. comm., 2003)

**Uhthoff-Kaufmann (1990d)

***British specimens (but also specimens originating from Continental Europe) reared or collected from *Frangula alnus*, *Salix* spp. but also other hosts (including some Rosaceae) will very likely belong to a sibling, as yet undescribed, *Tetrops* species (Rejzek, Lopez, Booth in preparation).

The biology of several species deserves further comment. *Pedostrangalia revestita* is very polyphagous but oak is preferred. Larvae of this species feed in dead red-brown and relatively moist wood (a product of a special decay process) in close contact with living tissue, such as the bases of dead branches surrounded by living callus. The pupal cell as well as the larval galleries are filled with typical long red wood fibres created by the larvae. Late instar larvae can easily be reared from vials but attempts to rear adults from young larval instars in their natural substrate usually fail.

Trinophylum cribratum Bates is an Indian species imported to England and now naturalised and occurring rarely in Worcestershire (Whitehead, 1990). In Whitehead's experience (pers. comm., 2003) of larvae working in logged cultivated apple wood *Malus domestica*, the galleries are distinctive but irregular and are cut in the superficial layers of the sapwood and in the bast. They do not conform to a recognisable pattern and are strongly etched by the larval mandibles. The larvae accept and are thought to prefer wood which is actively drying out, or dying slowly and have been found in covered sawn logs some 12 months after cutting. Three host genera are recognised in Worcestershire (as at January 2003), oak (*Quercus*), apple (*Malus*) and plum (*Prunus*). The biology of *Trinophylum cribratum* has also been described by Uhthoff-Kaufmann (1990d). According to this work the species seems to be fairly polyphagous on deciduous trees (*Malus*, *Fraxinus*, *Fagus*, *Betula*, *Carpinus*, *Quercus*, *Pyrus*, *Platanus*, *Pyracantha* and *Juglans regia* but prefers *Quercus robur* and *Quercus coccifera*) and conifers (*Larix* and *Pinus sylvestris*). Larvae feed in well-seasoned wood (timber or standing trees). The development lasts one year and the crepuscular adults emerge from mid-summer until as late as September.

Saperda populnea is the sole British cerambycid species whose larvae provoke the proliferation of host tissues, which results in the formation of a conspicuous protective gall. The larvae feed largely on the tissue by which the plant attempts to heal the wound. The gallery system of such species is then remarkably restricted. Rearing adults from larvae developing in living hosts is very difficult and should in general be avoided whenever possible. To some extent it can be successful when late-instar larvae are collected in their larval substrate (twigs, branches). The substrate after removal should be inserted in wet sand.

Larvae of *Phytoecia cylindrica* and *Agapanthia villosoviridescens* develop in herbaceous plants (cf. **Table 1**). Larvae of *Phytoecia cylindrica* are to be found in the roots of their hosts and the last-instar larva builds a pupal cell in the root, girdles the stalk several centimetres above ground level and secures the gallery by a wad of fibrous frass. The section of stalk above the girdle will break off sooner or later, exposing the characteristic wad. Through this method the larva ensures that the plant always breaks above ground level rather than below ground where the pupal cell is situated. The adults emerge by removing the wad. Larvae of *Agapanthia villosoviridescens* feed in stalks of their hosts. The pupal cell is built in the stalk rather than the roots although it may occur very close to ground level. Similarly to the previous species the last instar larva girdles the stalk above the pupal cell and secures the gallery at this point by a wad of fibrous frass. Interestingly, in contrast to *P. cylindrica* the emerging adult leaves the stalk by biting an oval emergence hole in its side, usually about 1 cm below the cut. Both species can be reared from larvae collected in their larval substrates. The moisture content of the natural substrates can be kept constant by inserting the roots or stalks into wet sand. Late instar larvae, however, will pupate and adults will hatch even in vials.

Last but not least, attention should always be paid to the accurate recording of collecting data. In the case of rearing adults from larvae kept in vials, each individual vial must bear detailed collecting data accompanied with the name of host plant if known. Substrate containing insect larvae should be kept in a separate well-labelled vessel and should never be mixed with substrates originating from different localities. Ideally each individual piece of wood containing larvae should be marked to avoid mistakes. Moreover, emerging adults must be prevented from escaping the rearing vessels to ensure correct recording of collecting data and host association.

Breeding Longhorn Beetles – Frequently it is possible to breed cerambycid species under laboratory conditions, rearing adults *ex ovo*. This is usually very successful because no parasitic insects are present and good control can be had over fungi harmful to the larvae. The method usually yields large numbers of adults that can either be used for further studies, or released in the place of origin. Rearing adults from larvae offers an even deeper insight into the insect's life history, providing information on larval substrate requirements, duration of development, exact date of hatching and adult emergence, adult life expectancy, maturation feeding, copulation and oviposition.

Most cerambycids will copulate immediately or shortly after emerging from their larval substrate. Copulation can usually be encouraged by providing the beetles with freshly cut substrate (usually natural host) into which they later oviposit. An artificial source of light (placed a reasonable distance from the rearing vessel to avoid over-heating) will help in the case of diurnal species; nocturnal species will copulate in darkness. Species requiring a maturation feed will have to be provided with plant tissue exactly matching their requirements.

Cerambycids usually oviposit in the cracks of bark or wood. The substrate should therefore be chosen accordingly. Shortly after oviposition most cerambycids die. The substrate should occasionally be sprayed with water to avoid desiccation. After about two weeks most larvae will hatch from the eggs and will start making their way into the preferred host tissues. Once a month or so the substrate should be submerged into water and allowed to remain there for about 1 hour. Most species will not tolerate either mould or over-desiccation. For at least two months the substrate should be kept outdoors and exposed to temperatures alternating between -5 to +5°C (a refrigerator is not convenient) and allowing the larvae to hibernate. The exposure to such temperatures also influences the ability of the insects to synchronise their emergence. The substrate is then brought back to room temperature and the adult beetles emerge shortly thereafter.

Callidium violaceum can serve as an example of a species easy to reproduce. The adults should be provided with freshly cut spruce branches (diameter 2-3 cm) and allowed to oviposit there. The larvae will soon consume most of the branches leaving only thin bark. Occasional cracks in the bark will expose typically broad and shallow larval galleries partially filled with frass. The next generation will hatch after two years of development.

As always, attention should be paid to the accurate recording of collecting data. Adults obtained by reproduction should bear the collecting data of the original animals and in addition be labelled with '*ex ovo*'.

Mounting – In order to avoid excess moisture damaging sensitive body pubescence and to obtain relaxed material, adult cerambycids should be collected in 100ml plastic (polypropylene) bottles half filled with very dry wood shavings and treated with only a very small amount of ethyl acetate (three to five drops) (plate 110). It is important to ensure that the plastic bottle and stopper are resistant to ethyl acetate. At the end of an excursion an additional amount of ethyl acetate should be added (10 drops) to the collecting bottle, mainly to prevent mould. The following day the material is ready to be mounted. With perhaps the exception of *Prionus coriarius* all British cerambycids should be card mounted. The body should be well glued down using the minimal amount necessary. The fragile appendages should be kept within the boundaries of the card to ensure their protection. Antennae should preferably be positioned backwards along each side of the body and should never obscure the pronotum and elytra. Antennae and tarsi should also be glued to the card surface. To avoid excessive amounts of glue, that after drying would show on the card surface, first use a pin and water to wet both the appendages to be glued and the card surface. Next gently touch the appendage with a pin previously dipped in a very dilute aqueous solution of the glue. Capillary action will distribute the solution evenly. If need be, remove any excess of this solution using a wet pin. Specimens prepared in this way will be less prone to damage.

Dried material should first be placed into hot water containing a small amount of a surfactant. After an hour or so the relaxed material should be placed between two layers of absorbent tissue and gently pressed (avoid rubbing off the fragile setae). The rich pubescence, frequent in cerambycidae, will become erected and separated again. Metallic beetles and also *Saperda scalaris*, however, will lose their colour when this method is applied.

In the case of British cerambycids the aedeagus is never an essential feature for identification. It does, however, become important when beetles of certain Continental genera are being identified. Most species can be sexed by longer antennae or by expanded basal segments of the front tarsi in males. The aedeagus is extracted by opening the apical abdominal segments, which can be pushed back under the elytra when the specimen is mounted. The parameres should be separated from the median lobe and both glued onto a card using DMHF. Recently, the internal sac and the female spermatheca have received more attention as they have proved to display valuable diagnostic characteristics in certain Continental species otherwise difficult to distinguish. They are, however, of little importance for identification of British cerambycids.

Larvae can be prepared according to Duffy (1953) or Švácha & Danilevsky (1987). They should first be submerged in approximately 90°C hot water for two to 20 seconds depending on their size. Using a pin, the larvae should then be perforated once or twice in lateral intersegmental zones to prevent osmotic deformation. Finally they should be kept in Pampel's fluid, which contains glacial acetic acid (four parts), distilled water (30 parts), 40% aqueous solution of formaldehyde (six parts) and finally 95% ethyl alcohol (15 parts).

References

Bense, U., 1995. *Bockkäfer, Illustrierter Schlüssel zu den Cerambyciden und Vesperiden Europas.* [*Longhorn Beetles. Illustrated Key to the Cerambycidae and Vesperidae of Europe*]. Weikersheim, 512 pp.

Booth, R.G., Cox, M.L. & Madge, R.B., 1990. *The Guides to Insects of Importance to Man, 3. Coleoptera.* University Press, Cambridge, 384 pp.

Böving, A.G. & Craighead, F.C., 1931. An illustrated synopsis of the principal larval forms of the order Coleoptera. *Entomologica Americana*, **11**:1-351.

Crowson, R.A., 1981. *The Biology of the Coleoptera.* Academic Press, London, pp. 694-698.

Danilevsky, M., 1979. Descriptions of the Female, Pupa and Larva of *Apatophysis pavlovskii* Plav. and Discussion of Systematic Position of the Genus *Apatophysis* Chevr. (Cerambycidae), *Entomologicheskoe Obozrenie*, **58**(4):821-828.

Demelt, C.v., 1966. *Die Tierwelt Deutschlands und der angrenzenden Meeresteile, II. Bockkäfer oder Cerambycidae, I. Biologie mitteleuropäischer Bockkäfer (Col. Cerambycidae) unter besonderer Berücksichtigung der Larven.* VEB Gustav Fisher Verlag Jena, 115 pp., 9 pls.

Duffy, E.A.J., 1952. Cerambycidae. Handbooks for the Identification of British Insects **5**(12), 18 pp. Royal Entomological Society, London,

Duffy, E.A.J., 1953. *A monograph of the immature stages of British and imported timber beetles.* British Museum (Natural History) Monograph, 350 pp.

Hambler, C. & Speight, M.R., 1995. Biodiversity Conservation in Britain: Science Replacing Tradition. *British Wildlife*, **6**(3):137-148.

Hanks, L.M., 1999. Influence of the Larval Host Plant on Reproductive Strategies of Cerambycid Beetles. *Annual Review of Entomology*, **44**:483-505.

Horion, A., 1974. *Faunistik der mitteleuropäischen Käfer, Band XII: Cerambycidae - Bockkäfer.* Überlingen - Bodensee, 228 pp.

Lawrence, J. F. & Newton, A. F., 1995. Families and subfamilies of Coleoptera (with selected genera, notes, references and data on family-group names). pp.779-1006 *in* Pakaluk, J. & Ślipiński, S. A. (eds.) *Biology, phylogeny, and classification of Coleoptera: Papers celebrating the 80th Birthday of Roy A. Crowson.* Muzeum I Instytyt Zoologii PAN, Warszawa.

Linsley, E.G., 1959. Ecology of Cerambycidae. *Annual Review of Entomology*, **4**:99-138.

Linsley, E.G., 1961. The Cerambycidae of North America. Part I: Introduction. *University of California Publications in Entomology*, **18**:1-97+35 pls.

Rejzek, M., 2004. Check-List of Cerambycidae (Col.) of the British Isles. *Entomologist's Monthly Magazine*: **140**: 51-57.

Sama, G., 2002. *Atlas of the Cerambycidae of Europe and Mediterranean area.* Part 1 : Northern, western, central and eastern Europe (British Isles and Continental Europe from France (excluding Corsica) to Scandinavia and Urals). Kabourek, Zlín, 177 pp., 36 colour plates.

Švácha P., & Danilevsky, M.L., 1987. Cerambycoid Larvae of Europe and Soviet Union (Coleoptera, Cerambycoidea). Part I. *Acta Universitatis Carolinae - Biologica*, **30**:1-176.

Švácha P. & Danilevsky, M.L., 1988. Cerambycoid Larvae of Europe and Soviet Union (Coleoptera, Cerambycoidea). Part II. *Acta Universitatis Carolinae - Biologica*, **31**:121-284.

Švácha P.,. & Danilevsky, M.L., 1989. Cerambycoid Larvae of Europe and Soviet Union (Coleoptera, Cerambycoidea). Part III. *Acta Universitatis Carolinae - Biologica*, **32**:1-205.

Švácha P., 2001. Überfamilie: Chrysomeloidea, Familie: Cerambycidae, Unterfamilie Lamiinae, pp. 248 - 298. In Klausnitzer B., (ed.): *Die Larven der Käfer Mitteleuropas*. 6. Band: Polyphaga, Teil 5. Spektrum Akademischer Verlag Heidelberg, Berlin, 309 pp.

Twinn, P.F.G. & Harding, P.T., 1998. *Provisional atlas of the longhorn beetles (Col., Cerambycidae) of Britain*. Huntingdon: Biological Record Centre. Henry Ling Ltd., Dorchester, 96 pp.

Uhthoff-Kaufmann, R.R., 1985. The genus *Obrium* (Col.: Cerambycidae) in Great Britain; a reappraisal. *Entomologist's Record & Journal of Variation*, **97**:216-223.

Uhthoff-Kaufmann, R.R., 1987. The distribution of the genus *Leptura* L. (Col.: Cerambycidae) in Great Britain. *Entomologist's Record & Journal of Variation*, **99**:195-202.

Uhthoff-Kaufmann, R.R., 1988. The occurrence of the genus *Strangalia* (Col.: Cerambycidae) in the British Isles. *Entomologist's Record & Journal of Variation*, **100**:63-71.

Uhthoff-Kaufmann, R.R., 1989. The occurrence of *Grammoptera* Serville and *Alosterna* Mulsant (Col.: Cerambycidae) in the British Isles. *Entomologist's Record & Journal of Variation*, **101**:97-103.

Uhthoff-Kaufmann, R.R., 1989a. The occurrence and distribution of the genera *Acmaeops* Lec. and *Judolia* Muls. (Col.: Cerambycidae) in Great Britain. *Entomologist's Record & Journal of Variation*, **101**:179-182.

Uhthoff-Kaufmann, R.R., 1989b. The genera *Rhagium* F. and *Stenocorus* Müll. (Col.: Cerambycidae) in the British Isles. *Entomologist's Record & Journal of Variation*, **101**:241-248.

Uhthoff-Kaufmann, R.R., 1990. The occurrence of the sub-family Aseminae (Col.: Cerambycidae) in the British Isles. *Entomologist's Record & Journal of Variation*, **102**:55-63.

Uhthoff-Kaufmann, R.R., 1990a. The genera *Clytus* Laich. and *Anaglyptus* Muls. (Col.: Cerambycidae) in the British Isles. *Entomologist's Record & Journal of Variation*, **102**:111-114.

Uhthoff-Kaufmann, R.R., 1990b. The occurrence of the Callidini tribe (Col.: Cerambycidae) in the British Isles. *Entomologist's Record & Journal of Variation*, **102**:161-166.

Uhthoff-Kaufmann, R.R., 1990c. The genera *Nathrius* Brèthes and *Molorchus* F. (Col.: Cerambycidae) in Great Britain. *Entomologist's Record & Journal of Variation*, **102**:239-242.

Uhthoff-Kaufmann, R.R., 1990d. The distribution of the genera *Trinophylum* Bates, *Gracilia* Serv., *Aromia* Serv., and *Hylotrupes* Serv. (Col.: Cerambycidae) in the British Isles. *Entomologist's Record & Journal of Variation*, **102**:267-274.

Uhthoff-Kaufmann, R.R., 1991. The distribution and occurrence of the Tanner Beetle, *Prionus coriarius* L. (Col.: Prionidae) in Great Britain. *Entomologist's Record & Journal of Variation*, **103**:3-5.

Uhthoff-Kaufmann, R.R., 1991a. The genera *Lamia* F., *Mesosa* Latr. and *Leiopus* Serv., (Col.: Lamiidae) in the British Isles. *Entomologist's Record & Journal of Variation*, **103**:73-77.

Uhthoff-Kaufmann, R.R., 1991b. The distribution and occurrence of the genus *Saperda* F. (Col.: Lamiidae) in Great Britain. *Entomologist's Record & Journal of Variation*, **103**:129-134.

Uhthoff-Kaufmann, R.R., 1991c. The distribution and occurrence of *Acanthocinus* Dej. and *Agapanthia* Serv. (Col.: Lamiidae) in the British Isles. *Entomologist's Record & Journal of Variation*, **103**:189-192.

Uhthoff-Kaufmann, R.R., 1991d. The genus *Pogonocherus* Zett. (Col.: Lamiidae) in the British Isles. *Entomologist's Record & Journal of Variation*, **103**:243-246.

Uhthoff-Kaufmann, R.R., 1992. The genera *Oberea* Muls., *Stenostola* Muls., *Phytoecia* Muls. and *Tetrops* Steph. (Col.: Lamiidae) in Great Britain. *Entomologist's Record & Journal of Variation*, **103**:243-246.

Uhthoff-Kaufmann, R.R., 1992a. Extinct and very rare British longhorn beetles (Col.: Cerambycidae and Lamiidae). *Entomologist's Record & Journal of Variation*, **104**:113-121.

Whitehead, P.F., 1990. Analysis of a Coleoptera fauna from Broadway, Worcestershire. *Entomologist's Monthly Magazine*, **126**:27-32.

MEGALOPODIDAE, ORSODACNIDAE & CHRYSOMELIDAE
(Plates 89 – 99)

by M.L. Cox

There have been major changes to the higher taxonomy of the Chrysomelidae since the last edition of the *Handbook*. Changes effecting the British fauna are tabulated below:

Family	Sub-family	Genus/genera
MEGALOPODIDAE		*Zeugophora*
ORSODACNIDAE		*Orsodacne*
CHRYSOMELIDAE	Bruchinae	*Bruchus, Bruchidius, Acanthoscelides, Callosobruchus* and *Zabrotes*
	Donaciinae	*Donacia, Plateumaris* and *Macroplea*
	Criocerinae	*Lema, Oulema, Crioceris* and *Lilioceris*
	Cassidinae	*Hypocassida, Pilemostoma* and *Cassida*
	Chrysomelinae	*Timarcha, Leptinotarsa, Chrysolina, Gastrophysa, Phaedon, Hydrothassa, Prasocuris, Plagiodera, Chrysomela, Gonioctena* and *Phratora*
	Galerucinae	*Galerucella, Xanthogaleruca, Pyrrhalta, Galeruca, Lochmaea, Phyllobrotica, Luperus, Calomicrus, Agelastica, Sermylassa, Phyllotreta, Aphthona, Longitarsus, Altica, Hermaeophaga, Batophila, Lythraria, Ochrosis, Neocrepidodera, Derocrepis, Hippuriphila, Crepidodera, Epitrix, Podagrica, Mantura, Chaetocnema, Sphaeroderma, Apteropeda, Mniophila, Dibolia* and *Psylliodes*
	Lamprosomatinae	*Oomorphus*
	Cryptocephalinae	*Labidostomis, Clytra, Smaragdina* and *Cryptocephalus*
	Eumolpinae	*Bromius*

In addition there have been several species added to the British List and others sunk as synonyms, see Cox, 2000.

They are surprisingly not a particularly popular group with the amateur, considering the frequency with which they are encountered in the sweep-net or on the beating-tray, and many are large and showy. However, about half the British fauna is represented by the flea-beetles which are small, often difficult to catch and often uniform in colour and thus present the most difficult problems as regards accurate identification. The lack of good up-to-date keys also partly explains why this still remains an unpopular family.

Specific identification in the majority of genera is relatively straightforward since many display good external characters. However, there are still 'tricky' groups in most sub-families since colour in these is variable and often external characters are only relative and unreliable. In such cases resort has to be made to dissection of the genitalia. Notably difficult genera include *Phratora, Phaedon, Galerucella, Longitarsus, Altica, Phyllotreta, Aphthona, Chaetocnema, Psylliodes* and *Cassida*.

MEGALOPODIDAE. *Zeugophora* adults are dull, either unicolorous orange-brown or bicolorous yellow-orange and black. They occur on the foliage of young poplars, aspens and sometimes willows from May to October. Their damage is characteristic since they cause netting of the leaves which become completely perforated by small holes and only the sclerotised veins remain intact.

The yellow oval ova are laid singly in small concavities eaten by the females in the upper epidermis and covered by a secretion. Oviposition probably commences from mid-May and the larvae mine the parenchyma of the leaves during June and July causing characteristic black blotch mines. Pupation occurs within a cell in the soil. New generation adults emerge from about mid-August, feed and move below into their over-wintering quarters from September. The *Zeugophora* species are univoltine.

Adults may be collected by beating young aspens and poplars especially those showing the characteristic damage of the adults and larvae. The three British species are readily separated on head and body colour and hence it is unnecessary to dissect their genitalia. The sexes are difficult to distinguish and apparently can only be separated on the shape of the apex of the final abdominal sternite.

ORSODACNIDAE. The pale-brown or black adults of *Orsodacne* species frequent the flowers of various herbaceous and woody plants from May until July. They feed upon the anthers and pollen and probably oviposit from June to July. The ova are white and very elongate with rounded ends. Incubation requires about two weeks and larvae are probably present in the field from mid-June. The position of feeding of *Orsodacne* larvae is not known with certainty. Morphological evidence suggests that they are external or internal root-feeders (Cox, 1981) and not miners in leaf petioles as proposed by Balachowsky (1963). In Britain it is possible that *Orsodacne* species overwinter as first instar larvae after a lengthy summer diapause. They emerge from hibernation in early spring, feed up and the adults emerge from May until July. It is not definitely known where the larvae hibernate and when and where they commence feeding.

It is possible that hibernation is above ground. *O. lineola* (Panzer) larvae were most commonly dissected from gizzards of blue tits, which obtain most of their food from twigs, buds and leaves, but rarely were they found in gizzards of ground-feeding great-tits (Betts, 1955, 1956).

The adults are best collected from areas in and around mature broad-leaved woodland by beating, especially rowan or hawthorn in flower, or by sweeping vegetation rich in flowering Apiaceae. *Orsodacne lineola* and *cerasi* (L.) are easily separated using external characters and should be mounted dorsally.

CHRYSOMELIDAE.

BRUCHINAE. These beetles, commonly called pea or bean weevils, but preferably seed beetles are small, oval, compact, sombre, usually brown or black and clothed with uniform or variously coloured recumbent pubescence. The adults are pollen-feeding and occur on the white flowers of Rosaceae, Fabaceae and Apiaceae and also the yellow flowers of broom and rock-roses, from which they may be beaten or swept. The adults overwinter in hibernation sites and oviposit in the spring when suitable host plants are available.

The eggs are usually flattened-oval, usually laid singly on the surface of a pod or seed and covered with exudate which hardens into a protective cover.

The larvae develop within the seeds of Fabaceae, either in storage (*Bruchus pisorum* (L.), *Callosobruchus* spp., *Acanthoscelides obtectus* (Say), *Zabrotes subfasciatus* (Boheman)), or in the field (*Bruchus* and *Bruchidius* spp.) and pupate within the larval feeding cell.

The males in most British genera furnish reliable characters in species recognition and therefore it is important to be able to sex bruchids. Males and females of *Bruchidius villosus* (F.) and *Acanthoscelides obtectus* can be separated on the shape of the pygidium (apical abdominal tergite) and apical abdominal sternite. Male *Bruchus* species have mesotibiae bearing one or two spurs

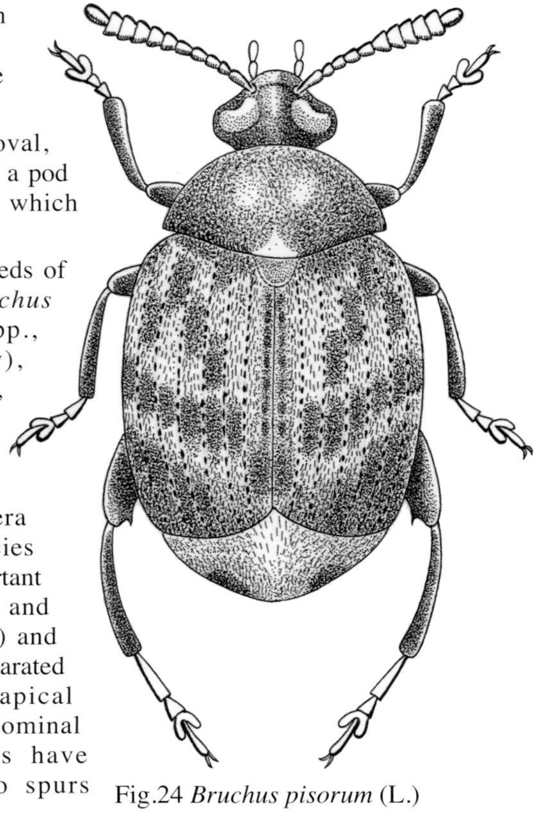

Fig.24 *Bruchus pisorum* (L.)

apically/pre-apically; these are absent in females. The antennae of male *Callosobruchus* species are sometimes pectinate or distinctly serrate, whereas in the females of the corresponding species they are weakly serrate. In *Zabrotes subfasciatus*, the males have uniform pale brown pubescence over a dark cuticle, whilst the females are strongly marked with white patterning on a dark, almost black, background. The median lobe of the aedeagus is useful in separation of *Bruchidius* and *Callosobruchus* species. For full details of sexual differences and keys to genera and species see Cox, 2001.

Most species are restricted to England and Wales and occur in a wide range of habitats. *Bruchidius olivaceus* (Germar) was associated with the arable crop sainfoin (*Onobrychis viciifolia*) and the last authenticated record is from Cothill, Oxfordshire in September 1923. *Bruchidius varius* (Olivier) was added to the British fauna by Hodge (1997) and is now quite widespread in southern England.

Since ventral characters are not used in specific determination, specimens are best mounted normally, with the legs, especially the mid-tibiae and hind femora well displayed. Specimens are easily sexed and resort may have to be made to dissection of the male genitalia. Apparently, the spermatheca is not useful in specific determination in the Bruchinae.

DONACIINAE. (See also under 'water beetles', page 16) These usually brightly metallic-coloured beetles frequent the foliage and flowers of a variety of aquatic plants during the summer. They eat strips from the leaves and also feed on pollen.

The ova are usually laid in groups, e.g. on the undersides of water lily leaves, and are covered by an opaque gelatinous substance. The larvae are aquatic and feed on the rhizomes and roots of their host plant and obtain oxygen through a pair of hollow caudal spines arising from the 8th abdominal segment which are inserted into the host plants tissues. Pupation occurs in tough brown cocoons attached to the roots of the host plant. Larvae and pupae may over-winter but adults of the new generation may emerge in the autumn, either remaining in the cocoon or emerging above water. They are univoltine.

Adults may be collected by sweeping vegetation in the littoral zone of lakes, lochs and small ponds or from the margins of streams and rivers. Larvae and pupae are obtained by pulling up clumps of uprooted water plants using a grappling iron.

Although the adults in this subfamily are moderately large, and sometimes with distinctive coloration and sculpture, they still prove quite difficult. This is because the body colour is sometimes very variable intra-specifically and also in those species with the hind femora armed, the sexes are often dimorphic as regards the size and position of the teeth. These teeth are usually a good specific character but are sometimes variable as in females of *D. versicolorea* (Brahm) where they may be large or greatly reduced (virtually absent). Attempts have been made to use the elytral depressions in species separation in *Donacia* but these are variable and not a valid character. The aedeagus is useful in separating males of *Macroplea appendiculata* (Panzer) and *M. mutica* (F.) since keys based on external characters are unreliable.

The sexes are usually easy to distinguish since the males have the anterior tarsi dilated; the size and arrangement of teeth on the hind femora are usually different and rarely the body colour is dimorphic (as in *Plateumaris affinis* (Kunze) in which the females are coppery and the males black with a purple or violet reflection). Since ventral characters are not used in specific determination, specimens are best mounted dorsally with the legs (especially the anterior tarsi and hind femora and tibiae) well displayed.

The British donaciines were keyed by Menzies and Cox, 1996.

CRIOCERINAE. The adults are typically narrow, elongate, with the pronotum much narrower at the base than, and often differently coloured from, the elytra which are rectangular and often with a bright metallic coloration. The adults, which usually stridulate using an elytra / abdominal apparatus, may be seen from April to September feeding on the leaves of their hosts (Poaceae, Asteraceae, Liliaceae). Those feeding on grasses and cereals (*Oulema* species) cause characteristic damage since they eat short longitudinal strips in parallel lines 1mm wide, completely perforating the leaf blade.

The ova, white (*Lema cyanella* (L.)), greenish-grey (*Crioceris asparagi* (L.)), yellow (*Oulema*) and red-brown (*Lilioceris lilii* (Scopoli)) are stuck by an adhesive substance in ones and twos on the under-surface of the leaves or sometimes deposited in grooves eaten by the adults. In *C. asparagi* they are fixed by their extremities, while in *L. lilii* they are stuck laterally. The larvae are strongly convex and cover themselves dorsally with a viscous layer of excrement which seems to protect them against predators. They feed openly, eating elongate strips 1mm wide from the upper layer of the epidermis, leaving the lower epidermis intact as a fine transparent membrane, especially in those feeding on Poaceae. There are four instars and pupation occurs in white cocoons constructed by the larvae either on the host plant or in the soil. New generation adults usually emerge from July to October and enter hibernation in grass tussocks and other sites from September onwards. *O. melanopus* (L.), *O. obscura* (Stephens) and *L. lilii* are probably univoltine whilst *L. cyanella* and *C. asparagi* may have two generations per year.

Adults are best obtained by sweeping their known hosts, especially if showing characteristic damage. *O. melanopus* and *O.obscura* also occur in grass tussocks during the winter. *Lema cyanella* occurs in broad-leaved woodland sites, clearings and edges, grassland, scrub etc. *Oulema erichsoni* (Suffrian) occurs in coastal areas and grassy places besides dykes and ditches;

C. asparagi is to be found on arable land, gardens and allotments on asparagus. *Lilioceris lilii* is associated with lilies in gardens, garden centres and the like.

Most criocerines are distinct and easily identified, except *Oulema obscura* and *erichsoni*. *O. obscura* is the smaller and more widespread and with elytra much less elongate than in *erichsoni*.

Sexes are difficult to recognise, except of *C. asparagi*, the males of which have longer more strongly curved anterior tibiae. Species should be mounted with the dorsal surface uppermost.

CASSIDINAE (treated by some authors in Hispinae). The adults of the 'tortoise beetles' are usually greenish dorsally and some have metallic golden bands (*vittae*) on the elytra. These are chemical colours which fade soon after death of the beetle but can be maintained if the beetle is preserved in 70% alcohol. They frequent the foliage of various low growing plants during the spring and summer. Their green colour provides good camouflage against the foliage of their host plants.

The elongate-oval ova are laid either singly, in pairs or in groups of up to 10 within brown oothecae, usually on the under-surface of leaves. They do not diapause, but hatch in from one to two weeks depending upon the prevailing temperature. There are five larval instars and the larvae are characterised by long lateral forked processes on the thorax and abdomen and a pair of usually long caudal processes which are bent forwards over the abdomen. Exuviae and faeces are sometimes retained on this caudal fork, which when held over the body acts to camouflage the larvae since they resemble droppings of birds. Pupation always occurs on the stem or foliage of the host plant. They are usually univoltine with the new generation adults emerging from the end of June until mid-September and overwinter in this stage. Adults are best collected by sweeping vegetation in the summer.

Specific habitats of some of the more local/rare species are as follows: *Pilemostoma fastuosa* (Schaller) – sweeping on grassland (downs), disturbed ground and probably also coastal habitats. *Cassida hemisphaerica* Herbst – sweeping on grassland, woodland rides, clearings and also coastal habitats. *Cassida murraea* L. – sweeping wetland, grassland, scrub and possibly disturbed ground. Formerly widespread in England (except north), Wales (except north). This species appears to have contracted its range to S.W. England, S. Wales and Dyfed/Powys where it is still common. *Cassida nebulosa* L. – sweeping on disturbed ground (arable farmland) probably also grassland and scrub. *Cassida nobilis* L. – sweeping in coastal habitats (England, Wales except N., S.W. Scotland). *Cassida prasina* Illiger – sweeping on grassland, disturbed ground and probably scrub. *Cassida vittata* de Villers – sweeping coastal shingle, maritime cliff, satmarsh and probably sand dunes.

Since dorsal (colour, puncturation, etc.) and ventral (colour) characters are used in specific determination, it is best to mount one specimen dorsally and one ventrally or to micro-pin a singleton. Resort does not have to be made to genital dissection and sexing is very difficult, the only apparent difference being the shape of the apex of the final abdominal sternite.

CHRYSOMELINAE. The majority of this subfamily are small to medium-sized, usually brightly coloured beetles frequently encountered in the spring and summer on the foliage of their host plants. The larvae and adults feed openly on the foliage of low herbaceous plants in various families (*Timarcha, Chrysolina, Gastrophysa, Phaedon, Hydrothassa*, and *Prasocuris*) or on bushes and trees, especially of *Salix* and *Populus* (*Plagiodera, Chrysomela, Gonioctena* and *Phratora*).

The white, yellow or reddish, oval or elongate-oval ova are usually laid singly or in batches on the undersides of leaves but sometimes covered with faeces and regurgitated food and deposited at the base of their host-plants or at a shallow depth in the soil as in *Timarcha* spp. In some *Phaedon* and *Hydrothassa* species they are laid in depressions cut into the leaves or actually laid inside the hollow stems of their host-plants. The larvae have three or four instars and usually have well-developed dark dorsal sclerotised tubercles. They also sometimes have paired eversible defensive glands on the thorax and abdomen. Pupation usually occurs in the soil but in *Plagiodera* and *Chrysomela* it takes place on the leaves and stems of their host plants. Most of the species are probably univoltine, the adults usually overwinter, rarely as diapausing eggs (*Timarcha*), non-diapausing eggs or possibly early instar larvae (*Chrysolina hyperici* (Forster) and *brunsvicensis* (Gravenhorst)) or diapausing late instar larvae (*Chrysolina herbacea* (Duftschmidt) for example).

Adults and larvae are best collected by sweeping or beating

Fig. 25 *Timarcha tenebricosa* (F.)

vegetation; some species may also be collected in grass tussocks during the winter (for example *Chrysolina staphylaea* (L.), *Chrysomela aenea* L., *Hydrothassa glabra* (Herbst), *Phaedon cochleariae* (F.) and *tumidulus* (Germar) and *Prasocuris junci* (Brahm)).

Specific habitats and distributions of some of the rarer species are as follows: *Chrysolina graminis* (L.) – sweeping *Tanacetum vulgare*, *Mentha* spp., on river flood plains or in marshes. *Chrysolina haemoptera* (L.) – sweeping maritime cliffs, coastal shingle, sand dunes and possibly salt marshes, also sometimes downland, inland. *Chrysolina marginata* (L.) – sweeping well-grazed short-turf grassland, river margins and dry sandy habitats, for example sandpits. *Chrysolina herbacea* – sweeping wetland mostly, but also grassland, wood edges, commons, gardens and disturbed ground. *Chrysolina oricalcia* (Müller) – sweeping woodland rides, clearings and woodland edges, also calcareous grassland, disturbed ground and possibly wetland. *Chrysolina sanguinolenta* (L.) – sweeping broad-leaved woodland rides, clearings and wood-edge, grassland and disturbed ground. *Chrysolina violacea* (Müller) – sweeping calcareous grassland, possibly also scrub and disturbed ground on base-rich soil, also chalk quarries. *Chrysolina brunsvicensis* – broad-leaved woodland rides, clearings and wood-edge, grassland and disturbed ground, also on disused railway lines and grassy places beside rivers. *Chrysolina cerealis* (L.) – Very rare and still only known from Snowdonia though apparently well established there. Adults occur on the host plant, thyme, in montane grassland between mid-April and September. They may also be found by grubbing at the base of the host plant or by searching under stones. *Chrysolina intermedia* (Franz) – maritime cliffs and probably dry sandy grasslands near the coast, N.W. and S.W. Scotland, Orkney and Shetland. *Phaedon concinnus* Stephens – saltmarshes, by sweeping or grubbing at plant roots. *Hydrothassa glabra* (Herbst) – sweeping grassland, wetland, quarries, downland, heathland, possibly also woodland edges and disturbed ground. Also by grubbing, tussocking and in flood refuse. *Hydrothassa hannoveriana* (F.) – sweeping wetland and semi-aquatic habitats. *Chrysomela aenea* L. – by beating alder. Under loose bark and in tussocks during winter. England except S.E. and E. Anglia, Wales, Scotland. *Chrysomela populi* L. – beating in broad-leaved woodland, wetland, particularly willow carr. Also recorded from wet flushes on heathland, dunes and cliff top. *Chrysomela tremula* F. – Formerly widespread in southern England but last recorded during the 1950's. Broad-leaved woodland. *Gonioctena decemnotata* (Marsham) – beating in broad-leaved woodland. *Gonioctena viminalis* (L.) – beating in broad-leaved woodland. *Phratora polaris* (Schneider) – a high altitude species usually collected by shaking out moss from below the host plants. So far only recorded from two vice-counties in N.W. Scotland and one in central Scotland but probably occurring in other similar montane localities in Scotland above 950m.

Sexing is usually accomplished by examining the anterior tarsi, the males *usually* having one or the first three segments dilated. In those species where this obvious difference is not exhibited, then the apical abdominal sternite should be examined. The apex of this is entire and usually rounded in females, but slightly indented or sinuous in males.

Specific determination in this subfamily usually relies upon colour or puncturation, antennal and leg (tarsal) characters. Specimens are best mounted with the dorsal surface showing and legs and antennae extended. However, in *Chrysolina* ventral characters are sometimes used so at least one specimen should be thus mounted.

GALERUCINAE. GALERUCINI – The soft-bodied adults are usually small to medium sized, dull brown, yellow or black, rarely metallic. They occur during the spring, summer and autumn on the foliage of low growing herbaceous plants (some *Galerucella, Galeruca, Phyllobrotica,* and *Sermylassa)* or shrubs and trees (some *Galerucella, Pyrrhalta, Luperus, Calomicrus* and *Agelastica).*

Galerucine ova are usually yellow or brown, approximately spherical, hemispherical or oval and are usually deposited singly or in batches on foliage or in the soil. Those laid in the soil often have the chorion with strong hexagonal microsculpture. Sometimes eggs are laid inside stems of the hostplant *(Pyrrhalta viburni* (Paykull)) or in oothecae attached to low vegetation (*Galeruca*). The ova in some cases diapause; in *Sermylassa halensis* (L.), *Pyrrhalta viburni* and *Galeruca tanaceti* (L.) this is probably obligatory, whilst in *Phyllobrotica quadrimaculata* (L.) it is facultative. These species oviposit in late summer/autumn whilst the other species oviposit in early or mid-summer.

First instar larvae lack egg bursters and hatch by biting their way from the egg. The larvae either feed openly exposed on foliage (*Galerucella, Pyrrhalta, Galeruca, Agelastica, Sermylassa*) or at the roots of their host plants *(Luperus, Calomicrus, Phyllobrotica)*. The adults and larvae of the former group feed on the same hostplant whilst the hosts of the adults and larvae of the latter group may differ. For example, adults of *Luperus* species feed on the foliage of *Betula* whilst their larvae probably feed on the roots of grasses (*?Molinia*). The larvae of *Lochmaea crataegi* (Forster) are unusual since they feed within the pulp of *Crataegus* haws whilst their adults feed on pollen from flowering hawthorn. Foliage-frequenting larvae usually have dark dorsal tubercles on thorax and abdomen and dark heavily sclerotised heads, whilst root-feeders have pale weakly sclerotised body tubercles and pale heads. Larvae of *Sermylassa* and *Agelastica* have paired segmental openings of defensive glands on the abdomen. Larvae have three instars. Pupation usually occurs in earthen cells in the soil, except in *Galerucella nymphaeae* (L.) and *sagittariae* (Gyllenhal) which pupate on the leaves of their aquatic and semiaquatic hostplants.

The adults of G. *tanaceti* have a period of summer aestivation (June to August) when they stop feeding and remain inactive in grass-tussocks (Donia, 1958; Siew, 1966). *Agelastica alni* also apparently aestivates from July to August.

The adults are best collected by beating or sweeping vegetation; *L. crataegi* especially from hawthorn, and *G. tanaceti* by tussocking from June to August.

Specific habitats and distributions of some of the local or rare species are as follows: *Galeruca laticollis* (Sahlberg) – was recently recorded from Sarlingham, East Norfolk during 1996 – dry heathland/grassland, areas of dry hillsides and broads. *Galeruca tanaceti* – grassland, maritime cliff, disturbed ground, heath, commons, quarries. *Phyllobrotica quadrimaculata* – wetland and semi-aquatic habitats; also wetland areas within woodland, alder carr. *Luperus flavipes* (L.) – beating bushes (birch, willow) in broadleaved woodland, parkland, scrub, heath etc. *Calomicrus circumfusus* (Marsham) – beating *Ulex* spp., *Genista* spp., broom on heathland, grassland, scrub, maritime cliff. *Agelastica alni* (L.) – beating alders in wetland, particularly old carr, also along river margins and in wet flushes. Presumed extinct in Britain since last recorded in 1910 at Devil's Punch Bowl, Hindhead, Surrey.

The sexes in the galerucines are usually distinct and easily recognisable. In *Luperus* the antennae of the males are as long or longer than the body and the apical segment is at least twice as long as the first. In the females the antennae are much shorter than the body and the apical segment is at most only slightly longer than the first. In addition, males have large, more prominent eyes, and the head broader than the pronotum, whilst in the female the head is narrower.

In *Calomicrus* the sexes can only be distinguished by the apical abdominal sternite, which in males has a deep apical notch on both sides and a large median depression; the apex is entire in the female and has no depression.

The sexes of *Sermylassa*, *Agelastica* and *Galeruca* can only be distinguished by the shape of the apex of the final abdominal sternite. In the males of the first two genera this is sinuous with a median lobe and in the females it is evenly rounded. In *Galeruca* the apex is deeply emarginate in the male and evenly rounded in females. In addition females of this genus when gravid show physogastry.

The sexes in *Phyllobrotica* differ in the following respects. The first tarsal segment of the fore-legs is dilated and as broad as segment three in males, whereas it is slender and narrower in the females. There are also distinct differences in the shape of the abdominal sternites 3, 4, and 5.

In *Galerucella tenella* (L.), *lineola* (F.), *pusilla* (Duftschmidt), *Pyrrhalta* and *Lochmaea* the mid-tibiae of the males bear a short spur whereas these are absent in the females. Males of *Galerucella calmariensis* (L.) have the hind tibiae armed with a spur. In this species the apical abdominal sternite is also sexually dimorphic. In addition in *Lochmaea* the males usually have the first segment of the metatarsus strongly dilated and the hind tibiae stouter and more sinuous than in the females. In *G. sagittariae* (Gyllenhal) tibial spurs are absent in both sexes and the only reliable difference in the sexes is the shape of the apex of the final abdominal sternite.

Galerucine species are easily separated except for those in *Galerucella*. Here resort has to be made to the dissection of the male aedeagi for accurate identification. Specimens are best mounted with the dorsal surface uppermost, except in *Galerucella* where at least one specimen should be mounted on its back since ventral characters are of value.

GALERUCINAE. ALTICINI – The 'flea-beetles' are small, dull or brightly-coloured beetles which have the ability of jumping. They often cause characteristic damage by shot-holing or completely perforating with numerous small holes the foliage on which they feed. This is particularly true of the genus *Phyllotreta* when they attack young brassicaceous plants. They usually occur in the spring, summer and autumn on low vegetation since the majority feed on herbaceous plants with the exception of *Altica brevicollis* Foudras and *Crepidodera* which are beaten from bushes and trees of *Corylus, Salix* or *Populus*.

The oval or elongate-oval, whitish, yellow or orange eggs are usually laid in the soil with the exception of *Altica, Phyllotreta nemorum* (L.), *Hippuriphila modeeri* (L.) and certain *Psylliodes, Mantura, Sphaeroderma* and *Apteropeda*, where they are deposited on the foliage. Those laid in the soil often have distinct hexagonal microsculpture on the chorion. Larvae are elongate and hatch by biting an escape hole in the egg or cut slits in the chorion using paired egg bursters, usually situated on the meso- and metathorax. The larvae of the majority of species feed in concealed positions within their host plants or in the soil, except for those of *Altica* which feed openly on foliage. Some mine leaves, others stems, but the majority live inside roots or at roots in the soil. There are usually three larval instars and pupation always occurs in cells in the soil.

Alticines are usually univoltine and the eggs apparently do not diapause except for certain *Longitarsus* species. The overwintering stage is usually the adult but occasionally the larva may also overwinter (*Sphaeroderma* – leaf-miner in thistles; *Noecrepidodera* – stem-miner in various plants; *Derocrepis* – root or stem-miner; *Podagrica* – root feeder; *Longitarsus gracilis* Kutschera, *flavicornis* (Stephens) and *luridus* (Scopoli) – root feeders). There is usually only one period of oviposition annually, either in Spring/early summer but in some species there is also oviposition in late summer/autumn (e.g. some *Longitarsus*) with some *Longitarsus* gravid in every month of the year. Brachypterous genera in the Alticini include *Derocrepis, Podagrica* and *Hermaeophaga* whereas the apterous genera are *Mniophila, Apteropeda* and *Batophila*. The genera *Longitarsus* and *Aphthona* include micropterous/ brachypterous/ macropterous species.

Adults are usually collected by sweeping and beating vegetation and secured with a pooter before they leap away! They may also be collected in Malaise traps (*Altica, Longitarsus*); flight interception traps, suction traps (*Chaetocnema concinna* (Marsham)); yellow water traps (*Phyllotreta*); in tussocks during winter (*Crepidodera, Chaetocnema, Sphaeroderma, Altica, Psylliodes* and *Longitarsus*) or by grubbing at the base of plants and shaking out short turf and moss (*Chaetocnema*).

Specific habitats of some of the more local/rare species are as follows:

Calcareous grassland (chalk or limestone, downland). *Aphthona* ?*atratula* Allard, *herbigrada* (Curtis); *Longitarsus agilis* (Rye), *curtus* (Allard), *anchusae* (Paykull), *dorsalis* (F.), *nigrofasciatus* (Goeze), *obliteratus* (Rosenhauer), *quadriguttatus* (Pontoppidan), *tabidus* (F.); *Batophila aerata* (Marsham) and *Mantura matthewsi* (Curtis). **Grassland.** *Aphthona nigriceps* (Redtenbacher); *Longitarsus anchusae* (Paykull), *ochroleucus* (Marsham), *parvulus* (Paykull), *pellucidus* (Foudras); *Psylliodes chalcomera* (Illiger) and *sophiae* Heikertinger. **Broadleaved woodland.** *Altica brevicollis* Foudras; *Crepidodera nitidula* (L.); *Epitrix atropae* Foudras; *Chaetocnema conducta* (Motschulsky) (N.E. Yorkshire only, doubtfully British) and *Mniophila muscorum* (Koch). **Saltmarsh.** *Longitarsus absynthii* Kutschera, *plantagomaritimus* Dollman; *Neocrepidodera impressa* (F.); *Chaetocnema sahlbergi* (Gyllenhal); *Podagrica fuscipes* (L.). **Wetland.** *Phyllotreta tetrastigma* (Comolli); *Longitarsus brunneus* (Duftschmidt), *ferrugineus* (Foudras), *holsaticus* (L.), *nigerrimus* (Gyllenhal) (S.England), *rutilus* (Illiger); *Lythraria salicariae* (Paykull)(England); *Chaetocnema aerosa* (Letzner) (England S. & S.E.), *arida* Foudras, *subcoerulea* (Kutschera) and *Apteropeda globosa* (Illiger). **Moorland.** *Altica longicollis* (Allard) and *ericeti* (Allard). **Arable land, disturbed ground, gardens, allotments.** *Phyllotreta punctulata* (Marsham), *cruciferae* (Goeze), *flexuosa* (Illiger), *striolata* (F.); *Psylliodes attenuata* (Koch) and *luteola* (Müller).

The specific identification in many genera is relatively straightforward, except in the large, intraspecifically variable *Longitarsus* and the small genus *Altica* (in which there are no reliable external characters). The spermatheca, notably the number of twists in the spermathecal duct, is especially important in *Longitarsus* for species recognition and in *Altica* for species-group recognition. The aedeagus is an extremely important specific character in both of these genera. Males are needed for accurate determination in *Altica*.

Sexing of beetles in these two genera, especially *Altica*, is important. In *Altica*, males usually with first segment of anterior tarsi wider than second; females with tarsal segments one and two sub-equal in width. In addition the males have the apical abdominal sternite with the posterior margin laterally incised, females with posterior margin entire, not laterally incised. In *Longitarsus* the first segment of the anterior tarsi is usually more strongly dilated in the male. However, sometimes this difference is not very obvious and if both sexes are not available for comparison it is difficult to decide which sex you have (especially in *L. succineus* (Foudras), *ochroleucus, membranaceus* (Foudras) and *parvulus* (Paykull)). In these cases the difference in the shape of the apical abdominal sternite is particularly useful. In addition the elytral shoulders in the males are often less strongly marked than in the females, but the antennae in some *Longitarsus* species are noticably longer in the males.

The antennae are modified in the males of several *Phyllotreta*. The fifth segment is very elongate and dilated in *P. ochripes* (Curtis) and *exclamationis* (Thunberg), the fourth and fifth are moderately dilated in *nemorum*, the third and fourth are strongly dilated in *nodicornis* (Marsham), whilst *consobrina* (Curtis) has the fifth segment elongate and slightly dilated.

LAMPROSOMATINAE. The small brassy adults of *Oomorphus concolor* (Sturm) are seldom seen openly on vegetation. Little is known concerning the biology of this species. Adults have been collected in all months of the year except March, with the greatest number of records for June. They probably overwinter, reappearing in April and ovipositing during May, June and July. Kolbe (1898) demonstrated that in this species hibernation is also undertaken in the larval stage. The adult fed upon the leaves of *Astrantia* and *Aegopodium* (Apiaceae) whereas the larva was polyphagous, feeding nocturnally on the stem and leaves of different plants on the ground and usually preferring the petioles. This work was substantiated by the observations of Kasap & Crowson (1976) who collected fully grown larvae at Morroch Bay, Scotland on 11th September 1974. The case-bearing larva fed nocturnally on the fresh green leaves and stems of *Hedera helix* (Araliaceae). Adults are best collected by evening sweeping, by grubbing at the roots of low plants, in water nets (October, November, December) or by beating in old ivy hedges. Adults and larvae are also extracted by collecting leaf litter and often detritus from *Hedera* and other vegetation. This species occurs chiefly in broadleaved woodland throughout England and Wales, but only rarely in Scotland. There are no obvious differences between the sexes.

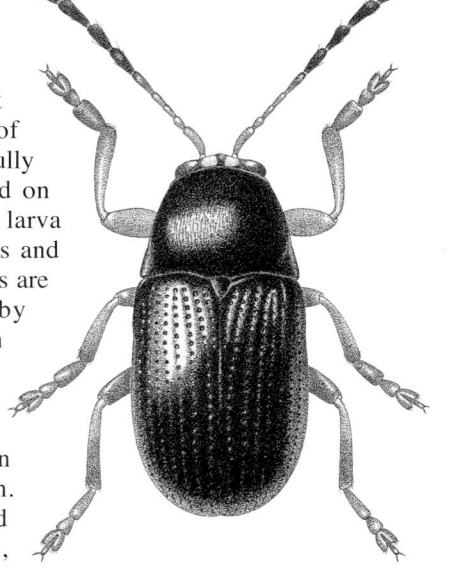

Fig.26 *Cryptocephalus exiguus* Schneider

CRYPTOCEPHALINAE. The adults of *Clytra, Labidostomis* and *Smaragdina* are usually encountered nibbling the leaves or young shoots of various bushes, including *Salix, Betula* and *Prunus*, or around the nests of ants, especially *Formica rufa* L. from April to October.

The eggs are covered with excrement by the females and either dropped to the ground (*Clytra, Smaragdina*) near the nests of *Formica rufa* or attached together by long pedicels of excrement to the leaves of bushes (*Labidostomis*). The ants move the eggs into their ant-hills possibly mistaking them for seeds or other plant debris. The larvae, after hatching, build a larval case onto the egg case and feed upon plant material in the ant-hill. The larval cases are very variable in form and furnish excellent generic characters. The larvae may be met with in ant-hills (*Clytra*) whilst *Labidostomis* larvae occur under stones near ant-hills, but not at the interior. Pupation occurs within the case but the entrance is completely closed by the final instar larva. According to Donisthorpe (1902) the cycle in *C. quadripunctata* (L.) is yearly and as eggs have an incubation period of about 20 days, the first instar larvae are likely to occur within the nest the third week of June. The larvae would probably feed during the summer, autumn and winter becoming fully developed the following spring, giving rise to a new generation during May and June.

Adults are best collected by beating bushes but those of *Clytra quadripunctata* have been taken in light traps. The adults of British clytrini are easily separated as they belong to three distinct genera. *Smaragdina affinis* (Illiger) is confined to the Wychwood area of Oxfordshire where it has been beaten from hazels in dedicuous forest (Holland, 1910; Atty, 1970). *Labidostomis tridentata* (L.) has been collected from young birches, usually in rough open ground in woodland during May, June and July (Massee, 1945); recorded mostly from eastern England, the most recent being the mid 1950's. *Clytra quadripunctata* is widespread and sometimes common, especially in woodland, throughout Britain. *Clytra laeviuscula* Ratzeburg is doubtfully British but there are possible old records for the Black Wood, Rannoch where it is possibly associated with *Formica sanguinea* Latreille and also chalk hills (Streatley, Berkshire, Chilterns) where it may be associated with *Lasius niger* (L.) and *L. alienus* (Förster) (Allen, 1976).

Sexing is fairly straightforward, especially in *Labidostomis* which has serrated antennal segments 5-10 broader in males, the mandibles, are longer and have a dorsal prominence, the anterior tibiae are longer and more curved in relation to the mid- and posterior tibiae, and tarsal segments 1 and 2 are longer in relation to the same segments on other legs. The females, as in *Clytra* and *Smaragdina*, have a deep median circular pit in the apical abdominal sternite. There is sometimes only a shallow broad smooth depression in the males of *Clytra*.

Cryptocephalus adults are variable in colour and may be almost entirely metallic green or blue; or entirely yellowish, black or brownish or a combination of these. They occur on the foliage of either bushes and trees or low growing herbaceous plants from June until September. During oviposition the female retains each egg between the posterior tarsi, turning them around and depositing on each a regular layer of excrement. They are then dropped to the ground where they may be collected by ants. Incubation of the eggs takes about three weeks and the larvae never abandon their cases but enlarge them by adding new pieces made from excrement. There is much controversy over the larval food of *Cryptocephalus* species, but in captivity *C. labiatus* (L.), *pusillus* F. and *parvulus* Müller larvae took *Betula* leaves although *parvulus* seemed to prefer leaves which had turned brown and had a fungus infection. There are four or more instars and before pupation the larvae close up their cases and turn to face the opposite end of the case. Pupation usually occurs with the case attached to the bark of trees or on the ground. Adults emerge by removing a small cap from the larval case about 1mm from the anterior end.

Cryptocephalus species overwinter as larvae and may complete their development the year following oviposition (e.g. *C. pusillus*). However, some species (e.g. *C. parvulus* and *C. labiatus*) probably require two years to achieve development to the adult. Larvae (probably of *C. pusillus)* may feed on leaf litter as three larvae (one penultimate and two final instar) were extracted from oak leaf litter on March 25th 1980. Donisthorpe (1908) suggested that they feed on lichens on trees. On April 16th, 1910 he found a larva of *C. fulvus* in a nest of *Lasius fuliginosus* (Latreille), the adult emerged on June 8th-9th and was eaten by ants. *C. sexpunctatus* (L.) has been recorded near a nest of *Formica rufa* L., and it was suggested that all *Cryptocephalus* species pupate within ant nests, but this is unproven and highly unlikely.

Cryptocephalus species occur in the following general habitat types (N.B. some species occur in more than one type of habitat): **Calcareous grassland** – *aureolus* Suffrian, *bilineatus* (L.), *bipunctatus* (L.), *hypochaeridis* (L.), *labiatus, moraei* (L.), *nitidulus* F., *primarius* Harold and *pusillus*. **Broad-leaved woodland** – *aureolus, bipunctatus, coryli* (L,), *decemmaculatus* (L.), *frontalis* Marsham, *fulvus, hypochaeridis, labiatus, moraei, nitidulus, parvulus, punctiger* Paykull, *pusillus, querceti* Suffrian, and *sexpunctatus*. **Wetland (bog, fen, carr)** – *biguttatus* (Scopoli), *exiguus, frontalis, labiatus, pusillus* and *?fulvus*. **Heathland** – *biguttatus, bipunctatus, fulvus, labiatus, moraei, parvulus* and *pusillus*. **Coastal sand dunes** – *aureolus, bilineatus, fulvus, labiatus* and *pusillus*.

More specific habitats of some rare species are as follows: *C. biguttatus* – swept from cross-leaved heath *(Erica tetralix)* and heather *(Calluna vulgaris)* on boggy heaths and moors. *C. coryli* – beaten from young birch and oak. *C. decemmaculatus* – beaten from dwarf sallows and birch. *C. exiguus* – beaten from *Betula* and *Salix cinerea* but the larval host plant is *Rumex*. *C. nitidulus* – beaten from young birches. *C. primarius* – swept from warm sheltered dry hillsides with grasses, *Hieraceum* and *Helianthemum*. *C. querceti* – oaks and hawthorn. *C. sexpunctatus* – from hazel, birch, aspen and crack-willow.

Sexing is readily achieved as all female *Cryptocephalus* have a deep median circular depression in the apical abdominal sternite whilst this is absent in the males. In addition the antennal segments are usually more elongate in the males, especially notable in *pusillus*. The males of this species often have the elytra dark-brown or black, whereas in the females they are yellow. Colour is also sexually dimorphic in *coryli* in which the pronotum of the males is black whilst in the females it is orange. There are also modifications of the abdomen in some species. In males of *sexpunctatus* abdominal sternite 2 bears a pair of lateral processes.

Cryptocephalus species are easily identified using colour, elytral plus pronotal puncturation, antennal and leg characters making resort to genital dissection unnecessary. Specimens are best mounted with the dorsal surface uppermost and with legs and antennae extended.

EUMOLPINAE. This subfamily is represented in Britain solely by the rare *Bromius obscurus* (L.). The dull black or brownish, pubescent adults occur on various willow-herbs, *Epilobium* spp. and especially the Rosebay Willow-herb *Chamaenerion angustifolium* (L.) Scop. On the Continent it was recognised as a pest of the grape vine *Vitis vinifera* L. as early as the 15th century.

According to Balachowsky (1963) this species is parthenogenetic and only the female is known. However, this is not true for N. American populations, in which males commonly occur (pers. observation). Adults start to emerge from the soil during May. Oviposition commences in early June and continues up to August and adults may survive for nearly three months. The bright yellow, 1.0 x 0.5mm ova are laid in batches of 20-40 either in the soil near the host plants or at the base of the stem, slightly above the root neck under the old sloughed-off epidermis. The white slightly 'C'-shaped larvae feed in groups on the roots, decorticating them and thus removing the epidermis and even sometimes the superficial wood. They eat either linear or irregular incisions, the latter resembling the damage caused by certain scarabaeid larvae. The larvae develop during the summer and early autumn and penetrate deep into the soil to overwinter. Pupation occurs in an earthen cocoon during March or early April and the adults emerge 20-30 days later. The adults feed on the foliage of the host plant.

This species seems to prefer light sandy soil alongside rivers. It is known as a single colony from only one site in Cheshire where it was first discovered during 1970. It is possibly a relict species since it was apparently very common in Britain during the mild phase about 11,000-12,000 B.P. [Before Present].

Stephens (1831) is perhaps the only British author to include this amongst the reputed British species. He refers to a specimen in the BM(NH) collection which was said to have been taken in Lincolnshire, but no subsequent record has since been reported.

The males and females are extremely difficult to distinguish with only a slight difference in the shape and puncturation of the apical abdominal sternite.

References

Allen, A.A., 1976. Notes on some British Chrysomelidae (Col.) including amendments and additions to the list. *Entomologist's Record & Journal of Variation*, **88**:220-225 & 294-299.

Atty, D.B., 1970. Gloucestershire beetles: a few records and an appeal. *Entomologist's Monthly Magazine*, **105**:199.

Balachowsky, A., (ed.) 1963. *Entomologie Appliquée a L'Agriculture*. Tome I. Coléoptères. (second volume). Masson et Cie., Paris.

Betts, M.M., 1955. The food of titmice in oak woodland. *Journal of Animal Ecology*, **24**:282-323.

Betts, M.M., 1956. A list of insects taken by titmice in the Forest of Dean (Gloucestershire). *Entomologist's Monthly Magazine*, **92**:68-71.

Champion, G.C., 1910. Some interesting British Insects (ii). *Entomologist's Monthly Magazine*, **46**:2.

Cox, M.L., 1981. Notes on the biology of *Orsodacne* Latreille with a subfamily key to the larvae of the British Chrysomelidae (Coleoptera). *Entomologist's Gazette*, **32**:123-135.

Cox, M.L., 2000. Progress report on the Bruchidae/Chrysomelidae recording scheme. *The Coleopterist*, **9**(2):65-74.

Cox, M.L., 2001. Notes on the natural history, distribution and identification of seed beetles (Bruchidae) of Britain and Ireland. *The Coleopterist*, **9**(3):113-147.

Donia, A.R., 1958. *Reproduction and reproductive organs in some Chrysomelidae (Coleoptera)*. Unpublished Ph.D. Thesis, London University.

Donisthorpe, H.St.J.K., 1902. The life history of *Clytra quadripunctata* L. *Transactions of the Entomological Society of London*, **35**:11-24 + 1 pl.

Donisthorpe, H.St.J.K. 1908. A few notes on Cryptocephali. *Entomologist's Record & Journal of Variation*, **20**:208-209.

Hodge, P.J., 1997. *Bruchidius varius* (Olivier) (Chrysomelidae) new to the British Isles. *The Coleopterist*, **5**(3):65-68.

Kasap, H. & Crowson, R.A., 1976. On systematic relations of *Oomorphus concolor* (Sturm) (Col., Chrysomelidae), with descriptions of its larva and of an aberrant cryptocephaline larva from Australia. *Journal of Natural History*, **10**:99-112.

Kevan, D.K., 1962. The British species of the genus *Haltica* Geoffr. *Entomologist's Monthly Magazine*, **98**:189-196 (43 figs).

Kevan, D.K., 1967. The British species of the genus *Longitarsus* Latreille. *Entomologist's Monthly Magazine*, **103**:83-110.

Kolbe, W. 1898. *Lamprosoma concolor* Sturm in biologischer Beziehung. *Z. Ent.* **23**:22-29.

Massee, A.M. 1945. Abundance of *Labidostomis tridentata* L. (Col., Chrysomelidac) in Kent. *Entomologist's Monthly Magazine*, **81**:164-165.

Menzies, I.S. & Cox, M.L., 1996. Notes on the natural history, distribution and identification of British reed beetles. *British Journal of Entomology and Natural History*, **9**:137-162.

Shute, S.L., 1975a. *Longitarsus jacobaeae* Waterhouse (Col., Chrysomelidae): identity and distribution. *Entomologist's Monthly Magazine*, **111**:33-39.

Shute, S.L., 1975b. The specific status of *Psylliodes luridipennis* Kuts. *Entomologist's Monthly Magazine*, **111**:123-127.

Siew, Y.S. 1965. The endocrine control of adult reproductive diapause in the chrysomelid beetle *Galeruca tanaceti* L. *Journal of Insect Physiology*, **11**:1-10.

British Key Works to Larval Identification

Cox, M L., 1981. Notes on the biology of *Orsodacne* Latreille with a subfamily key to the larvae of the British Chrysomelidae (Coleoptera). *Entomologist's Gazette*, **32**:123-135.

Cox, M.L., 1982. Larvae of the British genera of chrysomeline beetles (Coleoptera, Chrysomelidac). *Systematic Entomology*, **7**:297310.

Cox, M.L., 1991. The larvae of the british *Phaedon* (Coleoptera: Chrysomelidae, Chrysomelinae). *Entomologist's Gazette*, **42**:267-280.

Cox, M.L., 1998. The genus *Psylliodes* Latreille (Chrysomelidae: Alticinae) in the UK. *The Coleopterist*, **7**(2):33-65.

Emden, F.I. van., 1962. Key to species of British Cassidinae larvae. *Entomologist's Monthly Magazine*, **98**:33-36 + 10 figs.

Mann, J.S. & Crowson, R.A., 1981. The systematic positions of *Orsodacne* Latreille and *Syneta* Lac. (Col., Chrysomelidae), in relation to characters of larvae, internal anatomy and tarsal vestiture. *Journal of Natural History*, **15**(5):727-749.

Marshall, J.E., 1979. The larvae of the British species of *Chrysolina* (Chrysomelidae). *Systematic Entomology*, **4**:409-417.

Marshall, J.E., 1980. A key to some larvae of the British Galerucinae and Halticinae (Coleoptera: Chrysomelidae). *Entomologist's Gazette*, **31**:275-283.

CURCULIONOIDEA (WEEVILS)
(Plates 100 – 104)

by M. G. Morris

In world terms the Curculionidae constitute the largest family in the animal kingdom, based on the number of recognised species. Even in temperate regions, where they are outnumbered by the Staphylinidae, the group is very speciose, contributing substantially to their biodiversity. This abundance is the result of almost exclusive exploitation of one type of feeding strategy – the use of plants and plant material. Within this overall plan there is every type of relationship with the substrate or food that is utilised. Many weevils feed on a single plant species, or on those in one genus or family. These beetles may be described as stenophagous, with more restrictive terms (e.g. monophagous, oligophagous) for the various degrees of specialisation.

The dependence of weevils on plants has several implications for students of the group. In its most fundamental aspect, study of weevils involves working vegetation. It is convenient to consider this at three levels, assuming that access to the high canopy of trees is unlikely to be achieved frequently. Branches of trees and shrubs which can be reached from the ground may be worked by beating, using the conventional tray. Herbaceous vegetation and lower branches may be swept using a sweep net, though better results are often achieved by tapping such vegetation and holding the open net under the plants. At ground level weevils may be sought using the activity variously known as grubbing, grovelling or hand-collecting. For the last type of habitat, vacuum samplers are increasingly being used and are often very effective. The usual pattern of powered sampler is based on modified garden leaf-collectors, as described later see p. 215. Many weevils are fully aquatic and so may be found using a water-net, or by shaking or sieving water-weed or other aquatic vegetation.

None of these methods of finding weevils is specific to the group, of course, and species in other families will be found as well. What may apply only to weevils is the knowledge that particular species are associated with their own special host plants. Here, as with other groups of phytophagous beetles, such as *Meligethes* and Chrysomelidae, a working acquaintance with the species of vascular plants is an immense advantage. It is not an exaggeration to say that the successful student of weevils must be a competent field botanist. As well as being an aid to finding particular species, knowledge of host plants can be a valuable help in identification. For example, an '*Apion*' shaken from a clump of Dyer's Greenweed (*Genista tinctoria*) will almost certainly be *Exapion difficile* (Herbst). In the following account the names of vascular plants are those in Stace (1997). The *New Atlas of the British and Irish Flora* (Preston, Pearman & Dines, 2002) will be found invaluable in showing whether a food-plant is likely to be present in any particular area of the country.

Weevil life-histories are varied, but some general principles can be stated. Species which feed on aerial structures of particular plants usually have one generation a year and can be found as adults during winter, spring and autumn. This pattern can be modified if the weevil larvae feed in structures which occur early in the year (e.g. catkins) or in late-developing fruits and seeds. Some root-feeding species have similar types of life-history, but others have a long duration of larval life and occur as adults for relatively short periods in spring and early summer. Another group with short adult lives is the genus *Magdalis*; the larvae live for a long time under the bark of various trees, but there is a single generation a year. Other wood-feeders, such as Cossoninae, may be found in all developmental stages at any time of the year. Very few weevils have more than one generation a year, and those that do generally feed as larvae in or on long-lasting structures such as leaves. Some few species appear to have variable life-histories within the same population, or at least within the same area. Others vary in their times of appearance in relation to latitude, becoming adult later in northern Scotland than in southern England.

Most students of weevils are not overly concerned with their higher classification, which is just as well as there is little agreement among specialists working on the subject. This section of the *Handbook* follows the recent check list of Morris (2003). The arrangement of Lawrence & Newton (1995) is often followed, as it covers the whole of the Coleoptera, but has been superseded, as far as weevils are concerned, by the comprehensive world catalogue of weevil genera produced by Alonso-Zarazaga & Lyal (1999). This arrangement is followed by Morris (2003) but differs considerably from the more traditional treatment of Pope (1977) and the idiosyncratic classification of Joy (1932).

NEMONYCHIDAE-DOYDIRHYNCHINAE. There is only one British species in this group. *Cimberis attelaboides* F. is associated with Scots Pine, larvae feeding in the male flowers. The weevil is local but widely distributed and can be abundant in parts of the Scottish Highlands.

ANTHRIBIDAE-ANTHRIBINAE. None of our few anthribine species is very common and one or two are very rare. Some species are associated with dead wood, and may be worked for by setting faggot traps, or bundles of dry wood, in areas favoured by the species, and shaking or beating the faggots over a sheet. *Platyrhinus resinosus* (Scopoli) feeds on the fruiting bodies of the large black fungus *Daldinia concentrica* (King Alfred's Cakes) which grows mainly on dead ash but occasionally on other species of tree; this anthribid is one of the less rare species. Our two species of *Anthribus* are unusual in parasitizing scale insects (Coccidae etc.) on various trees. It is difficult to go out and work for Anthribinae and specimens fall to the collector by lucky chance in most cases.

ANTHRIBIDAE-CHORAGINAE. Our two species in this group contrast with each other. *Choragus sheppardi* Kirby is a small, uncommon weevil associated with dead Ivy, *Hedera helix*, while *Araecerus fasciculatus* (De Geer) is a cosmopolitan stored products pest which should be looked for in disused granaries and similar places.

ANTHRIBIDAE-URODONTINAE. Only one species in this group is known to inhabit the British Isles. *Bruchela rufipes* (Olivier) is a recently rediscovered weevil that seems to be spreading, particularly in eastern England. It is associated with Wild Mignonette, *Reseda lutea*, in the fruits of which the larvae develop.

RHYNCHITIDAE. Although we have only a small number of species the British Rhynchitidae include some particularly

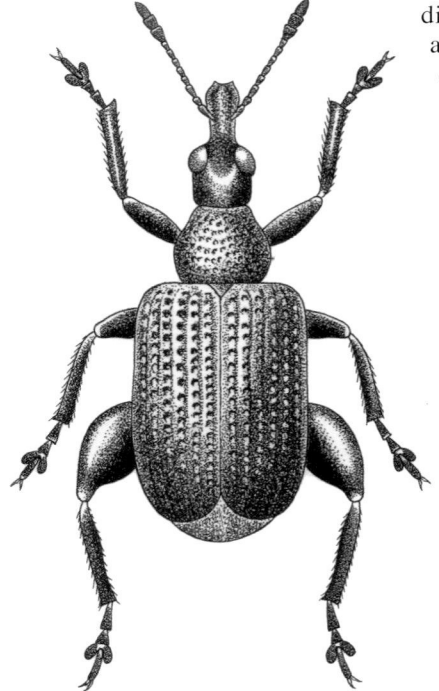

Fig.27 *Deporaus betulae* (L.)

attractive and interesting species. All are fairly specific in their plant hosts, but the feeding habits of the larvae and behaviour of egg-laying females are very varied. One well-known method of larval feeding, unique in our fauna to the rhynchitids (and attelabids), is for the ovipositing female to construct a 'cradle' from a leaf or leaves of the host, usually a tree. By rolling up the leaf lamina after laying an egg on it the female provides protection and nutriment for the developing larva. The conical, pendant leaf-rolls of the common *Deporaus betulae* (L.) are often abundant on Birch, Alder and Hazel in May and June. Other leaf-rollers include our two species of *Byctiscus*, both of which are rare, with *B. populi* (L.) a 'priority species' under the United Kingdom's Biodiversity Action Plan. *B. populi* is associated with poplars, particularly Aspen, and the less rare *B. betulae* (L.) with various trees, most commonly Hazel.

Other Rhynchitidae are bud-, petiole- and twig-feeding species as larvae, and some have very short periods of adult life. The most obvious of these is *Neocoenorrhinus aeneovirens* (Marsham), the larvae of which develop in buds of Oak. Adults are rarely found except in the months of April and May. *Deporaus mannerheimi* (Hummel) has larvae which develop in the petioles of Birch, but the adults are not so restricted in time, though more usually taken in late summer than at other times. Most Rhynchitidae do not overwinter as adults, but an exception is *Involvulus caeruleus* (De Geer). Known to fruit-growers as 'twig-cutting weevil', it is a minor pest of apple trees, but also attacks Hawthorn and other rosaceous trees and shrubs. The twigs are half-cut through by females after an egg has been laid in the severed portion. This eventually falls to the ground where the larva completes its development in the dead tissue.

Most of our Rhynchitidae are arboreal and adults are most easily obtained by beating appropriate trees and shrubs. However, because some species are large and brightly coloured they can be 'spotted' as they sit on foliage; examples are the metallic blue (male) or green (female) *Byctiscus betulae*. A non-arboreal species is the common *Neocoenorrhinus germanicus* (Herbst). It is associated with herbaceous Rosaceae, with females laying their eggs in the shoots, runners and stolons; the weevil can be a minor pest of strawberry plants.

ATTELABIDAE. The two British species in this group are both leaf-rollers, *Attelabus nitens* (Scopoli) producing distinctive, tightly-rolled cradles on Oak, and *Apoderus coryli* (L.) ones of different construction on Hazel. The latter species is a bizarre-looking red and black beetle which the beginner may not immediately identify as a weevil at all. Neither it nor its leaf-rolls are uncommon in southern England, where it can have two generations a year.

APIONIDAE. This is the most species-rich group so far considered, with 86 species recorded from the British Isles. Despite the number of species and their small size they are not very difficult to identify, many being distinctive. However, in some cases the two sexes differ considerably, so that this has to be borne in mind when attempting determination. As so often with weevils, the host plants are a valuable clue to identification. By working clovers, vetches and other Fabaceae, docks and knotgrasses (Polygonaceae), mallows (Malvaceae) and selected species of Asteraceae the beginner will accumulate many of the common apionids and perhaps some more local species as well.

Some hosts in other plant families support only one species of Apionidae. *Aizobius sedi* (Germar) feeds on species of *Sedum*, principally *S. anglicum* and *S. acre*. The adult weevils are most usually taken by 'grubbing', as the low-growing food-plants cannot easily be swept. Each of our species of Mercury is host to a species of *Kalcapion* and notable for being poisonous. The widely-distributed, but local, *K. pallipes* (Kirby) feeds as a larva in the stems of Dog's Mercury (*Mercurialis perennis*) and *K. semivittatum* (Gyllenhal) has the same habit on Annual Mercury (*M. annua*). The latter weevil was once considered rare, but appears to have increased its range in southern England in recent years. A closely related species, *Taeniapion urticarium* (Herbst), feeds on common Stinging Nettle (*Urtica dioica*) but is more local than its host and tends to occur on light soils such as sands, peats and limestones.

Species in the genus *Squamapion* feed on Lamiaceae: *S. atomarium* (Kirby) on *Thymus*, *S. flavimanum* (Gyllenhal) on *Origanum* etc., *S. cineraceum* (Wencker) on *Prunella* and *S. vicinum* (Kirby) on *Mentha* species. None of these weevil species is common, with the possible exception of *S. atomarium*.

Only three Apionidae are associated with trees in Britain, and indeed the arboreal habit is unusual in the group, at least in Europe. *Betulapion simile* (Kirby) feeds as a larva in the catkins of Birch and is of interest as being one of our Holarctic species, occurring in North America as well as Europe. *Melanapion minimum* (Herbst) is a rarer species, and indeed a priority one under the UK Biodiversity Action Plan. It has an unusual type of host, larvae being inquilines (harmless guests) in the galls induced by sawflies (Hymenoptera-Symphyta), and possibly some Diptera, on Sallows (*Salix* spp.). Quite recently another arboreal apionid has been discovered to occur in Britain. *Ixapion variegatum* (Wencker) feeds in

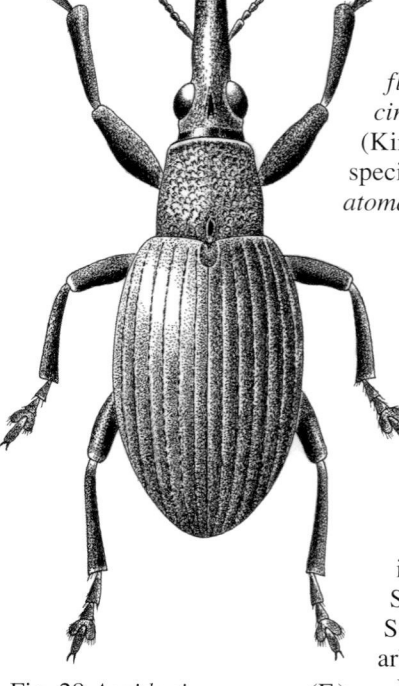

Fig. 28 *Aspidapion aeneum* (F.)

the stems of Mistletoe (*Viscum album*), growing as a parasite particularly on apple trees. The weevil has been found in the West Midlands where the host is particularly common. All our arboreal Apionidae are most easily collected by beating their hosts.

Many Apionidae are associated with Fabaceae. Species of *Protapion* feed as larvae mainly in the seed-heads of clovers (*Trifolium* spp.). Several of these weevils are very common and can be an introduction to learning about the whole group. Adults may be reared from collected seed-heads of the hosts. Some of the species in this genus are sexually dimorphic, the males differing considerably from the females; *P. dissimile* (Germar) and *P. difforme* (Germar) are particularly extreme examples. There are also legume feeders in several other genera of Apionidae. For instance, our two species of *Ischnopterapion* are often abundant on bird's foot trefoils (*Lotus* spp.).

Because Apionidae are restricted to particular hosts the best way of finding them is to work these plants. Some quite local species, such as Wild Liquorice (*Astragalus glycyphyllos*) or *Genista* spp. support uncommon apionids. Morris (1990) gives the hosts of all our Apionidae, as well as keying the species (giving attention to differences between the sexes), but the nomenclature is rather out of date. Since its publication, as well as *Ixapion variegatum* another species, *Helianthemapion acicularis* (Germar), has been added to the British list, showing that there is still scope for making new discoveries in our weevil fauna.

NANOPHYIDAE. Two species in this group, which is not always recognised as a separate family from Apionidae, are found in the British Isles. *Nanophyes marmoratus* (Goeze) is a common species, limited only by the distribution of its foodplant, Purple Loosestrife (*Lythrum salicaria*). *Dieckmanniellus gracilis* (Redtenbacher) is much more local, though often abundant where it occurs. The host is Water Purslane (*Lythrum portula*), an obscure little plant which grows in wet places and provides a challenge for coleopterists wishing to find the weevil. However, *D. gracilis* is probably commoner than published records suggest.

DRYOPHTHORIDAE. This is a small group in Britain which has only recently been given family rank. It includes cosmopolitan stored products pests of the genus *Sitophilus*, of which we have three species. Nowadays they are seldom found, but may be sought in old granaries and similar places. *Dryophthorus corticalis* (Paykull) is our only other representative of the family. It is a rare, wood-inhabiting weevil that was formerly included in the Cossoninae.

ERIRHINIDAE. This group has also only recently been classified as a family. It has been shown that two fundamentally distinct kinds of weevils were previously included in the curculionid subfamily Erirhininae. A restricted number of these constitute the separate family, though Morris (2002) retains the group as a subfamily, for practical reasons. These weevils are mostly waterside species. *Notaris*, *Tournotaris* and *Thryogenes* include species whose biology is imperfectly known, though most of the host plants recorded for them are bur-reeds, sedges and club-rushes. The weevils may be found in marsh litter and in tussocks on marshy ground as well as by sweeping. *Grypus equiseti* (F.) feeds on horsetails and is most usually found by grubbing at their roots. However, at evening, particularly after wet weather, the weevils often ascend their hosts and can be conspicuous. *Stenopelmus rufinasus* Gyllenhal, introduced accidentally from North America, is a poorly-adapted aquatic species which feeds on the Water Fern (*Azolla filiculoides*) and is well established in southern England. Two species of *Procas* are the most terrestrial erirhinid weevils. Their biology is very imperfectly known, though *P. granulicollis* Walton is usually taken where Climbing Corydalis (*Ceratocapnos claviculata*) is growing. This weevil was thought to be a British endemic species, but this is now known not to be the case.

RAYMONDIONYMIDAE. This is another group formerly classified within the Curculioninae in the broad sense, but which is now regarded as constituting a separate family. The species included are very interesting, small, eyeless, soil-inhabiting weevils. We have only one species in Britain, *Ferreria marqueti* (Aubé), and this was discovered only relatively recently, through the employment of specialised pitfall traps set deep in the soil.

CURCULIONIDAE. This family includes most of the beetles classified as weevils, i.e. the superfamily Curculionoidea. In the following account the system of subfamilies adopted by Morris (2003), following Alonso-Zarazaga & Lyal (1999) will be followed. The arrangement of Lawrence & Newton (1995) has had no currency in the British literature, either traditional or recent, and to avoid confusion references to their system will be kept to a minimum.

CURCULIONINAE. One beetle which many naturalists would quote as the typical weevil is *Curculio nucum* L., the Nut Weevil. This species, along with four others, gives its generic name to the whole superfamily. They, and three closely related species in *Archarius* (formerly *Balanobius*), have particularly long and slender rostra which are used for boring into nuts, acorns and other structures. In the case of species of *Archarius* these structures are the galls induced by various Hymenoptera in which the weevil larvae feed as inquilines. *A. salicivorus* (Paykull), feeding in the galls of sawflies on sallows, and *A. pyrrhoceras* (Marsham), which inhabits 'currant' galls on Oak, are both common species.

We have 13 species of *Anthonomus* in Britain (including one in subgenus *Furcipus*, often regarded as distinct) and the closely related *Brachonyx*. Most of these species are arboreal, the exceptions being *A. rubi* (Herbst), a very common species on herbaceous Rosaceae, and *A. brunnipennis* (Curtis), which is predominantly northern, with Tormentil (*Potentilla erecta*) one common host. Hawthorn, Blackthorn, fruit trees, Elm and Scots Pine are the main hosts of the arboreal species, the larvae feeding in flower, fruit and vegetative buds and in 'capped' blossoms. Because buds are attacked many species are active in winter. For example *A. bituberculatus* Thomson is common by beating the bare branches of Hawthorn from late November to March and is less frequently taken at other times. Care is needed in the identification of *Anthonomus* species, the host trees once again being a useful guide. In some old literature keys to the species are often not satisfactory, with some being omitted or included only as 'varieties'.

Cionus and *Cleopus* species are among our most familiar weevils. Like *Hypera* (dealt with later) they have ectophagous larvae, but these are very different in appearance from the caterpillar-like ones of *Hypera*. Feeding on Scrophulariaceae, particularly species of *Scrophularia* and *Verbascum*, they resemble small slugs, being covered in protective slime. The slime converts to a paper-like cocoon when the larvae pupate and adults are easily reared from collected cocoons. The adults of *Cionus scrophulariae* (L.) and *C. hortulanus* (Fourcroy), two of our commonest species, resemble the seed-capsules of *Scrophularia nodosa* (in particular).

Our single species of *Acalyptus*, which is placed in a separate tribe within Curculioninae, is arboreal, feeding as a larva in the buds of sallows. Also in a separate curculionione tribe (but very recently found to be molytine – a group dealt with below) are the two British species of *Anoplus*. Associated with Birch and Alder, their larvae mine the leaves of their hosts.

All the species of Ellescini, another curculionine tribe, feed on poplars or sallows (willows). Larvae of species of *Dorytomus* feed in the male or female catkins; consequently they are easily reared if damp sand or soil is provided for them to pupate in. Timing of the life-history depends on the appearance of the catkins of the host tree, which is usually early in the year. In winter many of the species can be found hibernating under bark or in hollow trees. *D. taeniatus* (F.) is the commonest species, feeding on *Salix* spp., while *D. tortrix* (L.) and *D. dejeani* Faust are the Aspen feeders which are most frequently encountered. Whereas *D. tortrix* oviposits in autumn and chooses only male catkin buds, *D. dejeani* waits until late winter or early spring and utilises both male and female ones, but mainly the latter. The rare *D. affinis* (Paykull), also an Aspen feeder, oviposits exclusively in female catkins or their buds.

Many of the British Mecinini are seed feeders. Species of *Miarus* and *Cleopomiarus* are associated with Campanulaceae. Because plants of this family often flower in late summer, *M. campanulae* (L.), for instance, is frequently found in association with other plants earlier in the year; it seems to favour yellow-flowered Asteraceae and rockroses and may eat their pollen. Species of *Mecinus* are associated mainly with plantains, some feeding as larvae in stems, others in fruits. *M. collaris* Germar may be reared from inconspicuous galls induced below the flower heads of Sea Plantain (*Plantago maritima*), but the larvae sustain high rates of parasitism. Most of our species of *Gymnetron* feed on Scrophulariaceae, particularly speedwells. The genus formerly included one of our commonest weevils, *G. pascuorum* (Gyllenhal), but this species has recently been transferred to *Mecinus*, following phylogenetic study of the group.

Most coleopterists know species of Rhamphini (=Rhynchaeninae) because of their ability to jump, indeed the common name for this group is 'flea weevils'. They are also notable for the larval leaf-mining habit, and as a group in which stridulation (by means of a file on the underside of the elytra and a scraper on the abdomen) has been studied. Most British species are arboreal, and we have ones associated with Oak, Elm, Hazel, Birch, Alder and Sallow. By far the commonest is the Beech weevil, *Orchestes fagi* (L.) (placed in *Rhynchaenus* in previous classifications). The larval mines can be very numerous and the adults very abundant. Only one British rhamphine is associated with a herbaceous plant; *Pseudorchestes pratensis* (Germar) mines the leaves of Common Knapweed (*Centaurea nigra*) and is rather local. We have three species of *Rhamphus*, one being a very recent discovery as a British species. They are minute jumping insects which need looking for, not because they are rare, but because of their small size and active habits. *R. oxyacanthae* (Marsham) is associated mainly with Hawthorn, *P. pulicarius* (Herbst) with Sallow, Birch and Bog-myrtle (*Myrica gale*) and *R. subaeneus* Illiger (the recent addition) with Apple.

Species of *Smicronyx* are associated with dodders (*Cuscuta* spp., Convolvulaceae) and Gentianaceae, mostly Common Centaury (*Centaurium erythraea*).

Orthochaetes are leaf-miners as larvae and mainly ground-living insects; both our species are parthenogenetic. *O. setiger* (Beck) is a common species but seldom collected if only beating and sweeping are used. Vacuum collecting has shown how abundant it can be, especially on chalk downland. Included with *Orthochaetes* in the tribe Styphlini is the rare *Pseudostyphlus pillumus* (Gyllenhal).

Species of *Tychius* feed exclusively on Fabaceae, while *Sibinia* spp. are associated with Caryophyllaceae. The larvae of both genera are seed-feeders. *T. picirostris* (F.) is the commonest species of *Tychius* and is easily recognisable (under the microscope) by the six-segmented antennal funiculus; all our other species have seven segments. Most *Tychius* are small weevils, *T. pusillus* Germar particularly so. *T. parallelus* (Panzer) is one of the largest; it feeds on Broom (*Cytisus scoparius*) and has a curiously disjunct distribution, being found locally in southern England and East Anglia and again in north-east Scotland. None of our species of *Sibinia* is very common and all are southern in their distribution. *S. pyrrhodactyla* (Marsham) is a weevil of particularly unpredictable occurrence as it tracks its foodplant, Corn Spurrey (*Spergula arvensis*) a ruderal plant of disturbed sandy ground that rarely persists for long in one place.

BAGOIINAE. Species of the interesting genus *Bagous* are all semi-aquatic and feed on water plants. Most are rare (*B. limosus* (Gyllenhal) and *B. alismatis* (Marsham) the least so) and at least three species on the British list are almost certainly extinct here. The extant ones can be taken by collecting water-weed, wringing it out to get rid of most of the water, and sieving or shaking it over a waterproof sheet. Alternatively, the weevils may be heat-extracted from the vegetation, using an ordinary desk lamp or other source of heat. In many of the areas where bagoines occur, particularly in drainage ditches, the vegetation is periodically cleared out, or roded. This vegetation can be worked for these and other aquatic weevils and on warm, sunny days the *Bagous* may be seen crawling back to the water. Bagoinae appear to be a relict group across Europe. Some species seem to be sporadic or uncertain in their appearance and much patience is needed to collect more than a few of the less rare ones.

BARIDINAE. This is a group which has been variously treated in higher classifications. In the British Isles there are only seven species, but the group is very speciose in other parts of the world. Only five species of *Baris* have been reliably recorded from Britain and they are found predominantly in southern England, where they feed on various dicotyledonous plants; *B. picicornis* (Marsham), for example, is fairly common on Wild Mignonette (*Reseda lutea*). In contrast, our two species of *Limnobaris* are associated with sedges and, though local, are found over a much wider area of our country.

CEUTORHYNCHINAE. In Britain this subfamily is one of the most speciose and diverse (in the associations and biology of the species, not their general appearance). Nearly all the species are closely tied to particular plants. In recent years the large genus *Ceutorhynchus* (s.lat.) has been 'split' and the constituent genera are distinguished by their host-plant relationships as well as by morphology.

Mononychus punctumalbum (Herbst) is a distinctive and taxonomically somewhat isolated species in that it feeds as a larva in the seed-pods of our native *Iris* spp., both Yellow Iris (*I. pseudacorus*) and Stinking Iris (*I. foetidissima*). It is most readily found by examining these pods in late September and October, but is very local, being practically confined to the coasts of Dorset and the south of the Isle of Wight.

Few ceutorhynchines are arboreal. *Rutidosoma globulus* (Herbst), a rare species, is associated especially with Aspen, while our three *Coeliodes* are found on Oak and our two *Coeliodinus* on Birch. Many of the species of *Zacladus*, *Stenocarus*, *Micrelus*, *Microplontus*, *Datonychus*, *Trichosirocalus* and *Ceutorhynchus* etc. can be found by general sweeping. However, this is a poor substitute for working the known food-plants of the species. All our *Ceutorhynchus* are associated with Brassicaceae, with the exception of *C. resedae* (Marsham) which feeds on the closely related Weld (*Reseda luteola*, Resedaceae). Many species are attached to only one food-plant, so that a good knowledge of this plant family is particularly valuable in finding the less common species of weevil. Not surprisingly, it is the very common species that tend to have the widest range of hosts; examples are *C. typhae* (Herbst) (formerly *floralis* (Paykull)), *C. pallidactylus* (Marsham) and *C. minutus* (Reich) (previously *contractus* (Marsham).

Many Ceutorhynchinae are waterside or aquatic species. Most aquatic of all is *Eubrychius velutus* (Beck), which can spend its whole life under water. It is beautifully adapted, both morphologically and physiologically, for the aquatic life. It should not be passed over when netting water beetles, but can also be found by sieving water-weed, particularly if the hosts, milfoils (*Myriophyllum* spp.) are present. Many of the semi-aquatic ceutorhynchines can swim on the surface of water, or underneath it. However, most of them are most readily found by working their known hosts in the usual way. Examples are the species of *Phytobius* (milfoils), *Rhinoncus* (Polygonaceae), *Pelonomus* (a range of different food-plants), *Amalorrhynchus* and *Drupenatus* (watercresses), *Tapeinotus* (Yellow Loosestrife, *Lysimachia vulgaris*) and *Poophagus* (both white- and yellow-flowered *Rorippa* spp.). A few species require special collecting methods. Thus *Rhinoncus albicinctus* Gyllenhal, not long known as a British species and still a rarity, can only readily be found by examining the floating leaves of its host, Amphibious Bistort (*Persicaria amphibian* f. *natans*), so requiring the use of waders or a boat.

One problem encountered when working food-plants for weevils, especially Ceutorhynchinae, is that the insects drop from the plants when approached. Certainly, care needs to be exercised, and any rough or sudden movement avoided. However, in the author's experience the ability of some weevils (e.g. *Tapeinotus sellatus*) to drop at the slightest provocation is exaggerated.

COSSONINAE. These weevils are a distinctive group of mainly wood-living species. They are seldom beaten or swept and have to be extracted from the dead wood in which they are breeding. Finding timber that is in the right condition is often difficult, as it is frequently either too fresh or too old. *Pselactus spadix* (Herbst) has the specialised habitat of dead wood which is periodically inundated by the sea; it is often plentiful in old groynes, wharves etc. and is sometimes found in driftwood. The Holly Weevil, *Rhopalomesites tardyi* (Curtis), despite its common name, actually feeds in many kinds of dead wood, even including Rhododendron. It is our largest cossonine, is sexually dimorphic, and found mainly in western Britain. Several Cossoninae are semi-synanthropic, occurring in old floorboards and similar structures indoors. An example is *Euophryum confine* (Broun). This species, which arrived from New Zealand about 1930, has spread throughout the British Isles and can now be found commonly in the open as well as in houses and other buildings.

CRYPTORHYNCHINAE. This is an important curculionid subfamily worldwide, but we have only four native species. *Cryptorhynchus lapathi* (L.) is a large, black-and-white weevil which resembles a bird-dropping. Larvae bore in the stems of various willows and are a pest of osiers grown for basket-making in some areas. Only three species of the very speciose genus *Acalles* (s. lat.) inhabit the British Isles; they are small, rather obscure weevils associated mainly with dead twigs of deciduous trees and dwarf shrubs.

CYCLOMINAE. The name of this curculionid subfamily is not very familiar to British coleopterists but was included as a group of 'true weevils' by Morris (2002). It contains only two species of *Gronops*, one a recent colonist, the other (*G. lunatus* (F.)) a native species associated mainly with spurreys (*Spergularia* spp.).

ENTIMINAE. The broad-nosed weevils (Morris 1997) are an important and species-rich group. The larvae are root-feeders and many adults are polyphagous. These include ground-living species but also many associated with herbaceous vegetation or trees. However, separation into groups occurring at different heights of vegetation may be misleading, as little account has been taken of the nocturnal habits of many of the species. With no necessity for long rostra or 'beaks' to penetrate plant structures the organ is not developed into the characteristic feature of other weevils, hence the name of 'broad-' or 'short-' nosed weevils.

The speciose genus *Otiorhynchus* includes arboreal, field-layer and ground zone species. Several are generalised pests, with *O. sulcatus* (F.) (Vine Weevil) one of the most notorious and prevalent ones. In recent years several species of the genus have been introduced into Britain, or have colonised the country, and have become established. Two of these are so recent that they are not included in the most recent keys to the species (Morris 1997). Few *Otiorhynchus* are restricted to particular plants. Several species are very common, but collecting the rarer ones often depends more on geography than botany. *O. arcticus* (Fabricius, O.), *O. nodosus* (Müller) and two particular rarities in Great Britain, *O. morio* (F.) (which may in fact be extinct) and *O. auropunctatus* Gyllenhal (which is common enough in Ireland) are all northern, especially Scottish, insects.

Caenopsis, *Trachyphloeus*, *Cathormiocerus* and their allies are ground-living weevils with many features of interest. Most of our *Trachyphloeus* are parthenogenetic, males being unknown, rare or restricted to small areas of continental Europe; *T. scabriculus* (L.) is an exception. *Cathormiocerus* are not only among our rarest weevils, but they are restricted to the extreme western edge of Europe, their 'headquarters' being the Iberian Peninsula and Morocco. Our species are all very local and restricted to the south coast of England, though one has recently been discovered to occur in south Wales as well. The Lizard Peninsula, Cornwall, is a favoured area for these weevils. They may be found by 'grubbing', vacuum collecting, or specialised methods such as setting moss-traps if a lengthy stay in the area is possible. Such traps consist of damp moss bundled up and secured by chicken-wire and set in suitable hollows and rabbit burrows or under heather or other plants. After a while the traps can be taken up, the moss sifted or examined, and any weevils therein captured. Short vegetation may also be collected and sieved, or turf and moss 'extracted' under a lamp or heat source until the weevils crawl out. Stone walls are often favoured by *Cathormiocerus* and other species and can often be profitably worked.

Broad-nosed weevils with very different habits include species of *Phyllobius* and *Polydrusus*, otherwise 'leaf weevils'. Although larvae of these species feed on roots the adults are mostly arboreal and often very abundant on deciduous trees in spring and early summer. In fact, there can hardly be said to be an uncommon *Phyllobius*, though some species (for example *P. vespertinus* (F.), *P. glaucus* (Scopoli) and *P. viridicollis* (F.)) are not found absolutely everywhere. There are also some very common *Polydrusus*, such as *P. cervinus* (L.) and *P. pterygomalis* Boheman, which often occur on such trees as Oak and Hazel along with species of *Phyllobius*. *Polydrusus pulchellus* Stephens is a saltmarsh species, feeding polyphagously, not specifically on Sea Wormwood (*Artemisia maritima*) as suggested in some text-books. Our other *Polydrusus* are all arboreal, although *P. confluens* Stephens can be found by grubbing under Dwarf Furze (*Ulex minor*) (wear gloves!) as well as by beating Gorse (*U. europaeus*). All adult 'leaf weevils' have limited seasons in spring and early summer, though some individuals may be found as late as September, particularly in northern Britain.

There are several genera of broad-nosed weevils which have only one or two species in them. *Philopedon plagiatum* (Schaller) is a common and well-known inhabitant of coastal dunes which is occasionally found inland in sandy areas. *Liophloeus tessulatus* (Müller) is frequently found in association with Ivy (*Hedera helix*) despite feeding as a larva on the roots of various Apiaceae. Our three species of *Tropiphorus* are all uncommon; their reported association with Dog's Mercury seems to be only partially true at best and the same is true of *Barynotus moerens* (F.), although the plant is certainly one host of the weevil. Species which are often classified in groups on their own include *Tanymecus palliatus* (F.), which is polyphagous, and *Graptus* (formerly *Alophus*) *triguttatus* (F.), which is associated with Ribwort Plantain (*Plantago lanceolata*).

Another group of broad-nosed species is specialised to feed on Fabaceae, specifically on the root nodules formed by the plants. This is the genus *Sitona*, of which we have about twenty species, including some well-known pests, for example the 'Pea and bean weevil', *S. lineatus* (L.). As in most groups, there are also uncommon species. *S. cambricus* Stephens is one of these less abundant weevils. Like other *Sitona*, it may often be detected by the semicircular feeding marks or notches on the edges of the leaves of its food-plant, in this case Greater Birds-foot Trefoil (*Lotus pedunculatus*). *S. gemellatus* Gyllenhal is a rare, or rather very local, species that was rediscovered as a British species in Dorset a few years ago and proves to be abundant on a short stretch of the sandy undercliff in the west of the county; it feeds principally on Common Restharrow (*Ononis repens*) but is also found in association with other legumes. A feature of *Sitona* species is that several exhibit alary polymorphism, or in simpler terms have individuals some of which are flightless and others which fly readily. Thus *S. lineatus* is fully winged, *S. sulcifrons* (Thunberg) brachypterous (short winged and flightless), while *S. hispidulus* may be either macropterous or brachypterous. The beginner is likely to find species of *Sitona* to be among the first weevils encountered, there being several common species which force themselves on the attention.

HYPERINAE. Included among the 'true weevils' by Morris (2002), this subfamily contains the two genera *Hypera* and *Limobius*. Many species of *Hypera* are quite common. They feed on various herbaceous Fabaceae, Caryophyllaceae, Apiaceae and Geraniaceae, and are most easily collected by working appropriate food-plants. Searching, tapping and sweeping are all effective. However, unlike most weevils, species of *Hypera* and *Limobius* can be reared from collected larvae. Again unlike most weevils, the larvae feed externally on host-plant foliage or, occasionally, on flowers or other parts. Weevil larvae are usually white or yellowish, inactive , comma-shaped grubs, but those of *Hypera* and *Limobius* are colourful (green, pink or purplish, often with longitudinal stripes) and they crawl actively over vegetation by means of body-lobes (pedal discs). The larger larvae are easily reared, rather like lepidopterous caterpillars. Like some moths, the larvae spin cocoons in which to pupate, and the duration of the pupal stage is usually short. After emergence from the pupa the callow adults take some days to mature and do so more effectively if allowed to feed. This is a

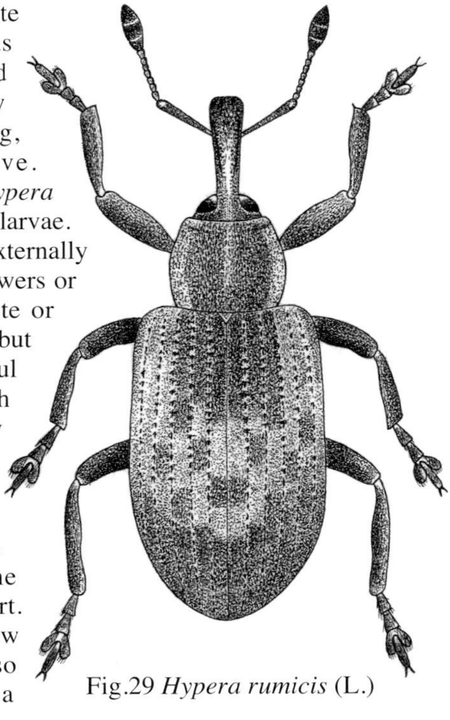

Fig.29 *Hypera rumicis* (L.)

good way to obtain specimens, particularly of the less common species, if an expedition to collect them is mistimed so that only larvae but no adults are present, as in midsummer.

Among the common species of *Hypera* are *H. plantaginis* (De Geer) (on *Lotus*, despite its name), *H. postica* (Gyllenhal) (on *Medicago*), *H. nigrirostris* (on *Trifolium*) and *H. rumicis* (L.) (on species of docks, *Rumex* spp.). Our species of *Limobius* are either rare (*L. mixtus* (Boheman)) or local (*L. borealis* (Paykull)); the former is associated with Storksbill (*Erodium cicutarium*), the latter with various species of cranesbills (*Geranium* spp.).

LIXINAE. This subfamily includes some of our largest and rarest weevils; indeed several species have become extinct during the past 150 years. Many more lixines are found in the Mediterranean region, where the species also tend to be much commoner. One of the least rare of the British representatives is *Cleonis pigra* (Scopoli), associated with thistles (especially Creeping Thistle, *Cirsium arvense*) but occurring only in sandy places. In favoured areas it can be fairly common and may be found sitting on bare sand near its hosts. Also associated with thistles are *Larinus planus* (F.) and *Rhinocyllus conicus* (Frölich). Both can be quite common in the extreme south of England, but they are rare elsewhere. Tapping thistles over a sweep net, or just spotting the weevils on the plants, are the best methods of collecting them.

Although there are six or seven species of *Lixus*, a genus of elongate, generally large, weevils on the British list, all are probably extinct except one added to it only in 1992. This species, *L. scabricollis* Boheman, is abundant at the site where it was first discovered and is undoubtedly spreading along the south and south-east coasts of England. Like other species in the genus, larvae of *L. scabricollis* feed in the stems of its hosts, which in this case are species of Chenopodiaceae. *L. paraplecticus* (L.) was once so abundant that it was anecdotally considered to be a danger to cattle in some areas because the weevils could get stuck in their throats when ingested with vegetation. The weevil is probably extinct in Britain, its demise having been brought about by intensive agriculture, industrial development and other land-use changes. Larvae fed in the stems of waterside Apiceae.

MESOPTILIINAE. This subfamily is much more familiar to British coleopterists as Magdalidinae, and indeed includes only the genus *Magdalis* in the British fauna. The larvae feed in cells under the bark of trees. The adults can be beaten from the hosts, since they do not live within the dead wood as do most Cossoninae. However, they generally have a very short season, being found mainly from late May to July. The species associated with Elm (*M. armigera* (Fourcroy)) has perhaps become particularly common since Dutch Elm disease has provided quantities of dead elm wood. Other species are associated with various rosaceous shrubs and fruit trees, Birch and Scots Pine.

MOLYTINAE. The species included in this subfamily are rather disparate. One, *Hylobius abietis* (L.), the Pine Weevil, is a notorious and abundant forestry pest. Its congener, *H. transversovittatus* (Goeze), in complete contrast, is a rare 'Red Data Book' insect which feeds in the rootstocks of Purple Loosestrife (*Lythrum salicaria*); it appears to be much commoner on the continent of Europe. *Liparus germanus* is one of our largest weevils and occurs only in the extreme south-east of England. Adults often give themselves away by their feeding activity on the foliage of the usual host, Hogweed (*Heracleum sphondylium*). Because its smaller and commoner congener, *L. coronatus* (Goeze), feeds mainly on Cow Parsley (*Anthriscus sylvestris*), which has finely divided leaves, its feeding activity is much less obvious. Both our *Liparus* may occasionally be found walking along country lanes and tracks. *Leiosoma* species are superficially like miniature *Liparus*. *L. deflexum* (Panzer) is a common species the presence of which, again, may be revealed through its feeding damage to the foliage of Wood Anemone (*Anemone nemorosa*) and common buttercups (*Ranunculus* spp.). It is ground-living and not often swept.

The exceedingly local *Anchonidium unguiculare* (Aubé), which is found in oakwoods around Gweek in Cornwall, but also on cliffs in South Devon, is a molytine which is most readily collected by sieving moss and leaf litter from its habitat. It is often overlooked, even when very small quantities of litter are painstakingly examined. Also included in the Molytinae is the Fern weevil, *Syagrius intrudens* C. O. Waterhouse. Although a member of an Australian genus, it has not (yet) been found there and is known only from the British Isles. Conservationists seem to disregard this fact and ignore the claims of the weevil to be included as a Red Data Book, or at least Notable, species.

OROBITINAE. The taxonomic affinities of *Orobitis cyaneus* (L.) are in doubt; although often included with Ceutorhynchinae its current position is in a subfamily of its own. Larvae feed in the seed capsules of various violets (Violaceae) and the adults are so similar in appearance to violet seeds that they are easily passed over.

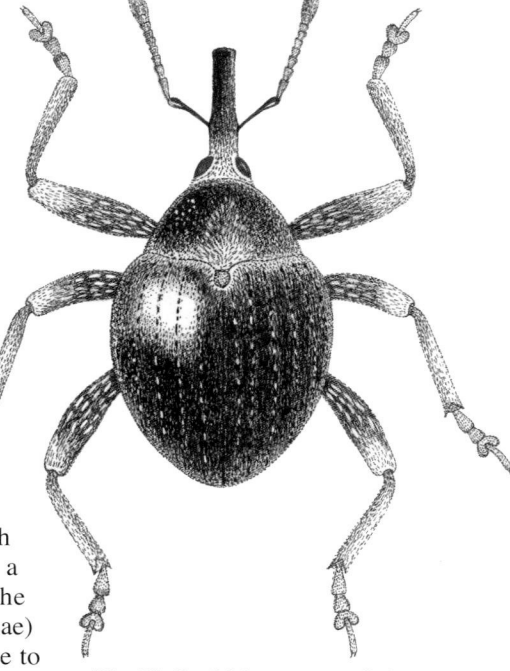

Fig.30 *Orobitis cyaneus* (L.)

SCOLYTINAE. For long regarded as constituting a separate family, the bark- and ambrosia beetles are currently classified as a subfamily of Curculionidae by most (though not all) specialists. All are associated with dead wood, or the dead stems of woody plants and many construct complex galleries in which the larvae feed. The ambrosia beetles, in particular, have a close symbiotic relationship with fungi that are a part of the decomposition cycle of dead wood.

Bark beetles are only occasionally swept or beaten, though abundant species such as *Leperisinus varius* (F.), *Phloeophthorus rhododactylus* (Marsham) and *Scolytus multistriatus* (Marsham) may be found in this way on their respective food-plants, Ash, Gorse (and Broom) and Elm. Similarly, *Xylocleptes bispinus* (Duftschmid) may be beaten from Traveller's-joy (*Clematis vitalba*). Several Scolytinae engage in maturation feeding, in which newly emerged adults eat green shoots and foliage; they can be collected more easily at this time than when in dead wood. If the timing is right, peeling back the bark of suitable trees may result in the discovery of mature adults; they are not normally found in such situations on the trees serving for maturation feeding. Though many species make superficial galleries under bark, others penetrate more deeply into the wood, for instance *Xyleborus dispar* (F.), which exhibits sexual dimorphism. Careful use of the chisel is needed to extract this species, which is most frequently found in the wood of Oak and fruit trees. Puffing smoke into the beetles' exit holes has sometimes been found to be effective in inducing them to emerge.

Many scolytines are associated with conifers rather than deciduous trees, in contrast to Cossoninae, which predominantly attack broad-leaved trees. Some species are forest and plantation pests, examples being *Xyloterus lineatus* (Olivier), *Ips cembrae* (Heer) and *Hylastes* spp. Several conifer-feeding bark beetles have established themselves in Britain during the last twenty years or so. One which is a particularly damaging pest is *Dendroctonus micans* (Kugelann). This species is exceptionally large for a scolytine and attacks spruces (*Picea* spp.).

TANYSPHYRINAE. It is perhaps rather unfortunate that, because of the 'nested alphabetical' sequence of categories in Morris (2003) this subfamily appears after the bark beetles, Scolytinae, often regarded as a separate family. *Tanysphyrus lemnae* (Paykull), our only representative of the group, is a small species which feeds on the floating plants of species of duckweeds (Lemnaceae). However, like the erirhinid *Stenopelmus*, it is unable to swim.

PLATYPODIDAE. Though considered to be only a subfamily of Curculionidae by Lawrence & Newton (1995) it is considered preferable here to treat the group as a distinct family, as do Alonso-Zarazaga & Lyal (1999). *Platypus* is the only British genus. These beetles are especially associated with tree stumps and can sometimes be detected from the fine sawdust they leave by their exit holes. Two species are on the British list but only *P. cylindrus* (F.) is at all widespread, though local. It is an elongate, cylindrical insect, the prothorax in particular being very long.

CURATING. The details of forming a collection of weevils, or of beetles more generally, are largely a matter of personal preferences, but also of experience. Consequently, some advice may be acceptable to beginners.

Weevils are best kept alive until reaching home and not killed in the field. In this author's view nothing beats ethyl acetate as a killing agent. Unlike many beetles, weevils do not relax well after killing if once they are allowed to stiffen. If the intention is to have a tidy, neat and aesthetically pleasing collection then well relaxed specimens will be needed. One good procedure is to stun the weevils in ethyl acetate (or other killing agent), set them, and then return them to the vapour of the killing agent for completion of the process. An excellent alternative to this procedure, particularly if time is short, or facilities limited (e.g. on a lengthy collecting trip or expedition) is to kill the weevils and store them (with adequate labels!) in tubes of weak acetic acid solution; 2-3% is usually recommended but some coleopterists prefer a stronger solution. This method leaves the insects well relaxed. They should not be kept too long in the solution, though weevils can last for several years in it; the exact period can depend on the size and robustness of the specimens. There are a few disadvantages to this method; abdomen and wings may get filled with fluid, though this can be expelled during setting, or 'preparation', and some species with metallic scales may change colour. However, acetic acid for temporary storage is far preferable to IMS or alcohol, which leaves weevils stiff and often unsettable (though they can be pinned). However, much depends on personal preference; many coleopterists do not particularly want to spend valuable time producing well-set specimens.

Nowadays considerable reliance is placed on the form and structure of the median lobe of the male genitalia as an aid to identification, especially for some groups of weevils. The student should get used to dissecting at least a specimen or two of any new species taken in order to use this important characteristic for determination. It is much more difficult to extract aedeagi from dry and long-dead specimens, though not impossible of course; museum taxonomists do it every day. The spermathecae of females are also often dissected out and may be helpful aids to identification in some cases. In general, however, they are less useful than the male organs.

Once 'hooked' on weevils the coleopterist may find interest in other beetles waning. But that is a risk worth taking!

References

Alonso-Zarazaga, M. A. & Lyal, C. H. C., 1999. *A World Catalogue of families and genera of Curculionoidea (excepting Scolytidae and Platypodidae)*. Entomopraxis, Barcelona.

Joy, N. H., 1932. *A Practical Handbook of British Beetles*. (2. vols.). Witherby, London.

Lawrence, J. F. & Newton, A. F., 1995. Families and subfamilies of Coleoptera (with selected genera, notes, references and data on family-group names). pp.779-1006 *in* Pakaluk, J. & Ślipiński, S. A. (eds.) *Biology, phylogeny, and classification of Coleoptera: Papers celebrating the 80th Birthday of Roy A. Crowson.* Muzeum I Instytyt Zoologii PAN, Warszawa.

Morris, M. G., 1990. Orthocerous Weevils. Coleoptera: Curculionoidea (Nemonychidae, Anthribidae, Urodontidae, Attelabidae and Apionidae). *Handbooks for the Identification of British Insects*, **5**(16): 1-108.

Morris, M. G., 1997. Broad-nosed Weevils. Coleoptera: Curculionoidea (Entiminae). *Handbooks for the Identification of British Insects*, **5**(17a): 1-106.

Morris, M. G., 2002. True Weevils (Part 1). Coleoptera: Curculionoidea (Subfamilies Raymondionyminae to Smicronychinae). *Handbooks for the Identification of British Insects*, **5**(17b): 1-149.

Morris, M. G., 2003. An annotated check list of British Curculionoidea (Col.). *Entomologist's Monthly Magazine* **139**; 193-225.

Pope, R. D., (Ed.), 1977. Kloet & Hincks. A check list of British Insects. Second Edition (completely revised. Part 3: Coleoptera and Strepsiptera. *Handbooks for the Identification of British Insects*, **11**(3): i-xiv + 1-105.

Preston, C. D., Pearman, D. A. & Dines, T. D., (Eds.), 2002. *New Atlas of the British and Irish Flora.* OUP, Oxford.

Stace, C., 1997. *New Flora of the British Isles.* Second Edition. CUP, Cambridge

Beetle Larvae
by M.L. Luff

1. INTRODUCTION

It is obvious to any serious entomologist, although perhaps not thought about by many collectors, that every adult beetle that they seek must first have passed through developmental stages of egg, larva and pupa. Unlike some other insects, many British beetles spend much of their total life (and in particular the winter) as adults. There are, however, common exceptions to this. Some beetles such as the larger wood-boring Cerambycidae, and soil inhabiting Elateridae may spend a year or more as larvae. Others may be more readily found as larvae than as adults. Larvae and adults of the same species frequently share the same habitat, perhaps because beetles, more than any other group of endopterygote insects, are adapted as adults to live in confined spaces and sheltered situations, where larvae also occur. One must be careful, however, not to assume that just because larvae and adults are found together, that they are necessarily of the same species; rearing out of the larvae may be needed to confirm this.

The study of beetle larvae has, however, been neglected by the majority of collectors: a few have reared larvae, but only to obtain perfect adults for their collections. The number who deliberately rear them in order to study their ecology, bionomics, physiology or taxonomy must be very small.

The aim of this chapter is not to direct coleopterists into a specialised channel, but merely to make them 'larva conscious'. Comparatively few life-histories of our British beetles have been worked out and many of those that have been are but superficially dealt with. Much still remains to be done in larval systematics, for although nearly all the family characters are known, there still remain many genera and species yet to be described and keyed; while many of the 'known' species need to be verified by breeding, because systematists are so frequently forced by lack of material to rely on questionable specimens and inadequate descriptions.

It must be emphasised that over-zealous searching for larvae, even more than for adults, may lead to habitat destruction and loss of rare species. The JCCBI Code for Insect Collecting applies as much or even more to collecting larvae (which are the future breeding stock of the species) than to adults which may already have bred. In particular, long-term habitats such as old timber should not be intensively worked, and as much material as possible should be left intact.

2. MORPHOLOGY

As the structure of beetle larvae is both more variable and less familiar to the general collector than that of the adults, it is appropriate to outline the general features of beetle larvae (Fig. 31.) This will both enable them to be recognised as such and serve as a starting point for more detailed identification, if this is to be attempted. It must be remembered that as beetle larvae grow by moulting several times during their life, the size of any species can vary considerably. Generally a fully-grown beetle larva is about 1.5 times the length of the corresponding adult. Most species have between three and five larval stages or instars, with a moult between each instar. In a few cases their morphology changes markedly between instars; this is called 'hypermetamorphosis.'

Beetle larvae are seldom brightly coloured; those living in the open tend to be dark brown or black, whereas the majority living in sheltered habitats are pale brown, cream or white, sometimes with a darker head. They generally have a distinct head capsule, unlike many fly larvae, with curved, opposable mandibles, and two sets of palpi (maxillary and labial) as in the adults. Up to six simple eyes or ocelli may be present on each side of the head. The antennae vary from minute two-segmented appendages to long filaments, but are usually three or four-segmented. The thoracic legs range from 5-segmented (plus one or two claws) to completely absent. The abdomen usually has from 8-10 obvious segments, and is without prolegs (in contrast to those found in Lepidoptera) but there may be smaller swellings (*ampullae*) or hooks on particular segments. The apical abdominal segments may bear paired cerci (technically known as *urogomphi*) which can be short and hook-like, or long and slender.

Fig. 31

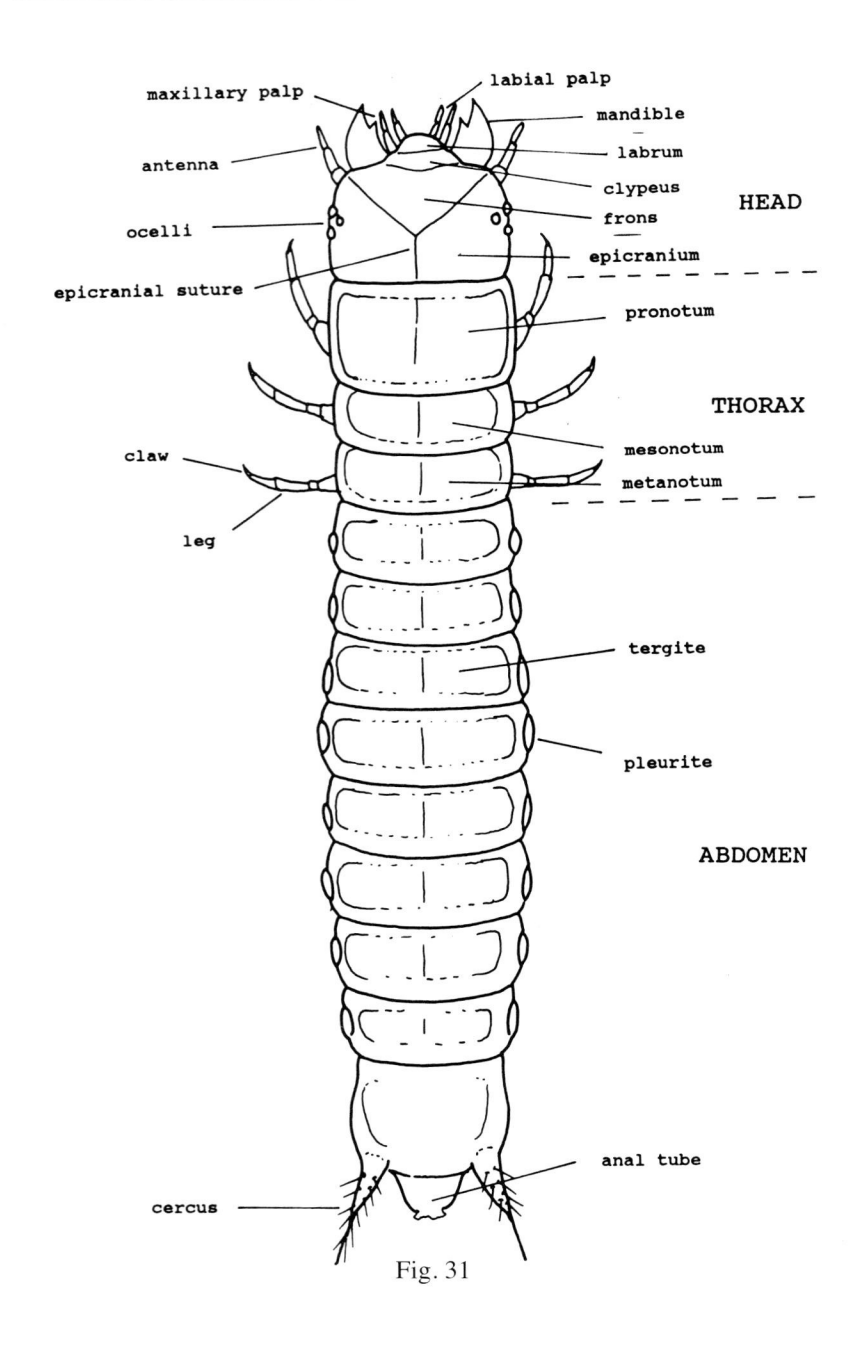

This general structure is usually modified into one of four main morphological types:

1. Campodeiform, as in Carabidae (plate 13) and Staphylinidae. These larvae are active, elongated specimens with long legs (and often long cerci). They hold the mouthparts forwards at the front of the head and are usually predatory.
2. Scarabaeiform, as in Lucanidae, Scarabaeidae (plate 46) and Byrrhidae. These are fat, usually C-shaped grubs, tending to lie on their sides within their larval habitat such as dung or soil. They have well-developed thoracic legs, but their movements are limited by the bulk of their bodies.
3. Eruciform, as in many Chrysomelidae. These are more 'caterpillar-like' larvae, with a small head capsule bearing ventrally-directed mouthparts, and short but functional thoracic legs, so that they can crawl over their substrate, see Coccinella larva (plate 66).
4. Apodous, as in Curculionidae. These are legless grubs, rather like the scarabaeiform type in their C-shaped body, but unable to move other than by bodily contractions; many are found within their food medium such as in plant tissue.

Intermediate forms occur between these main morphological types, but any one family usually tends to have a characteristic type of larva.

3. IDENTIFICATION

There is currently no complete and easily usable key for the identification, even to family level, of British beetle larvae. The 'classical' work is that of van Emden (1942) There are other papers in the same series on particular families, but all are now difficult to obtain. There are comprehensive works on larvae by Boving & Craighead (1931), Ghilarov (1964) and, more recently, a comprehensive volume, with many keys to genera, edited by Klausnitzer (1978). None of these are readily available to the amateur, however. At the time of writing, there is a volume on the identification of beetle larvae being prepared in the Royal Entomological Society's 'Handbooks for the Identification of British Insects' series; hopefully this will fill the gap. It would be nice, however, if a beetle larvae key was available in the A.I.D.G.A.P. series sponsored by the Field Studies Council, similar to that already available for the families of adult beetles.

Keys are available for a few particular families; the best is that to larvae of Coccinellidae, in the excellent book 'Ladybirds' by Majerus & Kearns (1989). Some volumes in the 'Fauna Entomologica Scandinavica' series (written in English) include keys to larvae, e.g. Bílý (1982). Details of other families are given in the earlier family sections of this Handbook.

4. FAMILY CHARACTERISTICS

The following brief descriptions outline some of the larval characters and habitats of selected families of British beetles. The larval characters of the various families mentioned are not a conclusive guide to their identification and, to facilitate identification in the field, preference has been given whenever possible to the more conspicuous characters. Such a procedure must inevitably leave uncovered many exceptions, but these can be verified by employing more specialised keys, after the beginner has grasped the fundamentals.

SUB-ORDER ADEPHAGA

Larvae usually campodeiform; legs 5-segmented, with one or two movable claws; antennae usually 4-segmented.

CARABIDAE (Fig. 32 and plate 13) including CICINDELIDAE (Fig. 32).

Abdomen 10-segmented (including anal tube), form elongate, slender, with paired sub-apical cerci; tarsi with one or two claws. Cicindelid larvae with prominent dorsal hooks on the 5th abdominal segment.

Habitat: On or in the soil. Larvae of *Nebria, Notiophilus* and fully-grown *Carabus* and *Pterostichus* are surface-active. Those of some *Harpalus* store seeds in vertical burrows. *Cicindela* are predatory in vertical burrows.

DYTISCIDAE (Fig. 32 and plate 2.)

Abdomen 8-segmented (no anal tube), form slender, tapering, with slender apical cerci; tarsi with two claws which are fringed with hairs; mandibles large and sickle-shaped.

Habitat: In pools, ditches, ponds, streams, etc. Predaceous on small soft-bodied animals. Larvae of *Ilybius fenestratus* (F.) and other Dytiscids may be dug up from the banks of ponds about March, prior to pupation.

HYGROBIIDAE (Fig. 32.)

Abdomen 8-segmented, the last segment with a long median process and slender cerci; form tapering with very large head and thorax; tufts of ventral gills on thorax and first three abdominal segments.

Habitat: The larva of *Hygrobia hermanni* (F.) is confined to muddy ponds where it feeds on midge larvae in the mud.

HALIPLIDAE (Fig. 32.)

Long, slender-larvae with 9 or 10-segmented abdomen; cerci apical if present; tarsi with a single claw; no gills.

Habitat: In mud or on vegetation in streams, where they feed on filamentous algae such as *Spirogyra*.

The Beetle Families

GYRINIDAE (Fig. 32.)

Elongated larvae with small head and prominent pairs of tracheal gills on each of the abdominal segments except the last, which has 4 small hooks.

Habitat: On the substrate or aquatic vegetation in the same habitat as the adults, where they are predatory.

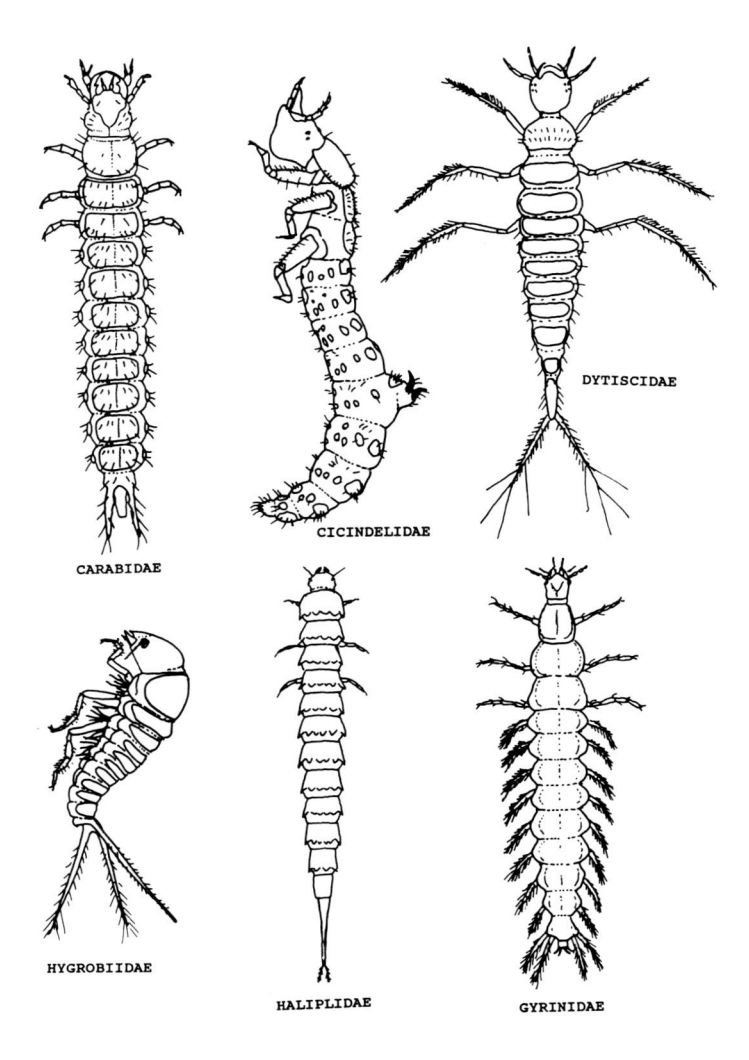

Fig. 32

SUB-ORDER POLYPHAGA

Legs with at most 4-segments, sometimes absent. Antennae at most 3-segmented. The body form can be any of the four main types.

HYDROPHILIDAE (Fig. 33.)

Variable in appearance, typically elongate, fleshy, with small head and legs (which are vestigal in *Cercyon*), and eight abdominal segments dorsally (the 9th and 10th segments are small and retracted into the 8th); cerci usually small, longer and segmented in *Helophorus* (which superficially resemble many carabid larvae), absent altogether in *Spercheus*.

Habitat: Aquatic species in ponds, etc., particularly where there is an abundance of floating weed; carnivorous. Terrestrial species in similar habitats to those of the adults, including dung and damp soil.

HISTERIDAE (Fig. 33.)

Parallel-sided, somewhat flattened and rather fleshy larvae, with mouth-parts directed forwards; labrum absent, mandibles with small retinaculum but no mola; cerci short, two-segmented.

Habitat: In dung, decaying animal and vegetable matter, nests, etc. The predaceous larvae of *Hister 4-maculatus* L. and *H. unicolor* L. feed on dipterous larvae in dung. Bird nests, especially those in hollow trees, are worth searching for histerid larvae; these should be examined as soon as the fledglings have flown. Larvae of *Dendrophilus punctatus* (Herbst) have been taken from a woodpecker nest.

SILPHIDAE (Fig. 33 and plate 34.)

Elongated, often flattened and onisciform (woodlouse-like), with broad thorax and small head. Mouthparts directed downwards; cerci short, two-segmented.

Habitat: Chiefly in soil beneath carcasses and other decaying animal matter. *Silpha* larvae usually feed underneath the carcass, but those of *Nicrophorus* live in the soil beneath on portions of flesh carried down by adults. Larvae of *Dendroxena quadripunctata* (L.) are predaceous on oak-feeding caterpillars in the spring, and may be beaten from oaks at night.

STAPHYLINIDAE (Fig. 33.)

Elongated, campodeiform larvae, often superficially resembling those of Carabidae (but separable by the 4-segmented legs, 3-segmented antennae). Mandibles usually slender, without mola. Cerci usually two-segmented, often long.

Habitat: In soil, dung, fungi, moss, decaying animal and vegetable matter, under bark, etc. Some of these larvae live in dung (*Ontholestes murinus* (L.),

The Beetle Families 185

HYDROPHILIDAE

HISTERIDAE

SILPHIDAE

LUCANIDAE

SCARABAEIDAE

STAPHYLINIDAE

Fig. 33

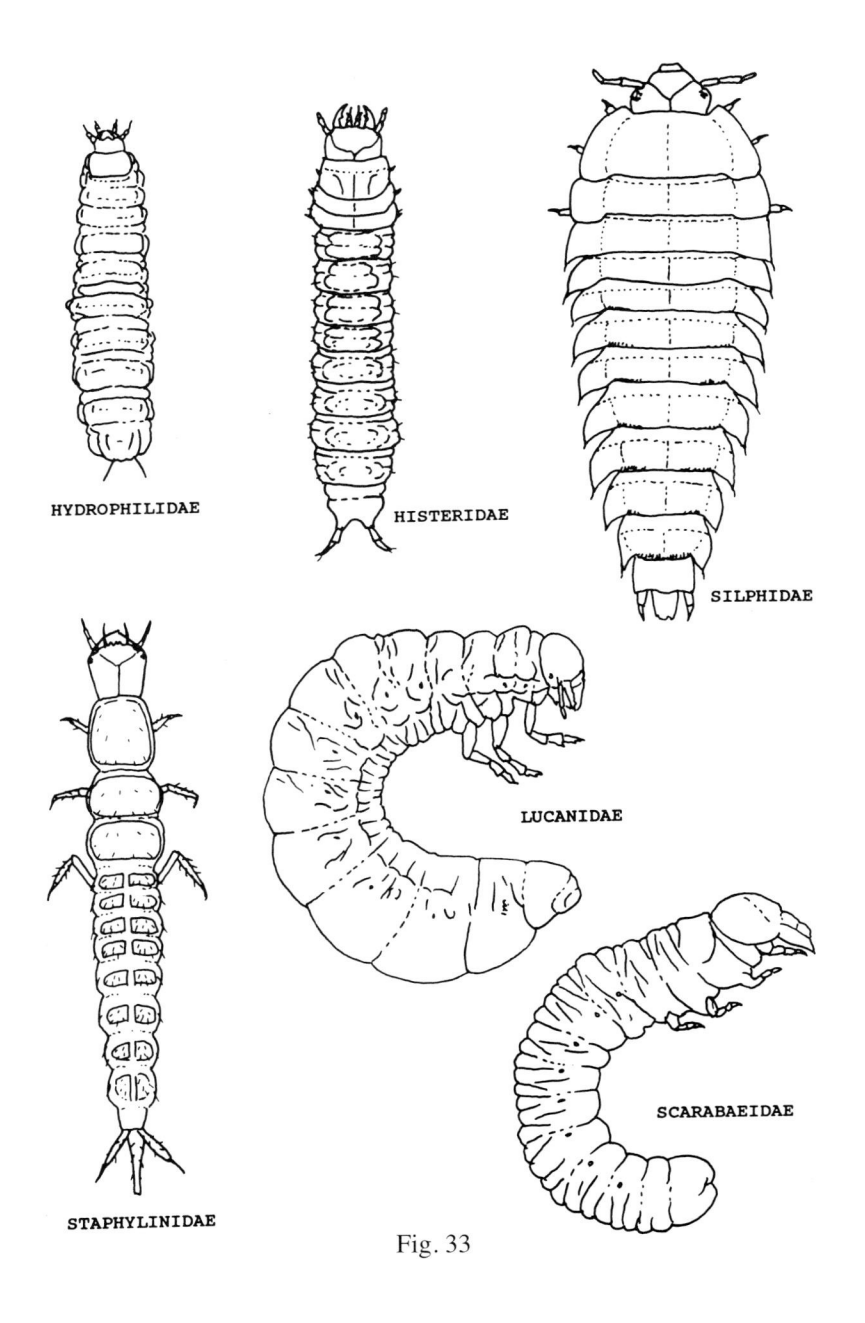

Philonthus spp.), others in saturated moss near brooks (*Lesteva*), in carrion (*Creophilus maxillosus* (L.)) and under bark (*Siagonium quadricorne* Kirby), which are predaceous on 'scolytid' larvae). Many of the smaller species are found in soil-living fungi. The nests of birds harbour the minute larvae of *Microglotta* spp. There are a number of parasitic species, too; for example, larvae of *Velleius dilatatus* (F.) are predatory on dipterous larvae, e.g. *Volucella* species hornets' nests; dipterous puparia (especially those in decaying vegetable matter) should be kept as they may prove to contain the parasitic *Aleochara* spp. Ant nests are particularly rich in staphylinid larvae and much investigation remains to be done. Mammal nests also harbour a number of species. Some of the larger Staphylinine larvae are active on the soil surface, and may be caught in pitfall traps.

LUCANIDAE (FIG. 33.)

Typical scarabaeiform larvae, fleshy and curved ventrally, abdomen fattest towards its apex; legs small and slender, with stridulatory organs on the mid coxae and hind trochanters.

Habitat: In rotten trees and stumps. The bases of old stumps particularly oak and elm, house the large, fleshy larvae of *Lucanus*. Those of *Dorcus* generally prefer rotten elm, ash and oak trunks. Rotten beech, ash and hawthorn stumps should be broken up for *Sinodendron* larvae.

SCARABAEIDAE (including GEOTRUPIDAE, etc.) (Fig. 33 and plate 46)

Similar to Lucanidae but abdominal segments with extra dorsal folds, making the number of segments not distinct. No stridulatory organ on the legs.

Habitat: In dung or at roots; those feeding in the latter mostly take several years to attain maturity. The majority of species (e.g. *Aphodius*) are coprophagous and it must be borne in mind that the adults of many species carry pellets of dung several inches down into the earth where they lay their eggs. Hence it is even important to examine the earth immediately beneath the dung. Deep holes, roughly 1-2cm diameter, near rabbit dung in sandy areas indicate the presence of adults of *Typhaeus typhoeus* (L.). Larvae will be found in those that appear to be freshly filled up. Deer dung is attractive to some species. The round plump larvae of *Melolontha* and several other species of 'chafer' are root feeders; these may be obtained by digging or by pulling up old stumps. Larvae of *Cetonia aurata* (L.) congregate in colonies beneath rotten elm stumps. *Gnorimus* larvae live in the soft wood mould in hollow oaks and fruit trees. *Trox* and *Saprosites* are exceptional in their habitat, the former living in dried carcasses and in the nests of owls and jackdaws in hollow trees and the latter under bark.

DASCILLIDAE (Fig. 34)

Elongate, rather cylindrical, fleshy larvae, slightly curved ventrally, with prominent V-shaped dorsal suture on head. Mandibles with mola.

Habitat: In soil, where they feed at roots. Larvae of *Dascillus cervinus* (L.) may easily be reared in pots of root fibre and damp sandy or loamy soil.

SCIRTIDAE (Fig. 34)

Small, flattened, tapering larvae, rather onisciform; antennae long with a multi-articulate thread-like terminal flagellum (which is apparently unique among all holometabolous insect larvae).

Habitat: Aquatic, in ditches, rot-holes in trees, etc. There are several not uncommon species in this group: larvae of some species may be found crawling on rotten submerged branches and leaves in rather shady, foul pools; those of others prefer water weeds and leaves in more open pools, or occur in rain-water in roots or hollow boles of beeches. Larvae of *Elodes minuta* (L.) live among mossy stones in streams.

BYRRHIDAE (Fig. 34)

Superficially rather like dascillid larvae, but apex of abdomen is more abruptly curved ventrally; the mandibles have no mola, and the dorsal suture of the head is Y-shaped.

Habitat: In soil under turf, mossy tree stumps, roots, etc.

DRYOPIDAE (Fig. 34)

Elongated, cylindrical, tapering to a point apically, where the ninth abdominal segment has a closeable operculum which covers the retractable 10th segment with its tracheal gills (these absent in some terrestrial *Dryops*).

Habitat: In similar aquatic habitats to the adults; some species are abundant in turf subject to waterlogging for extensive periods annually, as in the moist areas of fields.

ELMIDAE (Fig. 34)

Similar to Dryopidae, but more flattened, sometimes onisciform, larvae, with three tufts of gills usually protruding from the apical operculum.

Habitat: Crawling beneath stones, etc., in small but swiftly-flowing streams, occasionally in ponds. They are best collected by placing a pond net in a stream and stirring up the gravel or stones or, if roots are present, by shaking them violently so that the larvae are carried out by the current into the net.

DASCILLIDAE

SCIRTIDAE

BYRRHIDAE

DRYOPIDAE

ELMIDAE

BUPRESTIDAE

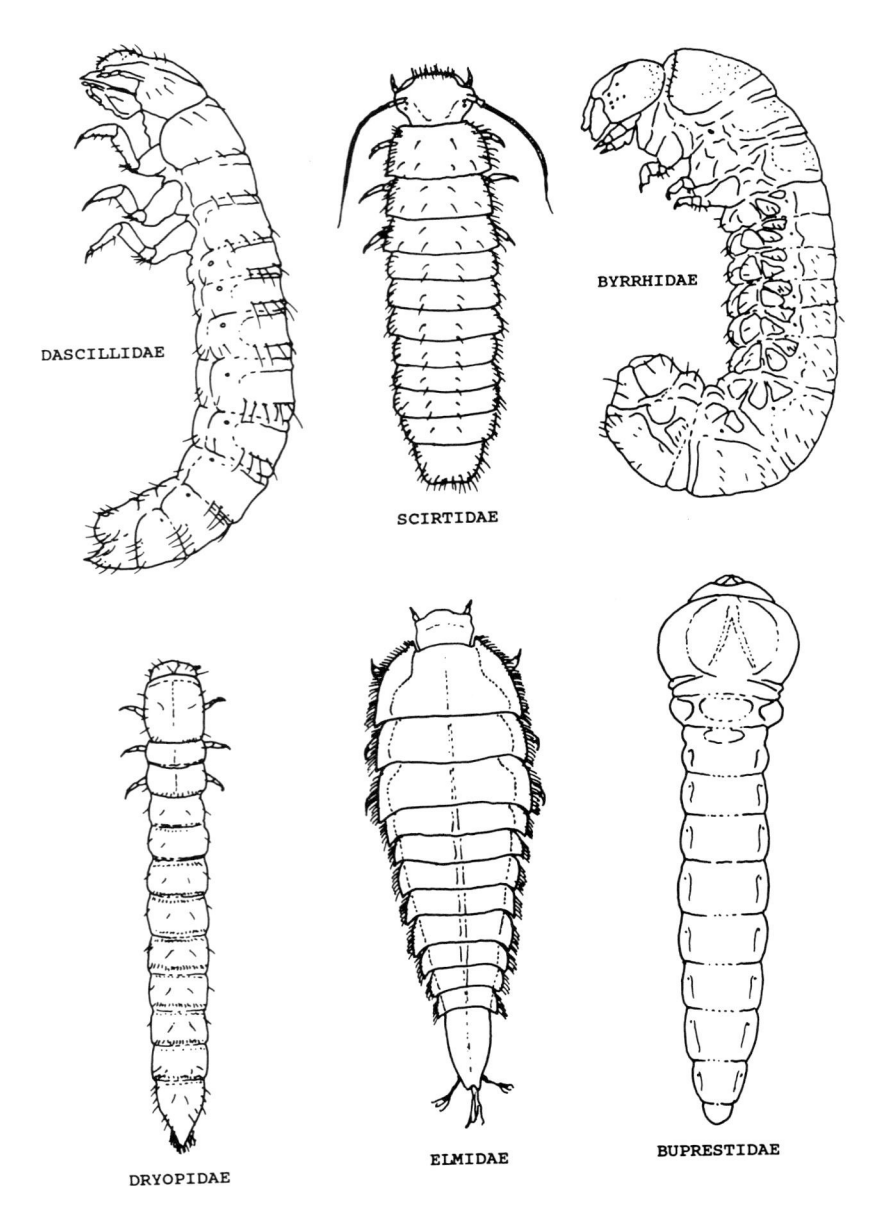

Fig. 34

BUPRESTIDAE (Fig. 34)

Cylindrical or flattened, tapering larvae, either head elongate and deeply retracted into prothorax (*Trachys*) or with thorax enlarged; legs absent or minute and labial palps fleshy and unjointed.

Habitat: In leaves and under bark, etc. Larvae of *Trachys* are leaf miners; those of *T. troglodytes* Gyllenhal mine leaves of *Succisa pratensis,* and *T.scrobiculatus* Kiesenwetter *Glechoma hederacea* and *Hyoscyamus niger.* It should be born in mind that these are the recorded hosts in Britain, continental literature quotes different hosts. *T.minutus* (L.), according to 'European' literature, mines *Salix, Corylus avellana, Ulmus campestris* and *Sorbus. Aphanisticus emarginatus* (Olivier) larvae occur low down in *Juncus* stems; *A.pusillus* (Olivier) feeds on *Schoenus nigricans* their larvae are also miners. *Agrilus* larvae develop subcortically in dying or freshly dead branches and/or trunks of various trees and shrubs (see 'Buprestidae' p. 58).

ELATERIDAE (Fig. 35 and plate 52).

Elongated, slender and cylindrical larvae, usually well-sclerotised and orange-brown in colour, giving them their popular name of 'wireworms'. Labrum fused to head capsule and toothed on its front margin. Apex of abdomen either subconical, or forming a flattened plate with two apical teeth or prongs.

Habitat: Many species feed at roots of plants and some are serious pests. A few, however, live in damaged or rotten trees where they are mainly predatory on other larvae, particularly those of the Cerambycidae. Larvae of *Denticollis linearis* (L.) occur in rotten stumps, under bark and in peat. Larvae of *Cardiophorus* are of an unusual appearance, being soft, flaccid and elongate like an extremely slender centipede. Full-grown larvae of *C. asellus* Erichson may be dug up in early spring in sandy soil at the roots of heather and pine (usually 10-13cm deep). Larvae of *Kibunea minuta* (L.) occur in the dry, mycelium-filled soil beneath 'fairy rings'. The sturdy blackish-brown larvae of *Stenagostus rhombeus* (Olivier) live under the bark of oak boughs and are more commonly seen than the nocturnal adults. The orange-brown larvae of *Melanotus villosus* (Geoffroy *in* Fourcroy) are abundant under the bark of pine stumps where they feed on other insect larvae etc.

CANTHARIDAE (Fig. 35).

Straight, elongated larvae, campodeiform with ventrally-grooved mandibles. Head with sutures indistinct. Body usually covered with velvety pubescence; no cerci.

Habitat: In loose soil, under logs, or crawling on pathways *(Cantharis,* etc.), where they are frequently seen in the winter even when snow has fallen; hence their popular name 'snow-worms'. Predaceous on worms, larvae, etc.

ELATERIDAE

CANTHARIDAE

LAMPYRIDAE

DERMESTIDAE

ANOBIIDAE

CLERIDAE

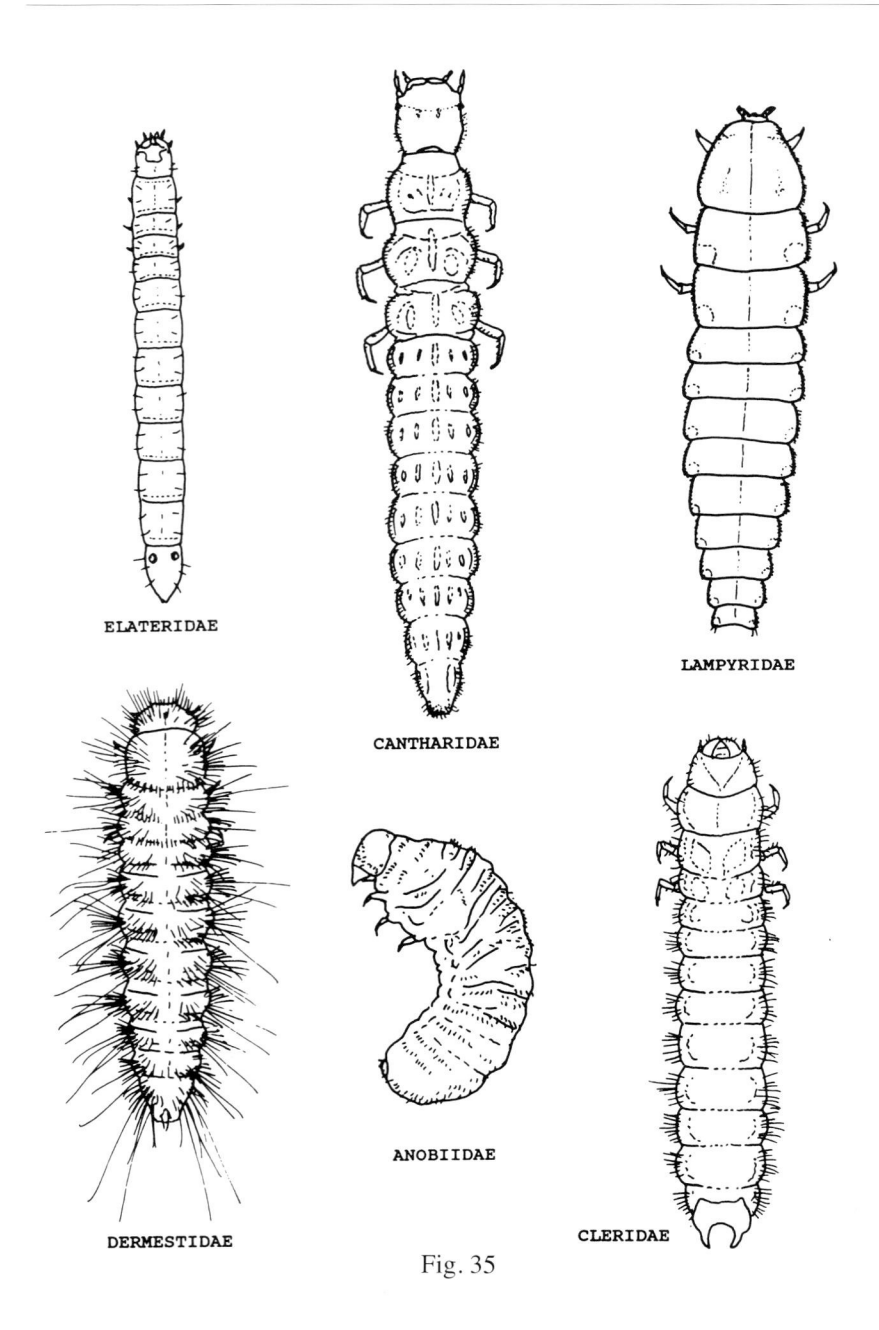

Fig. 35

LAMPYRIDAE (Fig. 35).

Similar to cantharids, but flattened and leathery; head covered dorsally by enlarged pronotum, with dorsal sutures distinct, and sides not meeting each other ventrally. Apex of abdomen with a brush-like tuft of hairs.

Habitat: On grass, pathways, etc., where they feed nocturnally on snails. The slightly luminous larva of *Lampyris noctiluca* (L.) closely resembles the adult female in general form and colour.

DERMESTIDAE (Fig. 35 and plates 58 and 114a).

Short larvae with a rather humped pronotum; body densely covered with usually long (and sometimes branched or barbed) hairs. Cerci small *(Dermestes)* or absent altogether.

Habitat: In dried carcasses, skins, woollens, stored products, etc. Larvae of *Dermestes* species mostly feed in dried carcasses, except *D. lardarius* L., which, as its name implies, is associated with foodstuffs. Old blankets are frequently reported to be infested with larvae of *Attagenus pellio* (L.). Larvae of *Megatoma* and *Ctesias* both occur under bark, the latter feeding on particles of insects ensnared in old webs and cocoons of spiders. The brown, hairy larvae of *Anthrenus* are unfortunately too well known to the collector who becomes careless in the maintenance of his collection.

ANOBIIDAE (Fig. 35)

Ventrally curved, C-shaped larvae, covered with minute spines. With a small head capsule bearing ventral mouthparts; thoracic legs present (unlike superficially-similar weevil larvae). Ocelli absent.

Habitat: In the hard wood of trees, old posts, etc. For example *Anobium punctatum* De Geer in rotten oak boughs (as well as in the timber structure of buildings, where it and *Xestobium rufovillosum* De Geer do serious damage); *Grynobius excavatus* (Kugelann) in holly, birch, hawthorn; *Ptilinus pectinicornis* (L.) in oak boles; *Xestobium rufovillosum* in rotten oak planks and posts in fields. Larvae of *Caenocara bovistae* (Hoffmann) feed in the spongy bases of puffballs.

CLERIDAE (Fig. 35)

Superficially resembling cantharid larvae, covered with rather longer, finer pubescence and with apical segment of abdomen forming a flat plate which bears two short horn-like cerci.

Habitat: Under bark, in carcasses, etc. The pink larvae of *Thanasimus* are predaceous on 'scolytid' larvae in the bark of pines etc. *Necrobia* larvae are to be found in old, dry carcasses.

NITIDULIDAE (Fig. 36)

Moderately elongate, somewhat depressed larvae, sometimes with thoracic or abdominal tubercles; cerci short or absent.

Habitat: Under bark (at sap), on flower-heads, in carcasses, etc. Larvae of *Nitidula* and *Omosita* often occur in old dry carcasses. Larvae of *Pityophagus* inhabit old borings of Scolytids. Those of *Glischrochilus* feed subcortically in freshly felled oaks. *Cossus* borings often contain numbers of *Cryptarcha* and *Soronia* larvae. Larvae of *Cychramus luteus* (Fairmaire) live in the spongy bases of puffballs. Those of *Epuraea aestiva* (L.) are to be found in nests of *Bombus*. Larvae of *Meligethes aeneus* (F.) and *M. viridescens* (F.) can commonly be found in the buds and flowers of Cruciferae, and are pests of oilseed rape.

COCCINELLIDAE (Fig. 36 and plate 66).

Short but active larvae; body surface usually bearing dorsal setiferous tubercles or spines; often brown or bluish-black, with yellow markings. Legs long, with small apical claw: no cerci.

Habitat: On plants and trees, e.g. heather, thistles, fir, sallow, etc.; mostly predaceous on aphids but a few feeding on mildews, etc. Many are easily reared, but must be provided with ample food if cannibalism is to be avoided.

TENEBRIONIDAE (Fig. 36)

Elongate, cylindrical larvae, often yellowish-brown and resembling larvae of Elateridae; but the labrum is free (not fused to clypeus) and not toothed along its front margin. Apex of abdomen either rounded or with short prong-like cerci.

Habitat: In damp cellars (*Blaps*) where they feed on rodent droppings, etc.; beneath bark of elm, fir, etc., where they are predaceous on other larvae; at roots of grass and heather (*Nalassus*); in granaries (*Palorus, Tribolium, Tenebrio,* etc.); in fungi (*Eledona,* etc.).

PYROCHROIDAE (Fig. 36 and plate 76).

Parallel-sided and flattened larvae, well-sclerotised; labrum distinct, mouthparts directed forwards. End of abdomen elongated, with a pair of parallel backwardly-directed prongs.

Habitat: Under bark. Larvae of *Pyrochroa coccinea* (L.) feed in rotten oaks and those of *P. serraticornis* (Scopoli) are to be found in various trees, particularly elm.

MELANDRYIDAE (Fig. 36) and TETRATOMIDAE

Elongate but rather fat and fleshy larvae, with large head and thorax; similar to Cerambycidae, but with well-developed legs and more deeply retracted mouthparts.

Plate 2. *Dytiscus marginalis* L. 'The Great Diving Beetle' larva *ca* 32mm © Roy Anderson.

Plate 4. *Carabus arvensis* ssp *silvaticus* Dejean *ca* 18mm © Roy Anderson.

Plate 1. *Rhantus exsoletus* (Forster) *ca* 12mm © Roger Key.

Plate 3. *Graphoderus zonatus* (Hoppe) *ca* 15mm © Roger Key.

Plate 5. *Carabus granulatus* ssp *hibernicus* Lindroth *ca* 20mm © Roy Anderson.

Plate 6. *Carabus monilis* F. *ca* 25mm © Roy Anderson.

Plate 7. *Carabus nemoralis* Müller *ca* 24mm © Roy Anderson.

Plate 8. *Carabus clathratus* L. *ca* 30mm © Roy Anderson.

Plate 9. *Carabus nitens* L. *ca* 15mm © Roy Anderson.

Plate 10. *Carabus problematicus* ssp *gallicus* Géhin *ca* 25mm © Roy Anderson.

Plate 11. *Carabus glabratus* ssp *lapponicus* Born *ca* 25mm © Roy Anderson.

Plate 13. *Carabus intricatus* L. (larva) *ca* 30mm © Clive Turner.
Plate 13a [inset] *Carabus intricatus* L. (ovum) *ca* 1mm © Clive Turner.

Plate 12. *Carabus intricatus* L. *ca* 28mm © Clive Turner.

Plate 15. *Nebria complanata* (L.) *ca* 20mm © Roy Anderson.

Plate 16. *Pelophila borealis* (Paykull) *ca* 11mm © Roy Anderson.

Plate 14. *Cychrus caraboides* ssp *rostratus* (L.) *ca* 18mm © Roy Anderson.

Plate 17. *Notiophilus aquaticus* (L.) *ca* 5.5mm
© Roger Key.

Plate 18. *Elaphrus riparius* (L.) *ca* 7mm
© Roy Anderson.

Plate 19. *Cicindela campestris* L. 'The Green Tiger Beetle' *ca* 15mm © Roy Anderson.

Plate 20. *Blethisa multipunctata* (L.) *ca* 11mm
© Roy Anderson.

Plate 21. *Dyschirius obscurus* (Gyllenhal) *ca* 4.5mm
© Roy Anderson.

Plate 22. *Aepus marinus* (Ström) *ca* 2.3mm
© Roger Key.

Plate 23. *Lasiotrechus discus* (F.) *ca* 5mm
© Roy Anderson.

Plate 24. *Asaphidion curtum* (Heyden) *ca* 4.2mm
© Roy Anderson.

Plate 25. *Bembidion* (*Peryphus*) *femoratum* Sturm
ca 5mm © Roy Anderson.

Plate 26. *Pterostichus aterrimus* (Herbst) *ca* 15mm
© Roy Anderson.

Plate 27. *Cillenus lateralis* Samouelle *ca* 4mm
© Roy Anderson.

Plate 28. *Laemostenus complanatus* (Dejean)
ca 15mm © Roy Anderson.

Plate 30. *Demetrias imperialis* (Germar) *ca* 5mm © Roger Key.

Plate 32. *Hydrophilus piceus* (L.) 'The Great Silver Water Beetle' *ca* 35mm © Roger Key

Plate 29. *Stenolophus mixtus* (Herbst) *ca* 6.5mm © Roy Anderson.

Plate 31. *Panagaeus bipustulatus* (F.) *ca* 7mm © Roger Key.

Plate 34. *Silpha tristis* Illiger (larva) *ca* 12mm © Roy Anderson.

Plate 36. *Nicrophorus investigator* Zetterstedt *ca* 20mm © Roger Key.

Plate 33. *Margarinotus* sp. *ca* 5.2mm © Roger Key.

Plate 35. *Dendroxena quadrimaculata* (Scopoli) *ca* 12mm © Roy Anderson.

Plate 38. *Lordithon thoracicus* (F.) *ca* 4mm © Roy Anderson.

Plate 40. *Staphylinus erythropterus* Linnaeus *ca* 15mm © Roger Key.

Plate 37. *Tachyporus hypnorum* (F.) *ca* 3.5mm © Roger Key.

Plate 39. *Ocypus olens* (Müller) 'The Devil's Coach-horse' *ca* 25mm © Roy Anderson.

Plate 42. *Aphodius rufipes* (L.) *ca* 12mm © Roger Key.

Plate 44. *Serica brunnea* (L.) *ca* 9.5mm © Roger Key.

Plate 41. *Typhaeus typhoeus* (L.) *ca* 16mm © Roger Key.

Plate 43. *Melolontha melolontha* L. 'The Cock Chafer' *ca* 25mm © Roy Anderson.

Plate 46. *Gnorimus nobilis* (L.) larva © Roger Key.

Plate 48. *Gnorimus nobilis* (L.) larval frass exposed in an ancient apple tree. © J.Cooter.

Plate 45. *Cetonia aurata* (L.) 'The Rose Chafer' *ca* 20mm © Roger Key.

Plate 47. *Gnorimus nobilis* (L.) *ca* 16mm © Roger Key.

Plate 49. *Agrilus biguttatus* (F.) *ca* 12mm
© Roger Key.

Plate 50. *Byrrhus fasciatus* (Forster) *ca* 8mm
© Roger Key.

Plate 51. *Ampedus cardinalis* (Schiödte) *ca* 11mm
© Roger Key.

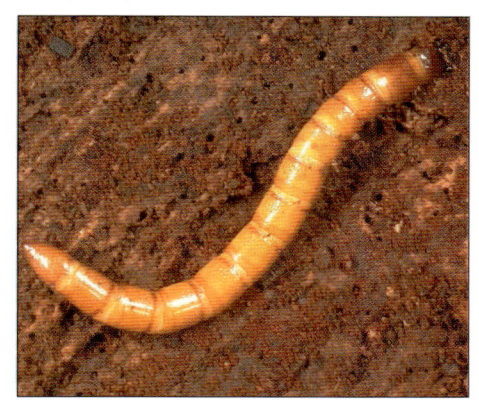

Plate 52. *Ampedus cardinalis* (Schiödte) larva
 ca 16mm © Roger Key

Plate 53. *Anostirus castaneus* (L.) *ca* 10mm
© Roger Key.

Plate 54. *Lampyris noctiluca* (L.) female. *ca* 17mm
© Clive Turner.

Plate 55. *Rhagonycha fulva* (Scopoli) male *ca* 8mm and female *ca* 10mm © Clive Turner.

Plate 56. *Malthinus flaveolus* (Herbst) *ca* 4.8mm © Roger Key.

Plate 57. *Malthodes marginatus* (Latreille) *ca* 4.2mm © Roger Key.

Plate 58. *Ctesias serra* (F.) larva *ca* 4mm © Roger Key.

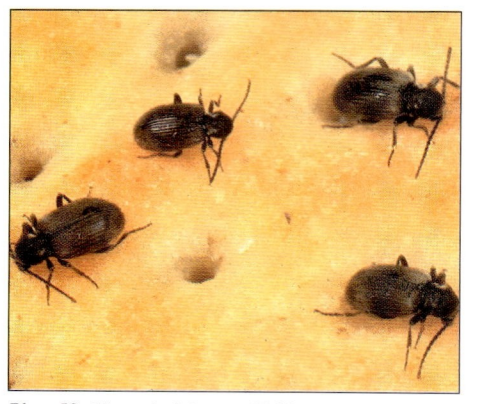

Plate 59. *Niptus hololeucus* (Faldermann) *ca* 3.8mm © Roger Key.

Plate 60. *Ptilinus pectinicornis* (L.) *ca* 5mm © Roger Key.

Plate 61. *Opilo mollis* (L.) *ca* 10.5mm
© Roger Key.

Plate 62. *Anthocomus rufus* (Herbst) *ca* 4.7mm
© Roger Key.

Plate 63. *Bitoma crenata* (F.) *ca* 3.2mm
© Roger Key.

Plate 64. *Triplax russica* (L.) *ca* 6.3mm
© Roger Key.

Plate 65. *Endomychus coccineus* (L.) *ca* 4.8mm
© Roger Key.

Plate 66. *Coccinella* sp. larva *ca* 4mm
© Clive Turner.

Plate 67. *Anatis occelata* (L.) *ca* 8.5mm
© Roger Key.

Plate 68. *Halyzia sedecimguttata* (L.) *ca* 6.2mm
© Roy Anderson.

Plate 69. *Mycetophagus quadripustulatus* (L.)
ca 5.5mm © Roger Key.

Plate 70. *Melandrya caraboides* (L.) *ca* 12.5mm
© Roger Key.

Plate 71. *Tomoxia bucephala* Costa *ca* 7mm
© Roger Key.

Plate 72. *Cteniopus sulphureus* (L.) *ca* 8.5mm
© Roger Key.

Plate 73. *Meloe proscarabaeus* L. *ca* 32mm © Roger Key.

Plate 74. *Pyrochroa serraticornis* (Scopoli) *ca* 12.5mm © Roger Key.

Plate 75. *Osphya bipunctata* (F.) *ca* 10mm © Roger Key.

Plate 76. *Pyrochroa serraticornis* (Scopoli) larva *ca* 18mm © Roger Key.

Plate 77. *Rhagium* sp. larva *ca* 15mm © Clive Turner.

Plate 78. *Leptura quadrifasciata* L. *ca* 16mm © Michal Hoskovec.

Plate 79. *Stictoleptura rubra* (L.) *ca* 18mm
© Michal Hoskovec.

Plate 80. *Judolia sexmaculata* (L.) *ca* 11.5mm
© Michal Hoskovec.

Plate 81. *Molorchus minor* (L.) *ca* 10.5mm
© Michal Hoskovec.

Plate 82. *Hylotrupes bajulus* (L.) *ca* 22mm
© Michal Hoskovec.

Plate 83. *Callidium violaceum* (L.) *ca* 15mm
© Michal Hoskovec.

Plate 84. *Pyrrhidium sanguineum* (L.) *ca* 12mm
© Michal Hoskovec.

Plate 85. *Clytus arietus* (L.) *ca* 13mm
© Michal Hoskovec.

Plate 86. *Pogonocherus hispudulus ca* 6.5mm
(Piller & Mitterpacher) © Michal Hoskovec.

Plate 87a. *Acanthocinus aedilis* (L.)
male *ca* 15mm © Michal Hoskovec.

Plate 87b. *Acanthocinus aedilis* (L.) female
ca 18mm © Michal Hoskovec.

Plate 88. *Saperda scalaris* (L.) *ca* 16.5mm
© Michal Hoskovec.

Plate 89. *Donacia versicolorea* (Brahm) *ca* 10mm
© Roger Key.

Plate 90. *Cryptocephalus decemmaculatus* (L.)
ca 4mm © Roger Key.

Plate 91. *Cryptocephalus coryli* (L.) larva
ca 4.5mm © Roger Key.

Plate 92. *Timarcha tenebricosa*
(F.) 'The Bloody-nosed Beetle'
ca 16mm © Roger Key.

Plate 93. *Chrysolina fastuosa* (Scopoli) *ca* 7mm © Roger Key.

Plate 94. *Galerucella sagittariae* (Gyllenhal) adult, *ca* 6mm and inset, ova © Roger Key.

Plate 95. *Psylliodes luridipennis* Kutschera *ca* 3mm © Roger Key.

Plate 96. *Pilemostoma fastuosa* (Schaller) *ca* 6mm © Roger Key.

Plate 97. *Cassida murraea* L. *ca* 8mm © Roger Key.

Plate 98. *Cassida rubiginosa* Müller larva *ca* 5.2mm © Roger Key.

Plate 99. *Cassida viridis* L. *ca* 8.2mm © Roger Key.

Plate 100. *Curculio glandium* Marsham *ca* 5.2mm © Roger Key.

Plate 101. *Magdalis carbonaria* (L.) *ca* 5.3mm © Roger Key.

Plate 102. *Sitophilus granarius* (L.) *ca* 2.5mm © Roy Anderson.

Plate 103. *Cionus scrophulariae* (L.) adult and larvae *ca* 4.2mm © Clive Turner.

Plate 104. *Apoderus coryli* (L.) *ca* 7.3mm © Roger Key.

Plate 105. The 'Stihl' Vacuum Sampler, demonstrated by Mike Morris, in his accustomed cliff top habitat, at Rinsey Head, Cornwall. Species collected on this occasion included *Opatrum sabulosum* (L.) and *Aizobius sedi* (Germ.), *Cathormiocerus britannicus* Blair, *C. maritimus* Rye, *Caenopsis waltoni* (Boheman), *Graptus triguttatus* (F.), *Barypeithes sulcifrons* (Boheman), *Neliocarus nebulosus* (Stephens) and *Trichosirocalus dawsoni* (Brisout) (Curculionoidea). © Max Barclay.

Plate 106. A rare photograph of an 'autocatcher' in use. Alas, the fine nylon mesh collecting bag at the left hand end of the apparatus is not visible.

Plate 107. Flight interception trap set up in a rather open situation. Note the ropes and bamboo canes under great tension keeping the net as tight as a drum. The collecting trays are resting on two short planks of wood and vegetation obstructing them has been cut. © J.Cooter.

Plate 107a. The same flight interception trap taken down prior to removal. © J.Cooter.

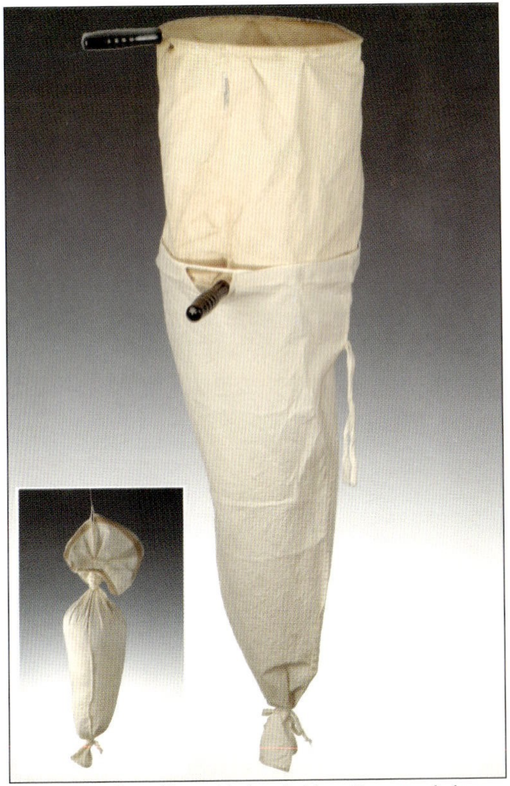

Plate 108. A bag sifter with detachable calico sample bag. The bag sieve is made of stout canvas. Welded stainless steel is used throughout. © Derek Evans Photography.

Plate 108a (inset). A sample bag removed from the sifter and tied off at the top. Another sample bag will be attached to the sifter and sieving continued. © Derek Evans Photography.

Plate 110. A 100ml polypropylene bottle with cork stopper into which the 'wood shavings' and dead beetles will be added. A 25ml polypropylene bottle with dropping neck containing ethyl acetate (see pages 140 and 313). © Martin Rejzek.

Plate 111. Rearing vials for Cerambycidae (and other wood inhabiting beetles). Left to right: standard tube; flat-headed pencil for rolling paper inserts; paper insert; pen with Indian ink for date recording on paper insert; forceps; metal box for up to 50 vials (the polypropylene caps have a small hole bored through for ventilation. See page 131. © Martin Rejzek.

Plate 109. A Winkler-Moczarski type extractor in use with three sample bags inside (this model can take five sample bags). Hanging beside the working extractor is another packed ready for transport. © Derek Evans Photography.

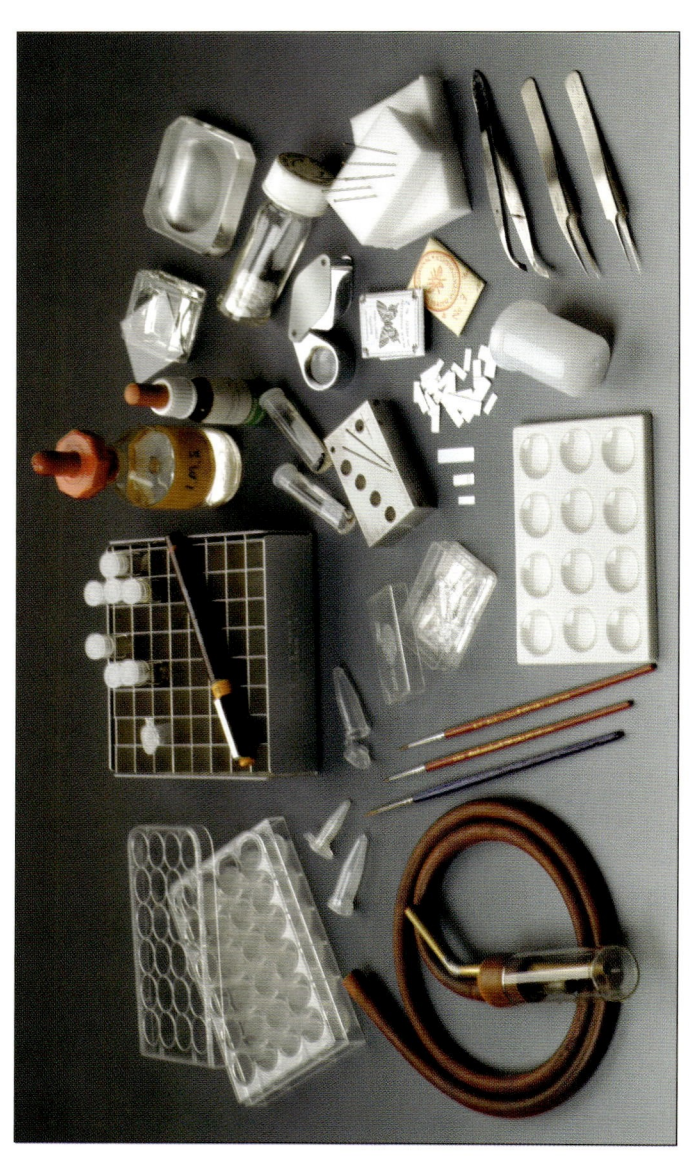

Plate 112. Some of the author's (JC's) essential equipment. Top left, clockwise: A sorting tray, page 315. Eppendorf tubes. A metal box with grid for keeping tubes of material in spirit upright and tidy. Bottle of 70% spirit with pipette dropper. Two tubes of beetles in spirit laying in front. A bottle of clove oil, page 359. Two excavated glass blocks, the smaller with a cover glass, page 316. Killing bottle with wadding at base to receive ethyl acetate. Page 312. X10 hand lens. Packets of pins, page 317. Squares of Plastazote. Pinning forceps with peg. Page 316. Watchmaker's stainless steel, non-magnetic fine pointed forceps, 5A at top, 5 below. Page 315. 35mm film cassette – numerous uses as containers for mounting cards, postal boxes, rearing containers and so on. Mounting cards, a small selection of different sizes, page 318. Above the cards – pinning stage. Biologists spot-testing palette, page 316. Every home should have one. Above palette, a box containing pre-cut plastic strips for mounting genitalia in Euparal, page 360. Three artists top quality sable hair brushed. Sizes 000, 00 and 0, page 315. Pooter, although a piece of collecting equipment, included here to show what a home made pooter can look like. This one has lasted me for 15 years, page 210. © Derek Evans Photography.

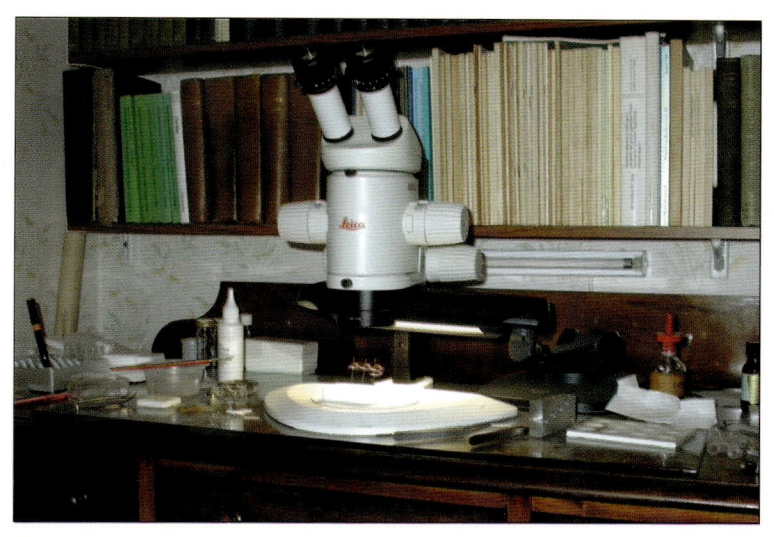

Plate 113. The author's (JC's) desk showing a typically crowded work area and most commonly used books within easy reach. The low energy fluorescent desk light has been modified to fit under the optics carrier of the microscope. © J.Cooter.

Plate 114a. A specimen of *Rutpela maculata* (Poda) being attacked by *Anthrenus verbasci* (L.) adults and larvae. Note also the characteristic small frass pellets. The beetle will be completely devoured in a matter of days.
© Roger Key.

Plate 114. Attack by *Anthrenus*. The transition from pristine insects to dust and fragment takes only a month or so; vigilance against pests has to be a priority. If correctly identified and with data attached, a collection in this state still has more value than one comprised of perfectly mounted entire insects without any data. Name of collection withheld at request of owner. © J.Cooter.

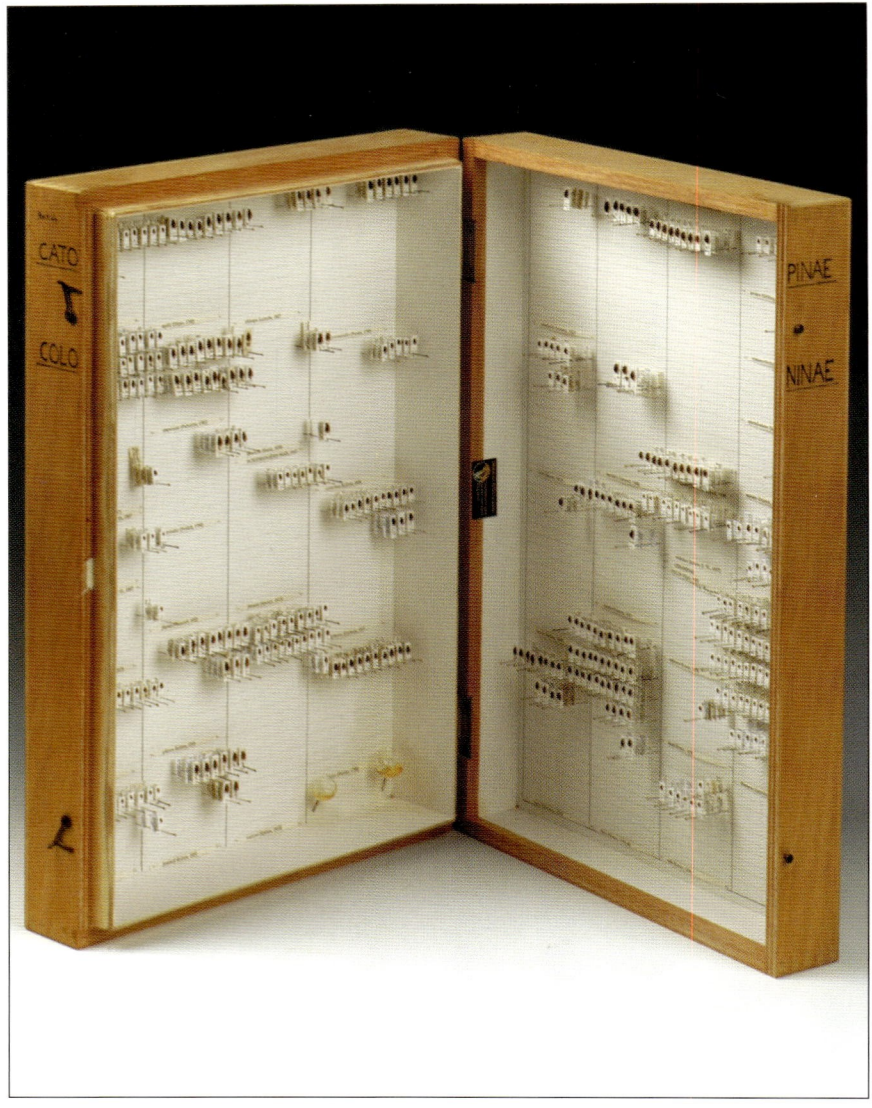

Plate 115. A (330mm x 440mm deep) store box from the author's (JC's) collection (six and a half columns of British Cholevinae and one and a half of British Coloninae) to some extent this betrays the age of the author – a cork/paper lined box. Part of a specialist collection hence the large amount of space allowed for each species, some blanks, not surprisingly, in Coloninae. Two fumigant cups pinned at the base containing essential oil of citronella. Note the marking out of right hand side with a 5mm border all the way around to prevent specimens being pinned in the area where the protruding 'seal' of the left hand side fits when closed. Store boxes of this type should always be stored vertically, with the beetles 'head up'.
© Derek Evans Photography.

Plate 116. Two 'continental carton' type store boxes, see page 330. Both are Plastazote lined, size 300mm x 230mm x 50mm. The box with removable lid is very useful, in use taking up half the desk space of the hinged lid model. Glass fumigant cups containing essential oil are in both boxes, see page 332. © Derek Evans Photography.

Plate 117. A 300mm x 230mm x 50mm 'carton' with hinged lid, part of Martin Rejzek's superbly curated Cerambycidae collection. The use of standard sized cards, standard mounting style, labelling and pin heights make for easy reference and comparison as well as being pleasing to the eye. © M.Rejzek.

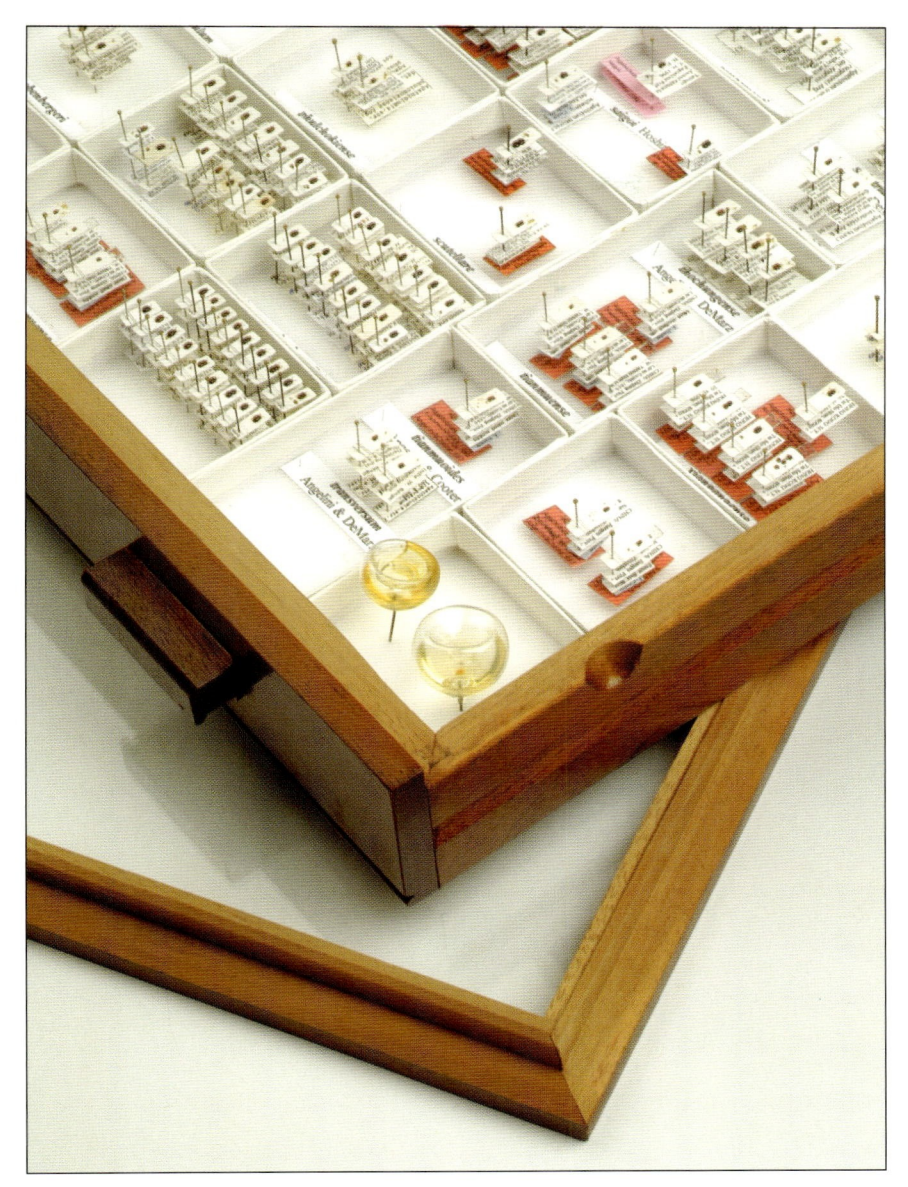

Plate 118. Part of the author's (JC's) specialist Leiodidae collection, in this case one of several drawers of genus *Agathidium*. The cabinet purchased brand new about 1994 marketed as a 'Hill type cabinet' i.e. not made to the Hill patent. The cabinet was ordered without camphor cells or cork / Plastazote lining to facilitate the use of unit trays.

Compared with plate 119 the drawer is of heavy construction, machine made and with relatively thicker sides. The lid is also of robust construction with deep lip extending so far into the drawer that it prevents insects being placed adjacent to any of the four sides, thus restricting the available space in the drawer. The lids themselves are in entomological cabinet terms, not a tight fit. © Derek Evans Photography.

Plate 119. A drawer from the author's (JC's) British collection housed in a 32-drawer cabinet. Note: the light construction of the drawers with raised fillet on the drawer top and rebate in the lid making an excellent seal (compare with plate 118). The drawers have a small camphor cell built in to the drawer front, but glass fumigant cups with essential oils are preferred by the author.

Points of curatorial interest worth mentioning are: two different sized mounting cards in the series of *Leptura quadrifasciata*; the very old flat mounted *Callidium violaceum*; the numerical code on some specimens – this started as a 'good idea' numbering each specimen and keeping note of the data numerically in a ledger. After a few years and hardly ever referring to the ledger the author realised time could be better spent studying beetles rather than documenting his collection. © Derek Evans Photography.

The Beetle Families

Habitat: In rotten wood and woody fungi. Larvae of *Conopalpus testaceus* (Olivier.) and *Orchesia undulata* Kraatz live in rotten, fallen oak boughs, but the pink larvae of *O. micans* Panzer infest the woody bracket fungi of oak and ash. Another fungus-feeding larva is the rare *Hallomenus binotatus* Quensel in the 'beef-steak' fungus of oak. Larvae of *Melandrya caraboides* (L.) often occur quite commonly in stumps of oak and birch.

TENEBRIONIDAE

COCCINELLIDAE

NITIDULIDAE

PYROCHROIDAE

MELANDRYIDAE

RHIPIPHORIDAE

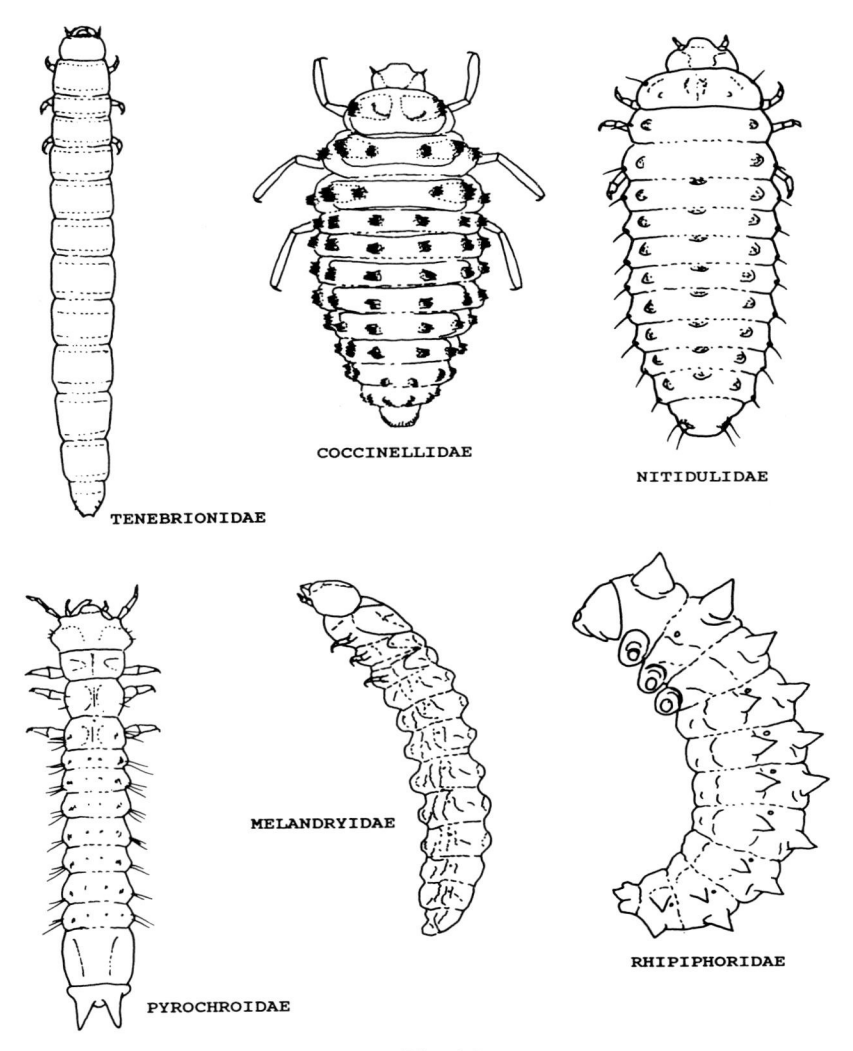

Fig. 36

RHIPIPHORIDAE (Fig. 36)

Soft, fleshy larvae, with reduced or vestigial appendages; dorsal surface with conical projections; legs with a single sucker-like claw. Undergoes hypermetamorphosis; the first stage (triangulin) larvae is campodeiform, with long legs, bristle-like antennae and cerci, and 4 or more ocelli on each side of the head.

Habitat: In cells of wasp nests, ectoparasitic on larvae of *Vespula, Metoecus paradoxus* L. is the only British representative of this family. Wasp nest should be examined in late summer for the curious parasitic larvae; the capped, less opaque cells are the most likely to contain the full-grown larvae. To 'take' a nest, one should stuff the entrance as deeply as possible with rags and saturate them with chloroform. This should be done at dusk when most of the wasps are inside, care being taken to drive off any late-comers. The nest should then be dug out early the following morning, and any wasps that show signs of activity killed.

MELOIDAE (Fig. 37)

Similar to Rhipiphoridae, but without dorsal projections on the body; triangulin stage with at most two ocelli on each side.

Habitat: In nests of Apidae where they feed on eggs, food masses, etc. The minute first-instar larvae, which have three claws on each side of their tarsi, rest on flower-heads until the latter are visited by bees, etc., whereupon they attach themselves to them and are carried away; naturally many fail to be successfully transported to the appropriate nests.

CERAMBYCIDAE (including PRIONIDAE, LAMIIDAE) (Fig. 37 and plate 77)

Fleshy, sub-cylindrical larvae; prothorax broader than remainder of the body; head retractable into the thorax; legs very short (not protruding beyond sides of body) or absent; sometimes with protruberances on the dorsal surface which assist them in moving along their galleries.

Habitat: In living or decayed wood such as the boles of trees, rotten stumps, fallen boughs, roots, posts, palings and twigs; some are confined to the stems of Umbelliferae, etc. For detailed and information see the chapter 'Cerambycidae' (pp. 121-142).

CHRYSOMELIDAE (Fig. 37, plates 91 and 98)

Fleshy larvae, body either slightly curved ventrally or straight; legs usually short, 3- or 4-segmented (or absent in leaf-mining species); mandibles without mola; cerci absent. They are the most 'caterpillar-like' of beetle larvae, some resembling lepidopterous or sawfly larvae, but they do not have ventral abdominal prolegs like these other Orders.

MELOIDAE

CERAMBYCIDAE

CHRYSOMELIDAE

CURCULIONIDAE

Fig. 37

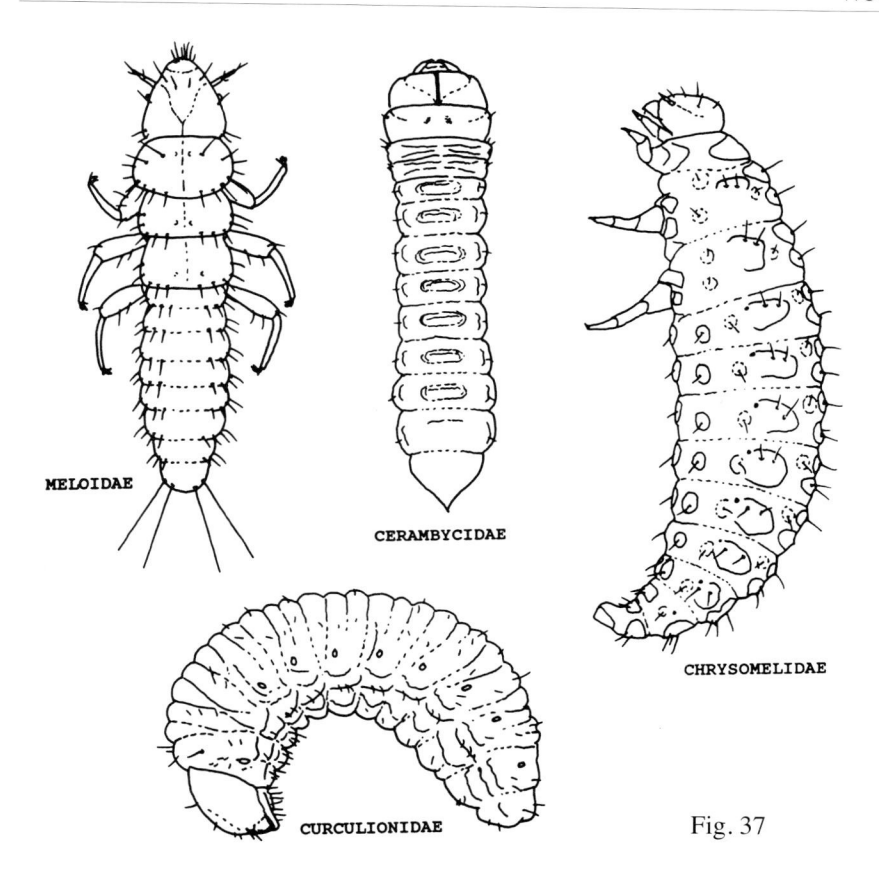

Habitat: Mostly on leaves, often large colonies. The lush semiaquatic flora provides good hunting-grounds, particularly the water dock for *Galerucella* spp. etc. Leaves of guelder rose are often reduced to a network through the ravages of larvae of *Pyrrhalta viburni* (Paykull). Poplar, sallow, hazel, aspen, broom and bedstraws are favourite hosts. The quaint *Cassida* larvae, which are short, oval, slightly depressed and distinctly spiny with a lateral fringe of processes, construct for themselves an 'umbrella' of frass and exuvia. They feed externally on leaves of thistles, goosefoot, water mint, etc. The aquatic *Donacia* larvae, which extract oxygen from submerged plants, are most easily procured in the autumn by pulling up these plants and examining their roots for the plump white larvae which make clusters of brown cocoons. Larvae of *Clytra* may be dug out (in their cases) from nests of *Formica rufa*. Larvae of many flea beetles (Alticinae) are miners in leaves, stems or roots of plants.

CURCULIONIDAE (Fig. 37 and plate 103).

Rounded and ventrally-curved apodous larvae with a slightly 4-lobed anal segment. The absence of legs distinguishes them from Anobiidae, etc. No characters are known which distinguish 'Scolytidae' from all other Curculionidae.

Habitat: In wood, seeds, roots, stored products, etc. The green larvae of *Hypera* are to be swept from *Ononis, Trifolium, Lotus* and *Rumex*. Leaves of *Scrophularia* and *Verbascum* should be examined for the slug-like larvae of *Cionus* which pupate in fibrous cocoons. Certain species are stem feeders such as *Mecinus* in *Plantago maritima*. Larvae of *Cleonis* should be searched for in thistles during August when brown withered stems will indicate their presence; the roots should be examined for swellings. Hazel nuts are often rendered inedible by the presence of *Curculio* larvae. Many '*Apion*' larvae infest seed pods of Leguminosae. Beetles of the genera *Attelabus, Apoderus,* and *Deporaus* roll up the leaves of oak, hazel and birch respectively and deposit their eggs therein; these leaves should be gathered in the autumn before they start to fall. Many species cause considerable damage to the roots of plants and trees; for example, the plump larvae of *Hylobius abietis* (L.) in roots of pine, *Sitona* larvae in clover and vetch, and larvae of *Leiosoma deflexum* (Panzer) in roots of *Ranunculus*. Larvae of *Cryptorhynchus lapathi* (L.) make large galls in willow stems. Many species feed subcortically in stumps and boles. The uncommon larvae of *Magdalis carbonaria* (L.) feed under the bark of birch stumps. Larvae of *Platypus cylindrus* (F.) occur locally in old oak stumps where they make vertical galleries in the heartwood. 'Scolytid' larvae form conspicuous and often intricate patterns beneath the bark of elm (*Scolytus scolytus* (F.) and *S. multistriatus* (Marsham)), oak (S. *intricatus* (Ratzeburg)) and of various fruit trees (S. *mali* (Bechstein), S. *rugulosus* (Ratzeburg)). Sallow catkins should be gathered in April; from these may be reared larvae of *Dorytomus,* etc.

5. PRESERVATION

Beetle larvae, as compared with those of Lepidoptera, are not particularly colourful and this, together with the fact that the majority are so small, renders them unsuitable for general exhibition purposes. Certain species, however, such as the large lucanid and chafer larvae and the more sclerotised elaterid and pyrochroid larvae, if well preserved, make useful and attractive additions to Coleoptera collections and add to their general interest; moreover, if pupae and cocoons are included, life-history series can be prepared illustrating the great diversity of various groups.

The larger, soft, fleshy larvae can be prepared by the method generally adopted for 'blowing' lepidopterous larvae, i.e., by enlarging the anus, pressing out the abdominal contents and inflating with hot air until rigid. If, in addition to this, they are then filled with suitably tinted melted wax and allowed to 'set' they present a very life-like appearance. Incidentally, cerambycid larvae, especially *Prionus,* cannot be 'blown' satisfactorily, because they turn black in the drying process. The wax-like, semi-transparent quality of the integument is best preserved by adopting the following procedure:

Place the larva in a test tube containing 30% alcohol (to which one or two drops of glacial acetic acid have been added) and heat at just below boiling point for twenty to thirty minutes; then store in 80% alcohol.

Another satisfactory method, most suited to the smaller, more sclerotised larvae, is as follows:

Immerse the larvae for one week (or two, if very large) in 95% alcohol. They should then be dehydrated by immersion for one to three weeks (according to size) in absolute alcohol which should be changed weekly. This should be followed by immersion successively for twelve hours in mixtures of one part xylol to two parts absolute alcohol, two parts xylol to one part absolute alcohol and pure xylol respectively. Finally they should be allowed to dry off and then be mounted in the same way as adults.

The foregoing methods are, of course, unsuitable if one wishes to form an extensive collection of larvae with a view to studying their taxonomy. For this each batch of larvae should be killed, preferably by immersion in a standard fixative such as Carnoy's fluid, made up from – glacial acetic acid 10 parts, 95% alcohol (industrial) 60 parts, chloroform 30 parts.

Details of this and many other useful preservation techniques are given in the book by Wagstaffe & Fidler (1955). If a fixative is not available, kill larvae in 70% alcohol, to which a few drops of ethyl acetate have been added. They should then be stored in 70% alcohol; the addition of 5% glycerol prevents the liquid from drying out, and keeps the larvae softer.

Larvae should be stored in individual small glass or polythene tubes, corked with plastic stoppers, or plugged with cotton wool. In the latter case they should then be kept inverted in a well-sealed bottle (such as a Kilner jar) which is itself filled with 70% alcohol, and checked regularly to prevent it drying out. If alcohol-preserved larvae do dry out and shrivel up, they can often be restored by immersion for up to 24 hours in cold 5% potassium hydroxide solution (followed by washing in water prior to re-storage in 70% alcohol).

Data, biological notes and any observations made during collecting larvae should be carefully recorded in a field notebook. An example might be:-

Jar 2   *Ilybius fenestratus* (F.) DYTISCIDAE
Tube 18   Surrey: Mytchett. 22.iii.1989
Dug up under grass, near lake; 2 specimens preserved.
Tube 274   Surrey: Ash Vale. 27.ii.1990
Dug up from banks of canal; two specimens and exuvia preserved; two being reared.

And on the other side of the page, observations thus:

Beaker 274. Larvae kept in damp soil in beaker; food, small worms. 20.iv, two larvae became inert and slightly curled having pressed out pupal cells. 26.iv, both pupated; exuvia preserved; soil moistened every third day. 11-12.v, adults emerged; took two days to become pigmented. Larvae apparently pupate several yards away from water.

All larvae should be labelled, using labels written either in indian ink, or with one of the readily available disposable micropoint nylon tipped drawing pens. But be sure to use one that is waterproof, and test the ink for alcohol-fastness before using it for labels. Each tube should have a label giving date as in the notebook referred to above, together with a unique number which relates the tube (and specimens in it) to that book.

For the serious study of larvae, their identification requires examination of cleared and mounted specimens under the high magnification of a compound microscope, using transmitted light. There are several standard methods available of making such slide preparations of larvae. One of the quickest is as follows:

Heat larvae gently on a hot plate in 5 or 10% potassium hydroxide, until the internal tissues have dissolved. Wash them in water, and then warm them (on a cooler hotplate) in chloralphenol (a solution of chloral hydrate in phenol). This renders them transparent; they can then either be examined on a slide in the same liquid, or mounted in Berlese's mountant, made from – distilled water 20ml, chloral hydrate 150g, gum arabic 15g, glucose syrup 10ml, glacial acetic acid 5ml.

This and similar mounting media are also available from commercial entomological or biological suppliers. It has the advantage of being reversible, that is the cover slip and larvae can be soaked off the slide by immersion in water, if needed at a later date.

FIGURES 31 - 37

The figures of whole larvae (Figs. 31-37) have been drawn specially from preserved specimens wherever possible. In a number of Families, however, larvae were not available, and some existing figures have therefore been re-drawn for this Handbook. These have been based mainly on the classical works of Boving & Craighead (1931), Ghilarov (1964) and Klausnitzer (1978).

The figure of a weevil larvae is modified from Crowson (1981) , and that of a buprestid from Bílý (1982). The drawings do not always represent any particular species, but have been modified so as to show 'typical' features of the family, sometimes based on more than one species.

References

Bílý, S., 1982. The Buprestidae (Coleoptera) of Fennoscandia and Denmark. *Fauna Entomologica Scandinavica* 10; i-x, 1-210. Brill Publication Co., New York.

Boving, A. G., and F. C. Craighead., 1931. *An Illustrated Synopsis of the Principal Larval Forms of the Order Coleoptera.* The Brooklyn Entomological Society, Brooklyn, New York. 351 pp. [Reprinted 1953].

Crowson, R.A., 1981. *The biology of the Coleoptera.* Academic Press, London, 802 pp.

Ghilarov, M.S., (ed.) 1964. *Keys for the identification of soil-inhabiting insect larvae* Moscow, 919pp. [In Russian].

Klausnitzer, B., 1978. *Ordnung Coleoptera (Larven).* Dr. W. Junk, The Hague. 378 pp.[In German].

Majerus, M.E.N. & Kearns, P.W.E., 1989. *Ladybirds*, Naturalists' Handbooks 10,. Richmond Publishing, Slough 103pp.

van Emden. F.I., 1942. Larvae of British Beetles. III. Keys to the Families. *Entomologists' Monthly Magazine,* 78: 206-272

Wagstaff, R., and J. H. Fidler., 1955. *The Preservation of Natural History Specimens. Vol. I. Invertebrates.* H. F. & G. Witherby, London. [reprinted 1961]

Collecting

COLLECTING EQUIPMENT

by J. Cooter

The novice can do no better than learn from the experience of others, either by asking advice or by careful field observation. Personal experience is a great teacher and one's methods and equipment become adapted to those habitats most often worked or those beetles most avidly sought. One lesson is very quickly learned – collecting equipment should be light but robust and minimal. My own collecting equipment generally consists of a few tubes, pooter, killing tube, sifter and sample bags, trowel and a dinner fork, occasionally a sweep net, rarely a beating tray (the two being mutually exclusive for my activities). The tubes, pooter and killing tube fit into pockets, the rest into a back pack along with lunch and a drink.

Most of the popular items of equipment can be purchased from suppliers. These are designed for general entomological use and their efficiency for collecting beetles can be enhanced by simple modifications if desired. The best equipment is that which you make yourself or which is made by an entomologist on a small scale, often at request, for favour or sale. In the following notes various modifications have been shown and it is hoped the diagrams will serve to make these self-explanatory (none are beyond the capabilities of anyone interested in DIY or model making; 'evening classes' at high schools or technical colleges might offer an opportunity for equipment conversion or manufacture, especially if 'model engineering' is offered or facilities are available).

Collecting equipment falls into two categories:
 1. That needed for finding beetles.
 2. That needed for securing their capture.

The equipment described here is not exhaustive, only the most essential items are described.

Sweep-net

The main requirement for a good sweep-net is a light-weight frame and a bag of strong light-weight fabric deep enough to fold over the frame to prevent any active beetles taking flight.

My own is a lepidopterist's cane-framed kite net with a light canvas home made bag reinforced around the mouth with heavier quality canvas. It has stood the rigours of 35 or more year's sweeping and I find a handle unnecessary. My experience of the specially designed shop-bought sweep-net with central bar/handle has not been very good. I find the bag is too shallow and the centre-tube is a confounded nuisance. They have considerable weight making their continued use on evening sweeping quite tiring. However, this might be personal prejudice as they are widely used, and some friends of mine with an interest in weevils swear by them, as their weight and centre-tube allow easy sweeping of quite thick scrubby vegetation. A light aluminium/nylon framed net, similar to my cane net described above is also available commercially. I have no experience of the latter but often hear colleagues express the desire that 'someone one day will make a good sweep net'.

Abroad, especially in tropical, sub-tropical and hot dry regions, a net with very deep bag affixed to a very long pole (ca 5m +) is invaluable for collecting Buprestidae and Cerambycidae. The net, of tough light-weight nylon, should be approximately 1m diameter and 2-2.5m deep. It is used to 'sweep' the peripheral crown branches of forest trees. The results can be astonishing, for example in temperate China my colleague 'swept' deciduous trees and found as well as the expected Cerambycidae and Buprestidae, Melyridae etc, Pselaphinae, Cryptophagidae, Latridiidae, Elateridae and many other families in numbers and variety. The same 'small fry' were otherwise very difficult to obtain by the traditional sieving/extracting methods; indeed at the location in question, during a whole week I found no Cryptophagidae or Latridiidae in the daily renewed sieved leaf litter samples placed in the eleven extractors I was using.

Beating tray

A sweep-net can double as a beating tray, but the tray cannot be used for sweeping; beating is after all a vertical variation of sweeping. The traditional 'Bignall Pattern' tray is readily available with white, or upon request, black cloth. The white cloth tends to give a high degree of glare on a sunny day, but many beetles are scarcely visible against a black background. To dislodge insects, one or two sharp taps with a stout stick or vigorous shaking will suffice; most of those dislodged should fall onto the tray, some will take flight quickly, others will feign death and remain still for a long time, others will miss the tray completely.

For beating higher branches, a method I have used successfully when beating large masses of mistletoe high up in standard fruit trees for *Ixapion variegatum*, is to use a decorator's dust sheet. Place this under the part of the tree to be beaten and use a suitable long pole to beat. The large area of sheet helps to cope with any insects that might 'travel' either as a result of impact or by being carried on the breeze during their fall.

Water nets

For collecting water beetles a suitable net is essential. The net must have a strong enough head to deal with the weight of water and material that enters it, a strong handle and the net must be of a material which will allow the water to drain through quickly without releasing the smallest beetle being sought. Home made and adapted nets are usually not much use and it is better to buy a purpose-built net which should give years of service and not let you down while on a trip.

The 'standard net' available from Educational Field Equipment UK Ltd (contact details at end of chapter) is almost universally recommended by water beetle specialists and other types of fresh water biologist. It is of robust construction and has a proven track record of reliability.

The net contents are washed into a white dish and the object of the hunt transferred to suitable containers. Some people prefer to suck the beetles up into a glass tube with a rubber bulb at one end; others prefer fine flexible forceps.

NB. Water nets are discussed at length in the *Balfour-Browne Club Newsletter* number **13** (July 1979).

Sifter and tray/sheet

Coleopterists with an interest in Staphylinoidea and various 'small fry' will use a sieve more frequently than a sweep net or beating tray; indeed the latter two may not figure at all amongst their equipment. The most useful type of sifter is illustrated in plates 118 and 118a.

A sample bag is attached to the sieve and the bottom tapes are tied, leaf litter or rotten wood is gathered into the upper section until it is about half full and then sieved. The finer debris and insects fall through the mesh into the sample bag. After a good shake the top section is emptied by tipping and shaking, more refuse added and sieved. When sufficient debris has accumulated in the sample bag, its top is tied and the bag removed and another bag put in its place. This process continues until debris sufficient to fill all of one's extractors or a sufficient quantity of the litter / wood sample has been processed.

The sieved material is taken home and placed in an extractor (see below).

A simpler, but less efficient sieving method is simply to use a small garden sieve and a tray (or a normal sized garden sieve and a very large tray). Litter is placed into the sieve and sieved onto the tray. Good trays for this purpose are the larger sized photographic print developing trays readily available from photographic shops. These are robust and have suitably high sides. A useful and lighter alternative is a tray of the type used in a butcher's shop for displaying cuts of meat. Again heavy quality but with lower sides; a rigid tray is preferable to a plastic sheet which develops creases and always has to be placed on the ground where it never lies flat – a tray can be held conveniently while standing and can double as a small beating tray for bracket fungi etc. The tray should be large enough to just fit into your equipment bag.

However, a sheet does have its uses, being much larger than a tray, piles of refuse can be placed upon it. A sheet is useful if investigating grass tussocks.

Experience has found it better to rough-sort bulky material on a sheet, remove large stones and sticks etc then tip the remains into the sieve to reduce the bulk and concentrate the insects.

Extraction funnel

Outside of research institutes, this equipment has largely been superseded by the extractor described below. It can be made by anyone with an interest in DIY. All that is needed is a large funnel – this can be purchased from a laboratory supply company or home-made from thin sheet aluminium. A large diameter plastic funnel can have its diameter increased by fixing a plastic bowl to the top (first cut out the bottom of the bowl) with a strong synthetic adhesive. If purchased, the spout has to be cut short. A mesh grid is placed inside at the bottom of the funnel cone, this will prevent excessive amounts of debris entering the collection jar. The whole will need support in a home-made cradle. A conical lid or cover with light bulb is also required.

A large tube or glass jar collecting vessel is attached to the tube of the funnel making sure there are no gaps between funnel 'spout' and collecting vessel. Preservative can be added to the vessel if needed, but inspection of the collecting vessel two or three times a day will suffice. A screw of paper in the dry collecting vessel will afford the inhabitants some protection. Often a predator will damage a number of its fellow occupants and a spider or two will cause havoc by spinning threads. Extraction may take several days.

Debris is placed into the funnel, the lid placed on and the light bulb turned on. It is thought by some that the gentle heat causes the debris to dry out from the top, forcing the inhabitants to the lower more moist regions, and eventually into the collecting jar. It is a good idea to examine the jar regularly and so deal frequently with small batches, rather than sort one large mass.

Extractor

The extractor is nowadays becoming popular in Britain after many years of use by our neighbours across the Channel. Basically the extractor consists of an oblong box-like cage of cloth supported by wires, tied at the top to prevent escapes and at the bottom tied to a collecting jar. Inside are a number of net bags. The whole is suspended from a rafter in the garage, hotel wardrobe rail or similar support, see plate 109.

A jar is fitted to the extractor, debris is placed in the net bags, top tied. The beetles and other inhabitants will crawl out and fall or otherwise make their way into the collecting jar. With the larger surface areas exposed, this extraction method is much quicker than the funnel. Generally, the bulk of the inhabitants will have been extracted within a couple of days. It is generally assumed that the beetles in the litter filled bags dislike disturbance so become active and fall into the collecting jar. The litter in the bags can be 'disturbed' regularly by giving each a stir with the hand. NB when emptying the extractor bear in mind a very few beetles will somehow climb up to the top. Having found one *Procas granulicollis* Walton by sieving in the field, I spent an hour or so sieving a bulk sample for extraction at home. After five days in the extractor the sample was bone dry and no weevils had appeared in the jar. When I untied the extractor five *Procas* were sheltering in the folds at the top where the cloth was bunched just below the tie.

Details for making a home made extractor have been given by Owen, 1987.

Traps (General)

It is a good idea to encourage beetles to catch themselves by using a range of traps. These can be quite simple and offer the beetles a concentration of a biotope otherwise scarce in that locality. In literature one can read of bunches of flowering *Heracleum* being placed in Scottish woodlands attracting a range of scarce Highland beetles – a very hit and miss method but it would seem to produce good results if weather and other conditions are favourable (the sort of 'trap' worthy of employment if based near a Caledonian pine forest remnant for any period of time).

Piles of fungi and decaying fruit/vegetables can be left and regularly sieved over the weeks. I once had good results by placing a dead cat on top of a large biscuit tin filled with sand. As the corpse decayed a succession of beetles came and went. By sifting through the sand a very rich fauna was sampled which, if on bare earth, would have been scattered in a much larger volume.

Lawn mowings can be placed in paper sacks or plastic bags for a few days to rot a little. This is then taken into woodland and placed in piles at bases of trees. After a week or so and at intervals thereafter, the pile can be sifted. Such rotting vegetation attracts a wide variety of uncommon Staphylinidae (including Pselaphinae), Ptiliidae, Scydmaenidae, Cryptophagidae and other small-fry.

Faggot piles – bundles of twigs made up in woodlands and stacked to dry out, although largely a product of past woodland management, can be made. They are untied and beaten over a sheet or beating tray. Often uncommon species have been produced in this way.

A 'bird-nest' trap was described by J.A. Owen (1976) who used them with success in many parts of Britain. They can be left and inspected at long intervals or more regularly if time and travel permits. A variant of this – a large plastic bottle with 'windows' cut in the top, quarter filled with beer and suspended high up in a tree was described to me when working the Natural History Museum, Geneva by Dr Claude Besuchet. One such trap in a wooded area we visited was contained a number of *Cerambyx cerdo* L.

Each year when digging up my potato crop, I inspect the rotten seed potato planted during the spring. Invariably I find *Parabathyscia wollastoni* (Janson) and *Anommatus duodecimstriatus* (Müller) present in varying numbers, often accompanied by a range of small Staphylinidae. It would seem logical that vegetable peelings buried on purpose and left for a few months (say April to August/September) might produce similar results. A plastic mesh bag of the type citrus fruit is sold in might form a useful receptacle for the bait.

Some types of trap can be used with success throughout the year, especially unbaited pit-fall traps, which can produce good results through November to March. With a glycol-based fluid (car antifreeze) in the trap the contents will not freeze, but the traps should still be inspected regularly, ideally each week.

Although flight interception traps seem to be most productive from spring to autumn when the bulk of beetles are active, they continue to function well throughout the winter.

Pit-fall traps

This type of trap can be used with a bait or without. When used with a bait, the vast majority of the beetles captured will of course have been attracted to the bait. Without bait, the catch is smaller but more varied and often some quite rare species might be found.

As with all trapping methods some places give better results than others, often an apparently 'likely' spot will give disappointing results. Fortunately traps can be moved from place to place quite easily and pit-fall traps can be set out in some numbers in various places. Be warned though, it takes a lot of time to service traps and to sort the accumulated catch; use only sufficient that can be dealt with.

A basic pit-fall trap is a plastic container of the type and size of a yoghurt pot or drinks dispenser beaker. It is sunk in the ground so that its top is level with the surface. If inspected regularly no preservative fluid or simply water plus a few drops of detergent need be added. If inspected at weekly intervals, use a glycol-based car antifreeze solution; fill the pot about a quarter full and make a small hole in the 'trap' about one-quarter from the top to allow excess rain water to escape. The trap should be placed away from the curious passer-by or vandal (often the two are synonymous). It is advisable to place a 'roof' over each trap, this might be a large piece of bark or a flat stone, anything that will keep rain water out. The 'roof' should be raised a little above the top of the trap. Some wastage will occur as a result of attention from foxes and other scavenging animals. A series of traps is best set out in any one locality, either randomly or in a line. Remember their location, as vegetation grows during the year the precise sites might become difficult to locate.

Baited traps have been used with great effect for special purposes. Welch (1964) describes a method for collecting insects from rabbit burrows. He sunk a jam-jar at arms length inside the burrow and used dead roach (handily available at the time) as bait. From March to late June 1963 he thus captured beetles belonging to seventeen families, including 1,182 Leiodidae from 13 traps in one warren. Another remarkable capture of Leiodidae was achieved by Ashby (see Cooter, 1989) using unbaited traps.

In tropical or sub-tropical regions a square of wire mesh placed over the trap will prevent ingress of larger beetles and land crabs which will cause damage to the smaller more delicate species. Very often in warmer climates larger Scarabaeidae are commonly trapped in this way along with Silphidae. A larger container, say a 1-litre ice cream tub or a small plastic bucket might be preferable to a 0.5-litre yoghurt carton.

A variation on the normal pit fall trap has been described by Turner (1997). This consists of setting a number of un-baited traps in a circle and in the middle of that circle suspending a bait. The beetles are attracted to the bait and those walking to it fall into the traps. The advantage of this method is that the 'catch' having not come into contact with a suppurating bait is a lot cleaner, and the bait itself is protected from foxes and similar vertebrate scavengers.

Subterranean pit-fall traps

We are way behind coleopterists on mainland Europe in employing these traps. Possibly necessity is the key here, in Britain we lack the rich subterranean fauna found south of the Alps and in eastern Europe.

Details of a servicable subterranean pit-fall trap have appeared in the entomological journals (Owen, 1995). Use of such traps has revealed some subterranean species, such as *Rhizophagus oblongicollis* Blatch & Horner and *Ferreria marqueti* (Aubé) to be much more widespread than previously believed.

Trapping in scree

This is involves burying the pit-fall trap at about arm's length in the scree, it is a method pioneered by Růžička (see for example Růžička, 1999). The same method can be employed in caves with a floor composed of rubble. Both methods are very productive abroad, especially in cave entrances – interesting cave dwelling Leiodidae: Leptodirini can be obtained by this method using a bait of meat or cheese. Due to the unstable nature of scree and cave rubble the arm's depth hole ends up with very large top diameter. Place a large stone over the trap vessel then carefully re-fill the hole with the rubble just excavated. Be sure to mark the location of scree traps Some years ago I set several scree pit-falls in the Kazakhstan Tien Shan relying on visual markers for location. Alas, upon return the micro-landscape had changed due to snow melting and seepages drying up. A pile of a few large stones is a suitably unobtrusive marker.

I use this method in caves when on holiday in France. Servicing the trap requires all the rubble in-fill to be carefully removed, but over a holiday fortnight one can expect good results after one week, thus only one burial / excavation cycle is required. Cheese bait in caves works well, unlike fish or meat it does not decompose into a sticky goo which will drown or coat the insect inhabitants. A small screw of paper inside the trap will offer protection; I suspend the bait from a bent wire resting across the top of the trap. The bait situated on the bend will hang down into the trap and remain an effective attractant, but one that the insects cannot reach.

Flight interception traps

This simple trap, used with great success for a number of years by American and Canadian entomologists, especially when collecting in the tropics, is now quite regularly used in Britain. Suitably located, it can trap a range of beetles not normally encountered, or only rarely, by other collecting methods.

An excellent flight interception trap was manufactured and marketed by R. George, Marris House Nets. I have used 'Marris House' flight interception traps with great success in England, France and China.

The traps are best set up on private land away from public gaze and vandalism. Always get the owner's permission and take time to select a likely site – a woodland ride or forest edge or some natural corridor seem likely places, but make sure these are not frequented by deer or other large animals (I had a trap ruined by a deer one autumn after being in position for seven months). Always avoid places where domestic animals roam. A trap set adjacent to a fallen tree can collect a large array of dead wood species.

The flight interception trap (see plates 107 and 107a) is set up as follows:

After selecting a suitable site roughly at right angles to the prevailing wind direction and with ground roughly horizontal, push a 2m long bamboo cane vertically into the soil. Roughly level the ground over the length of the trap – an easy way to do this is to use a single plank about 10cm shorter than the net. The plank can sit on a number of tent pegs hammered into the soil so that the plank is horizontal. Place one fluid container at one end of the plank adjacent to the bamboo cane. Slide the bottom corner loop on one side down the cane until the net touches the fluid container. Slide a screw adjusting hose clip down the cane and tighten it so that the net fits onto the top of the fluid container, but not pressing it down. Slide a second hose clip onto the cane, slide the top corner net loop over the cane. Pull the net vertically so that the side is taught. Screw up the hose clip. Thus one side of the net is set up. Insert the second bamboo cane through both loops, slide two hose clips onto the cane. Pull the net tight and at the other end of the plank push a second bamboo cane into the soil. Fix this side of the net as previously. The net should now be held firmly between the two canes. Insert a tent peg through the central bottom loop of the net and hammer it into the soil so that the bottom of he net in the centre is flush with the top of the collecting trays. Place two guy ropes over one bamboo cane such that their looped ends fit over the top hose clip. Hammer two tent pegs into the soil about 3m from the cane and fix the free ends of each guy to the peg such that and angle of about 45 degrees is formed between the two ropes. Repeat on other side. Adjust the guy ropes so that the net is stretched tight. Take a long guy rope and fix one end to a stout tree trunk or base of a branch about 2m above the top of the net. Thread the rope through the top loop of the net and secure the end to a similar stable point (another tree trunk for example) on the opposite side of the net. Tighten this guy rope so that the top of the net is pulled upwards. A length of cotton thread can be tied between the two end canes about 3-5cm above the net (this simple addition stops birds perching on the net and defecating onto the net or into the fluid containers). Place the rest of the fluid containers under the net and clip them together with large paper clips. (I use the transparent plastic half-sized seed tray propagator lids. I cut the flange off the longer sides to enable then to clip closely together. They are quite deep so the fluid will not totally evaporate in hot weather and rarely overflow during periods of rain) Other workers I know favour a pair of white plastic 'window boxes', or, especially in the tropics where weight and disposability are important considerations, aluminium 'takeaway food' containers. Fill each container about one-quarter full with water, add a tea spoon of chloral hydrate (bactericide) and one drop of washing up liquid (to break the surface tension) to each container. Make a final adjustment of the guy ropes – the net should be stretched as tight as a drum. The trap is now running and has probably already caught a couple of beetles. (N.B. chloral hydrate solution will corrode metal, prolonged use will puncture aluminium containers.)

In Britain flight interception traps are generally used with a roof to keep rain out of the containers.

Beware, one trap requires regular servicing. I once made the mistake of locating one about an hour's drive from home. Two hours drive, a 30-minute walk to and from the trap site plus 20 minutes to service the trap. It was only in position one month!

Servicing involves removing the accumulated 'catch' and re-charging the fluid filled trays. If located some way from a water supply you can save a lot of effort carrying water by adopting the simple technique of taking an extra tray with you.

Unclip the trays, pick out any slugs, twigs and leaves (give the twigs and leaves a 'wash' in the fluid to dislodge any adhering insects). Line a large funnel with nylon mesh – net sold for Lepidoptera breeding cages or butterfly nets is ideal. Hold the funnel over the empty (spare) tray, empty the first trap tray by pouring it into the funnel and draining the fluid into the 'spare' tray. Take the next container, place the funnel over the container just emptied, pour through funnel. Carry on until you come to the last container, refill that one with the fluid from the first one emptied (in the 'spare'). This re-use of solution can be carried out two or three times only, eventually fresh water and chemicals are required. Re-clip the containers, check the guys are tight. If on private ground, you might be able to leave all the servicing kit on site.

You now have a funnel with the contents of the trap resting in a nylon mesh square. Carefully bunch up the corners of the mesh and empty the mass of insects and debris into a suitable large mouthed container (one with a watertight lid). Wash the nylon mesh with ethanol or industrial spirit and add sufficient spirit to cover the 'catch'. Seal the container.

The 'catch' is best sorted a little at a time in a petri dish under the microscope at home.

Beware: one trap can easily produce sufficient material to occupy all your free time. Some species will be in large numbers, but there will always be singletons, most of which represent scarce species and often minute ones.

Pooter

This is perhaps the most essential item of equipment; it is a simple device for the rapid collection of small insects. Coleopterists tend to prefer the tube-type pooter (Fig. 38a and plate 112) rather than the cylinder model (Fig. 38b) much favoured by our near relatives, the dipterists. The tube-type does indeed have advantages when used for capturing insects that walk rather than fly. The head can be removed and quickly replaced with a stopper, the head being fitted to a fresh empty tube. Emptying a cylinder-type pooter on a hot sunny day requires a swiftness of hand more associated with illusionists; remember dipterists kill or anaethsetise their catch in the pooter before emptying. A spider inadvertently caught should be removed at once as invariably its silken thread causes problems.

The shop-bought tube pooter can be easily modified (see Fig. 38) but it is probably far easier to make one from scratch, based upon this diagram. The main points of modification are: replacing the mouth-piece tube with a shorter length of larger diameter tube; replacing the bulky cloth filter with a neat discof metal or nylon gauze; replacing the inlet tube with a longer length of metal tube available from model shops. Aluminium tube is easier to bend cold but care must be taken as it will 'kink' if bent into too tight an arc. Brass is much stronger but will not bend, it can however be cut at an angle and soldered (silver-soldered if possible). Replace the rubber tube with a length of a slightly larger diameter, a little longer than one's arm. The reasons for these changes are: the cloth filter traps beetles, the inlet tube is stronger and projects longer into the pooter, the larger inlet arrangement permits a more powerful 'suck' to be generated and the whole can be used at arm's length.

Collecting

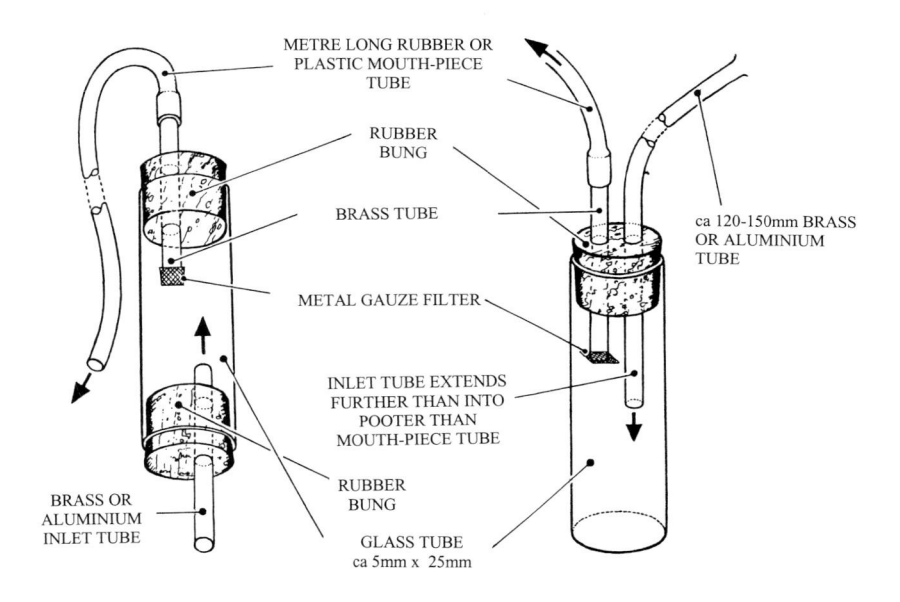

Fig. 38. Cylinder pooter (left) and tube pooter (right).

Many Coleopterists studying smaller beetles make a 'micropooter' of similar design, but based on a tube about 40-45mm long x 12-15mm diameter with inlet and mouth-piece tubes made from lengths of electrical insulation tubing stripped from the copper wire.

Manufacture of your essential pooter might require a degree of lateral thinking – where can I get one or two rubber bungs? Where can I get the holes bored into them? (a drill is completely useless). How can I get the brass tube cut and soldered? See what evening classes are available, speak nicely to the science teacher or technician at your local school/college (a small payment to 'school fund' can work wonders). If you are setting out to make a pooter, make more than one and keep the 'spares' safe as an investment for the future. The glass tubes break regularly, but in 35 years I have only ever lost one 'pooter head' – a complete mystery which an hour of searching on hands and knees in the small area where it must have fallen failed to solve.

A variant of the suction pooter relies upon a strong blow, but has never gained popular use. The various filters help to weaken the 'blow' and prolonged use is likely to be exhausting. However, if its efficiency can be improved it would be a useful device for collecting at carrion, dung or in dusty places. It should go without saying that the use of a suction pooter in such places is fraught with health hazards.

The 'spider pooter' is a simple tube expanded at one end beyond the filter. It is used for holding a specimen whilst it is examined in the field. It would seem to have minimal application for working Coleoptera. However, there are commercially available 'spider pooters' sold at some large DIY stores. They are designed to pick up and transport a large *Tegenaria* from the lounge carpet to garden. They are battery operated and would seem to offer potential to the coleopterist in place of the 'blow' pooter (above).

Tubes

Nowadays the most widely available glass tubes have push-on polypropylene caps, the corked tube going out of fashion and possibly manufacture long ago. These tubes are not as thick walled as the older cork stoppered ones and sometimes break when the cap is pushed in. Some beetles used to eat their way out of the cork, or burrow into it; cork does however, permit the tube to 'breathe', the plastic has an air-tight fit. Cork used to shrink and the stopper work loose, cork often contained passageways through which small beetles could effect their escape – I once looked on bemused as several *Hypebaeus flavipes* made their way from the glass tube through the network of tunnels and holes in the cork stopper to freedom.

Tubes are available in many sizes, the most popular being in the 40-50mm long and 10-20mm diameter range. A few of larger size are useful, say, 50mm x 25mm and 75mm x 25mm for larger beetles, which double up for replacement pooter tubes and make very useful killing tubes.

Some form of box should be made or used to hold the tubes as they are easily broken if left loose in the collecting bag, and a box will also protect the beetles against sunlight and the consequent formation of moisture inside the tube.

Many entomologists are now using 50ml 'centrifuge tubes' made of modern plastics that are not attacked by ethyl acetate. These have a screw on lid that can be tightened to seal, or attached loosely to allow the tube to breathe. They usually come with stickers on the lid or side, for easy labelling, and are light and unbreakable.

Tins etc

A tin, plastic box, cloth bag or even polythene bag is very useful for bringing home various oddments, especially larvae with a supply of their pabulum. My son once captured two *Prionus coriarius* as they blundered in flight during the early evening. No tube large enough to hand, they were put into two supermarket carrier bags tied in half.

Knife

In these politically correct days a knife of suitable size and strength is likely to not only raise a few eye-brows but also attract unwanted attention, it is illegal to carry a knife with a blade over 2.5 inches in length. A woodworker's firmer chisel of about 25-30mm width is a useful substitute and ably handles a wide a variety of tasks, most usually lifting bark and working rotten wood, removing bracket fungi etc. It can even be used for digging under dung in the absence of a trowel. Many coleopterists are now purchasing a tiling hammer – this has a pick-like point at one end of the head and a chisel-shaped point at the other. For working rotten and hard dead wood this tool is superior to a knife or chisel, the levering action of the handle enabling relatively easy 'digging' into even hard wood (that said, I broke the handle of my hammer when trying to lever some hard rotten wood).

Suppliers

Water nets:

Educational Field Equipment UK Ltd., (trading as EFE & GB Nets), P.O. Box 1, Bodmin, Cornwall, PL31 1YJ (web site, http://www.efe-uk.com)

Nets, Malaise and interception traps:

Marris House Nets, has recently been bought by B&S Entomological Services, 37 Derrycarne Road, Portadown, BT62 1PT (www.entomology.org.uk).

References

Cooter. J., 1989. Some notes on the British *Leiodes* Latreille (Col., Leiodidae) *Entomotogist's Gazette*, **40**(4):329-355.

Manser & Goulet, 1981. A new Model of flight interception trap for some Hymenopterous insects. *Entomological News*, **92**(5):199-202.

Owen, J.A., 1976. An Artificial nest for the study of bird nest beetles. *Proc. Trans. Br. Ent. Nat. Hist. Soc.*, **9**:34-35.

Owen, J.A., 1987. The 'Winkler Extractor.' *Proc. Trans. Br. Ent. Nat. Hist. Soc.*, **20**: 129-132.

Owen, J.A., 1995. A pitfall trap for repetitive sampling of hypogean arthropod faunas. *Entomologist's Record and Journal of Variation*, **107**: 225-228.

Růžička, J., 1999. Beetle communities (Insects: Coleoptera) of rock debris on the Boreč hill (Czech Republic:České středohoří mts). *Acta Soc. Zool. Bohem.*, **63**: 315-330.

Turner, C.R., 1997. A simple, effective and cheap baited pitfall design. *British Journal of Entomology and Natural History*, **10**:31-32.

Welch, R.C., 1964. A simple method of collecting insects from rabbit burrows. *Entomologist's Monthly Magazine*, **100**:99-100.

AUTOCATCHER (AUTOKÄTSCHER) OR 'CAR NET'

by Alex Williams

Plate 106

Evening walkers in country lanes in Hertfordshire in the 1930's might have been surprised to see a cyclist approaching with a large net fitted firmly to his bicycle; this would have been the distinguished coleopterist B.S. Williams experimenting with an early version of an aerial net. I have no doubt he caught many interesting beetles in this unconventional way, which would be dangerous with today's road traffic.

A car net is its modern equivalent, it is safer to use and a more efficient way to catch flying beetles, easy to make and very simple to use. The main body of the net needs to be made from cotton or nylon sheet and a piece of curtain netting fixed to the narrow end with the complete net tied to two bamboo canes which are then fixed to the vehicle. The cotton allows the net to fill out as it is driven along whilst the netting allows the air to pass through, sucking the insects in. In the days when car windows incorporated opening quarter-lights it was a simple matter to tie the canes to the window frame, with modern cars it may be more difficult and it may be necessary to use a roof rack with certain models of car. The results will certainly repay any trouble involved.

With the net in place, see plate 106, the car is driven slowly around country lanes on warm evenings, if the road runs through woodland so much the better, a closed canopy over the road is an advantage. It is sometimes possible to drive through an area that contains old trees, a good example is the New Forest, where with suitable permission beetles can be collected that would be hard to find by more conventional methods. Good map reading and regular emptying of the net is necessary if recording is being carried out as it is easy to cover quite a wide area in a motor car in a very short time.

On a good evening the amount of insects caught in this way can be enormous and it becomes necessary to pooter them up quickly before they fly away (it is invariably the 'best' ones that are lost) particularly as they cling to the cotton part of the autocatcher as well as the end net section.

The last run of the evening is often the most productive and the net can be taken down and carefully rolled up for examination the next morning.

VACUUM SAMPLERS

by M.V.L. Barclay

Plate 105

Ecologists have used vacuum-based devices for invertebrate sampling for many years, predominantly in grassland habitats. Among the best known of these devices are the 'D-vac' and 'Vortis' samplers. These purpose built machines can be expensive, very heavy and difficult to obtain, and have never gained much currency outside of mainstream ecology. Nowadays, however, more and more entomologists are investing in portable, petrol driven garden 'leaf suckers', which can easily be converted for invertebrate sampling. In performance tests, these 'leaf suckers' have compared very favourably with the D-Vac (Stewart & Wright, 1995) for collecting invertebrates. Their widespread use in recent years has added at least one new species to the British list, the tiny cantharid *Malthodes lobatus* Kiesenwelter (Barclay & Kopetz, 2003), and has done much to extend the known distributions of many taxa that were formerly under-recorded. They may be the most important addition to the Coleopterist's toolkit since the development of Flight Interception Trapping.

The conversion of a leaf 'blow & vac' is simple, and is outlined in detail by Stewart & Wright (1995), who also provide much interesting data, such as air velocity, dry weights of catches etc. A white net bag, about 40cm long, is attached to the end of the wide inhalant tube, using bungee cords or a length of elastic. Most users also place a barrier of chicken wire inside the tube in case the bag becomes detached, to prevent it from being sucked into the fan. The net used should be strong, but with very fine mesh, such as that used for aerial butterfly nets, not a calico bag, which creates too much air resistance, especially when wet.

Various commercial brands of vacuum are available from hardware stores, garden centres etc, including *McCulloch*, *Stihl* and *Sabre*, all costing around £100-£200. The basic principal is similar, and I have used all three models above with good results. A sieve (a 15 inch circular potting sieve with 1cm square holes is ideal), and some white trays (e.g. standard 16 x12 x3.5 inches) are also required for sorting the catch.

The vacuum sampler can be used almost anywhere, though in woodland it may collect excessive quantities of leaf litter which reduces its efficiency and makes the catch difficult to sort. In reed-beds it will give excellent results, but care should be taken not to vacuum any water, or the contents of the net will become a soggy mess. The tool is at its best on short to moderate sward, such as chalk grassland, cliff-tops (see plate 105), meadows and roadside vegetation. A great deal of folklore already exists as to the optimal technique, but I have observed friends with techniques completely different to my own produce very similar results; there is much scope for experimentation. Certainly, though, the nozzle end of the inhalant tube must make contact with the ground. After a few minutes of operation, the net bag can be inverted into the sieve, and sieved into a tray.

Vacuum samplers catch hundreds of individuals from a wide variety of invertebrate groups, but the bulk of the catch is usually Collembola and arachnids. Numerous Hemiptera, Hymenoptera (mainly ants), Diptera and a great diversity of beetles, especially representatives of Carabidae, Ptiliidae, Staphylinidae (Tachyporinae, Aleocharinae, Steninae, Paederinae and Pselaphinae), Leiodidae, Scymaenidae, Byrrhidae, Latridiidae, Nitidulidae, Cryptophagidae, Coccinellidae, Corylophidae, Chrysomelidae, Apionidae and Curculionidae will also be caught. It has been suggested that larger beetles, about 1cm or over, may avoid being collected, presumably by holding on to or burrowing into the vegetation mat. If this is so, it is a result of the beetle's behaviour rather than simply a lack of suction power, as quite large stones, and animals such as newts and snails are commonly sucked up.

Once placed in the tray, fast moving species, such as Carabidae, alticine Chrysomelidae and Coccinellidae can be quickly collected with a pooter or stork-bill forceps, and the remaining substrate watched carefully for movement of the less obvious species. Some entimine and bagoine weevils may remain still for considerable lengths of time, especially in colder weather. Some workers, especially dipterists, prefer to first invert the net into a white sweep net for preliminary investigation, so that fast moving flies can be pootered while escaping up the walls; I have found this method useful for *Longitarsus* spp. and *Bruchidius varius* (Olivier) (Chrysomelidae), which, like flies, are apt to escape from a tray (Barclay & Ismay, 2001).

The sampler is particularly effective for Curculionoidea, Chrysomeloidea and Nitidulidae when its use is combined with reasonable plant-recognition skills. It is good for collecting cryptic, nocturnal and/or ground living weevils, which are otherwise frequently overlooked; for example, in suitable areas one can quite easily collect members of the weevil genera *Nanophyes*, *Caenopsis*, *Trachyphloeus*, *Cathormiocerus*, *Smicronyx*, *Bagous*, *Orthochaetes*, *Baris*, *Orobitis*, *Ceutorhynchus*, *Sibinia* and *Tychius* in reasonable numbers. Grubbing and sweeping, though much more laborious and destructive, generally produce fewer specimens and species. The sampler can also be used on vertical surfaces, such as vegetated walls and cliff faces, or large pieces of dead wood; mossy logs may yield *Amauronyx maerkelii* (Aubé) (Pselaphinae) (D.J. Mann pers. comm.). R.A. Jones (2002) has recently demonstrated its use on planted roof-gardens in urban London. Vacuuming on, under and around cowpats, carrion or piles of beached seaweed is very productive in terms of Staphylinidae (aleocharinae, *Philonthus* etc), Hydrophilidae, Ptiliidae and the like.

The vacuum sampler is a powerful tool, and is capable of collecting very large numbers of specimens, often from more obscure groups, in a comparatively short time. Woodcock et al (in prep) recorded 3064 specimens of 97 spp. in 90 suction samples of 10 seconds each on a grassland site. The numbers of specimens and species collected makes it exceptionally useful for site surveys, and for adding to lists for already well-recorded sites. However, it is worth being aware that the beetles collected in this way almost all belong to the 'more difficult' half of the Coleoptera, and the catch from a few hours may take several weeks or months to mount and identify. For this reason, unless one is doing quantitative or comparative work, it is not recommended to take the whole catch home 'to sort out later'; a series of trays in the field usually reveals most species. If examination of the total catch is desired, sievings can be bagged up and placed in a Tulgren funnel or Winkler extractor.

Naturally, the vacuum sampler has its disadvantages. It is heavy, bulky, noisy and smelly. Manufacturers even recommend ear and eye protection. For those without a car, it is cumbersome and difficult to carry on a bicycle or public transport, especially along with the sieve, trays, and other equipment you may need. The use of a petrol powered vacuum sampler may disrupt the pleasant tranquillity of a summer's afternoon in the field, not just for the entomologist but for other site-users as well, and it is a good idea to get permission from the site owner beforehand! Once, at Box Hill, while collecting *Orobitis cyaneus* Germar (Curculionidae), I was accosted by hostile members of the public who accused me of pumping poison gas into rabbit warrens!

The vacuum sampler is also a very useful teaching tool. Natural History societies and school groups are fascinated by the vast number and diversity of creatures that can be sucked up from a patch of the most ordinary looking ground, and it serves as a useful illustration of the 'struggle for life' as the spiders and predatory beetles immediately begin feasting on the Collembola. It also lends itself to undergraduate ecology projects, especially when combined with a quadrat, and there are still many unanswered questions which groups of students could very usefully address. It is a tool that will require parts of the Red Data Books to be rewritten, and as its use becomes more widespread, will make a great contribution to our understanding of the beetle fauna of these islands.

References

Barclay, M.V.L. and Ismay, J.W., 2001. *Bruchidius varius* (Olivier, 1795) (Chrysomelidae: Bruchinae) new to Oxfordshire. *The Coleopterist* 10.

Barclay, M.V.L. and Kopetz, A., 2003. *Malthodes lobatus* (Kiesenwetter) (Cantharidae) new to Britain. *The Coleopterist* **12**: 97-100.

Jones, R.A., 2002. *Tecticolous invertebrates: a preliminary investigation of the invertebrate fauna of ecoroofs in urban London*. 34pp. English Nature, Peterborough.

Stewart, A.J.A. and Wright, A.F., 1995. A new inexpensive suction apparatus for sampling arthropods in grassland. *Ecological Entomology.* **20**:98-102.

Woodcock, B. A., Mann, D. J., Mirrielees, C., McGavin, G. C., and McDonald, A. W., *in press*. Re-creation of a threatened lowland flood-plain meadow: management implications for invertebrate communities. *Journal of Insect Conservation*.

COLLECTING – SAMPLE HABITATS

by G. B. Walsh – Revised by J. Cooter

Introduction

Initially the beginner will find plenty of interesting beetles by turning over stones and logs (remember to examine the crevices in the bark of the overturned log) and by general use of the beating tray, sweep- and water-net. The more experienced coleopterist will also find these methods productive on the first visit to a new area. After a while more specialised collecting methods will have to be adopted and it becomes a good idea to make a search for a particular, usually a rare, species. During the course of searching for the desired quarry, a wide range of other species will be encountered as a matter of course.

Nearly all coleopterists will be able to recount 'beginner's luck' stories, and the occasional rarity will continue to crop up unexpectedly throughout your collecting. However, the aim should be to try for consistently good results and to achieve this an amount of simple research is necessary. Advice from other coleopterists, reading species notes in the journals, the study of maps are all essential preparation. If a beetle is phytophagous, its precise requirements might be known — does it feed on the leaf, in the stem, at roots for example. If its biology is unknown, be careful to note its habits when you eventually find it. Quite often the coleopterist embarks upon a spot of 'trophy hunting' – visiting a well-known locality with the sole aim of securing the rarity that place is famous for. Such collection often involves a long journey from home followed by an assiduous search in the precise place worked by others. The upshot, while the desired species may well be secured, is that the rest of the site is unworked and species lists for such places (often SSSI's or nature reserves) are small and strongly biased towards the rarities. Try to spend some time working other parts of these sites – such time is rarely wasted and pleasant surprises are often made.

Many people have an interest in more than one insect Order; often Hemiptera and certain families of Diptera are paired with the Coleoptera. It is generally best to embark upon a day's collecting with one of these Orders in mind, but keeping an eye open for desirable species of the other Order which might be turned up during the day.

As stated in the first edition of this *Handbook*, patience or better *PATIENCE* is needed and hard work (which is in itself enjoyable and productive) has to be put in. Many beetles are restricted to very small areas of what to us looks like suitable and identical habitat. Courage is often required – I recall the late Eric Gardner telling me of a search he made during the 1960's in the New Forest for *Velleius dilatatus* (F.) in a hollow tree with a hornet's nest. After accidentally disturbing the hornets and receiving so many stings on his bald head that he had to visit hospital, he returned to the site, continued working the debris under the nest and 'eventually got my series'.

The following notes are arranged seasonally at times when optimum results can be expected. They can of course be followed at any time with varying results, a different range of species can be expected at different seasons.

1. Winter

Winter collecting is usually very disappointing if the temperature is at or below freezing-point or if the night has been very cold.

- *Flood-refuse:* This is a most productive method of collecting at any time of the year and will often yield many thousands of specimens. Long continued rain in the winter (or a few days' heavy fall in summer) will cause rivers to overflow their banks and bring down quantities of refuse. With luck, one may find a bridge that spans the stream or other safe location and fish out with a net the material that is being carried down with the flood water, or we may search the banks to find a place where the refuse has been piled up in rows or in little heaps. In the winter, beetles will remain in the refuse for days or even weeks; but in the summer most of them will have flown in 24 hours. Great care should be taken when physically close to a river in flood, never under estimate the power of the flow or the effect the low temperature of the water has on the human body. A step too far or a collapse of the bank can spell disaster. Never carry out collection of flood refuse from close to a river in flood alone.

  The best flood-refuse is the fine material formed of small pieces of grass, stubble, horse-dung and general refuse of the banks; twigs, large grass-tufts and coarser material generally contain few specimens. The material is lifted up from the bottom and put on the waterproof sheet and is then rough-sieved through the fingers to remove the bigger useless material. The remainder can be sieved and the concentrated sample brought home in cloth bags and placed in one or more extractors.

  Alternatively it can be worked through piecemeal, a small handful of sieved refuse being placed on a tray and searched assiduously. The sample bag can be fastened up at the top and put into a cool place to be examined again later. If we can identify the beetles we have got, it will help us to make a selection when next we sieve the material. The unwanted material including the sievings should, ideally, be returned to or near the place from where it was collected. Some species, e.g. *Pseudopsis, Micropeplus* and *Monotoma,* do not move for some time and a special look-out should be kept for them; their presence is best recognised when they run about. A fairly large reading-glass is of great help here.

  In summer, the same method is used, but the beetles are much more active and perhaps more numerous, and hand-sorting of the sieved sample must be carried out quickly. Use of extractors is thoroughly recommended.

- *Damp refuse* from the bottom of hay-ricks and straw-stacks can be treated in the same way. We are particularly fortunate if we can find a hay-stack that has been standing neglected for some years and is beginning to ferment at the bottom. The 'hot-pockets' that form and the material round these will repay examination – we may find *Pseudopsis* as well as a large range of other Staphylinidae, Cryptophagidae and other interesting small-fry. Older literature mentions 'bottoms of old haystacks.' I have always had problems with this habitat as 'haystacks' are an extremely uncommon feature of British farming, and how does one sample the decaying hay at the bottom without demolishing the haystack? Possibly it was a case of 'being there at the right time' when the haystack collapsed or the farmer sought to remove it. Whatever the case it is a rather academic point in the 21st century, and the coleopterist should now look for piles or bales of hay decomposing in the fields. The 'hot-pockets' that form within the material will repay examination. Permission from the farmer is best sought as it might be necessary to use a garden fork to dig into the warmer parts of the heap; always 'rebuild' the heap after investigation.

    Bales or piles of straw do not give such good results, but are worthy of attention. A good indication of the age of a straw pile are the weeds growing on the top. Piles of straw rich farm manure left in a field prior to 'muck spreading' are another potentially rich source of beetles, especially in the warm dry parts with white fungal hyphae present.

- *Big tufts of feathery moss* should be squeezed free of much of their water, roughly sifted and the residue packed into cloth bags and brought home for examination as before. The best results are obtained from mosses such as *Mnium* and *Hypnum;* close growing mosses such as *Leucobryum*, *Polytrichum* and similar types are only poor. In November *Mniophila muscorum* (Koch) is to be found in this way, as well as swarms of other small beetles, e.g., *Tachyporus, Sepedophilus,* etc.

- *Tussocks of coarse grass:* The Rev. E.J. Pearce (1948) gives the detailed instructions for working tussocks. However, the methodology described, although somewhat dated, does contain elements essential today. I advocate thorough sieving of the tussock in the field and bringing the residue home for extraction rather than spending time in the field achieving the same end in far from ideal circumstances. Whichever method is adopted, it is necessary to cut the base of the tussock just above root level quickly and cleanly. Before doing so, the surrounding vegetation needs to be pressed back and a sheet ready to take the severed tussock.

    Once cut the tussock is placed on the sheet. It should be broken by hand into pieces of suitable size for sieving, each piece being placed in the top section of the sieve and broken up into individual stems then thoroughly sieved. After the entire tussock has been so treated, the residue is transferred to a cloth bag.

Isolated tufts at the edges of fields and woods yield the best results. This is a fine haunt for species of *Sepedophilus, Cypha* and *Tachyporus*.

- *Bark* can be torn off with a chisel or equivalent tool. Under the loose bark of living trees such as sycamore and willow we may find beetles that have gone into hibernation there, as well as truly subcortical species. *Carpophilus sexpustulatus* (F.) occurs under the bark of freshly felled oaks. The loose bark of stacked timber, firewood logs and palings can be removed as well as that of dead trees. The best results are achieved when the bark peels off fairly easily, and underneath it the wood is still sappy; here we may hope to find species of *Rhizophagus* and *Cerylon, Agathidium, Anisotoma*, Staphylinidae, Nitidulidae, *Salpingus, Triplax,* etc. Very old bark with woodlice, the hyphae of fungi, and worms, usually has very little underneath it except perhaps hibernating Carabids. Both moss and bark are usually less productive after a night's frost.

- *Bracket fungi* on trees, especially birch, may be examined for *Cis, Triplax, Tetratoma*, etc., and if larvae are present in the fungus, it can be taken home and kept in a tin for breeding purposes.

- *Stems:* In the early spring the dead leaves of reed mace *(Typha)*, carefully pulled apart, reveal hibernating beetles especially *Alianta incana* (Erichson), *Telmatophilus typhae* (Fallén). Good results are often found during the summer in stems which are bored by the moth *Nonagria typhae* (Thunberg).

- *Pit-fall traps.* Unbaited pit-fall traps can be used with success throughout the winter. Less frequently encountered species of Leiodidae can be caught this way, often in numbers from autumn to mid-January. Fish-baited pit-fall traps placed as far as possible inside the mouths of rabbit burrows can produce numbers of Cholevinae including *Catopidius depressus* (Murray) and a variety of Staphylinidae (see Welch, 1964).

    Carrion-baited pit-fall traps can be generally used during mild spells in the winter, but better results are obtained in the spring and summer. There are no particular 'winter species' to trap for.

- *Herbaria, etc:* In old and perhaps neglected herbaria, especially in museums, there may occasionally be found the slender Latridiid *Dienerella filum* (Aubé); and in stored tobacco *Lasioderma serricorne* (F.) may sometimes be found, occasionally becoming almost a pest. Beware: it was general practice in Victorian times and later to use herbarium paper impregnated with mercuric chloride. Seek advice prior to handling such material. Fortunately it is unlikely that any insect would survive in an herbarium thus treated.

- *Beating:* Some species are at large during the winter. Beating suitable trees during mild spells in late winter may produce species of *Anthonomus*. Old oaks should be beaten for *Phloiophilus edwardsi* Stephens in late autumn through to mid-winter, its host fungus being *Peniophora* spp.

2. Spring

With the coming of Spring, more and more outdoor work may be done. The winter methods may still be followed, with the following new ones.

- *Ground beetles:* This is an excellent time of the year for turning over loose stones, clods, logs etc – every type of habitat should be tried – edges of fields, the coast, moorlands, hillsides, woods especially forest glades, edges of streams. On moorlands, pieces of peat can be turned over; on the edges of fields, tufts of grass and clods of earth uprooted by the plough; on the coast the jetsam thrown up by the tide.

    The stone, log, etc., should always be replaced, so that it can be utilised again and continue to give protection to the other invertebrates present.

- *Moorland beetles:* Good results can be obtained in the spring by lying down near old patches of heather and searching at their roots, or the material can be put on a tray and examined there.

- *Nests of mammals and birds:* Dr. N.H. Joy first pointed out this method of collecting from mammals' nests. The easiest mammals' nest to find is that of the mole to which at least nine species of beetles are specially attracted. Among a number of the smaller mole heaps there is frequently a much larger one, possibly containing about four to six times the amount of soil. A good place to look for such mounds is under scrub and hedges at a field edge. Winter offers advantages for this – the vegetation has died down making it easier to locate a nest mound and also, the cold weather will act to subdue the inhabitants of the nest, which will include very large numbers of fleas.

    The topsoil is cleared away, and then the trowel is used to dig down through the middle. The nest usually lies about 15-20cm below the surface of the ground, and is generally a sphere 10-20cm diameter. Personal experience has shown the use of 'marigold gloves' pulled over one's cuffs to be advantageous as even during winter the fleas soon become active. The hole is enlarged sufficiently to allow the nest to be lifted out entire with both hands. It is then placed in the sieve, teased apart, shaken and sieved, the residue is then placed in a bag for extraction at home. The sieved material can be gathered into a ball and replaced in the hole after which the soil is replaced. This at least gives the mole the wherewithal to reconstruct the nest if so desired.

    Dr. Joy says 'moles' nests are made of grass, leaves or sedge, the structure depending on the situation in which the nest is placed. I have found those made of leaves as a rule most profitable, and the grass ones least so.' This is certainly true in my experience of 'mole nesting'. Apparently there are more beetles to be found in nests in the south than in the north, and in the winter and early spring than in the summer.

The nests of badgers, rabbits, mice, etc., can also be examined, if we are lucky enough to find them but do bear in mind some mammal species are protected by law and it is an offence to disturb their nests. This advice applies equally to the following section; bird nests should only be examined after the brood has left and advice sought as to the legality or sensibility of attempting to examine a raptor nest (which might have been in use for year and its continued use likely if left undisturbed).

Dry birds' nests are too dry for beetles, but nests in holes in trees and other damp nests are often very productive. Starlings' nests, owls' nests, buzzards' nests, all yield beetles of particular interest: these are best examined just after the young have left. It is good practice to bring home the refuse which contains beetle larvae; it can be put in a tin and kept moist, and so we may perhaps breed the species. Trees damaged by storms and gales can often give access to nest material if the hollow centre of the trunk is revealed. Generally such instances should be regarded as 'good luck' and the hollow interior of the fallen tree examined at once. The habitat has, effectively, been destroyed by the tree breaking open and the occupants of the accumulated nest material will soon seek refuge in their preferred habitat. After a mid-summer storm I once chanced upon an oak tree which had not only been blown over by the wind, but its trunk had split revealing a hollow cavity of which roughly 2.5m were filled with accumulated nesting material, remains of food (bones etc) and other debris. It was a rare occasion when I did not have my sieve with me, the purpose of my field trip being to secure certain species by beating. By shaking handfuls of nest material over my beating tray I secured several Histeridae, Staphylinidae and a couple of *Trox scaber* (L.) which was present in large numbers. I returned a day or two later with sieve only to find the material virtually barren.

The following are some of the species that may be found:

Gnathoncus spp. in birds' nests, especially of owls; *Peranus bimaculatus* (L.) in moles' nests; *Leptinus testaceus* Müller, in nests of mice and other small mammals; *Atheta nigricornis* (Thomson.) – general; *Haploglossa villosula* (Stephens) – starlings, tits, sandmartins and other small birds; *Haploglossa nidicola* (Fairmaire) – sandmartins; *Haploglossa picipennis* (Gyllenhal) – buzzards. *Aleochara cuniculorum* Kraatz at the entrance to the burrows and in rabbit burrows. *Quedius puncticollis* (Thomson), *Q. nigrocoeruleus* Fauvel – mole nests; *Quedius brevicornis* (Thomson) and *Q.truncicola* Fairmaire & Laboulbène and *Trox scaber* (L.) in bird nests in hollow trees; *Ptinus fur* (L.) and *Ptinus sexpunctatus* Panzer in nests in trees; *Atomaria morio* Kolentai – bird nests, especially jackdaw, in trees. This list shows merely a sample of species likely to be encountered.

- *Catkins*. These can be beaten for species of *Dorytomus, Ellescus bipunctatus* (L.) etc.

- *Sub-littoral species:* These are best sought for on a coast with rocks which can be easily split with a cold chisel and a hammer. The Liassic rocks and estuarine series are very suitable. In this way, we may hope to get *Aepus, Micralymma,* etc. The edges of the rocks are to be worked along to about mid-tide depth, looking for places where it will be reasonably easy to prise off slabs of rock, not necessarily of large size. We are likely to find great numbers of the Collembolan *Anurida maritima* (Guérin), on which the beetles probably feed, and it is probable that we shall find larvae, pupae and imagines of the beetles.

- *Carrion traps:* Carrion traps do well in the spring, species appearing then that later seem to disappear. Fish seems to form the best bait; a small fish or head/guts left over from preparing a fish for the table, deposited under leaves in a wood, will speedily have a large population of sarcophagous beetles.

    Traps, as described in the foregoing chapter on Field Equipment, should be laid in open woodland, on the moors, in the open country, in places on the coast and river margins where the water is unlikely to be a problem.

    Keepers' trees or lines should be examined whenever possible and the corpses beaten over a beating tray; there is a succession of species until everything edible has been consumed. Carrion traps can be used all through the year. It is well to keep the specimens alive in a suitable receptacle with moss or tissue for a day or two so as to get them clean and to enable them to empty their intestines.

- *Damp fallen leaves* in woods can be rough-sieved in the open and the fine material brought home for extraction. The productive leaf litter will have white fungal hyphae present and the accumulated litter at the bases of trees or in hollow stumps can be particularly productive.

- *Moss* may yield us *Bradycellus* spp., *Trichocellus* spp., *Otiorhynchus ovatus* (L.), *O. desertus* Rosenhauer, *Byrrhus* spp., and many others. Any moss on old walls may be turned over for *Ocys quinquestriatus* (Gyllenhal). In mid-May moss at the roots of trees, especially ash, in damp places may yield *Caenoscelis ferruginea* (Sahlberg) and *Atomaria fimetarii* (Herbst).

- *Moss in waterfalls:* The thick moss, even under small waterfalls, has its own special beetle fauna. The moss is pulled off and squeezed, and then examined on the sheet. Some of the species likely to be found are *Stenus* spp., especially the red-spotted species S. *biguttatus* (L.), *S. comma* LeConte, S. *guttula* Müller, S. *bimaculatus* Gyllenhal, S. *guynemeri* Jacquelin du Val., and also *Dianous coerulescens* (Gyllenhal), *Quedius umbrinus* Erichson, *Q. maurorufus* (Gravenhorst.), *Q. auricomus* Kiesenwetter, etc.

- *Spring flights:* On favourable days, Coleoptera fly in the spring in great numbers of both species and individuals, so that it is possible to use the voile net. The beetles seem to prefer to alight on light objects, such as new fences, white paint, sea walls etc. When spring bursts suddenly with a few particularly warm days after a long period of cold weather the swarms of flying beetles (and other winged insects) are particularly noticeable.

- *Beach collecting:* During early spring, often around April 14th but sometimes earlier, only very rarely later than late April, sandhills near the sea are amongst the most productive places for beetles. Many species, responding to the increased temperatures, become active and, whether flying or crawling, blunder into and become trapped in small depressions in the bare sand. When dry, the sand is so loose that the beetles cannot negotiate the slope and remain trapped. After rain, however, the sand is more consolidated and the beetles can escape. The ideal weather for this collecting method is bright and sunny with moderate breeze (a stiff wind makes conditions for the coleopterist unpleasant as well as causing the dry sand to blow freely, covering the depressions and their entrapped fauna). The sand must be dry underfoot. A search should be made of the small hollows left by passing walkers as well as the large expanses of breached dune, especially the basal halves of such features. By getting down on hands and knees and crawling along the paths even the smallest sand-coloured beetle will not escape detection. However, with experience, one needs only to walk slightly stooped concentrating on the nature and texture of the sand. Most beetles will not be moving and a few might even be dead (possibly from heat or dehydration). This method is often highly productive and can produce beetles which otherwise are not easily encountered by other collecting methods. For example, in two visits to adjacent sand-dune areas on the Welsh coast in early April, I captured (amongst a lot of more common species) *Psammodius asper* (F.), *Dicronychus equisetioides* Lohse, *Aphanisticus pusillus* (Olivier), *Catapion seniculus* (Kirby), *Squamapion atomarium* (Kirby), *Hypera dauci* (Olivier), *Glocianus punctiger* (Gyllenhal), *G. pilosellus* (Gyllenhal), *Ceutorhynchus atomus* Boheman and *C. erysimi* (F.). On a return visit to the same Welsh dunes in May, none of the above mentioned beetles were evident, and the hollows in sand were virtually barren. Possibly bunkers on golf links might act to trap beetles in a similar way.

    Some beetles must be sought in these places by other methods and in other habitats – *Cicindela maritima* Latreille & Dejean likes bare sand, areas of dune not yet fixed by vegetation, rather than the dune slopes. *Hypocaccus rugifrons* (Paykull) can be found at rest on the dune slopes, *Licinus depressus* (Paykull) under refuse, wood, etc., *Phytosus balticus* Kraatz under drift wood on damp sand above the high tide mark – look on the underside of the wood as well as minutely examining the sand.

- *Grubbing and Searching*: Many species of phytophagous beetle appear early in spring, often as early as March, and frequent the tender young growth of their host plants. By April or May at the latest, these species are no longer evident. Grubbing at roots as well as painstaking searching the foliage will produce good results. Another advantage from collecting early in the year is that many of the plants are not crowded out by the more varied and vigorous vegetation of summer. Several species of *Longitarsus* can thus be found and with this genus it is most important to record the host plant. *Longitarsus fowleri* Allen occurs in spring on teasel plants. It seems logical that such a diminutive species would attack this plant only when it is young and tender.

Searching: After a time, indiscriminate beating and sweeping cease to add new species to the collection, though there will always be the odd newcomer, even in the most thoroughly worked locality. Physically searching plants for beetles is often very rewarding and, at the same time, often frustrating if care is not taken to prevent beetles falling from the plant and being lost in the mass of roots and young growth – weevils are particularly infamous for this, while most alticines will readily jump and avoid capture. With experience, however, escapes will become less frequent. I recall finding, and indeed nearly passing it over as a seed, my first specimen of *Baris analis* (Olivier) – the first instance of it being found outside the Isle of Wight. Knowing its host plant to be fleabane, I searched the plants growing in the vicinity and found the beetle eating into the stem of the plant just at the point where the leaf joins the stem. None were on the leaves or stems, all half buried with only the dark red elytral apex projecting. A search of the literature, including this *Handbook* will provide much information as to the host plants and more precise habits of particular species. However, it must be borne in mind that the biology of many species is still not known and there is plenty of scope for the coleopterist to make a valuable contribution to our knowledge. Some beetles will frequent leaves, others the stem, roots, flower or seed. Sweeping therefore will probably not produce a range of root-feeders, other than by capturing the odd specimen straying from its normal biotope.

Looking for mines and galls and feeding holes in leaves, flower heads and stems all help in our quest. Difficult genera such as *Longitarsus* and certain sections of Apionidae become less problematic if the host plant is accurately recorded. Do remember though that a species might be on a particular plant by pure chance, especially isolated specimens and not necessarily breeding.

Many species of Ceutorhynchinae will be over by the end of June, but others can be found later in the year.

Grubbing: When grubbing, it is good practice to first tap the erect part of the plant first over the collecting tray or into a sweep net. Next, carefully lift the leaves and growth resting on the ground, the dead leaves are parted and the ground surface carefully examined. This job is greatly facilitated by the use of a dinner fork, both the pronged end and handle can be used. It has the advantage of being very much smaller than a hand, so does not get in the way so much and can get under small roots more easily and pull them to one side. It can of course be stuck in the soil to hold vegetation apart while the pooter is employed. After this, some soil can be scraped away and roots examined. *Hylobius transversovittatus* (Goeze) makes galls in the roots of purple loosestrife these can be dug up during autumn and winter and carefully opened to reveal the dormant adult beetle inside. Alternatively, if larvae are present in the galls, sections of plant with galls can be removed and potted up at home and placed in a suitably airy cage with the intention of breeding the adult through. This is a more certain way of obtaining this elusive beetle than by 'grubbing' amongst dense vegetation which very often includes stinging nettles.

3. Summer

The activities of spring and winter can be carried out during the summer months and new methods can be brought into practice as conditions permit. Generally, early summer is a most productive time. By the end of July or early August most coleopterists are of the opinion that the season is in decline until the autumn species become evident, usually during September. However, there will be plenty to do throughout.

- *Beating:* This is often extremely productive of common species. Certain trees such as oak, hazel, birch, sallow, pine are generally productive, others such as horse-chestnut, elder, yield very little. Specimens from the same species of tree should be kept separate for ease of identification. Over-mature trees, especially oaks, are particularly productive and may produce species that inhabit the rotten mould of the interior. Ancient woodlands/parklands harbour some very rare species: *Hypebaeus flavipes* (F.), *Trixagus*, *Anaspis septentrionalis* Champion, *Scraptia*, *Aderus*, anobiids including *Dorcatoma*, Ciidae etc.

- *Sweeping*: This too, gives plenty of material and once again we should sweep in as great a variety of places as possible. Very often in the heat of the day, from 12.00 to 16.00hrs there is little to get other than phytophagus species, but as evening draws on, species come out of shelter again and work once more becomes worthwhile. The best places are low, swampy meadows, meadows surrounded by woodlands or on the slopes of hills, undergrowth in open woodland, and neglected fields. Well-grazed pastures are very unproductive.

    In the daytime we may obtain Cantharidae, Chrysomelidae, Curculionidae, and in the late afternoon and more especially in the short interval from just before sunset to dark, Leiodidae (especially Leiodinae and

Colon), Pselaphinae and Scydmaenidae, may be swept from the tips of grasses, *Colon* spp. apparently preferring the grass Melic (*Melica uniflora* Retz.). Sweeping can be carried out from sundown until it is too dark to see and if conditions are right (a warm still humid evening or with only a light breeze being ideal) evening sweeping will produce a wide variety of beetles leaving their daytime shelter to disperse or search for a mate. The majority will climb up grass and other plants before launching into flight. The last captures can be put into a calico bag for examination in daylight, but it is far better to wear a caver's head-mounted torch and process the final sweeps in the field.

- *Searching*: Searching continues during summer much the same as during spring (above), but different species are likely to be encountered. Advice from a botanist is often very helpful in locating less common plants even on one's home territory.

  In the case of *Meligethes* spp., we may refer to a list of the food plants and their associated beetles, and then search in the corollae is far more likely to give us our desiderata than sweeping; the beetles often get so far down the corolla-tube that the net fails to dislodge them.

- *Flowers*: Certain plants when in flower are very attractive to beetles, e.g. hawthorn, guelder-rose, sycamore, hogweed; the beating tray may be used, or in the case of the umbellifers and very low growing plants, search is far better. The best results are obtained in a year when there is little blossom or where the necessary flowers are few and far between, for then the beetles are congregated in greater abundance in one place. Special flowers sometimes attract certain definite beetles, e.g. *Cimberis attelaboides* (F.), may best be found on male flowers of Scots pine, and species of *Donacia* must be sought on water plants; *D. obscura* Gyllenhal, may be found in June on the Norfolk Broads, on the flowers of *Carex rostrata* Stokes; the flower-heads may be moved and the water beneath examined. In the last week in July and the first week in August, flower-heads of bog asphodel (*Narthecium*) may be examined for *Phalacrus substriatus* Gyllenhal. In May hawthorn flowers may be beaten for a wide variety of beetles.

- *Dead or dying trees*: These give good results until the wood becomes thoroughly dry, which usually takes place in large trees in two or three years after the death of the tree, and in less time with smaller ones. The bark of such trees is the best collecting place for Histeridae, Cucujidae, Colydiidae, Scolytinae, etc., and the shady side is more productive than the sides exposed to the sun. Some species must be cut out, others may be found crawling over or resting on the bark in the bright sunshine, while the crepuscular and nocturnal species may be found on the tree towards evening or after dusk, when a torch is needed. A sharp lookout must be kept for beetles which take to flight almost at once, e.g. *Ernobius nigrinus* (Sturm).

Particularly good results may be obtained by beating the lopped branches of pine where the needles are just beginning to die. We may thus obtain *Sphaeriestes castaneus* (Panzer), *Pogonocherus fasciculatus* (De Geer), *Hylobius abietis* (L.), *Pissodes pini* (L.), *P. castaneus* (De Geer). Beating the dead twigs in the crowns of oak trees that died or fell the previous autumn/winter can produce *Poecilium alni* (L.).

- *Timber piles and cut stumps*: These should be examined, especially in the early morning, when the sap is still wet and fragrant. Even timber-stacks on dock sides yield good results at times, though it must be admitted that the beetles are, as a rule, not British.

- *Old manure heaps, compost heaps*: When old manure heaps and compost have lost their smell, they can be pulled to bits for species of *Monotoma, Myrmechixenus, Atomaria, Cryptophagus, Anthicus,* etc.; these may be sieved out. *Philonthus jurgans* Tottenham is most likely to be found at the very heart of a manure or compost heap. Older literature mentions 'hotbeds' as a good source of beetles. Today 'hotbeds' are rarely if ever made, they were a particular horticultural feature, by the means of which the decaying vegetable matter produced a gentle heat that enabled tender plants, usually vegetables, to flourish. They were confined and had movable glass lids. It was in the condensation on the glass that the beetles were to be found (Fowler often mentions 'hotbeds.' Their approximate equivalent today is likely to be the compost heap, and its examination requires a more invasive approach with sieve and extractor).

- *Dung*: All kinds of dung can be examined, either by turning it over, pulling it to bits over a sheet, or throwing it into water, when after a time, the beetles struggle to the surface and can be skimmed off. It is good policy to grub amongst the soil and roots where the dung has stood and even dig up about 5cm of soil and sieve this over a tray. Many species are local and occur only in certain types of locality. The dung beetles belong to families Geotrupidae (all British genera) and Scarabaeidae (subfamilies Coprinae (*Copris* and *Onthophagus*) and Aphodiinae (*Euheptaulacus, Heptaulacus,* and *Aphodius*). Some members of the Hydrophilidae, Staphylinidae, etc., also occur in dung. Dung, like carrion, is best worked when one is young, though perhaps the comments of the unenlightened are likely to be most irritating then.

  Dry horse dung which shows external evidence of fungal attack is well worth sieving for Ptiliidae.

- *Cut grass and reed refuse*: This often yields excellent results especially if the top of the cut swathe is quite dry but the underside is still damp; tossed or dry hay is practically useless. Handfuls of the grass should be carefully lifted with as much of the damp materials as possible, put into the sieve given a thorough shake and the grass returned to the pile, a fresh sample added to the sieve. The accumulated sievings bagged up and taken home for extraction. This is best done with mixed herbage, and gives better results in damp seasons than in very dry ones.

- *Shore refuse*: This is best when it has been lying for some time out of the reach of the waves. It is examined in the usual way over the tray (a special watch being kept for the very small beetles) or by sieving with the view to extract later. A dead porpoise, dead gull or other large animal tossed up on the shore may give us coastal carrion beetles. *Nebria complanata* (L.) is found under large items resting on wet sand near the high tide mark.

- *Seaweed*: Decaying seaweed thrown up by storms well above high-water marks should be sieved and extracted; alternatively sieving and inspection in the field is at times advantageous at least to form an opinion of what might be present. The best results are obtained when the weed is still damp; dry weed seems to yield little of interest. Many species are to be found, some adventitious visitors, but many in their normal habitat, among them being *Broscus cephalotes* (L.), *Aepus marinus* (Ström) (under heaps of decaying *Zostera), Cercyon litoralis* (Gyllenhal), *C. depressus* Stephens, *Ptenidium punctatum* (Gyllenhal), *Omalium rugulipenne* Rye, *O. laeviusculum* Gyllenhal, *O. riparium* Thomson, *Alaobia trinotata* (Kraatz), *Adota immigrans* (Easton), *Atheta triangulum* (Kraatz), *Thinobaena vestita* (Gravenhorst), *Halobrecta* species, *Aleochara grisea* (Kraatz), *A. obscurella* (Gravenhorst), *Actocharis readingii* (Sharp), *Phytosus* spp., *Arena tabida* (Kiesenwetter), *Diglotta mersa* (Haliday), *Anotylus* spp. *Philonthus fumarius* (Gravenhorst), *Cafius* spp., *Heterothops binotatus* Gravenhorst, *Cantharis rufa* L., *Corylophus cassidoides* (Marsham), *Corticaria punctulata* (Marsham), *C. crenulata* (Gyllenhal), *C. impressa* (Olivier), *Corticarina fulvipes* (Comolli), *Phaleria cadaverina* (F.), *Anthicus antherinus* (L.), *A. angustatus* Curtis, *Cassida flaveola* Thunberg.

- *Seaside timber*: At least two species of beetles are associated with timber that is washed by the sea in seaports and by the coast. *Nacerdes melanura* (L.) (Oedemeridae) can be caught flying in the sunshine and on large pieces of driftwood such as tree trunks, railway sleepers etc, while *Pselactus spadix* (Herbst) (Curculionidae) can be found crawling on posts by the seashore.

- *Clay cliffs*: Great care should be taken when working slumped clay, mud and sand cliffs. Very often there are 'quicksand' areas, undetectable to the eye which have been known to claim lives. In more vegetated areas of cliff there are usually deep cracks caused by the gradual movement of the slumped mass seaward; these are often concealed by vegetation and can easily cause injury. The possibility of 'new falls' should also be considered. Having made your 'risk assessment' you will most likely find such places support a very rich and diverse fauna.

Very often, and if the run of cliff is sufficiently long, there will be a wide variety of habitat to investigate. This will range from freshly slumped and fallen cliff to stable areas covered in scrub. Ruderal areas, those with vegetation just becoming established after a cliff fall can be very productive. Usually the cliff will be made up of a variety of rock types with clays and sands and perhaps the occasional limestone band, the varied geology itself gives different biotopes which are exploited by different species.

Under desiccated flakes of mud and clay we might find a variety of Carabidae, including *Nebria livida* (L.), *Asaphidion* species, *Bembidion saxatile* Gyllenhal, *B. stephensi* (Crotch), *B. genei* ssp *illigeri* (Netolitsky), *Tachys micros* (Fischer von Waldheim), *Chlaenius* species, and various Staphylinidae. On near vertical areas, after carefully removing desiccated flakes, it is worthwhile digging into the cracked damper clay.

Seepages and runs of water are also productive, the damp mud/sand at their sides should be carefully examined for signs of burrowing (usually evidenced by slightly raised, slightly drier runs of the clay/sand). If these areas are gently disturbed with a dinner fork or by stepping on top, after a minute or so, the beetles, *Dyschirius* and *Bledius* species will appear. The same can be carried out at the top of a sandy beach where fresh water runs off the cliff, here in addition we might find *Bembidion pallidipenne* (Illiger).

Although possibly too shallow for Hydradephaga, it is the margins of transitory pools on slumped cliff benches that hold interest. Those with tussocks of sedge growing in the water and in recently dried areas and with deep moss in places are prime targets for the coleopterist. The sedge tussocks need not be cut and sieved, but if bundles of the stems are gripped in each hand and pulled apart, the beetles sheltering at the base of the tussock will run up the stems to escape the disturbance; in this way *Drypta dentata* (Rossi) can be found quite easily in suitable areas of the south coast. Thick moss at the pool edges will produce a range of small Staphylinidae and in Dorset is the home of *Sphaerius acaroides* Waltl, along with *Thinobius brevipennis* Kiesenwetter. *Georissus crenulatus* (Rossi) can be found around the edges of drying pools and seepages, it is invariably covered in mud/clay and only detected when it moves; look for minute walking blobs of mud rather than a black beetle!

Moderately vegetated parts of slumped cliffs in Dorset and the Isle of Wight are home to *Cicindela germanica* L., this species avoids capture by running, unlike its relatives (in Britain) which readily take to flight.

Plants on slumped cliffs are worth investigating, for example *Mononychus punctumalbum* (Herbst) can be found in places on the south coast on plants of *Iris pseudacorus* and *I. foetidissima*. Restharrow growing in large clumps on bare mud is the haunt of *Sitona gemellatus* Gyllenhal. Many more chrysomelids and weevils are to be found on unstable sea cliffs.

- *Moors*: By walking among sphagnum in moist (not too wet) spots, one may, if lucky, stir up *Carabus nitens* L. Possibly the best way to collect on moors at this time of the year is to find a bare patch covered with damp felted algae. This is turned back and we are likely to find moorland Carabids, e.g. *Bembidion nigricorne* Gyllenhal, *B. mannerheimii* Sahlberg and *B. humerale* Sturm; under peat blocks we may find *Miscodera artica* (Paykull), *Nebria rufescens* (Ström), *Pterostichus lepidus* (Leske) and *P. adstrictus* Eschscholtz.

- *Rivers (swift flowing, stony beds)*: In the beds of swift flowing rivers with occasional falls, several methods of collecting may be used.

  The moss from waterfalls is pulled to bits over the sheet, and we shall probably find *Dianous* and desirable species of Aleocharines, *Stenus* and *Quedius*.

  Shingle-beds are often rich hunting grounds; stones can be turned over or the fine shingle can be disturbed with a trowel in plough-like action. A sharp lookout should be kept for minute species of *Bembidion, Lesteva, Zorochrus* and *Fleutiauxellus maritimus* (Curtis), the later preferring rather dry areas of very fine shingle/coarse sand away from the water's edge. *Lionychus quadrillum* (Duftschmidt) is found in this zone too, but prefers coarser gravel/shingle. In spring (April typically) shingle banks are the home of *Amara fulva* (Muller) and *Negastrius sabulicola* (Boheman).

  Water may be splashed on the shingle-beds or on the steeper earth banks to dislodge ground beetles and *Bledius* spp., *Heterocerus*, smaller Staphylinids, etc. Close to the water's edge, the gravel can be pushed with one's boot to make a depression which slowly fills with water, the beetles disturbed during this operation float on the surface and are captured with a tea strainer. This is an excellent way to collect beetles such as *Thalassophilus longicornis* (Sturm) and species of *Hydraena* and *Hydrosmecta*. The tea strainer is useful for collecting small swimming Hydradephaga from small pools and ruts in the shingle, *Bidessus minutissimus* (Germar) occurs in this habitat, to the eye it looks, when swimming, like a water flea.

  Stones should also be turned over in the almost dried-up beds; many specimens collect here when there is still a certain amount of moisture.

  The following is an excellent way of collecting small aquatic 'clavicorns' and 'palpicorns'. The net is placed where there is a very swift current of water, or a loosely woven cloth may be stretched across and through the stream. One then stands a little way up stream and turns over the bigger stones, rubs their undersides where the beetles collect, and stirs up the shingle with a stick or boot. These small beetles are washed down into the net and cling to its meshes.

Species of *Dryops* can be found by lifting floating logs and branches carefully from the water and examining their under surface, in shaded reaches *Orectochilus villosus* (Müller) and *Pomatinus substriatus* (Müller) can be found this way. *Heterocerus* species are to be found by disturbing damp sand and mud, an action that might also produce *Dyschirius* and *Bledius* species. By removing loose bark of partly submerged timber, especially alder stumps, we may find *Cyanostolus aeneus* (Richter). Caution should be taken while working in river beds in hilly country during stormy weather for 'freshets' (flash floods) may come down unexpectedly and in great volume. Only by providential good fortune did the writer of the first edition of this *Handbook*, G.B. Walsh, once escape death in the river Swale when a sudden flood 1m high wall of swirling water flooded the place in the river bed where he had been standing only 10 seconds before. Three years afterwards two boys were drowned by a freshet in the same river.

- *Slow flowing streams and ponds*: The banks of slow flowing streams and of ponds and mud flats, especially in salt marshes, will repay examination. Water should be splashed on the sides and we may tap the mud or walk over it to make the beetles appear on the surface. In this way *Panagaeus cruxmajor* (L.) has been taken at Wicken Fen, and it is an excellent way of obtaining Carabidae such as *Elaphrus, Clivina, Bembidion* and *Tachys* as well as Staphylinidae (*Tachyusa, Carpelimus, Stenus, Philonthus*, etc.) and from their galleries we may wash out species of *Dyschirius, Bledius* and *Heterocerus*.

- *Netting in flight*: Coleopterists of the early/mid-20th century have tried with great success the almost indiscriminate use of a butterfly net with lightweight bag, even carrying one whilst cycling. Many common species can be taken by this method as they fly during the early evening, but there are also many choice things to be so obtained. Freude, Harde and Lohse (1965:109), describe the 'autocatcher' – a large tapered net fitted above the roof of a car. This is effectively a mobile flight interception trap and is capable of capturing a range of beetles only rarely encountered by more conventional collecting methods. In Britain, Alex Williams, is one of the few to have used this device (see page 214).

- *Dry bark*: Dry bark may be removed in the hope of finding and larvae or adults of *Ctesias serra* (F.) found which are feeding on spider-webs; the larvae can be reared on this pabulum.

- *Mountains*: A characteristic 'boreal' fauna is to be found on mountains, especially if we can get above the 800m (2000ft) level. There we may find under stones *Leistus montanus* Stephens and *Nebria rufescens* (Ström) and *Patrobus* species; in dung, *Aphodius lapponum* Gyllenhal, *A. constans* Duftschmidt and in moss Staphylinidae such as *Geodromicus longipes* (Mannerheim), *Anthophagus alpinus* (Paykull), *Oxypoda soror* Thomson and other aleocharines. At the top of the highest mountains in Scotland one may find *Nebria nivalis* (Paykull).

- *Fungi and 'slime moulds'*: The large powdery fruiting bodies of myxomycetes on forest trees are home to species of *Anisotoma* and *Agathidium*, *Enicmus testaceus* (Stephens), *Aspidiphorus orbiculatus* (Gyllenhal) and *Symbiotes latus* Redtenbacher.

    Damp wood with fungoid growth on it may yield *Sphindus dubius* (Gyllenhal), *Biphyllus lunatus* (F.) and *Anommatus duodecimstriatus* (Müller), etc.

- *Burnt areas*: A restricted but interesting beetle fauna is associated with burnt areas and burnt timber. *Pterostichus quadrifoveolatus* Letzner and *Sericoda quadripunctata* (De Geer) are associated with burnt woodland areas, especially conifers. *Dromius angustus* Brullé can be found under bark on burnt trees. The buprestid *Melanophila acuminata* (De Geer), in Britain restricted to the Surrey and Berkshire heaths, is attracted to burnt areas by the infrared radiation caused by the forest/scrub fire, not by the scent of pine resin/oils given off during burning (see Bílý, 1982). *Ernobius mollis* (L.), *pini* (Sturm) and *gigas* (Mulsant & Rey) have all been recorded by beating burnt pines, as have *Sphaeriestes ater* (Paykull) and, in Finland, *Cartodere constricta* (Gyllenhal). At old fire-sites the small fungus *Pyronema confluens* is the host of the diminutive histerid *Acritus homoeopathicus* Wollaston and under pieces of charred twigs, etc. we might find *Micropeplus tesserula* (Curtis).

- *Vegetable garden*: When the potatoes are dug up, one may examine the inside of the shells of the seed potatoes planted during the early spring for *Parabathysicia wollastoni* (Janson), *Anotylus insecatus* (Gravenhorst), *Anommatus duodecimstriatus* (Müller) and *Langelandia anophthalma* Aubé; generally other small aleocharines are present too (see Wood, 1896). It might be worth mentioning I find *Parabathysicia* and *Anommatus* in the remains of seed potatoes in my garden in Hereford. The husk of the seed potato is best removed as soon as the potato plant is lifted and placed on a tray for examination.

- *Sawpits*: Many beetles associated with timber can be found in saw mills and timber yards especially in rural settings. The stacked timber as well as nearby flowers and vegetation are likely to harbour a variety of timber-associated beetles. Cerambycids will be in evidence and a variety of other families can be found at the cut ends of logs, exuding sap and under the bark of uncut trunks. Older accumulations of sawdust mixed with chippings and small off-cuts of wood are well worth searching and digging into, especially in the Scottish Highlands for *Dictyoptera aurora* (Herbst).

- *Moss in woods, etc.*: Moss in woods and along hedgerows, near violet plants, may yield *Orobitis* in August and September; it looks much like a seed. Moss growing on the tops of stone walls in shaded areas in the Peak District should be sieved for *Otiorhynchus porcatus* (Herbst).

- *Lepidopterists' sugar*: Many beetles come to this e.g. *Rhagium mordax* (De Geer), *Helops caeruleus* (L.), *Sermylassa halensis* (L.) but it is scarcely worth while sugaring for beetles as they can usually be captured more easily in other ways.

- *Cossus burrows*: The unpleasant-smelling sap round the wounds in trees in which the larvae of *Cossus* (the 'goat moth') are feeding is very attractive to some species of Coleoptera, for example *Phloeostiba plana* (Paykull), *Silusa rubiginosa* (Erichson), *Bisnius subuliformis* (Gravenhorst), *Epuraea guttata* (Olivier), *E. fuscicollis* (Stephens), and *E. thoracica* Tournier, *Soronia grisea* (L.) and *S. punctatissima* (Illiger), *Thalycra fervida* (Olivier), *Rhizophagus ferrugineus* (Paykull) and *R. parallelocollis* Gyllenhal. A similar beetle assemblage can be found around 'sap runs' on oak trees, and Max Barclay notes that suitable trees are most easily identified by looking for the large numbers of hornets *Vespa crabro* L. which they attract.

- *Salt marshes:* A number of species are restricted to coastal marshes that have soils of varying degrees of salinity. Some species are dependent upon certain plants that grow in these areas (see the beetle/host plant list, p. 270); some aquatic beetles inhabit brackish waters. The species listed here are more closely related to marsh litter, soil and related habitats. They may be collected by a variety of methods including sieving strand-line refuse and turning driftwood. Bare sand and mud showing signs of burrows will repay examination; often walking over the area will, after a few minutes, bring species to the surface. An alternative method is to scrape the surface to about 2cm depth and see what comes to the surface after a few minutes. [I had an amusing experience of this whilst collecting *Dyschirius* species, *Bembidion pallidipenne* (Illiger) and *Bledius* and species at Berrow, Somerset – a beach surreptitiously frequented by naturists. On a hot sunny day, with heavy oiled jacket, collecting bag, garden sieve, sheath knife with its 20cm blade exposed and wearing wellingtons, I crawled along the bare wet sand with pooter in mouth occasionally stabbing and scraping the bare sand. Approaching slowly towards a reed bed, two nudists stood up, gave me a filthy look, dressed and left; better safe than sorry I guess. Finding *Bembidion pallidipenne* (Illiger) and *Dyschirius* spp. and three species of *Bledius,* my mind was closed to all else].

Beetles typical of this saline habitat include:

Dyschirius nitidus (Dejean), *D. politus* (Dejean), *D. extensus* Putzeys, *D. salinus* Schaum; *Trechus fulvus* Dejean; *Bembidion aeneum* Germar, *B. ephippium* (Marsham), *B. minimum* (F.), *B. normannum* Dejean; *Pogonus chalceus* (Marsham); *P. littoralis* (Duftshmid), *P. luridipennis* (Germar); *Anisodactylus poeciloides* (Stephens); *Dicheirotrichus gustavii* Crotch, *D. obsoletus* (Dejean); *Polistichus connexus* (Fourcroy); *Hypocaccus metallicus* (Herbst); *Bledius spectabilis* Kraatz, *B. tricornis* (Herbst), *B. unicornis* (Germar), *B. furcatus* (Olivier), *B. bicornis* (Germar), *B. fuscipes* Rye, *B. fergussoni* Joy; *Carpelimus foveolatus* (Sahlberg), *C. halophilus* (Kiesenwetter); *Tasgius ater* (Gravenhorst); *Heterocerus flexuosus* Stephens, *Heterocerus obsoletus* Curtis, *Heterocerus maritimus* Guérin-Méneville; *Phylan gibbus* (F.); *Melanimon tibialis* (F.;) *Opatrum sabulosum* (L.); *Phaleria cadaverina* (F.); *Crypticus quisquilius* (L.).

- *Keepers' lines*: The dead animals to be seen on a gamekeeper's pole or line are well worth tapping over a tray from the time when they are first hung up till they become mere dry skins. In addition to the ordinary Histeridae, Cholevinae and Silphidae which we expected to get in corpses, we may get *Necrobia* spp., *Korynetes, Osmosita* spp., *Dermestes* spp., and *Trox* spp. In furs and dried skins may be found *Dermestes* spp., *Attagenus* spp., *Megatoma undata* (L.) and *Anthrenus* spp.

  It is often advantageous getting to know the gamekeepers working on estates where you collect insects. In my experience they are helpful, no longer regard you as a trespasser and are interested in what you are doing. A 'phone call to ask if it will be in order to visit on a certain day is more than good manners. You might arrive to find a shoot in progress in which case you will not be allowed onto the estate. You might arrive to find the game keeper culling deer – the first you might know about this is meeting the irate keeper having just frightened the deer he had been stalking all morning, or worse, the ricochet of a high-power rifle bullet which missed its target.

- *Submerged logs*: Partially floating timber and submerged branches and roots in rivers should be carefully examined for *Macronychus quadrituberculatus* Müller, and *Cyanostolus aeneus* (Richter). Both species are local; but the latter is fairly widely distributed. With them may be found *Pomatinus subtriatus* Müller and other related beetles.

- *Spider-webs*: Spider-webs would not seem to be a very nourishing pabulum, but at least two species of Coleoptera are associated with them - *Trinodes hirtus* (F.) and *Ctesias serra* (F.), their larvae feed on webs under dry bark, probably to a large extent on webs of the spider *Segestria senoculata* (L.).

- *Light*: In the tropics this seems to be a very good way of attracting many species of Coleoptera. Those recorded as coming to light in Britain are usually common species but some quite scarce beetles have been captured regularly at light, including the majority of British specimens of *Odonteus armiger* (Scopoli). It is not worth investing in a light trap for collecting in Britain, but making friends with lepidopterists, and giving them a supply of collecting tubes for any beetles they may encounter, can be productive.

- *Manure heaps*: These provide rich hunting grounds for many species of beetles, the best time being from early July to late September when the heap is dry and gives off little smell. The following beetles have been recorded; but there are many others to be found: *Cercyon* spp.; *Smicrus filicornis* (Fairmaire & Laboulbène) and other Ptiliidae; *Euconnus fimetarius* (Chaudoir); *Euplectus signatus* (Reichenbach); various Oxytelinae; species of *Lithocharis*, *Philonthus*; *Monotoma*; *Cryptophagus*; *Atomaria* and '*Anthicus*'.

4. Autumn

- *Sweeping*: Sweeping can be continued in favourable seasons right into November, especially in open woodland and along the glades and borders. We may hope to find *Kalcapion pallipes* (Kirby) on *Mercurialis perennis* ; evening sweeping may produce species of Leiodinae and Coloninae.

- *Fungi*: As Autumn advances fungi become more abundant. Every kind should be pulled to bits and sieved over the tray or through a Winkler sieve for extracting at home. Make sure the fungi are not too young, make an inspection of one or two before embarking on wholesale investigation. 'Toadstool' fungi, especially those growing on rotting stumps and cut logs will almost certainly produce species of *Lordithon* and *Cychramus luteus* (F.). Invariably a variety of small alerocharine Staphylinidae will be present too, amongst which will be *Gyrophaena* and *Agaricochara latissima* (Stephens), it is advisable to collect a good series as there will be the likelihood of more than one species present. Puffballs can give good results, e.g. *Cryptophagus lycoperdi* (Scopoli), *Pocadius ferrugineus* (F.). *Lycoperdina bovistae* (F.) has been recorded from *Lycoperdon bovista*, *perlatum* and *pyriforme*, the latter growing on dead wood. In *Lycoperdon gemmatum* and *caelatum* puffballs in the Suffolk Breck we may be lucky to find the rare *Lycoperdina succincta* (L.); adults of which have been known to overwinter in decaying puffballs. Fungi may be gathered in woods and put into little heaps in the hollows at the bases of trees; these should be examined at frequent intervals.

- *Nests of bees, wasps and hornets*: These nests can be dug out and examined, though the task is an unpleasant one so bear in mind wasps' nests give results even after the wasps have all died in late autumn. If the wasps are still present in small numbers, the mouth of the nest can be covered with turf and the returning wasps be suitably dealt with. If ethyl acetate is poured over the nest, it will kill the inhabitants, both wasps and Coleoptera. We may thus take *Leptinus testaceus* Müller and *Metoecus paradoxus* (L.) in subterranean nests of *Vespula vulgaris* (L.), along with *Cryptophagus pubescens* Sturm and *C. populi* Paykull. In nests of *Vespula germanica* (F.) in trees we may take *Cryptophagus micaceus* Rey after the wasps have died.

  The following species are associated with the nests of bees: *Cryptophagus pubescens* (*Bombus terrestris* (L.)); *Cryptophagus populi* (nests of *Colletes daviesana* Smith, F.); *Antherophagus* spp. (nests of *Bombus*); *Ptinus sexpunctatus* Panzer. (bred from nests of *Osmia rufa* (L.)); *Meloe* spp. (*Andrena* and *Anthophora* species).

  Donisthorpe (1906) used a trap consisting of an ordinary jam-pot buried up to its neck in the ground at the foot of a tree in which was a hornets' nest. The jam-pot was charged with a small quantity of lepidopterist's sugaring mixture which was frequently inspected and re-charged. *Velleius dilatatus* (F.) were attracted by this and fell into the jar where they became trapped.

  Metoecus paradoxus is parasitic in the nests of *Vespula vulgaris*. It seems to prefer nests in banks with rough herbage and woodland margins rather than those in open. It is often reported from nests in roof spaces, the adults appearing on the inside of windows during late summer/early autumn trying to escape.

- *Coastal sand-dunes*: Good results are obtained on the east coast hills when a south-westerly wind blows towards the middle or end of October. With good fortune *Hydnobius punctatus* (Sturm) and *Leiodes* species (especially *furva* (Erichson), *rufipennis* (Paykull)) are found on the leeward sides of the dunes, often struggling up the slopes and if very lucky, accumulating in small numbers at a spot where, one assumes, a hypogeal fungus is buried. Late one October on the dunes at Seaton Sands, Co.Durham, I found in this way an accumulation of about 40 *H. punctatus*, elsewhere in the vicinity were several *Leiodes rufipennis* and a few *obesa* (Schmidt); by sweeping the grassy area between the dunes and the high tide mark I found a number of *Sodga suturalis* (Zetterstedt). Sweeping in the dune areas was unproductive despite the obvious numbers of Leiodinae at large. This is the 'Hartlepool' locality mentioned in 'Fowler' (see for example Fowler, 1889:40) where Gardner collected *Sogda suturalis* freely over the years. It persists 130 years later in an area heavily polluted from the nearby chemical works and oil refineries, my sweep net was heavily blackened and remained so after several washes.

- *Dung*: Certain species of onthophagous beetles are most common in the autumn, for example *Aphodius paykulli* Bedel and *A. conspurcatus* (L.) in sheep dung in early November.

- *Cellars and out buildings*: The insects to be found in these places depend of course on the goods stored there and the degree and standard of 'house keeping'. In neglected cellars and out buildings traditionally *Sphodrus leucophthalmus* (L.), *Laemostenus terricola* (Herbst.), and *Blaps* spp. have all been found. Alas, such architectural features are these days few and far between and very often 'converted' for other uses.

References

Bílý, S., 1982. The Buprestidae (Coleoptera) of Fennoscandia and Denmark. *Fauna Entomologica Scandinavica,* **10.** Scandianavia Science Press, Klampenborg. 109pp. (ISBN 87-87491-42-7).
Donisthorpe, H. St. J. K., 1906. Traps for Coleoptera. *Entomologist's Record and Journal of Variation,* **18**: 186.
Fowler, W.W., 1889. *The Coleoptera of the British Islands,* **3.** Reeve & Co., London. 399pp.
Freude, H., Harde, K.W. & Lohse, G.A., 1964. *Die Kafer Mitteleuropas,* **1**. Goeke & Evers, Krefeld. 214pp.
Pearce, E.J., 1948. The Invertebrate Fauna of Grass-Tussocks. *Entomologist's Monthly Magazine,* **84:**169-174.
Welch, R.C., 1964. *Catopidius depressus* (Murray) (Col., Anisotomidae) in large numbers in a rabbit warren in Berkshire. *Entomologist's Monthly Magazine,* **100:**101-104.
Wood, T.W., 1898. Coleoptera at Broadstairs. *Entomologist's Monthly Magazine,***32**:258-9.

COLLECTING ABROAD

by J. Cooter

This book is not large enough to cover all aspects of collecting abroad; it would need to include details about wildlife legislation in each country, advice for collecting and preserving all families of Coleoptera, not just those in Britain, collecting methods for various habitats and climatic zones and lots more. In my experience there are two main factors to consider – the legality of your proposed activities and weight of equipment.

In these days of cheap flights and (once again) direct rail links to mainland Europe more and more coleopterists are travelling and some are finding the richer fauna across the Channel and further afield to be stimulating. At the same time, the authorities in many 'foreign countries' are enforcing often strict legislation to protect their native fauna and flora, or at least to stop its export. Make no mistake, the consequences of falling foul of a nation's legislation can be alarming and can involve a period in prison while your belongings are confiscated and whilst your family back home make arrangements (or petition) for your repatriation or while the nation you have offended makes arrangements for your deportation. If officialdom were not bad enough, in some parts of the world there is a risk from armed gangs, terrorist groups and robbers.

Many 'third world' countries hold the view that their natural history heritage has been steadily removed by westerners for many years. They fail to mention that until recently they had no expertise in their own country or suitable institution for the deposition and study of this material. That statement might be a generalisation, and it is no use using it as an argument when trying to exit a country with your freshly collected insects.

With heightened airport security the contents of your luggage will nowadays be more rigorously inspected. I was once held up briefly *on arrival* at Tashkent airport, the hand-luggage x-ray machine operator mistaking my box of 100 50mm x 12mm glass tubes with polypropylene caps for a box of bullets. Fortunately all I had to do was to unpack the bag and show the offending article to the official. So far so good, I was then asked why I was carrying the glass tubes. It might just be me, but I find when questioned by a uniformed official carrying a handgun (or as I encountered in Morocco, taking it from the holster, cocking the trigger and pointing it at me) my ability to communicate in a pidgin version of the native language vanishes and by now appearing too nervous for anyone on innocent business the situation takes a downward spiral. It is then I play my trump card. At Tashkent, I produced my letter of invitation written in Russian on official Kazakhstan Government agency (I was in transit to Kazakhstan via Tashkent, Uzbekistan) headed paper which explained not only was I an invited guest but why I was carrying some unusual equipment.

Where ever possible I try to make contact with entomologists in the country I propose to visit. This can have great advantage; you will be met on arrival, helped to your destination, accompanied in the field by someone who knows the region and can speak the local language; you no longer have to rely upon public transport or hire car (if available) and you will be charged 'local rates' rather than 'tourist rates' for food, accommodation etc. You can easily obtain letters on headed paper and in the *lingua franca* to show to officials that you encounter and thus explain your strange activities or presence off the usual tourist routes. An offer to return a percentage of your material to that country's national museum (or equivalent) cements the friendship and makes return visits so much easier. Very often a friendship will develop with the likelihood of return trips (or exchange visits) and fresh material arriving via the post in the future. On one occasion when leaving to fly home, my host arranged for me to be met at the airport by a colleague working in animal quarantine, I was then taken via the Diplomatic gate to the head of each queue and to the departure lounge a full 20 minutes before any of my fellow passengers had completed the necessary immigration formalities; I was not asked any questions and got home with my whole collection intact; several New Species were later described.

On the other hand, travelling 'kitted out' with all the necessary permits and letters can have disadvantages. You will not be able to argue the toss if an animal quarantine, customs or immigration official, after inspecting your 'catch' says, 'your permits states you are here to collect Carabidae and Scarabaeidae, but you also have other families in your collection.' You are now standing on the threshold of having your entire catch confiscated. (Bear this in mind; colleagues of mine collecting without permits in China came to the notice of local police. They were questioned and a couple of days later an officer from Beijing arrived with books and microscope and identified to family and logged every specimen they had collected. They were, however, allowed to continue collecting/travelling and had none of their catch confiscated).

Make no mistake adverse and terrible events do happen. I know of an East European coleopterist shot dead by an armed gang in the rain forest in Laos and have a Russian friend who has been deported from China after having been 'caught' collecting without the necessary permits. I know of two Slovak entomologists who were imprisoned in Palawan and were only repatriated (less all belongings) after the intervention of their Ambassador and after paying a hefty fine.

At the 'simply annoying' end of the scale is the assiduous or officious customs officer. When returning home from France to the Newhaven ferry terminal in the 1980's at a time when Customs and Immigration were taking industrial action and 'working to rule' I passed a car with an elderly couple being given the 'once over' by Customs. The couple had packed a picnic which included a quarter salami. This being a prohibited import, the customs sought to investigate what other contraband might be in their car. They had emptied every item from their luggage onto the parking bay, removed all the car seats, carpets and everything from the boot and were not yet finished. I at once declared my few dead, dry insects and requested the appropriate forms if they thought this necessary. Luckily they did not and waved me on my way. (It is important to stress 'dry' insects as 'spirit' may be charged excise duty. If leaving for home with material in spirit, always tell the officer that it is low concentration spirit. Too much flammable liquid inside the plane is frowned upon, though 'duty free allowance' alcohol seems immune to this).

How the individual regards and interprets the legal requirements of the country being visited is a personal matter. The risk has to be weighed up by the individual, but do bear in mind 'ignorance of the Law is no defence.' I will not make suggestions other than to say co-operation and abiding by the laws of the county you visit is generally preferable to arrest, imprisonment and deportation; a *'persona non grata'* stamp in your passport will attract attention in every other country you visit. For information and advice about insect collecting, contact the embassy of the country you intend to visit. In addition, staff at our National museums (South Kensington, Cardiff, Edinburgh, Liverpool) should also be able to give advice.

With limited luggage allowance on airlines and charges in the region (2005) of £20 for each kilo of excess baggage careful consideration must be given to the amount and type of equipment taken on foreign trips. My personal preference is to take only what I need and with the option of abandoning some prior to my return if needs be say, if airline luggage allowance is likely to be exceeded on the journey home. This is not the same as taking all my equipment – for example, I have never taken a beating tray abroad because phytophagous beetles are not of much interest to me and anyway, in many countries 'beating' is likely to produce little else but ants, and in alpine zones things to beat are few and far between. Decide which families of beetles or which habitats you want to investigate and gather the necessary equipment. Remember too that you will need suitable containers and packing for bringing the catch home.

As with collecting at home, there is no need to be over burdened with equipment. Unless especially interested in aquatic beetles, why take a water net when a tea strainer will serve when opportunity arises? Buy a cheap 99p gardening trowel that will probably last as long as your trip abroad and can be left behind, ditto chisel. There are of course the real essentials and these will be different for each coleopterist, for me they comprise a bag sifter with at least six spare sample bags, two to six Winkler-type extractors and a few empty yoghurt cartons (pit-fall traps). There are also 'universal' items such as pooter, tubes, killing fluid/bottle, hand lens and forceps. If you are due to arrive at your foreign destination late in the day (after the shops and cafés have shut – if indeed there are likely to be any in your chosen area) take a few 'pot-noodles' with you. Hardly haute cuisine, but they do provide a filling meal and have the bonus of being packaged in a pit-fall trap. If only everything could double up in its use!

I always take ethyl acetate and other liquids in thick-walled glass tubes with plastic 'medicine bottle' click caps. Ethyl acetate is or can be used in the production of heroin, thus a 250ml bottle might attract attention if not for its heroin association, then as a flammable substance. Read the notices at airport check in desks, flammable substances are prohibited . . . but 'duty free' alcohol in fragile glass bottles is permitted. I thus take my IMS or ethanol preservative as part of my 'duty free' allowance and have on occasion even poured it into an empty vodka bottle to add credibility to the masquerade (5% acetic acid in the alcohol will prevent accidental 'personal use'). In many countries outside Europe 'alcohol' is readily available from pharmacies and general stores, though it can be of dubious quality for entomological purposes. I always pack my tube of ethyl acetate in my toilet bag where it is less likely to attract attention if searched.

How to pack the catch for export is important. Day light, being a valuable commodity, is best used for collecting, so invariably preparing the catch for export takes place after dark. Provision of a light source might be a consideration. Based at 2800m in the Tien Shan, Kazakhstan in a yurt, my light source was a candle and the two small Maglite torches I brought with me. I purchased a head mounted 'caver's torch' upon return to UK. I have taken it abroad ever since and find it invaluable for 'curatorial' work after dark, night field work and a range of necessary domestic duties as well as for visiting caves and setting pit-fall traps therein.

Like so many things, how the catch is packed is a personal matter. Mine divides between smaller beetles in spirit and larger ones 'dry' between layers of cellulose wadding in strong cardboard boxes (each about 10cm x 15cm maximum and of a depth of about 8cm). Cardboard absorbs some moisture, the contents less likely to become mouldy as will be the case with plastic or metal boxes. Pack data with each tube and each layer of dry beetles, it might require a lot of label writing but it represents time well spent. Never simply number each tube/layer and record the field data once in your diary. If the diary is lost or damaged (water and fire are as safe bets in the field as lost luggage is on your way home) your catch will be virtually useless, especially so if you have been travelling widely while abroad). Always use a sharp HB pencil, it is preservative and water proof, cheap and unlikely to be stolen, and if it is, then it can easily be replaced. Packing beetles in small sized heavy-gauge transparent polypropylene self-seal bags ('Whirl-Pak' bags, see Bioquip website catalogue) with a little alcohol and field data can save a lot of space; the bags are themselves packed into stout polypropylene food boxes (generically referred to as 'Tupperware') which have near air-tight (ie alcohol proof) seals.

Before packing material in layers of wadding, make sure it is dry of all surface moisture (a short spell in full sun should do the trick (make sure everything really is dead as well!!)). Mould can be a real problem even during short trips (2-3 weeks) in sub-tropical and tropical regions. Dry material is also particularly vulnerable to ants and/or termites in warmer countries. (For other preserving methods see the chapter 'Curatorial' 'Killing' section, page 311).

Correctly recording the location of collecting areas when abroad can be difficult, maps might be very difficult to obtain and many countries will not have any form of map with places marked in English. A GPS device is invaluable when abroad. I recall poring endlessly over a scrap of an old Soviet military map to locate my collecting localities in the Aksu-Dzhabagly State Nature Reserve, Kazakhstan, the map covered only part of the Reserve. All the collecting sites were large distances from any named settlement and the settlements were indicated in Cyrillic and using names that were no longer current (thus if I could read Russian, Dzhabagly village appeared on the map as Novo Nicholaiovich). I purchased a GPS before my next trip abroad. During my subsequent visits to China I have only very rarely glimpsed a map of the any of the areas I intended to visit, generally these are large maps mounted on an office wall and of course with no English notations. However, the GPS and altimeter gave me the necessary data which when combined with the information supplied by the Chinese entomologist accompanying me gave detail of the magnitude I record when in the UK and using OS maps.

If desperate for a hard copy map of the area you are to visit, Stanfords, generally regarded as the UK's most comprehensive map shop will doubtless be able to help (visit < http://www.stanfords.co.uk > shops in London (Stanfords, 12-14 Long Acre, Covent Garden, London, WC2E 9PL Tel: 0207 836 1321).

GPS devices have altimeter functions but these (currently) are not known for their accuracy. I recommend a small hand held aneroid altimeter (available from good out door specialist shops). For example once when in France picnicking at a 'col' with altitude marked on a sign post, my hand held altimeter was within 3m but the GPS was out by 400m.

Another extremely valuable function of a GPS is the ability to 'way mark' and record routes. Without GPS my completely disorientated companion and I would have been totally lost after a day's field work in the vast featureless closed-canopy forest that extends over 50km from Bai He town to the North Korean border, Jilin Province, north-eastern China. A 'way mark' at the start of the day's activity enabled us, after realising we had no idea which way was 'home' to get back to our starting point with comparative ease but not without some anxiety as dusk approached.

If collecting in the Balkans, the eastern fringes of geographical Europe or outside Europe, it is worth bearing in mind that a percentage of the material you have collected will be New to Science, notice I say 'will be' rather than 'may be.' This being so, it is unpardonable not to make the effort to get as much of the collection as possible studied by specialists with the appropriate expert knowledge. Staff at our National museums should be able to help in providing names of these experts and some information might be obtained via the British Beetles e-group (beetles-britishisles@yahoogroups.com), there are 'foreign' subscribers and many of the UK natives have exotic (entomological) interests and suitable contacts.

This leads us on to what to do with type material and your duplicates. The person(s) identifying material for you will expect to be able to retain 25% of the duplicates; you might occasionally be asked for extra specimens too. If this does arise, bear in mind the expert who has just given up a fair amount of time to help you has seen a specimen of great interest that will be very useful in a specialist collection. It is a good idea to allow the expert to retain that (those) specimens too. (The 25% rule is a universally and internationally accepted 'gentleman's agreement').

I hold the view that type material belongs to Science and has no monetary value, therefore deposit it in a major museum or major specialist collection where it will be available to all. Of foreign beetles I maintain a specialist collection of Leiodidae, Agyrtidae and Silphidae. I give away to colleagues and friends all material belonging to other families; they have specialist collections of their particular family interests; from time to time these friends send me batches of foreign Leiodidae, much of which is New to Science. A bonus of such a 'system' is the friendships that it creates with the resulting occasional collaboration and offers of accommodation.

With 'political correctness' and other mores pervading every aspect of life, entomology has by no means escaped. For the average British entomologist working with the restricted, well-known and relatively well-documented British fauna in often threatened habitats, a very conservative attitude to collecting has, quite rightly, developed. For example, a British coleopterist having swept a woodland ride would never capture all 40 examples of *Grammopotera ruficornis* present in the net. However, when abroad such restraint can be counter-productive. Abroad and especially in the tropics and sub-tropics those 40 specimens of the same species in your net might be the only examples ever seen by Science and stand a good chance of representing a species New to Science. You capture the lot, not just select two or three which might turn out to be female specimens and unidentifiable.

Another foible that curses the British collector abroad is the applauded habit when in UK to select the two or three 'best' examples of a prolific species and mounting them with perfect symmetry before submitting them to an authority for identification. If this is adopted abroad those select two or three might be or are likely to be, if not New to Science, then representatives of a very little known species ... and you collected only three out of the 40 in your net, 40 that resulted from one sweep in one restricted area on one day. Far better that you leave behind, after your death or the cessation of your hobby, a wealth of well documented even unmounted material for others to work on than you leave it behind in the field where habitat destruction and other constraints mitigate against finding many species again, ever.

I can give an example of taking a good sample from personal experience, very often when beetles are slow to show themselves or when I've filled my back pack and arms with bags of sieved forest litter my attention is taken by other Orders. Whilst in the Wuyi Shan, Fujian Province, China during 2001 I saw a medium-sized 'conopid' fly settle on large cut bamboo (*Phylostachys pubescens*) stacked at the track-side. I captured the fly as I knew someone back home would be interested. Then another landed, I captured that and so on along the track, eventually netting about 15-20 specimens. Back home these flies were identified by Dr John Deeming, National Museum of Wales, who informed me they were not conopids but Syrphidae: Cerioidini genus *Sphiximorpha*. On a visit to the Natural History Museum, London where John attempted to identify them to species he found in the National collection a single specimen labelled 'China 1840' which they matched, but given the age of the specimen and the lack of modern literature could not place the fresh material to species with certainty. So not New to Science, but 15-20 fresh examples of a taxonomically obscure fly represented in UK by one ancient specimen, one sex, with inadequate data. As another example of mine I cite collecting in 1999 a number of bibionid flies in Western Tien Shan, Kazakhstan. These were identified by John Deeming as *Diplophus obscuripennis* Lundstöm, a species according to the literature to hand at the time, known only from the female holotype captured in the 1920's and deposited in the Natural History Museum, Budapest. The Diptera collection at Budapest took a direct hit from a Soviet tank shell in 1956 so we thought my series, both sexes present, potentially represented neotype material. However after further investigation and perusal of Soviet and Russian entomological literature, it became obvious that the species had been collected between 1920's and 1999. Also on that trip, I collected approximately 50 specimens of *Oodescelis transcaspica* Kaszab, an abundant ground living tenebrionid, I subsequently gave the lot to various museums and was told by Max Barclay, Natural History Museum, London, that the species was hitherto represented in the National collection by only the holotype; they now have a short series and both sexes.

Earlier I touched upon collecting time. Some people mount their daily catch whilst in the field but to me this is simply a way to waste time that should be spent amassing as a large a catch as possible during the brief time I am abroad. Rarely, if at all, can (Coleoptera) material be prepared in the field as well as it can once back at home with better facilities and full range of equipment and chemicals. In addition 'mounting' creates a bulky and quite fragile mass to add to your luggage and increases risk when travelling and again when exiting through customs and immigration. (NB some non-Coleoptera Orders, Neuroptera for example, are perhaps best mounted in the field, but this is beyond the scope of this *Handbook*).

In addition to making good use of the daylight hours, in the sub-tropics and tropics light is an excellent source of insects, more so if in a reasonably remote location where electric lights are few and far between. A light, even of low wattage electricity or a paraffin pressure lamp, on a white-washed wall or white sheet will attract a wealth of insects and can be sampled regularly, though often the best results are to be had for two or three hours after sun-set. Remember to examine the far dim edges of the pool of light and the ground under the light itself.

Searching tropical and sub-tropical forest trees with a torch at night can produce a number of beetles that are difficult to find or not seen by day time collecting methods (Elateridae, Eucnemidae, Anthribidae etc). Though do remember to check your bearings every couple of minutes as circling this tree then walking over to that one, circling it too, then off to another tree, in pitch darkness with no sounds of human activity nearby, no sodium glare or other 'markers' has huge potential for disorientation. This is true even in day light, see comments about GPS 'way marks' above.

Adequate travel insurance is an utmost necessity; always buy a policy with top-range medical cover including the use of a dedicated air ambulance plus doctor/nurse to fly you home or to a more 'civilised' location for treatment. It is also a good idea to check that your policy actually covers you for insect collecting. Whilst renewing my annual travel policy recently, in casual conversation with the clerk I mentioned visiting remote places to collect insects, only to be told the policy I had did not cover 'that sort of activity.' I wrote to 'head office' explaining I am pursuing my hobby, am not being paid nor will I be selling anything I collect, nor are my overseas entomological activities part of my work. I received a letter telling me that the policy *did* cover my entomological activities, I keep that letter with my policy documents.

There are some apparently eccentric features on some insurance policies. For example I read in my own policy that horse riding required extra cover. Realising I would be using a horse, I spoke with 'head office' and found that if I was riding for pleasure and extra premium would be payable, but as I was riding because there were no roads in the region, no extra cover was required as this was the 'normal mode of transport' (nor did they insist I wear a riding helmet).

If collecting with friends resident in the country you are visiting, it makes good sense to tell them you have good insurance cover and where your policy is kept (indeed tell your field companions whatever their nationality). If you are incapacitated, they will find the insurance and administer things for you.

Acquisition of a pack of sterile surgical necessities (containing needles, thread, hypodermics, sutures etc. available at high street pharmacists and good 'out door' shops) is very sensible if visiting remote 'third world' countries where disposable equipment is not as widely available as in Europe.

Your GP will, if requested, doubtless prescribe a broad spectrum anti-biotic for use if necessary. A word of caution here: make sure it really is necessary before taking as broad spectrum anti-biotics will destroy 'bad' and 'good' bacteria in your digestive system. An 'upset stomach' is almost to be expected once abroad, often this is a simple but inconvenient and sometimes embarrassing 'adjustment' to local conditions and all will be well in two or three days. I ignored my own advice when in a remote area of Heilongjiang Province, north east China in 2004. My health deteriorated slowly but markedly over about five days with various symptoms exhibiting. I took my anti-biotics and was fine after a week. Two weeks later I was *really* ill, my resistence to bacteria by then minimal thanks to the broad spectrum anti-biotics. Chinese medicine addressed the symptoms and I was thus able to get out into the field for a few hours each day (in 30°C heat and steep rough terrain). I was still not back to normal two months later and by then back home so I visited my GP, had a few tests all of which proved 'normal' and things eventually cleared up after about five months.

Some months before leaving for your trip abroad check with your GP which immunisations or booster injections you will require. Also enquire which other immunisations whilst not necessarily obligatory would be 'a good idea' – in this category I would include rabies as well as Lyme disease and tick-born encephalitis if going to an area known for either of these.

Your GP or pharmacist will also be able to recommend the correct anti-malarial; read the instructions as it is important to begin taking anti-malarial medication before leaving and after returning home. A bottle of Tea Tree essential oil is always useful in any travellers medical kit, amongst other things it will help combat fungal infections.

Further Reading

McGavin, G.C., 1997. Expedition Field Techniques. INSECTS and other terrestrial arthropods. Royal Geographical Society, London (ISBN 0-907649-74-2). 94pp + 28 figs.

BEETLES ASSOCIATED WITH ANTS AND ANT NESTS

by J. Cooter & D. Mann

(Based upon G.B.Walsh chapter in the first edition of this *Handbook*)

The standard work on collecting myrmecophilous Coleoptera was written by Donisthorpe (1927) and from time to time copies may be had in secondhand bookshops or from specialist book dealers; it is well worth acquiring a copy if the chance arises. The notes that follow originate from a paper also by Donisthorpe (1896) slightly modified and with nomenclature brought in line with current use (2005). It is worth bearing in mind a quote from Donisthorpe (1896) '*Patience and perseverance, with a determination not to be discouraged by frequent failures, are necessary, as it often happens that the day and situation you select as most propitious prove the contrary, and you return home without a single specimen.*' In addition, it is good practice to collect a few ants with the beetles and to mount some on a card on the same pin as the beetle and above the data label. Bolton & Collingwood (1975) or Skinner & Allen (1996) are inexpensive and useful texts for identifying the British ants.

Spring and autumn are the best periods for working ants' nests and the preferred time is in the morning, as after the sun has been shining and heating up the outer part of the nest the beetles tend to seek the cooler interior. Ants are also more active at that time, makes them harder to sort out.

The three most productive species of ants to work, the ones with which most beetles are found are:
> *Formica rufa* L.
> *Lasius fuliginosus* (Latreille)
> *Lasius flavus* (F.)

However, beetles are to be found with other species too, most notably:
> *Lasius niger* (L.)
> *Formica fusca* L.
> *Formica sanguinea* Latreille
> *Myrmica ruginodis* Nylander
> *Myrmica rubra* (L.)
> *Myrmica scabrinodis* Nylander
> *Tapinoma erraticum* (Latreille)

Some ant-nest beetles are exceedingly constant in keeping to their own hosts; others seem a little more cosmopolitan.

Formica rufa L. Generally several nests can be found in the same area and it is good practice to find a group that can be easily worked. Early in the spring one or two bricks or flattish rough stones and a piece of wood or thick branch can be placed on the uppermost sides and top of the nest mound. A bunch of long grass, twisted, can be pushed into a hole in the side of the nest. After setting these 'traps' the nest should be inspected regularly for best results, perhaps weekly or fortnightly, but in warmer weather more regularly as the ants' constant re-working of the nest material will result in the 'traps' soon becoming buried and lost to the coleopterist.

Inspection begins by setting out the necessary equipment and having the pooter ready, some people tie their cuffs or put cycle clips around their ankles to prevent too many ants getting under their clothing. This is one occasion when a sheet is perhaps more useful than the tray, its larger area and the fact that it does not have to be hand held are advantageous. Dark sheets are good, as the ants will hide underneath and be a little less active. The bricks, stones and wood are then quickly but carefully picked up and dropped onto the middle of the sheet. The tunnels and passageways under the 'trap' are inspected for beetles and any that are seen can be collected (with luck and great speed) by the pooter. Next the 'traps' are given a sharp tap to dislodge loose material from their under-surfaces. This material and the undersurface of the 'trap' is inspected, again any beetles are collected at once with the aid of the pooter. Some ant nest beetles are very slow to move, so a long and careful look through a seething mass of ants is necessary. It is good practice, once the bulk of the ants have left the collecting sheet, to sift the debris over a new sheet or tray and examine the siftings for small-fry and sluggish movers (e.g. histerids and *Monotoma*) or to take this residue home in a cloth bag for extraction.

Next the grass twist is bodily removed from the nest and unwound on the sheet and the debris similarly inspected.

The paths and runs used by the ants around the nest should also receive attention, leaf litter and moss sieved and a search made under any nearby logs or stones as such places are the haunt of, for example, *Zyras humeralis* (Gravenhorst).

We have used these methods in the Wyre Forest, Worcestershire, and found it a very good way to get *Quedius brevis* Erichson – a species that, in my limited experience, frequented the underside of wood or log 'traps' placed on the nest mound. *Clytra quadripunctata* (L.) is to be found by searching or sweeping/beating any overhanging vegetation – the females' drop their eggs onto the nest mound, but take care to keep off the ant-infested mound themselves lest they become food for the nest. *Zyras humeralis* is not difficult to find by sieving litter.

A second method, we have employed in the New Forest and in Glen Tanar, and always with a friend, is also described by Donisthorpe. It involves the wholesale removal of handfuls of nest material for sieving and inspection over a sheet. The ants are antagonised by such methods, so there is perhaps a greater need to guard against these active little creatures running up one's sleeves and trouser-legs or down one's neck. Doing this work with a friend has advantages as each worker can divide time between looking on the sheet and observing the progress of the ants on one's companion. The double-handfuls are dropped into the sieve and rapidly sifted. The material in *Formica rufa* nests, which is about an elbow's length plus in depth in the centre of the nest and warm to the touch the best material to sieve, onto a dark sheet. Ants can then be removed from one's person, the residue placed near the nest and a little time allowed for the bulk of the ants to clear the sheet. The debris is inspected, again keeping a sharp look out for small and sluggish species and those beetles that mimic ant locomotion. The debris is then re-sieved and re-inspected or taken home for extraction, but knowing what good results can be had purely by inspecting the nest siftings in the field, it is perhaps unnecessary to risk frayed tempers resulting from the release of large ants back at home. The residue can be tipped back onto or near the nest and the ants will soon make good any damage. By using this method in Warkwickshire and Devon, we have had both *Monotoma angusticollis* Gyllenhal and *conicicollis* Aubé, histerids and various staphylinids. In the Scottish Highlands we have found nearly full-grown larvae of *Protaetia cuprea* (F.) in nest mounds, and have successfully reared adults. The method is simple, only the largest larvae are taken and placed in a stout box or tin with a quantity of nest material. The whole is taken home and left for a few weeks. Careful inspection will reveal the progress, and after a while the adults can be removed.

Donisthorpe gives the following myrmecophilous Coleoptera associated with *Formica rufa*:
> *Myrmetes piceus* (Paykull)
> *Dendrophilus punctatus* (Herbst)
> *D. pygmaeus* (L.)
> *Hetaerius ferrugineus* (Olivier)
> *Ptenidium formicetorum* Kraatz
> *Ptilium myrmecophilum* (Allibert)
> *Acrotrichis montandonii* (Allibert)
> *Stenichnus bicolor* (Denny)
> *S. godarti* (Latreille)
> *Batrisodes venustus* (Reichenbach)
> *Oxypoda haemorrhoa* (Mannerheim)
> *O. formiceticola* Märkel
> *O. recondita* Kraatz
> *Thiasophila angulata* (Erichson)
> *Haploglossa pulla* (Gyllenhal)
> *Dinarda maerkeli* Kiesenwetter
> *Amischa analis* (Gravenhorst)
> *Amidobia talpa* (Heer)
> *Notothecta confusa* (Märkel)
> *N. flavipes* (Gravenhorst)
> *Lyprocorrhe anceps* (Erichson)
> *Alaobia sodalis* (Erichson)
> *Aleochara ruficornis* Gravenhorst
> *Drusilla canaliculata* (F.)
> *Zyras humeralis* (Gravenhorst)
> *Platydracus latebricola* (Gravenhorst)
> *P. stercorarius* (Olivier)
> *Heterothops* spp. (? *niger* Kraatz)
> *Quedius brevis* Erichson
> *Leptacinus formicetorum* Märkel
> *Othius myrmecophilus* Kiesenwetter
> *Gyrohypnus atratus* (Heer)
> *Cetonia aurata* (L.) – larvae and pupae
> *Protaetia cuprea* (F.) – larvae and pupae
> *Monotoma angusticollis* Gyllenhal
> *M. conicicollis* Aubé
> *Coccinella magnifica* (Redtenbacher)
> *Clytra quadripunctata* (L.) – larvae and pupae.

Donisthorpe lists special habitats for the following species:

Aleochara ruficornis – a rare species, taken in moss and by sweeping near the nest. *Dinarda maerkeli* is almost always found by means of the clump of wood placed on the nest (though Walsh has found it by sifting the nest contents).

Lasius fuliginosus (Latreille) – This is a tree-nesting ant, generally utilising oak or beech, often those that are dead or dying. Locating a nest can be difficult because the ants use set runs, which extend some distance from the nest entrance. In Britain, this ant can be positively identified by the strong, sweet lemony smell given off by a specimen crushed between the fingers (M.V.L. Barclay, pers. comm.). Having discovered the nest, one can observe the ants entering the interior of the tree by means of holes at the roots or in the trunk. All such cavities should be 'plugged' with grass twists, and if the tree is hollow, a large twist can be placed inside. A piece of old damp wood loosely wrapped in paper or an old bone placed in the nest entrance or cavity is often productive. After some days, the grass and any other 'traps' can be carefully removed and shaken over a sheet. Chinks in the tree should be carefully inspected and any loose bark carefully lifted and placed on the sheet.

The runs used by the ants will also repay examination and any stones or logs, litter and leaves near them ought to be examined or sifted. Tussocks growing near the tree can be cut and shaken over the tray or sieved, and it is often profitable to sift or otherwise inspect any 'saw-dust' ejected from the tree by the ants.

The following beetles are found with *Lasius fuliginosus* –

Dendrophilus punctatus (Herbst)
Ptenidium formicetorum Kraatz
P. gresneri Erichson
Batrisodes venustus (Reichenbach)
Amauronyx maerkelii (Aubé)
Oxypoda haemorrhoa (Mannerheim)
O. vittata Märkel
Homoeusa acuminata (Märkel)
Thiasophila angulata (Erichson)
T. inquilina (Märkel)
Ilyobates bennetti Donisthorpe
Amarochara bonnairei (Fauvel)
Haploglossa gentilis (Märkel)
H. pulla (Gyllenhal)
Amischa analis (Gravenhorst)
Notothecta confusa (Märkel)
Plataraea brunnea (F.)
Liogluta nitudula (Kraatz)

Enalodroma hepatica (Erichson)
Acrotona consanguinea (Eppelsheim)
Aleochara ruficornis Gravenhorst
Drusilla canaliculata (F.)
Zyras cognatus (Märkel)
Z. funestus (Gravenhorst)
Z. laticollis (Märkel)
Z. limbatus (Paykull)
Z. lugens (Gravenhorst)
Z. haworthi Stephens
Z. humeralis (Gravenhorst)
Heterothops spp. (? *praevius* Erichson)
Quedius brevis Erichson
Othius myrmecophilus Kiesenwetter
Gyrohypnus atratus (Heer)
Amphotis marginata (F.)

The species of *Zyras* have the habit of rolling up when disturbed and may remain motionless for some time, so may be easily overlooked. *Amphotis marginata* is often taken under loose bark but has been taken by sieving dead leaves in the ants' runs. *Leptinus testaceus* Müller is sometimes found with this ant.

Lasius flavus (F.) – nests either under a stone or in a mound of earth. With earth-mound nests little can be done and in general this type of nest seems quite unproductive. Those built under a stone are more productive. The stone should be quickly and carefully lifted and placed on the sheet, in a cloth bag or on a tray. The tunnels and passageways under the stone should then be inspected. Often the ants seize beetles and take them down into the nest as they do with their own larvae and pupae. The underside of the stone can then be inspected and afterwards carefully replaced over the nest.

The following species occur with *Lasius flavus* –
Hetaerius ferrugineus (Olivier)
Amauronyx maerkelii (Aubé)
Claviger testaceus Preyssler
Lamprinodes saginatus (Gravenhorst)
Homoeusa acuminata (Märkel)
Amischa analis (Gravenhorst)
Drusilla canaliculata (F.)
Zyras limbatus (Paykull)
Sunius bicolor (Olivier)
Platydracus stercorarius (Olivier)
Othius myrmecophilus Kiesenwetter

Formica fusca L. nests either under stones or in old posts and stumps; the former should be treated in the same manner as *L. flavus* and in the latter the stump or post can be broken up over a sheet and sieved. The following beetles have been taken with *Formica fusca* –

>*Hetaerius ferrugineus* (Olivier)
>*Neuraphes carinatus* (Mulsant)
>*Batrisodes venustus* (Reichenbach)
>*Amauronyx maerkelii* (Aubé)
>*Lamprinodes saginatus* (Gravenhorst)
>*Homoeusa acuminata* (Märkel)
>*Dinarda pygmaea* Wasmann
>*Aleochara ruficornis* Gravenhorst
>*Drusilla canaliculata* (F.)
>*Zyras limbatus* (Paykull)
>*Lomechusa emarginata* (Paykull)
>*L. paradoxa* Gravenhorst
>*Opatrum sabulosum* (L.)

These associations between other species of ants and beetles have been noted:

Formica pratensis Retzius:
>*Ptenidium formicetorum* Kraatz
>*Acrotrichis montandonii* (Allibert)
>*Oxypoda formiceticola* Märkel
>*O. haemorrhoa* (Mannerheim)
>*Leptacinus formicetorum* Märkel
>*Protaetia cuprea* (F.)
>*Monotoma angusticollis* Gyllenhal

Formica exsecta Nylander:
>*Dendrophilus punctatus* (Herbst)
>*Oxypoda haemorrhoa* (Mannerheim)
>*Dinarda hagensi* Wasmann
>*Amischa analis* (Gravenhorst)
>*Notothecta flavipes* (Gravenhorst)
>*Lyprocorrhe anceps* (Erichson)
>*Drusilla canaliculata* (F.)
>*Zyras limbatus* (Paykull)
>*Othius myrmecophilus* (Kiesenwetter)

Formica sanguinea Latreille:
    Hetaerius ferrugineus (Olivier)
    Lamprinodes saginatus (Gravenhorst)
    Oxypoda haemorrhoa (Mannerheim)
    O. recondita Kraatz
    Notothecta flavipes (Gravenhorst)
    Drusilla canaliculata (F.)
    Zyras limbatus (Paykull)
    Lomechusoides strumosus (F.)
    Quedius brevis Erichson
    Othius myrmecophilus (Kiesenwetter)

Lasius niger (L.):
    Claviger testaceus Preyssler
    C. longicornis Müller
    Homoeusa acuminata (Märkel)
    Drusilla canaliculata (F.)
    Zyras limbatus (Paykull)
    Opatrum sabulosum (L.)

Lasius alienus (Förster):
    Claviger testaceus Preyssler
    Drusilla canaliculata (F.)

Lasius brunneus Latreille: This is another tree nesting ant; its nests should be treated in the same manners as in *Lasius fuliginosus* (above).
    Dendrophilus punctatus (Herbst)
    Ptenidium formicetorum Kraatz
    P. turgidum Thomson
    Acrotrichis montandonii (Allibert)
    Euconnus pragensis (Machulka)
    Stenichus bicolor (Denny)
    S. godarti (Latreille)
    Eutheia formicetorum Reitter
    Batrisodes venustus (Reichenbach)
    B. delaporti (Aubé)
    B. adnexus (Hampe)
    Oxypoda recondita Kraatz
    Haploglossa gentilis (Märkel)
    H. pulla (Gyllenhal)
    Ilyobates propinquus (Aubé)
    Amischa analis (Gravenhorst)

Alaobia sodalis (Erichson)
Liogluta nitidula (Kraatz)
Aleochara sanguinea (L.)
Drusilla canaliculata (F.)
Zyras limbatus (Paykull)
Euryusa optabilis Heer
E. sinuata Erichson
Tachyusida gracilis (Erichson)
Quedius scitus (Gravenhorst)
Othius myrmecophilus Kiesenwetter
Leptacinus formicetorum Märkel
Xantholinus angularis Ganglbauer
Symbiotes latus Redtenbacher
Ptinus subpilosus Sturm
Dryophthorus corticalis (Paykull)

Lasius umbratus (Nylander):
Acrotrichis montandonii (Allibert)
Claviger longicornis Müller
Acrotona consanguinea (Eppelsheim)
Zyras humeralis (Gravenhorst)

Lasius mixtus (Nylander):
Claviger longicornis Müller
Homoeusa acuminata (Märkel)

Ponera coarctata (Latreille):
Tychobythinus glabratus (Rye)
Lamprinodes saginatus (Gravenhorst)
Drusilla canaliculata (F.)

Myrmica rubra (L.):
Drusilla canaliculata (F.)
Zyras collaris (Paykull)
Lomechusa emarginata (Paykull)
L. paradoxa (Gravenhorst)
Platydracus stercorarius (Olivier)

Myrmica ruginodis Nylander:
Lamprinodes saginatus (Gravenhorst)
Drusilla canaliculata (F.)
Lomechusa emarginata (Paykull)
Platydracus stercorarius (Olivier)
P. latebricola (Gravenhorst)

Myrmica sulcinodis Nylander:
    Drusilla canaliculata (F.)
    Lomechusa emarginata (Paykull)

Myrmica scabrinodes Nylander:
    Batrisodes venustus (Reichenbach)
    Amischa analis (Gravenhorst)
    Drusilla canaliculata (F.)
    Zyras limbatus (Paykull)
    Lomechusa emarginata (Paykull)
    Platydracus stercorarius (Olivier)
    Othius myrmecophilus Kiesenwetter

Leptothorax acervorum (F.):
    Drusilla canaliculata (F.)

Tetramorium caespitum (L.)
    Drusilla canaliculata (F.)
    Platydracus stercorarius (Olivier)

    When collecting beetles from and around ant nests abroad, it is of prime importance to take a small number of the ants and to preserve these with the beetles (for example mounting an ant on a card beneath the beetle mount). This will allow for confirmation of the identity of the ant, should there be any doubt or if the taxonomy of the taxa is under question. JC's personal experience of collecting ants abroad is largely confined to China. In the sub-tropical Wuyi Shan Special Protection Area spanning the borders of Fujian and Jiangxi Provinces he captured a variety of Pselaphinae (Batrisini and Clavigerini) in *Lasius* sp. nests in rotten trees. The bulk of these were New to Science. Leaf litter around ant nests in Fujian Province and north of Beijing produced many species of Batrisiine Pselaphinae, plus *Zyras, Pella, Homoeusa, Thaiosophila, Falagria* and *Tmesiphorus* including several species New to Science. In Zhejiang Province, again with *Lasius* ants, but under a stone in rough grassland at the top of the Tianmu Shan produced two species of Pselaphinae including a remarkable New Species of *Batristilbus* (Hlavač *et. al.*, 2002) the first record of this genus in China. DM collected a clavigerine under a concrete sewerage pipe laid against the curb, in the car park of the National Museum of Namibia, this pipe was rolled over, and several specimens of a clavigerine were collected, these were on the underside of the pipe, and were often picked up and carried away by the ants. This represented a new subfamily to Namibia and an undescribed genus, two years later DM collected more from the under the same pipe.

    It is worth bearing in mind that unidentified ants abroad should be treated with a great deal more respect than our comparatively harmless native species.

References etc.

Bolton, B. & Collingwood, C.A., 1975. Hymenoptera, Formicidae. *Handbk. Ident. Br. Insects.* **6**(3c): 34 pp. Royal Entomological Society, London.

Donisthorpe, H. St.J. K., 1896. Hints on collecting myrmecophilus Coleoptera. *Entomologist's Monthly Magazine,* **32**:44-50.

Donisthorpe, H. St.J. K., 1927. *The guests of British ants.* Routledge and Sons, London. xxiii +244pages, 16plates + 55figs.

Hlavač, P., Sugaya, H. & Zhou, H-Z., 2002. A new species of the genus *Batristilbus* (Coleoptera: Staphylinidae: Pselaphinae) from China. *Entomological Problems,* **32**(2):129-131.

Skinner, C.J. & Allen, G.W., 1996. *Ants.* New Naturalists' Handbook No. 24. Richmond Publishing Co., Slough. 83pp.

BEETLES ASSOCIATED WITH STORED FOOD PRODUCTS

by John Muggleton

Introduction

Not unreasonably, the average coleopterist usually confines his activities to searching for, and studying, those beetles found outdoors. In doing so he ignores the 500 or so species, worldwide, that have been found associated with stored food products or with other manufactured goods of plant or animal origin. I hope that this account will stimulate interest in these beetles by drawing attention to their existence, indicating where they may be found and how they may be bred. These beetles are of special interest as many are important cosmopolitan pests, destroying and spoiling food throughout the world, often in those countries where it can be spared the least. Although a number of species have been the subject of detailed laboratory investigations, there remains much to be learnt from studies of their distribution, habits and life histories outside the laboratory.

The term 'stored food products' is used to include any item of plant or animal origin which is stored in man-made premises. It includes unprocessed items such as cereals and other grains, for example rapeseed and legumes, together with dried fruit, herbs and nuts. It also includes processed items such as manufactured foods, animal feedstuffs, tobacco, animal skins, wool and textiles. These commodities may be stored for long periods in warehouses, mills, grain stores, barns, maltings, food factories, seed merchants, shops and, of course, the coleopterist's home. They are also temporarily held in the ships (often in containers) and vehicles used to transport them between and within the countries of the world.

Stored product beetles have been associated with man probably for as long as he has been storing food; the first sowing of cereals has been traced back to Syria in c.9000BC. Certainly the association has been in existence for several thousand years; *Tribolium* spp. have been recorded from Egyptian tombs dating from c.2500BC and the tomb of Tutankhamun (c.1340BC) contained at least six species all associated with stored food products today, including *Oryzaephilus surinamensis* (L.), at present the most important beetle pest of stored grain in the United Kingdom. With such a long association with man it is perhaps not surprising that many of these beetles are rarely found away from stored commodities. Some of them are confined to one product or to a group of products. Bruchids, for example, occur in the seeds of legumes, while *Sitophilus* and *Rhyzopertha* show a marked preference for whole grain. Others, such as *Oryzaephilus, Lasioderma* and ptinids, are more general feeders.

Habitats and species

I have on one occasion found a single specimen of *O. surinamensis* crawling up a grass stem in a deep Cotswold valley far from any grain store. On another occasion an evening meal outside a hotel restaurant in the Loire valley was interrupted by a succession of stored product beetles landing on the table. Such were their numbers that a surreptitious visit to my bedroom for a collecting tube was necessary. Investigation after the meal revealed that the source was a nearby grain silo. On a third occasion a number of *Cryptolestes ferrugineus* (Stephens) found on a table in a pub garden in Berkshire were most likely to have fallen from a succession of grain lorries passing by on the adjacent road. Such events are the exception and in general the coleopterist wanting to study stored product beetles will have to search the stored goods, the buildings in which they are stored and the vehicles in which they are transported, if he is to have any success in finding them.

Grain stores provide one of the most obvious habitats and the most frequently found species are *O. surinamensis, C. ferrugineus, Sitophilus granarius* (L.), *Ahasverus advena* (Waltl) and *Typhaea stercorea* (L.). Less frequently found are *Tribolium castaneum* (Herbst), *T. confusum* Jacquelin du Val, *Sitophilus oryzae* (L.), *S. zeamais* Motschulsky and *Rhyzopertha dominica* (F.). Access to grain stores is unlikely to be easy and the owners will not want to admit to having pests. Nevertheless, if the opportunity arises they are well worth investigating and will harbour many other species in addition to the stored grain specialists. However, as I have related above, warm weather will encourage the beetles to leave the stores and so the vicinity of stores will repay investigation. In such circumstances leaving bait bags (see below) for few days in a dry location near the store should yield results, but remember that the bags are likely to be attractive to badgers and squirrels.

Animal feed meals have their own beetle fauna which in addition to the grain store species mentioned above includes *Ptinus tectus* Boieldieu, *P. fur* (L.), *Stegobium paniceum* (L.), *Tenebrio molitor* L. and *Trigonogenius globulus* Solier. Again such places are unlikely to be accessible but their products are and can be found in the premises of animal feed merchants, in pet shops and stables. Places where the turnover of feeds is slow are likely to yield the best results. One can find *S. paniceum* in dog biscuits, ptinids in cereal based petfoods and *Dermestes maculatus* De Geer and *D. peruvianus* Laporte de Castelnau in dry petfood with a meat content.

Some of the beetles found with stored food may also occur in other situations, such as haystacks, bird's nests (a well-known reservoir for infestations), bee and wasp nests and cellars. The extent to which stored product species occur in natural habitats in the British Isles is unclear. Both *O. surinamensis* and *C. ferrugineus* have been found in the wild under bark in Britain (Halstead, 1993) which is rather unexpected for species of apparently tropical origin and which are primary grain pests. There is, however, some suspicion the latter species may be a mould feeder as well as a grain feeder. Certainly a number of secondary grain store pests, such as *T. stercorea*, which are principally mould feeders will be found in the wild. The coleopterist's home will, of course, provide a habitat for some stored product species. However, greater hygiene and the use of insecticides in stores and warehouses means that the number of the species reaching the customer is far less than at the time the first edition of this handbook was written. The coleopterist will, of course, also be aware that a number of stored product species (e.g. species of *Anthrenus* and *Ptinus*) are attracted to collections of dried insects and other animals, and species new to Britain such as *Reesa vespulae* (Milliron) and *Anthrenus olgae* Kalik have been found associated with collections in recent years (Peacock, 1993; Adams, 1988).

Collecting methods

Grain stores and warehouses are not the natural habitat of the coleopterist and therefore I have felt it useful to include some information which will aid the coleopterist when searching such *terra incognita*.

Permission: Permission must always be sought from the owner or manager before entering any premises to collect insects.

Safety: Great care must be taken when visiting grain stores, as heaps of grain are inherently unstable. Keep well clear of the edges of such heaps and the walling supporting them, never enter bins or silos storing grain. Some grain is stored admixed with insecticide or may have a surface treatment of insecticide dust, so you must be aware of the hazards of contact with organophosphorus and other insecticides. Grain handling and processing machinery can also be dangerous, so avoid searching in places where such machinery is installed, even if it does not appear to be working at the time. When an infestation with rats is observed or suspected there is a possibility of infection with Weil's disease, so gloves should be worn, and cuts and abrasions should be covered.

Searching: In any premises the best form of pest control is hygiene, so it follows that the best collecting grounds are those mills or warehouses that are in constant use and are not kept very clean. In such places accumulations of food are allowed to remain in odd corners and cavities for long periods and infestations are common. The best places to find specimens are among small heaps of debris and floor-sweepings, especially in damp and dark corners; behind the boarding around walls; at points where sacks or other containers of food are in contact with one another, or with the walls and floor; in cracks and crevices; and in the vicinity of accumulations of rodent and bird droppings or nest material. Many dead specimens can be found in cobwebs and on window-ledges. Finally, of course, one should search in and around the stored products themselves. Some species are active when it is dark and can be more easily obtained then.

Sieving: Beetles can be separated from various stored products, floor sweepings and debris, by sieving. A 'nest' of sieves is best used for this as it will allow the material to be passed through a range of meshes in a single operation. In practice the combination of a sieve with a 2mm aperture with another of 0.7mm aperture will separate most beetles from grain and smaller debris, with most beetles passing through the 2mm mesh and being retained by the 0.7mm mesh. The larger beetles, grain or other coarse material will be retained by the 2mm mesh. The contents of the sieve can be examined by emptying them onto a white tray or sheet and any beetles present can be removed using a pooter.

Trapping: Trapping will often reveal the presence of species which, because of their low population density, are not found by searching or sieving. Various baits or traps can be used and some are now marketed commercially for use in grain stores, but only two types, pitfall traps and bait bags, are likely to be of use to the amateur who will be able to make them for himself. Pitfall traps can be used in grain and other commodities. A plastic container such as a pint-size plastic beer 'glass' is ideal and should be inserted in the grain with the rim level with the surface of the grain. Surface dwelling beetles will fall in, but as many species can climb shiny surfaces it is necessary to put some water with a few drops of washing-up liquid into the bottom of the container to trap the beetles. The pitfall traps must be placed on a level area or else they will rapidly fill with grain. Pitfall traps made of glass should not be used because of the potential hazard of broken glass being left in the grain.

Baiting with food is another method that has been exploited commercially, and bait bags can be made quite easily. These bags should be made of nylon mesh (around 2mm mesh size) and measure about 100 x 160mm. The bags should be filled with a mixture of broken carob (locust bean), wheat grain and peanuts, and then sealed. Bait bags can be placed around the wall or floor margins, in large crevices, on ledges or on the surface of the stored commodity. They can be examined by shaking the bag and its contents over a white tray or sheet, and any beetles falling out can be collected up using a pooter. The bags will attract a wide range of species and can be used in grain stores, mills and any other situations where the presence of stored product beetles is suspected. Obviously one could experiment with the content of the bait bags to find that most effective for certain species, or for the widest range of species in a particular situation. Both pitfall traps and bait bags should be left in place for seven to ten days to be most effective. In commercial stores and farm premises care should be taken to ensure that they are put somewhere where they will not be disturbed. When placing traps on grain it is useful to tie them to a one metre marker cane in case the grain moves and buries the trap, and care should be taken to ensure that all the traps are removed from the stored commodity at the end of the visit. The bait bags are also very attractive to rats and mice and so should not be put in areas where there is obvious rodent activity.

Moisture, as well as food, will attract stored product beetles. This attraction can be exploited by placing pieces of wet sacking or small piles of moistened food on the floor and then examining them after a few days when a variety of species may be found.

Identification

The species most frequently found in the British Isles can be identified using the standard identification works. A key to the common stored product species can be found in the Natural History Museum publication *Common insect pests of stored food products* (Mound, 1989). More specialist keys to the genera of stored product beetles and to three families, the Laemophloeidae, Passandridae and Silvanidae, have been produced by Halstead (1986, 1993). There is also the detailed but much older work by Hinton (1945; reprinted 1963) dealing with many of the important pest genera. Unfortunately only the first part of this study was completed. Details of species first recorded from Britain in recent years can be found by looking under the entries for Dermestidae, Anobiidae, Nitulidae, Silvanidae, Laemophloeidae and Tenebrionidae in Hodge & Jones (1995). Although many of the species are cosmopolitan, with the result that even the most exotic foods may be infested by commonplace species, from time to time unusual species, that cannot be keyed out in the standard British or European works, will be found. These can be sent to the Central Science Laboratory, Sand Hutton, York, YO4 1LZ or to the Department of Entomology, The Natural History Museum, Cromwell Road, London, SW7 5BD, for identification; a charge, however, is usually made for this service, so you should enquire first.

Breeding

The Study of Life Histories. The discovery of larvae or pupae of stored product beetles will be followed by the need to breed them through to the adult stage in order to find out to which species they belong, but the study of life histories is also, of course, an activity in its own right and one that appeals to many coleopterists; it is to these to whom the following notes are especially addressed.

Although many of these beetles have been studied in the laboratory, there are still plenty of opportunities for the amateur to make novel observations while rearing them. Studies can be made on the range of temperatures at which the species will breed and on the length of the life cycle at various temperatures. They will be even more useful if it is possible to measure or control the humidity. The effects that differing photoperiods may have on life cycles and productivity may well produce interesting observations. The acceptability of various foods and the extent to which the life cycle can be completed on them is another area worthy of investigation. Whatever investigations are made it will be necessary to keep notes and it is particularly important to record the temperature and humidity at which the cultures are kept, the date on which the culture was set up and the number of adults, or other stage, used to start the culture. The other information required will depend on the nature of the investigation. Much of this information can be recorded on a label attached to the culture jar, but a permanent record must also be made and, if all this is done, it should be possible to produce useful contributions to our knowledge of this diverse group of beetles. Sokoloff's (1974) book on the genus *Tribolium*, while dealing with only a single genus, shows what might be studied, in less well-known species and genera, in relation to ecology, biology and nutrition.

A unique advantage of studying this group of beetles is that the situations in which they are found are usually man-made. Thus breeding and life history studies are easily carried out under conditions very close to those in which the beetles are normally found. Rearing stored product beetles is particularly easy provided that a few simple rules are followed in relation to the temperature and moisture of the food, its nature and the hygiene of the cultures. Although the rearing methods which follow were designed for use in the laboratory, for the most part they use readily accessible materials which should not pose a problem for the amateur coleopterist.

A General Culturing Method. The most convenient containers for cultures of stored product beetles are glass jars. The 1lb (454g) and 2lb (908g) sizes sold for storing jams and other preserves are the most useful, but good results can be obtained with most sizes and shapes of jars, although tall, narrow jars where the surface area of the food will be small in relation to its depth, are best avoided. Sufficient food should be put into the jar to occupy one third to one half of its volume. Different species have different requirements; *O. surinamensis* rarely penetrates more than the first 25mm of the standard culture medium, whereas the superficially (in sizes and shape) similar *C. ferrugineus* will burrow right to the bottom of the food. A piece of crumpled kitchen paper should be placed on top of the food. This paper provides a larger surface area for the adults and larvae and thus helps to reduce contact between individuals, and lessens the chances of cannibalism and reduces other effects of overcrowding. The paper also provides additional egg-laying and pupation sites for some species. Where necessary a pad of damp paper or a water tube can be added to the food at this stage (see below).

Although some species will pupate in the food itself, others require special pupation sites which often serve to protect them against the cannibalistic tendencies of their own adults. Members of the genera *Tenebrio, Tenebriodes* and *Dermestes,* in particular, will attack their own uncovered pupae, especially when they are kept under too dry conditions, but many members of other genera will also exhibit cannibalism if the cultures become overcrowded. To overcome such problems it is advisable to supply large corks (e.g. for *Dermestes* and *Tenebriodes*), lumps of cotton wool (e.g. for *Gnatocerus cornutus* (F.) and *Tribolium destructor* Uyttenboogaart) or small rolls of sacking or corrugated cardboard. These should be placed on top of the food in the jars, where they will provide adequate pupation sites. Only experience will tell if such measures are necessary in the particular conditions in which you are rearing your beetles.

Once the culture jars are prepared the adult beetles (or larvae) can be introduced. The number put into each jar will depend on the size of the adults and the numbers available. For most genera (e.g. *Sitophilus, Tribolium* and *Oryzaephilus*) up to 100 adults in a 2 lb jar is acceptable, but for the larger beetles (e.g. *Tenebrio*) there should be perhaps no more than 10 in a 2 lb jar. These are maximum numbers to avoid overcrowding; it is, of course, perfectly possible to start cultures with far fewer adults; a single female *O. surinamensis* will produce several hundred offspring in a couple of generations. The jars then need to be sealed both to keep out unwelcome guests and to keep in the inhabitants, as some stored product beetles can fly and others can climb glass. Glass-climbing ability can differ between members of the same genus, thus the *O. surinamensis* strains kept at the Slough Laboratory can climb glass whereas the *O. mercator* strains cannot. This is an important point to bear in mind when handling the beetles, as those that cannot climb glass are much easier to contain. The most effective method of closing the jars is to seal a filter paper or blotting paper circle to the rim of the jar with melted wax, and then as an added insurance to cover the top of the jar with a piece of cloth held in place by a rubber band (two are safer!). This method contains the insects but allows gaseous exchange through the paper. If wax and filter or blotting paper are not available then cloth held in place by rubber bands would be an alternative, but the bands must be very tight as many of the beetles can pass through very small holes. Similarly the normal screwtop of the jar could be used, but this would need to be punctured with holes that would allow air, but not the occupants, to pass through. I have known *O. surinamensis* to escape by walking along the screw thread of a closed screw-topped bottle. If the cultures are overcrowded some species (e.g. *Sitophilus* spp.) will chew their way out through the covering material.

Once set up, the culture jars should be placed in a warm place. It is worth remembering that a local microclimate will be set up in the jar itself and it is unlikely that this will be same as that in the room in which the beetles are being kept. Thus one's notes should state that the culture has been kept at, say 25°C and 70% R.H., rather than that the insects have been reared under these conditions. Jars should not be stored in direct sunlight, but may, otherwise, be put anywhere else. Cultures will breed perfectly well if kept in total darkness, and, indeed, many of their natural habitats will be totally dark.

The Importance of Temperature and Humidity

Temperature: Up to a certain point the warmer the environment the faster an insect will develop. There is, however, an optimum temperature at which development is quickest and mortality lowest. In laboratory studies the optimum temperature for development generally lies between 25°C and 30°C. However, stored product beetles will develop quite well at room temperature, and it is interesting to speculate just how often temperatures of 25°- 30°C will be found in the unheated premises in which food and grain is often stored in Great Britain. Even species like *O. surinamensis* will survive several days in a domestic refrigerator. The optimum temperature for one species will not necessarily be that for another. If the temperature rises above the optimum, the usual results are increasing sterility of the adults with rise in temperature, and the death of all stages at the lethal temperature zone. If the temperature is lower than the optimum, the development of the insect is longer, provided all other conditions are the same, until a temperature is reached at which development can barely take place and at which no eggs are laid. This is the threshold of development, and again differs for different species. As laboratory workers have tended to work with beetle strains which have become acclimatised to laboratory conditions, the amateur can make useful contributions to our knowledge of stored-product beetles by using populations freshly collected from the field to investigate the highest and lowest temperatures at which breeding will occur. A maximum-minimum thermometer hung near the cultures and read daily is the easiest means of recording the temperature of the surrounding air. It must be remembered, however, that the temperature inside the culture jar may be rather different and a thermometer inserted into the culture medium will give a better idea of the actual temperature of the culture.

Humidity: The next most important factor is moisture, not so much visible water, as the amount in the air and the food medium. That in the air may be measured as the relative humidity (R.H.) and that in the food as the moisture content; both these measurements are expressed as percentages. The humidity of the air can be found by means of a whirling hygrometer, wet and dry bulb thermometer, paper hygrometers and other apparatus. The moisture content of the food is difficult to determine accurately without access to a special drying oven.

Many species will breed poorly or not at all if the R.H. is below 50%, and an R.H. of 60-70% seems to be the optimum for most species and ideally the moisture content of the food should be brought into equilibrium with this (see below). Breeding will occur at higher humidities but in such situations the food rapidly becomes mouldy and unsuitable for many beetles. For a sealed culture kept at 70% R.H. and 25°C mould should not become a problem for 10 weeks or more. If mould growth appears earlier this is an indication that the conditions in the culture jar are too humid, or that the moisture content of the food was too high to start with. If the food goes mouldy rapidly it is best discarded, and the culture started afresh with new food. As mentioned earlier, for some species (e.g. *T. stercorea*) mould may be an important part of their diet and, for such species, the best results are obtained when the food supports a vigorous growth of mould. This can be achieved by adding a dampened pad of paper or cotton wool to the food when the culture is set up.

Without access to a laboratory, achieving the correct moisture content of the food is difficult. A simple way would be to store the food in a place where the relative humidity of the air was known and was close to that which you wished to achieve. After one or two weeks the food would equilibrate to the R.H. of the surrounding air. This method has the disadvantage of exposing the food to infestation by other organisms. Alternatively the food can be left for one to two weeks in an airtight container in which is placed, in a separate open container, some water saturated with an appropriate chemical salt. Care must be taken to ensure that the solution does not come into contact with the food. The principle of this method is that the food will absorb moisture until it is in equilibrium with the air surrounding it, and that when air is enclosed over certain salt solutions it has a fairly constant vapour pressure, depending on the strength of the solution. Thus at 20°C, and over a saturated solution of potassium carbonate, the R.H. of the air would be about 45%; over calcium or magnesium nitrate it would be about 55%; over sodium nitrite or ammonium nitrate about 65%; over sodium chloride about 78%, and over distilled water about 98%. At lower temperatures the R.H.s will mostly be a little higher. The purest salts obtainable should be used and the food should be spread in a thin layer. These are the simplest solutions to use; however when accurate work involving a range of humidities is contemplated, it would be necessary to use the appropriate strength potassium hydroxide solutions (Solomon, 1951) which cannot be easily produced by the amateur coleopterist.

Many of the stored product beetles will drink water eagerly and, although not essential, its availability may prolong their life and increase productivity. Among the beetles that might benefit from having water available are *Carpophilus hemipterus* (L.), *Tenebrio molitor* and many of the Ptinids. Water can be provided in a corked glass tube with a blotting paper wick. The water is put into the tube together with a strip of blotting paper which should reach to the bottom of the tube and pass between the cork and the side of the tube before projecting for about 25mm from the top of the tube. As long as there is water in the tube and the cork is not too tight a fit, the exposed portion of the blotting paper will remain damp. The water tube should be pushed down into the food in such a way that the blotting paper does not touch the surface of the food as, if it does, the food will go mouldy.

Food requirements

As with any other group of beetles the food requirements of stored product beetles will differ from species to species and this will apply to artificial, as well as natural, diets. The food must also be in an appropriate form to allow development to take place. Thus, *Sitophilus* spp. require whole grains of wheat for successful breeding as the larvae develop inside the wheat grain; other species, with free-living larvae, can be given ground wheat. Some species (e.g. *Dermestes*) require food of animal origin, but for most species an entirely vegetarian diet is acceptable.

There are two approaches to feeding, one is to provide the beetles with whatever food they were found on, the other is to provide a standard culture medium. The former approach will be rewarding as there is often little information on the range of natural foods on which development can be completed, and this could be a study in itself. The disadvantage is that such foods collected with the beetles, or directly from their habitat, may be infested with other organisms, including mites, bacteria, protozoans and moulds, detrimental to the beetles. The nature of the food may also make it difficult to handle; sticky dates or Turkish Delight infested with *Oryzaephilus* can be rather messy! If, however, one's principal concern is not with the range of natural foodstuffs, but with convenience, hygiene or rearing large numbers then a standard culture medium can be employed. Table 1 gives examples of some culture media and the species that will feed on them. The list is neither exhaustive nor exclusive and some species may be reared on more than one medium. As some grain is treated in store with insecticides, care needs to be exercised in choosing a supplier of grain; if possible find a supply that is said to be insecticide free. The sudden death, or greatly reduced productivity, of a whole culture on transfer to a fresh supply of food can often be due to contamination with insecticides. Several of the media require the use of whole ground wheat and this can be obtained by putting wheat grains through a coffee grinder, using a fairly fine setting. Dried yeast seems to be a useful addition to most media, especially to those with a low nutritive value, such as plain flour.

Culture hygiene

The cultures are a rich source of food to other organisms and are susceptible to infestation by moulds, mites and booklice, as well as by other species of beetle. All of these unwanted guests will compete for the food and may make conditions in the culture jar unsuitable for its intended occupants. Mites of the genera *Tyrophagus* and *Suidasia* will outcompete many of the stored product beetles, and *Caloglyphus* species may also eat the beetle eggs, as will the predatory mite species. Except when rearing species such as *Typhaea stercorea* and *Ahasverus advena*, if the food is too damp mould growth will soon make it unsuitable for consumption by the beetles. Booklice can build up to large population densities in culture jars, but appear to have only nuisance value. Different species of beetle will compete with each other for the food and even eat each other's eggs, larvae and pupae, so it is important to keep single species cultures. The problems of overcrowding have already been mentioned as this may lead to cannibalism and the spread of disease. It is worth noting here that some genera, in particular *Tribolium* and *Cryptolestes*, appear vulnerable to sporozoan infections.

Heat sterilisation of the culture medium at 70°- 90°C is often advocated as a method of preventing the infestation of cultures and the spread of disease. This does, of course, mean that the food will become very dry and need moistening, and there is the possibility that some ingredients will be denatured by the heat. The writer prefers storage for at least two weeks in a domestic deep freeze (at -18°C) as an alternative. (N.B. The freezing compartment of a refrigerator is not sufficiently cold for this purpose.) This method has been used for many years without problems and may even be a more effective way of killing sporozoan spores. It also has the advantage that the food can be prepared beforehand and stored for long periods until required. The food must be brought to room temperature before the beetles are introduced into it. Mould growth can be inhibited by ensuring that the jars are completely dry before use and that the food is not too moist. The danger from mite infestation is considerable, but can be overcome by scrupulous hygiene. As well as being in the food, mites may also be on the working surfaces and floating in the air, so culture jars should be stored upside down and rinsed thoroughly with hot water before use and then dried. Once set up, culture jars should be left open for as short a time as possible. Food should be stored in a closed container, preferably in a freezer. The prevention of mite infestation is another good reason for sealing the tops of the jars with filter or blotting paper, rather than using some other covering with gaps through which mites and booklice can creep.

When the food supply is exhausted or the culture becomes overcrowded it will be necessary to move some adults to fresh food. When this is done, it is important to transfer only the beetles and not the old culture medium which may contain fungal spores and frass.

Table 1. Some useful culture media and the species that feed on them.

Constituents	Ratio of constituents by weight	Species
Whole wheat	1:0	*Rhyzopertha dominica* and *Sitophilus* spp.
Whole ground wheat, rolled oats and yeast	5:5:1	*Oryzaephilus* spp. *Cryptolestes* spp. *Ahasverus advena Tenebrio* spp.
Fishmeal, yeast and bacon ends	16:1:4	*Dermestes* spp.
Fishmeal, yeast (and a piece of felt)	16:1	*Anthrenus verbasci* (L.), *Attagenus pellio* (L.)
Whole ground wheat, fishmeal and yeast	8:4:1	*Ptinus* spp., *Alphitobius diaperinus* (Panzer), *Gnatocerus cornutus* (F.) and *Tribolium destructor* Uyttenboogaart
Whole ground wheat and yeast	10:1	*Lasioderma serricorne* (Fabricius), *Stegobium paniceum* (L.), *Latheticus oryzae* Waterhouse and *Palorus ratzeburgi* (Wissman)
Wholewheat flour and yeast	12:1	*Tribolium castaneum* (Herbst) and *Tribolium confusum* Jacquelin du Val.

References

Adams, R.G., 1988. *Anthrenus olgae* Kalik new to Britain (Coleoptera: Dermestidae with notes on its separation from *A. caucasicus* Reitter. *Entomologist's Gazette*, **39**:207-210.

Halstead, D.G.H., 1986. Keys for the identification of beetles associated with stored products. 1 - Introduction and key to families. *Journal of Stored Products Research*, **22**:163-203.

Halstead, D.G.H., 1993. Keys for the identification of beetles associated with stored products. 2 - Laemophloeidae, Passandridae and Silvanidae. *Journal of Stored Product Research*, **29**:99-197.

Hinton, H. E., 1945. *A Monograph of the Beetles Associated with Stored Products, Vol. 1.* British Museum (Natural History), reprinted 1963, Johnson Reprint Company, London.

Hodge, P.J. and Jones, R.A., 1995. *New British Beetles*. British Entomological &Natural History Society, Reading.

Mound, L., 1989. *Common Insect Pests of Stored Food Products, a guide to their Identification, 7th edn*. British Museum (Natural History), London.

Peacock, E.A., 1993. Adults and larvae of hide, larder and carpet beetles and their relatives (Coleoptera: Dermestidae) and of Derodontid beetles (Coleoptera: Derodontidae). *Handbk. Ident. Br. Insects.*, **5**(3):144pages. Royal Entomological Society, London.

Sokoloff, A., 1974. *The Biology of Tribolium, Vol. 2*. Clarendon Press, Oxford.

Solomon, M.E., 1951. The control of humidity with potassium hydroxide, sulphuric acid, and other solutions. *Bulletin of Entomological Research*, **42:**543-554.

VASCULAR PLANTS AND THE BEETLES ASSOCIATED WITH THEM

by Eric G. Philp

Many species of beetle are closely associated with certain vascular plants and it will be of great help to the coleopterist to gain a working knowledge of the wild flowers of the countryside in order to search out some of the more elusive species of his quest. Beetles will be associated with certain plants because they or their larvae feed upon the plant or parts of the plant such as the roots, fruits, stem etc., each of which might have to be in a certain condition, young, ripe, dying etc. Or they may be present there because they feed upon other insects or fungi that are associated with that plant.

BEETLES AND THEIR HOST PLANTS

The following list of some of the British beetles, in Checklist order, also gives the major host-plants on which these beetles are found. A large number of 'common' species of beetle feed upon a wide range of plants or are attracted to almost any plant in flower, and these will be found by general beating or sweeping and are not dealt with here. By first seeking out the host-plant as listed here the reader will have a much better chance of finding the specific beetle that they seek.

CARABIDAE

Most species of Carabidae feed upon a mix of both animal and vegetable matter, and many are often found resting around the base of various plants, i.e. mats of *Stellaria media*, although this is more often for cover rather than for food. Most species of **Amara** and **Harpalus** will often climb herbaceous plants in search of pollen or seeds.

Calosoma inquisitor; *Quercus robur* and other species of oak. **Zabrus tenebrioides**; in cultivated cornfields on *Triticum aestivum, Secale cereale* etc. **Amara infima**; *Calluna vulgaris*. **Curtonotus aulicus**; Mayweeds such as *Tripleurospermum inodorum* and *Matricaria recutita*. **Curtonotus convexiusculus**; *Atriplex* and *Chenopodium* species. **Harpalus rufipes**; *Fragaria* x *ananassa*. **Bradycellus caucasicus**; *Calluna vulgaris*. **Trichocellus cognatus**; *Calluna vulgaris*. **Odacantha melanura**; *Phragmites australis, Typha* spp. **Demetrias imperialis**: *Typha* spp. **Demetrias monostigma**; *Ammophila arenaria, Leymus arenarius* and occasionally other sand dune grasses.

HYDROPHILIDAE

Helophorus nubilus; roots of *Triticum aestivum* and other plants. **H. porculus** and **H. rufipes**; at the roots of *Brassica rapa* and other Brassicaceae.

HISTERIDAE
Plegaderus dissectus; in wet decaying wood of *Ulmus* and *Quercus*. ***P. vulneratus***; under bark of *Pinus sylvestris*. ***Kissister minimus***; at the roots of *Rumex acetosella*. ***Paromalus flavicornis***; under bark of *Fagus sylvatica, Quercus* spp. etc.

SILPHIDAE
Dendroxena quadrimaculata; on *Quercus robur* and occasionally other trees where it feeds upon Lepidopterous and other larvae.

STAPHYLINIDAE
Eusphalerum primulae; on flowers of *Primula vulgaris*.

LUCANIDAE
All three species feed upon dead or rotten wood of various tree species, particularly those listed.
    Sinodendron cylindricum; *Fraxinus excelsior, Fagus sylvatica, Salix* spp., *Betula* spp., *Malus* spp., *Tilia* spp., *Castanea sativa*. ***Lucanus cervus***; *Quercus* spp., *Ulmus* spp. ***Dorcus parallelipipedus***; *Fagus sylvatica, Fraxinus excelsior, Salix* spp., *Ulmus* spp.

SCARABAEIDAE
Most of the Chafers develop underground, feeding upon the roots of various plants. The adults are frequently found at flower heads, particularly those of *Carduus* spp., *Cirsium* spp, *Rubus* spp. and *Heracleum sphondylium*. The larvae of ***Gnorimus nobilis*** develop in wood mould in stumps of *Prunus domestica* and *Malus domestica*, and those of ***G. variabilis*** in old *Quercus* spp.

BUPRESTIDAE
Melanophila acuminata; in and under bark or scorched and burnt *Pinus* spp., *Picea* spp. and occasionally *Betula* sp. ***Anthaxia nitidula***; larvae under bark of *Prunus spinosa* and other *Prunus* spp., adults on *Crataegus* spp, *Viburnum opulus*. ***Agrilus angustatus***; *Quercus* spp., *Corylus avellana*. ***A. laticornis***; *Quercus* sp., *Salix* spp., *Corylus avellana*. ***A. biguttatus***; *Quercus* spp. ***A. sinuatus***; on mature specimens of *Crataegus*. ***A. viridis***; mature or coppiced *Salix cinerea* and other *Salix* spp., stunted or damaged *Quecus* spp. ***Aphanisticus emarginatus***; on species of *Juncus*, particularly *J articulatus*. ***A. pusillus***; *Schoenus nigricans, Juncus* spp. ***Trachys minuta***; *Salix* spp., *Carpinus betulus*. ***T. scrobiculatus***; *Glechoma hederacea*. ***T. troglodytes***; *Succisa pratensis*.

EUCNEMIDAE
Microrhagus pygmaeus; *Fagus sylvatica, Quercus* spp. ***Melasis buprestoides***; *Fagus sylvatica*. ***Hylis olexai***; *Fagus sylvatica*. ***Eucnemis capucina***; *Fagus sylvatica*.

ELATERIDAE

The wire-worm larvae of the click beetles feed upon the roots of herbaceous plants, or in dead or decaying wood in trees (although some are predaceous upon other larvae in these situations).

Lacon querceus; *Quercus* spp. **Diacanthous undulatus**; *Betula* spp., *Pinus sylvestris*. **Calambus bipustulatus**; *Salix* spp., *Quercus* spp., *Alnus glutinosa*. **Paraphotistus impressus**; *Pinus sylvestris*, *Betula* spp. **P. nigricornis**; *Salix* spp. in swampy parts of woods. **Ampedus balteatus**; *Pinus* spp., and occasionally *Quercus* spp. and *Betula* spp. **A. cardinalis**; *Quercus* spp. **A. cinnabarinus**; *Fagus sylvatica*, *Betula* ssp., *Quercus* spp. **A. elongantulus**; *Quercus* spp., *Pinus* spp. **A. quercicola**; *Betula* spp., *Fagus sylvatica*, *Quercus* spp. **A. pomorum**; *Betula* spp. **A. rufipennis**; *Fagus sylvatica*. **A. sanguinolentus**; *Pinus* spp., *Betula* spp., adults often at roots of *Calluna vulgaris*. **A. tristis**; *Picea* spp. **Ischnodes sanguinicollis**; *Ulmus* spp., *Quercus* spp., *Acer* spp. **Procraerus tibialis**; *Quercus* spp., *Fagus sylvatica*. **Megapenthes lugens**; *Ulmus* spp., occasionally *Fagus sylvatica*. **Melanotus punctolineatus**; *Ammophila arenaria*. **Elater ferrugineus**; *Fagus sylvatica*, *Ulmus* spp. **Dicronychus equisetioides**; *Ammophila arenaria*.

DERODONTIDAE

Laricobius erichsonii; *Abies* spp., *Pseudotsuga menziesii*.

BOSTRICHIDAE

Rhyzopertha dominica; in decayed *Quercus* and other trees.

ANOBIIDAE

Dryophilus anobioides; *Cytisus scoparius*, *Rubus fruticosus* agg. **D. pusillus**; *Picea* spp., *Abies* spp., *Larix* spp. **Ochina ptinoides**; *Hedera helix*. **Xestobium rufovillosum**; in old trees, particularly species of *Quercus* and *Salix*. **Ernobius angusticollis**; species of *Picea* and *Pinus*. **E. mollis**; *Pinus sylvestris* and other coniferous trees. **E. nigrinus**; *Pinus* spp. **Gastrallus immarginatus**; *Acer campestre*. **Anobium inexspectatum**; *Hedera helix*. **Hadrobregmus denticollis**; in old *Quercus* and *Crataegus* spp. **Xyletinus longitarsis**; *Quercus* spp., *Ulmus* spp. **Anitys rubens**; in decayed trunks of *Quercus* spp.

LYMEXYLIDAE

Hylecoetus dermestoides; *Betula* spp., *Fagus sylvatica*, *Pinus* spp., *Quercus* spp. **Lymexylon navale**; *Quercus* spp.

TROGOSSITIDAE

Ostoma ferrugineum; *Pinus sylvestris*. **Thymalus limbatus**; *Fagus sylvatica*. **Nemozoma elongatum**; under bark of *Ulmus* spp.

CLERIDAE

Tillus elongatus; *Fagus sylvaticus*. ***Opilo mollis***; *Quercus* spp., *Crataegus* spp. ***Thanasimus formicarius***; *Pinus sylvestris* and other conifers. ***T. femoralis*** ; *Pinus sylvestris* and other conifers.

MELYRIDAE

Hypebaeus flavipes; *Quercus* spp. ***Sphinginus lobatus***; *Quercus* spp. ***Malachius barnevillei***; *Ammophila arenaria*.

KATERETIDAE

Brachypterolus antirrhini; *Antirrhinum majus*. ***B linariae***; *Linaria* spp., particularly *L. vulgaris*. ***B. pulicarius***; *Linaria* spp., particularly *L. vulgaris*. ***B. vestitus***; *Antirrhinum majus*. ***Brachypterus glaber***; *Urtica* spp., particularly *U. dioica*. ***B. urticae***; *Urtica* spp., particularly *U. dioica*. ***Kateretes pedicularius***; *Carex* spp. ***K. pusillus***; *Carex* spp. ***K. rufilabris***; *Juncus* spp.

NITIDULIDAE

Pria dulcamarae; *Solanum dulcamara, S. nigrum*. ***Meligethes aeneus***; on just about any flower of any species of plant, but particularly Oil-seed Rape *Brassica napus* and other species of Brassicaceae. ***M. atramentarius***; *Lamiastrum galeobdolon*. ***M. atratus***; *Rosa* spp. ***M. bidens***; *Clinopodium vulgare*. ***M. bidentatus***; *Genista tinctoria*. ***M. brevis***; *Helianthemum nummularium*. ***M. brunnicornis***; *Stachys sylvatica*. ***M. carinulatus***; *Lotus corniculatus*. ***M. coracinus***; *Brassica* spp. and other yellow Brassicaceae. ***M. corvinus***; *Campanula trachelium*. ***M. difficilis***; *Lamium album*. ***M. erichsoni***; *Hippocrepis comosa*. ***M. exilis***; *Thymus polytrichus*. ***M. flavimanus***; *Rosa* spp. ***M. fulvipes***; *Sinapis arvensis*. ***M. gagathinus***; *Mentha arvensis*. ***M. haemorrhoidalis***; *Lamium album*. ***M. incanus***; *Nepeta cataria*. ***M. kunzei***; *Lamiastrum galeobdolon*. ***M. lugubris***; *Thymus polytrichus*. ***M. morosus***; *Lamium album*. ***M. nanus***; *Marrubium vulgare*. ***M. nigrescens***; *Trifolium repens*. ***M. obscurus***; *Teucrium scorodonia*. ***M. ovatus***; *Glechoma hederacea*. ***M. pedicularius***; *Galeopsis tetrahit, G. bifida*. ***M. persicus***; *Stachys officinalis*. ***M. planiusculus***; *Echium vulgare*. ***M. rotundicollis***; *Sinapis arvensis, Sisymbrium officinale*. ***M. ruficornis***; *Ballota nigra*. ***M. serripes***; *Galeopsis angustifolia*. ***M. solidus***; *Helianthemum nummularium*. ***M. subrugosus***; *Jasione montana, Campanula glomerata*. ***M. umbrosus***; *Prunella vulgaris, P. grandiflora*. ***M. viridescens***; *Sinapis arvensis* and other yellow species of Brassicaceae. ***Pityophagus ferrugineus***; *Pinus* spp.

LAEMOPHLOEIDAE

Laemophloeus monilis; *Fagus sylvatica*, under bark. ***Cryptolestes spartii***; *Cytisus scoparius*, in dead branches. ***Leptophloeus clematidis***; *Clematis vitalba*.

PHALACRIDAE

(Note. *Phalacrus* are associated with the mentioned plants when these are infected by Smut Fungi, *Ustilaginales*). ***Phalacrus caricis****; Carex* spp. ***P. corruscus****;* cereal crops, *Triticum, Hordeum* etc. ***P. fimetarius****; Brachypodium pinnatum.* ***P. substriatus****; Carex* spp., *Narthecium ossifragum.* ***Olibrus aeneus****; Tanacetum vulgare, Matricaria* spp., *Artemisia* spp. ***O. affinis****; Hypochaeris radicata, Tragopogon pratensis.* ***O. corticalis****; Conyza canadensis, Senecio* spp. ***O. flavicornis****; Leontodon autumnalis.* ***O. liquidus****; Pilosella officinarum.* ***O. millefolii****; Achillea millefolium.* ***O. pygmaeus****; Crepis* spp., *Filago vulgaris, Leontodon* spp. ***Stilbus oblongus****; Typha latifolia.*

CRYPTOPHAGIDAE

Telmatophilus brevicollis*; Sparganium erectum.* ***T. caricis****; Sparganium erectum.* ***T. schoenherri****; Typha angustifolia.* ***T. sparganii****; Sparganium erectum, S. emersum.* ***T. typhae****; Typha angustifolia, T. latifolia.*

BYTURIDAE

Byturus tomentosus*;* adults on many flowers, particularly those of *Rubus idaeus.*

BIPHYLLIDAE

Biphyllus lunatus*; Fraxinus excelsior,* in the fungus *Daldinia concentrica* on dead branches. ***Diplocoelus fagi****; Fagus sylvatica,* under bark.

COCCINELLIDAE

Coccidula scutellata*; Typha* spp. ***Clitostethus arcuatus****; Hedera helix.* ***Scymnus nigrinus****; Pinus sylvestris.* ***S. auritus****; Quercus* spp. ***S. limbatus****; Salix* spp., *Populus* spp. ***S. suturalis****; Pinus sylvestris.* ***Nephus quadrimaculatus****; Pinus sylvestris, Hedera helix.* ***Chilocorus renipustulatus****;* on trunks of *Fraxinus excelsior, Salix* spp., *Alnus glutinosa.* ***Exochomus quadripustulatus****;* on trunks of *Pinus sylvestris, Fraxinus excelsior, Acer pseudoplatanus* etc. ***Aphidecta obliterata****; Larix* spp., *Picea abies, Pseudotsuga menziesii,* and occasionally other conifers. ***Coccinella hieroglyphica****; Calluna vulgaris.* ***Harmonia quadripunctata****; Pinus sylvestris, P. nigra.* ***Myzia oblongoguttata****; Pinus sylvestris.* ***Anatis ocellata****; Pinus sylvestris.* ***Epilachna argus****; Bryonia dioica.*

MELANDRYIDAE

Anisoxya fuscula*; Fagus sylvatica, Fraxinus excelsior, Salix* spp., *Acer campestre.* ***Abdera biflexuosa****; Quercus* spp., *Fraxinus excelsior.* ***A. flexuosa****; Alnus glutinosa, Salix* spp. ***A. quadrifasciata****; Carpinus betulus, Quercus robur,*

Fagus sylvatica and occasionally other trees. ***A. triguttata***; *Pinus sylvestris*. ***Phloiotrya vaudoueri***; *Carpinus betulus, Fagus sylvatica, Quercus* spp., *Fraxinus excelsior*. ***Hypulus quercinus***; *Quercus* spp., *Corylus avellana*. ***Zilora ferruginea***; in the fungus *Trichaptum abietinum* on *Pinus, Larix, Abies* and other species of conifer. ***Melandrya barbata***; *Fagus sylvatica, Quercus* spp. ***M. caraboides***; *Betula* spp., *Quercus* spp., *Ulmus* spp., *Fagus sylvatica*.

MORDELLIDAE

Mordellistena acuticollis; *Cirsium arvense*. ***M. nanuloides***; *Seriphidium maritimum*. ***M. parvula***; *Achillea millifolium, Artemisia vulgaris*. ***M. pumila***; *Cirsium* spp. (Note. Members of this family are attracted to the flower heads of Apiaceae, in particular to *Heracleum sphondylium*.).

COLYDIIDAE

Cicones variegata; under bark of *Fagus sylvatica, Carpinus betulus*. ***Aulonium ruficorne***; under bark of *Pinus* and *Larix* and occasionally other species of conifer. ***A. trisulcum***; under barks of *Ulmus* spp.

TENEBRIONIDAE

Prionychus ater; in decaying wood of species of *Quercus, Salix, Fraxinus, Ulmus, Malus* and *Prunus*. ***P. melanarius***; *Fagus sylvatica*. ***Pseudocistela ceramboides***; *Quercus* spp. ***Mycetochara humeralis***; in decaying wood of species of *Quercus, Fagus, Acer* and *Prunus*. ***Bolitophagus reticulatus***; in the fungus *Fomes fomentarius* growing on *Betula* spp. (Scotland only). ***Diaperis boleti***; in the fungus *Piptoporus betulinus* on *Betula* spp. ***Scaphidema metallicum***; *Ulmus* spp. ***Corticeus bicolor***; *Ulmus* spp., *Fraxinus excelsior*. ***C. fraxini***; *Pinus* spp. ***C. linearis***; *Picea* spp., *Pinus* spp. ***C. unicolor***; *Fagus sylvatica, Quercus* spp., *Betula* spp.

OEDEMERIDAE

Chrysanthia nigricornis; *Pinus sylvestris* (Scotland only).

MELOIDAE

Lytta vesicatoria; *Fraxinus excelsior, Ligustrum* spp.

PYTHIDAE

Pytho depressus; *Pinus sylvestris* (Scotland only).

PYROCHROIDAE

Pyrochroa coccinea; *Quercus* spp. ***P. serraticornis***; *Fagus sylvatica, Quercus* spp., *Ulmus* spp. ***Schizotus pectinicornis***; *Betula* spp.

SALPINGIDAE
Sphaeriestes castaneus; *Pinus sylvestris*.

ADERIDAE
Aderus oculatus; *Quercus* spp., *Tilia* spp. ***A. populneus***; *Quercus* spp., *Tilia* spp.

SCRAPTIIDAE
Species of ***Anaspis*** are attracted to flower blossom, particularly that of *Crataegus monogyna*.

CERAMBYCIDAE
Larvae of Longhorn Beetles feed upon the woody parts of various trees and shrubs and it is often the nature and condition of this, such as thickness of bark, moisture content, or state of decay that is more important than the actual species of tree on which the eggs are deposited. Adults will be found at the breeding sites or on nearby flowerheads, particularly those of *Rubus fruticosus* and *Heracleum sphondylium*. The following list is not exhaustive but just gives the major hosts (see also the Cerambycidae chapter in section on pages 121-142).

> ***Prionus coriarius***; *Quercus* spp. ***Rhagium mordax***, *Quercus* spp. ***Grammoptera ustulata***, *Quercus* spp. ***G. abdominalis***, *Quercus* spp. ***Lepturobosca virens***, *Betula* spp., *Pinus* spp., *Picea* spp. ***Anastrangalia sanguinolenta***, *Pinus* spp., *Picea* spp. ***Stictoleptura rubra***, *Pinus* spp., *Picea* spp., *Laric* spp., *Abies* spp. ***S. scutellata***, *Fagus sylvatica*. ***Anoplodera sexguttata***, *Quercus* spp. ***Strangalia attenuata***, *Pinus* spp., *Quercus* spp., *Tilia* spp., *Betula* spp. ***Asemum striatum***, *Pinus* spp., *Picea* spp., *Abies* spp., *Larix* spp. ***Tetropium castaneum***, *Picea* spp., *Abies* spp., *Pinus* spp., *Larix* spp. ***T. gabrieli***, *Larix* spp., *Pinus* spp. ***Arhopalus rusticus***; *Pinus* spp., *Picea* spp., *Abies* spp., *Larix* spp. ***A. ferus***; *Pinus* spp., *Picea* spp. ***Trinophylum cribratum***; *Quercus* spp., *Pinus sylvestris*, *Larix* spp. ***Gracilia minuta***, *Salix* spp. ***Molorchus minor***, *Picea* spp. ***Aromia moschata***; *Salix* spp. *Populus* spp. ***Pyrrhidium sanguineum***; *Quercus* spp. ***Phymatodes testaceus***, *Quercus* spp. ***P. alni***; *Quercus* spp. ***Agapanthia villosoviridescens***, *Heracleum sphondylium*. ***Lamia textor***; *Salix* spp., *Populus* spp. ***Pogonocherus fasciculatus***; *Pinus sylvestris*. ***Acanthocinus aedilis***, *Pinus sylvestris*. ***Saperda carcharias***; *Populus* spp., *Salix* spp. ***S. scalaris***, *Quercus* spp. ***S. populnea***; *Populus tremula*, *Salix* spp. ***Stenostola dubia***, *Tilia* spp., *Fraxinus excelsior*. ***Phytoecia cylindrica***, *Anthriscus sylvestris*, *A. caucalis*, *Daucus carota* and other Apiaceae. ***Oberea oculata***; *Salix* spp. particularly *S. caprea*. ***Tetrops praeustus***; *Malus* spp., *Crataegus* spp. and other species of Rosaceae ***T. starkii***; *Fraxinus excelsior*.

MEGALOPODIDAE

Zeugophora flavicollis; *Populus tremula*. ***Z. subspinosa***; *Populus tremula*. ***Z. turneri***; *Populus tremula* (Scotland only).

ORSODACNIDAE

Orsodacne cerasi; *Crataegus* spp., *Prunus* spp. ***O. lineola***; *Crataegus* spp.

CHRYSOMELIDAE

Bruchus atomarius; *Vicia cracca, V. sativa, V. sepium*. ***B. loti***; *Lathyrus pratensis*. ***B. rufimanus***; *Vicia* spp., particularly *V. faba, Lathyrus pratensis*. ***B. rufipes***; *Vicia* spp., particularly *V. sativa, Lathyrus* spp. ***Bruchidius cisti***; *Lotus corniculatus* and regularly found on the flowers of *Helianthemum nummularium*. ***B. olivaceus***; *Onobrychis viciifolia*. ***B. varius***; *Trifolium* spp., particularly *T. pratense*. ***B. villosus***; *Cytisus scoparius*. ***Donacia aquatica***; *Carex acutiformis* and other emergent vegetation at the edge of rivers, ponds and lakes where other members of this genus will also be found. ***D. bicolora***; *Sparganium erectum*. ***D. cinerea***; *Typha* spp. ***D. clavipes***; *Phragmites australis*. ***D. crassipes***; *Nymphaea alba, Nuphar lutea*. ***D. dentata***; *Sagittaria sagittifolia*. ***D. impressa***; *Schoenoplectus lacustris, Carex* spp., particularly *C. acutiformis* and *C. paniculata*. ***D. marginata***; *Sparganium erectum*. ***D. obscura***; *Carex rostrata*. ***D. semicuprea***; *Glyceria maxima*. ***D. simplex***; *Sparganium* spp. ***D. sparganii***; *Sparganium* spp. ***D. thalassina***; *Carex* spp., particularly *C. acutiformis* and *C. rostrata*. ***D. versicolorea***; *Potamogeton natans*. ***D. vulgaris***; *Sparganium* spp., *Typha* spp., *Schoenoplectus lacustris*. ***Plateumaris affinis***; *Carex* spp. ***P. braccata***; *Phragmites australis*. ***P. discolor***; *Carex* spp. ***P. sericea***; *Sparganium erectum, Iris pseudacorus*. ***Macroplea appendiculata***; *Myriophyllum* spp., *Potamogeton* spp. ***M. mutica***; *Potamogeton pectinatus, Ruppia* spp., *Zostera* spp. ***Lema cyanella***; *Cirsium arvense*. ***Crioceris asparagi***; *Asparagus officinalis*. ***Lilioceris lilii***; cultivated species of *Lilium*. ***Pilemostoma fastuosa***; *Inula conyza*. ***Cassida hemisphaerica***; *Silene vulgaris, S. uniflora*. ***C. murraea***; *Pulicaria dysenterica, Inula helenium*. ***C. nebulosa***; *Atriplex* spp., *Chenopodium* spp., *Beta vulgaris*. ***C. rubiginosa***; *Cirsium* spp., *Carduus* spp. ***C. sanguinosa***; *Achillea ptarmica*. ***C. vibex***; *Cirsium* spp., *Centaurea* spp., *Carduus* spp. ***C. viridis***; *Mentha aquatica*. ***C. vittata***; *Spergularia* spp. ***Timarcha goettingensis***; *Galium* spp. ***T. tenebricosa***; *Galium* spp. particularly *G. aparine*. ***Chrysolina americana***; *Lavandula* spp., *Rosmarinus officinalis*. ***C. banksi***; *Ballota nigra, Plantago lanceolata*. ***C. brunsvicensis***; *Hypericum* spp. ***C. cerealis*** *Thymus polytrichus*. ***C. fastuosa***; *Galeopsis tetrahit*. ***C. haemoptera***; *Plantago* spp. in sandy coastal areas. ***C. herbacea***; *Mentha aquatica*. ***C. hyperici***; *Hypericum* spp. ***C. oricalcia***; *Anthriscus sylvestris*. ***C. polita***; various species of Lamiaceae. ***C. sanguinolenta***; *Linaria* spp. ***C. varians***; *Hypericum* spp. ***Gastrophysa polygoni***; *Persicaria* spp., *Rumex* spp. ***G. viridula***; *Rumex* spp. ***Phaedon armoraciae***; *Veronica beccabunga*. ***P. cochleariae***; *Brassica* spp., *Rorippa* spp. ***P. concinnus***; *Cochlearia anglica*, *Triglochin maritima*. ***P. tumidulus***; *Heracleum sphondylium*. ***Hydrothassa glabra***; *Ranunculus* spp. ***H. hannoveriana***; *Caltha palustris*. ***H. marginella***; *Ranunculus* spp., *Caltha palustris*. ***Prasocuris junci***; *Veronica beccabunga*. ***P. phellandrii*** *Oenanthe aquatica*, *O. crocata*. ***Plagiodera versicolora***; *Salix alba*, *S. fragilis*, *S. viminalis*. ***Chrysomela aenea***; *Alnus glutinosa*. ***C. populi***; *Populus* spp., *Salix* spp. ***C. tremula***; *Populus tremula*, *Salix* spp. ***Gonioctena decemnotata***; *Populus tremula*, *Salix* spp. ***G. olivacea***; *Cytisus scoparius*. ***G. pallida***; *Corylus avellana*, *Sorbus aucuparia*. ***G. viminalis***; *Salix* spp. ***Phratora laticollis***; *Populus* spp. ***P. polaris***; mountain species of dwarf *Salix*. ***P. vitellinae***; *Populus* spp., *Salix* spp. ***P. vulgatissima***; *Salix* spp., *Populus* spp. ***Galerucella calmariensis***; *Lythrum salicaria*. ***G. lineola***; *Salix* spp. particularly *S. viminalis*. ***G. nymphaeae***; *Nymphaea alba*, *Nuphar lutea*. ***G. pusilla***; *Lythrum salicaria*. ***G. sagittariae***; *Sagittaria sagittifolia*. *Rumex hydrolapathum*. ***G. tenella***; *Filipendula ulmaria*.

Fig. 39 top *Galerucella sagittariae* (Gyllenhal) adult feeding on *Polygonum amphibium*. Bottom – *Pyrrhalta viburni* (Paykull) larval feeding on Guelder rose.

Pyrrhalta viburni; *Viburnum lantana, V. opulus*. ***Lochmaea caprea***; *Salix* spp. particularly *S. caprea*. ***L. crataegi***; *Crataegus* spp. ***L. suturalis***; *Calluna vulgaris*. ***Phyllobrotica quadrimaculata***; *Scutellaria galericulata*. ***Luperus flavipes***; *Betula* spp., *Salix* spp., *Alnus glutinosa, Corylus avellana*. ***L. longicornis***; *Betula* spp., *Salix* spp., *Alnus glutinosa*. ***Calomicrus circumfusus***; *Genista tinctoria, Cytisus scoparius, Ulex* spp. ***Agelastica alni***; *Alnus glutinosa*. ***Sermylassa halensis***; *Galium* spp. ***Phyllotreta atra***; *Brassica* spp. ***P. consobrina***; *Brassica* spp., *Sinapis* spp. ***P. cruciferae***; *Brassica* spp., *Sinapis* pp. ***P. diademata***; *Cardamine* spp., *Cochlearia* spp. ***P. nemorum***; *Brassica* spp. ***P. nigripes***; *Brassica* spp. ***P. nodicornis***; *Reseda lutea*. ***P. ochripes***; *Alliaria petiolata, Cardamine* spp., *Rorippa* spp. ***P. punctulata***; *Brassica* pp., *Sinapis* spp. ***P. striolata***; *Raphanus* spp., *Brassica* spp. ***P. tetrastigma***; *Cardamine* spp., *Rorippa* spp. ***P. undulata***; *Brassica* spp. ***Aphthona atrocaerulea***; *Euphorbia* spp. ***A. atrovirens***; *Origanum vulgare*. ***A. euphorbiae***; *Euphorbia* spp. ***A. herbigrada***; *Helianthemum nummularium*. ***A. lutescens***; *Lythrum salicaria*. ***A. melancholica***; *Euphorbia* spp. ***A. nigriceps***; *Geranium pratense*. ***A. nonstriata***; *Iris pseudacorus*. ***Longitarsus absynthii***; *Seriphidium maritimum*. ***L. aeneicollis***; *Lithospermum officinale*. ***L. aeruginosus***; *Eupatorium cannabinum*. ***L. agilis***; *Schrophularia* spp. ***L. anchusae***; *Echium vulgare, Cynoglossum officinale*. ***L. atricillus***; *Trifolium* spp. and other Fabaceae. ***L. ballotae***; *Ballota nigra, Marrubium vulgare*. ***L. brunneus***; *Aster tripolium, Thalictrum* spp. ***L. curtus***; *Echium vulgare, Symphytum* spp., *Pulmonaria* spp. ***L. dorsalis***; *Senecio* spp. ***L. exoletus***; *Echium vulgare, Cynoglossum officinale*. ***L. ferrugineus***; *Mentha* spp. ***L. flavicornis***; *Senecio* spp. ***L. fowleri***; *Dipsacus fullonum*. ***L. ganglbaueri***; *Senecio* spp. ***L. gracilis***; *Senecio jacobaea, S. vulgaris*. ***L. holsaticus***; *Pedicularis* spp. ***L. jacobaeae***; *Senecio jacobaea*. ***L. kutscherae***; *Plantago* spp. ***L. longiseta,*** *Veronica officinalis*. ***L. luridus*** is very common and found upon many plants, particularly species of Asteraceae. ***L. lycopi***; *Lycopus europaeus, Clinopodium* spp., *Mentha* spp., *Nepeta cararia*. ***L. melanocephalus***; *Plantago* spp. ***L. membranaceus***; *Teucrium scorodonia*. ***L. nasturtii***; *Symphytum* spp. and other Boraginaceae. ***L. nigerrimus***; *Utricularia* spp. ***L. nigrofasciatus***; *Verbascum* spp., *Scrophularia* spp. ***L. obliteratus***; *Thymus* spp., *Origanum vulgare*. ***L. ochroleucus***; *Senecio* spp. ***L. parvulus***; *Linum* spp. ***L. pellucidus***; *Convolvulus arvensis*. ***L. plantagomaritimus***; *Plantago maritima*. ***L. pratensis***; *Plantago* spp. ***L. quadriguttatus***; *Cynoglossum officinale, Echium vulgare*. ***L. reichei***; *Aster tripolium, Stachys palustris*. ***L. rubiginosus***; *Calystegia sepium*. ***L. rutilis***; *Scrophularia auriculata*. ***L. succineus***; *Achillea* spp., *Eupatorium cannabinum, Leucanthemum vulgare, Artemisia* spp. ***L. suturellus***; *Senecio* spp. ***L. tabidus***; *Verbascum* spp. ***Altica brevicollis***; *Corylus avellana*. ***A. ericeti***; *Erica tetralix*. ***A. helianthemi***; *Helianthemum nummularium, Sanguisorba minor*. ***A. longicollis***; *Calluna vulgaris, Erica* spp. ***A. lythri***; *Epilobium hirsutum, Lythrum salicaria*. ***A. oleracea***; *Calluna vulgaris, Erica* spp., *Epilobium* spp.,

Oenothera spp. ***A. palustris*** *Lythrum salicaria, Epilobium hirsutum, E. parviflorum.* **Hermaeophaga mercurialis**; *Mercurialis perennis.* **Batophila aerata**; *Rubus* spp. **B. rubi**; *Rubus* spp. particularly *R. idaeus.* **Lythraria salicariae**; *Lythrum salicaria.* **Ochrosis ventralis**; *Solanum dulcamara, Matricaria* spp. **Neocrepidodera ferruginea**; *Cirsium* spp., *Carduus* spp. ***N. impressa***; *Limonium vulgare.* ***N. transversa***; *Cirsium* spp. **Derocrepis rufipes**; *Vicia* spp. **Hippuriphila modeeri**; *Equisteum* spp. particularly *E. palustre.* **Crepidodera aurata**; *Salix* spp., *Populus* spp. ***C. aurea***; *Populus* spp., *Salix* spp. ***C. fulvicornis***; *Salix* spp., *Populus* spp. ***C. nitidula***; *Populus* spp., *Salix* spp. ***C. plutus***; *Salix* spp., *Populus* spp. **Epitrix atropae**; *Atropa belladonna.* ***E. pubescens***; *Solanum dulcamara.* **Podagrica fuscicornis**; *Malva sylvestris* and other species of Malvaceae. ***P. fuscipes***; *Malva sylvestris* and other species of Malvaceae. **Mantura chrysanthemi**; *Rumex acetosella.* ***M. matthewsi***; *Helianthemum nummularium.* ***M obtusata***; *Rumex* spp. ***M. rustica***; *Rumex* spp. **Chaetocnema concina**; *Persicaria* spp., *Rumex* spp. **Sphaeroderma rubidum**; *Cirsium* spp., *Carduus* spp. ***S. testaceum***; *Cirsium* spp., *Carduus* spp., *Centaurea* spp. **Dibolia cynoglossi**; *Galeopsis* spp., *Ballota nigra, Stachys* spp. **Psylliodes affinis**; *Solanum dulcamara.* ***P. attenuata***; *Humulus lupulus.* ***P. chalcomera***; *Carduus* spp. ***P. chrysocephala***; *Brassica* and other species of Brassicaceae. ***P. cucullata***; *Spergula arvensis.* ***P. cuprea***; *Brassica* and other species of Brassicaceae. ***P. dulcamarae***; *Solanum dulcamara.* ***P. hyoscyami***; *Hyoscyamus niger.* ***P. laticollis***; *Rorippa nasturtium-aquaticum.* ***P. luridipennis***; *Coincya wrightii.* ***P. luteola***; *Solanum dulcamara.* ***P. marcida***; *Cakile maritima.* ***P. napi***; *Brassica* and other species of Brassicaceae. ***P. picina***; *Lythrum salicaria.* ***P. sophiae***; *Descurainia sophia.* **Labidostomis tridentata**; *Betula* spp. **Smaragdina affinis**; *Corylus avellana.* **Cryptocephalus aureolus**; adults on heads of *Pilosella officinarum* and other yellow Asteraceae. ***C. coryli***; *Corylus avellana.* ***C. decemmaculatus***; *Salix* spp. ***C. exiguus***; *Salix* spp. ***C. hypochaeridis***; adults on heads of *Pilosella officinarum* and other yellow Asteraceae. ***C. labiatus***; *Quercus* spp., *Betula* spp., *Corylus avellana.* ***C. moraei***; *Hypericum* spp. ***C. nitidulus***; *Betula* spp., *Corylus avellana.* ***C. parvulus***; *Betula* spp. ***C. punctiger***; *Betula* spp., *Salix* spp. ***C. pusillus***; *Betula* spp. ***C. querceti***; *Quercus* spp. ***C. sexpunctatus***; *Betula* spp., *Corylus avellana.* **Bromius obscurus**; *Chamerion angustifolium.*

NEMONYCHIDAE

Cimberis attelaboides; *Pinus sylvestris.*

ANTHRIBIDAE

Platyrhinus resinosus; *Fraxinus excelsior* infested with the fungus *Daldinia concentrica.* **Tropideres sepicola**; *Quercus* spp., *Carpinus betulus, Fagus sylvatica.* **Choragus sheppardi**; *Hedera helix.* **Bruchela rufipes**; *Reseda lutea.*

RHYNCHITIDAE

Involvulus caeruleus; *Crataegus* spp., *Prunus* spp., *Malus* spp. and other woody Rosaceae. *I. cupreus*; *Sorbus aucuparia*. **Lasiorhynchites cavifrons**; *Quercus* spp. *L. olivaceus*; *Quercus* spp. **Neocoenorrhinus aeneovirens**: *Quercus* spp. *N. aequatus*; Roseaceous trees and shrubs particularly *Crataegus* spp. *N. germanicus*; herbaceous Rosaceae particularly *Fragaria vesca* and *Rubus* spp. *N. interpunctatus*; *Quercus* spp. *N. pauxillus*; *Crataegus* spp., *Prunus spinosa*. **Temnocerus longiceps**; *Salix* spp., *Betula* spp., *Pyrus* spp. *T. nanus*; *Betula* spp. *T. tomentosus*; *Salix* spp., *Populus tremula*. **Byctiscus betulae**; *Corylus avellana*, *Betula* spp. *B. populi*; *Populus* spp. particularly *P. tremula*. **Deporaus betulae**; *Betula* spp., *Alnus glutinosa*, *Corylus avellana*. *D. mannerheimi*; *Betula* spp.

ATTELABIDAE

Attelabus nitens; *Quercus* spp. **Apoderus coryli**; *Corylus avellana*.

APIONIDAE

Apion cruentatum; *Rumex acetosa*. *A. frumentarium*; *Rumex* spp. particularly *R. obtusifolius* and *R. hydrolapathum*. *A. haematodes*; *Rumex acetosella*. *A. rubens*; *Rumex acetosella*. *A. rubiginosum*; *Rumex acetosella*. **Aizobius sedi**; *Sedum acre, S. anglicum*. **Helianthemapion aciculare**; *Helianthemum nummularium*. **Perapion affine**; *Rumex acetosa*. *P. curtirostre*; *Rumex* spp. particularly *R. acetosa* and *R. acetosella*. *P. hydrolapathi*; *Rumex* spp. particularly *R. hydrolapathum* and *R. obtusifolius*. *P. marchicum*; *Rumex acetosella*. *P. violaceum*; *Rumex* spp. particularly *R. hydrolapathum* and *R. obtusifolius*. *P. lemoroi*; *Polygonum aviculare* agg. **Pseudaplemonus limonii**; *Limonium* spp. particularly *L. vulgare*. **Aspidapion radiolus**; *Malva sylvestris* and other species of Malvaceae. *A. soror*; *Althaea officinalis*. *A. aeneum*; *Malva sylvestris*. **Acentrotypus brunnipes**; *Gnaphalium* spp. and *Filago* spp. particularly *L. vulgaris*. **Ceratapion onopordi**; *Carduus* spp., *Cirsium* spp., *Centaurea* spp., *Arctium* spp. *C. carduorum*; *Carduus* spp., *Cirsium* spp. particularly *C. arvense* and *C. vulgare*. *C. gibbirostre*; *Carduus* spp., *Cirsium* spp. **Diplapion confluens**; *Tripleurospermum* spp., *Matricaria* spp. *D. stolidum*; *Leucanthemum vulgare*. **Omphalapion beuthini**; *Anthemis cotula*. *O. hookerorum*; *Tripleurospemum* spp. *O. laevigatum*; *Anthemis* spp., *Matricaria* spp. **Exapion difficile**; *Genista tinctoria*. *E. fuscirostre*; *Cytisus scoparius*. *E. genistae*; *Genista anglica*. *E. ulicis*; *Ulex europaeus*. **Ixapion variegatum**; *Viscum album*. **Kalcapion pallipes**; *Mercurialis perennis*. *K. semivittatum*; *Mercurialis annua*. **Melanapion minimum**; *Salix* spp. **Squamapion atomarium**; *Thymus* spp. *S. cineraceum*; *Prunella vulgaris*. *S. flavimanum*; *Origanum vulgare*. *S. vicinum*; *Mentha* spp. particularly *M. aquatica*. **Taeniapion urticarium**; *Urtica dioica*. **Malvapion malvae**; on Malvaceae particularly *Malva sylvestris*. **Pseudapion rufirostre**; *Malva* spp. particularly *M. sylvestris*. **Cyanapion spencii**; *Vicia* spp. particularly *V. cracca*. **C. afer**; *Lathyrus pratensis*. **C. gyllenhali**; *Vicia* spp. particularly *V. cracca*. **Eutrichapion ervi**; *Lathyrus pratensis*. **E. viciae**; *Lathyrus pratensis, Vicia cracca*. **E. vorax**; *Vicia* spp. **E. punctigerum**; *Vicia* spp. particularly *V. sepium*. **Hemitrichapion reflexum**; *Onobrychis viciifolia*. **H. waltoni**; *Hippocrepis comosa*. **Holotrichapion ononis**; *Ononis repens, O. spinosa*. **H. pisi**; *Medicago* spp. **H. aethiops**; *Vicia cracca, V. sepium*. **Oxystoma cerdo**; *Vicia cracca*. **O. craccae**; *Vicia* spp. **O. pomonae**; *Vicia cracca, V. sepium, Lathyrus pratensis*. **O. sabulatum**; *Lathyrus pratensis*. **Pirapion immune**; *Cytisus scoparius*. **Catapion curtisi**; *Trifolium* spp. **C. pubescens**; *Trifolium* spp. **C. seniculus**; *Trifolium* spp. **Ischnopterapion loti**; *Lotus corniculatus, L. glaber*. **I. modestum**; *Lotus pedunculatus*. **I. virens**; *Trifolium* spp. **Protopirapion atratulum**; *Cytisus scoparius*. **Stenopterapion intermedium**; *Onobrychis viciifolia*. **S. meliloti**; *Melilotus* spp. **S. tenue**; *Medicago* spp. particularly *M. lupulina*. **S. scutellare**; *Ulex* spp. particularly *U. minor*. **Synapion ebeninum**; *Lotus pedunculatus*. **Betulapion simile**; *Betula* spp. **Protapion apricans**; *Trifolium* spp. particularly *T. pratense*. **P. assimile**; *Trifolium* spp. **P. difforme**; *Trifolium* spp. **P. dissimile**; *Trifolium arvense*. **P. filirostre**; *Medicago* spp. **P. fulvipes**; *Trifolium* spp. particularly *T. repens*. **P. laevicolle**; *Trifolium* spp. **P. nigritarse**; *Trifolium campestre, T. dubium*. **P. ononidis**; *Ononis repens, O. spinosa*. **P. ryei**; *Trifolium pratense*. **P. schoenherri**; *Trifolium* spp. **P. trifolii**; *Trifolium pratense, T. medium*. **P. varipes**; *Trifolium pratense*. **Pseudoprotapion astragali**; *Astragalus glycyphyllos*.

Fig. 40. *Protapion fulvipes* (Geoffroy) (= *dichroum* (Bedel)) adult feeding on white clover.

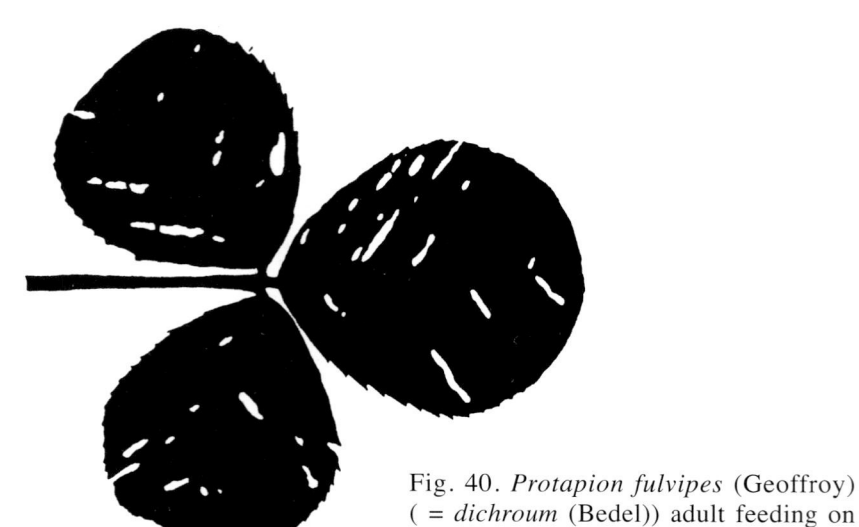

NANOPHYIDAE

Nanophyes marmoratus; *Lythrum salicaria*. ***Dieckmanniellus gracilis***; *Lythrum portula*.

ERIRHINIDAE

Grypus equiseti; *Equisetum arvense, E. palustre*. ***Notaris acridulus***; *Glyceria maxima*. ***N. aethiops***; *Sparganium erectum, Carex* spp. ***N. scirpi***; *Carex* spp., *Typha* spp. ***Procas granulicollis***; *Ceratocapnos claviculata*. ***Thryogenes fiorii***, *Carex* spp. ***T. nereis***; *Eleocharis palustris*. ***T. scirrhosus***; *Sparganium erectum*. ***Tournotaris bimaculatus***; *Phalaris arundinacea, Phragmites australis, Typha* spp., *Carex* spp. ***Stenopelmus rufinasus***; *Azolla filiculoides*. ***Tanysphyrus lemnae,*** *Lemma* spp.

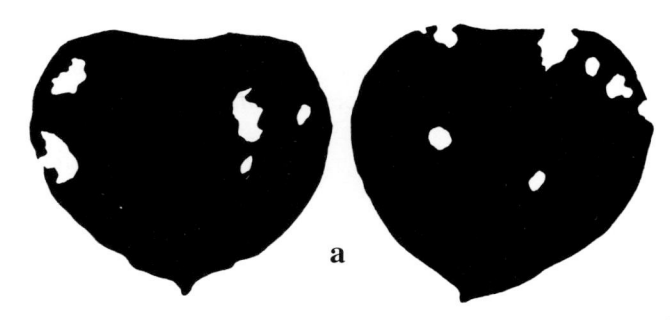

Fig.41. **a.** *Anthonomus brunnipennis* (Curtis) adult feeding on flower petals of *Potentilla erecta*. **b.** *Cionus scrophulariae* (L.): larval feeding on *Scrophularia nodosa*. **c.** *Cleopus pulchellus* (Herbst): larval feeding on *Scrophularia nodosa*.

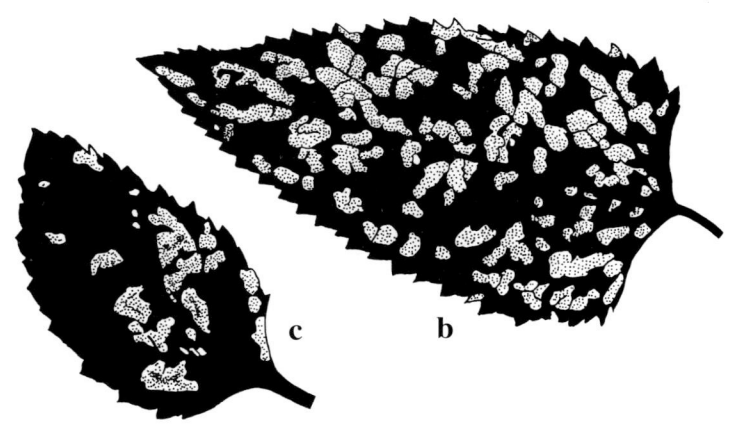

CURCULIONIDAE

Archarius pyrrhoceras; *Quercus* spp. *A. salicivorus*; *Salix* spp. *A. villosus*; *Quercus* spp. **Curculio betulae**; *Betula* spp. *C. glandium*; *Quercus* spp. *C. nucum*; *Corylus avellana. C. rubidus*; *Betula* spp. *C. venosus*; *Quercus* spp. **Acalyptus carpini**; *Salix* spp. **Anoplus plantaris**; *Betula* spp. *A. roboris*; *Alnus glutinosa*. **Anthonomus bituberculatus**; *Crataegus* spp. *A. brunnipennis*; *Potentilla erecta. A. chevrolati*; *Crataegus* spp. *A. conspersus*; *Sorbus aucuparia. A. humeralis*; *Malus* spp. *A. pedicularius*; *Crataegus* spp. *A. piri*; *Malus* spp. *A. pomorum*; *Malus* spp. *A. rubi*; various Rosaceae, particularly *Rosa* spp., *Rubus* spp., *Fragaria* spp. *A. rufus*; *Prunus spinosa. A. ulmi*; *Ulmus* spp. *A. varians*; *Pinus sylvestris. A. rectirostris*; *Prunus padus*. **Brachonyx pineti**; *Pinus sylvestris*. **Cionus alauda**; *Scrophularia auriculata, S. nodosa* and occasionally *Verbascum* spp. *C. hortulanus*; *Scrophularia auriculata, S. nodosa, Verbascum thapsus* and occasionally *Buddleja* spp. *C. longicollis*; *Verbascum* spp. particularly *V. thapsus*. *C. nigritarsis*; *Verbascum* spp. particularly *V. nigrum*. *C. scrophulariae*; *Scrophularia auriculata, S. nodosa, S.*

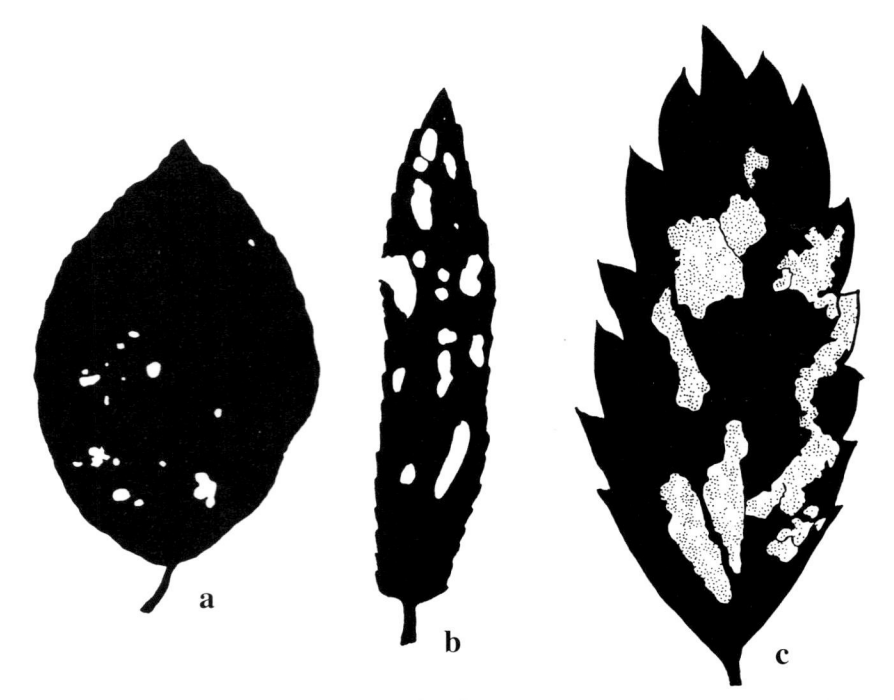

Fig. 42 **a.** *Orchestes fagi* (L.): adult feeding on *Fagus*. **b.** *Thamiocolus viduatus* (Gyllenhal): adult feeding on *Stachys palustris*. **c.** *Phytobius comari* (Herbst): larval feeding on *Potentilla palustris*.

scorodonia and occasionally *Buddleja* spp. ***C. tuberculosus***; *Scrophularia auriculata, S. nodosa*. ***Cleopus pulchellus***; *Scrophularia auriculata, S. nodosa, Verbascum thapsus*. ***Ellescus bipunctatus***; *Salix* spp. ***Dorytomus affinis***; *Populus tremula*. ***D. dejeani***; *Populus tremula*. ***D. filirostris***; *Populus* spp. particularly *P. nigra*. ***D. hirtipennis***; *Salix alba*. ***D. ictor***; *Populus nigra* agg. ***D. longimanus***; *Populus* spp. particularly *P. nigra* agg. ***D. majalis***; *Salix caprea, S. cinerea, S. aurita*. ***D. melanocephalus***; *Salix caprea, S. cinerea, S. aurita, S. repens*. ***D. rufatus***; *Salix caprea, S. aurita*. ***D. salicinus***; *Salix caprea, S. cinerea, S. aurita*. ***D. salicis***; *Salix repens*. ***D. taeniatus***; *Salix* spp. particularly *S. caprea, S. cinerea, S. aurita*. ***D. tortrix***; *Populus tremula*. ***D. tremulae***; *Populus tremula, P. alba*. ***Cleopomiaris graminis***; *Campanula* spp. ***C. plantarum***; *Campanula* spp. ***Gymnetron beccabungae***; *Veronica beccabunga*. ***G. melanarium***; *Veroniva chamaedrys*. ***G. rostellum***; *Veronica chamaedrys*. ***G. villosulum***; *Veronica anagallis-aquatica, V. catenata*. ***Mecinus circulatus***; *Plantago* spp. particularly *P. lanceolata*. ***M. collaris***; *Plantago maritima, P. coronopus*. ***M. janthinus***; *Linaria vulgaris*. ***M. labile***; *Plantago lanceolata*. ***M. pascuorum***; *Plantago lanceolata*. ***M. pyraster***; *Plantago* ssp particularly *P. lanceolata*. ***Miarus campanulae***; *Campanula* spp. ***Rhinusa antirrhini***; *Linaria vulgaris*. ***R. collina***; *Linaria vulgaris*. ***R. linariae***; *Linaria vulgaris*. ***Isochnus foliorum***; *Salix* spp. ***I. populicola***; *Salix* spp. ***Orchestes alni***; *Ulmus* spp. ***O. iota***; *Myrica gale*. ***O. pilosus***; *Quercus* spp. ***O. rusci***; *Betula* spp. ***O. signifer***; *Quercus* spp. ***O. testaceus***; *Alnus glutinosa*. ***O. fagi***; *Fagus sylvatica*.

Pseudorchestes pratensis; *Centaurea nigra*. ***Rhampus oxyacanthae***; *Crataegus* spp. ***R. pulicarius***; *Salix* spp, *Populus* spp. ***R. subaeneus***; *Malus* spp. ***Tachyerges decoratus***; *Salix purpurea*. ***T. pseudostigma***; *Salix* spp., *Betula* spp., *Corylus avellana*. ***T. salicis***; *Salix* spp. ***T. stigma***; *Salix* spp., *Betula* spp., *Corylus avellana*. ***Smicronyx coecus***; *Cuscuta epithymum*. ***S. jungermanniae***; *Cuscuta epithymum*. ***S. reichi***; *Centaurium erythraea*. ***Pachytychius haematocephalus***; *Lotus corniculatus*. ***Pseudostyphlus pillumus***; *Matricaria recutita*. ***Sibinia arenariae***; *Spergularia* spp. ***S. pyrrhodactyla***; *Spergula arvensis*. ***S. primita***; *Spergularia* spp., *Sagina* spp. ***S. sodalis***; *Armeria maritima*. ***Tychius crassirostris***; *Melilotus alba*. ***T. junceus***; *Medicago* spp. ***T. lineatulus***; *Trifolium* spp. ***T. meliloti***; *Melilotus* spp. ***T. parallelus***; *Cytisus scoparius*. ***T. picirostris***; *Trifolium* spp. ***T. polylineatus***; *Trifolium* spp. ***T. pusillus***; *Trifolium* spp. ***T. quinquepunctatus***; *Lathyrus* spp., *Vicia* spp. ***T. schneideri***; *Anthyllis vulneraria*. ***T. squamulatus***; *Lotus corniculatus*. ***T. stephensi***; *Trifolium* spp. ***T. tibialis***; *Trifolium* spp. ***Bagous brevis***; *Ranunculus flammula*. ***B. frit***; *Menyanthes trifoliata*. ***B. limosus***; *Potamogeton* spp. ***B. longitarsis***; *Myriophyllum* spp. ***B. lutulosus***; *Juncus* spp. particularly *J. bufonius*. ***B. nodulosus***; *Butomus umbellatus*. ***B. subcarinatus***; *Ceratophyllum submersum*. ***B. lutulentus***; *Equisetum fluviatile*. ***B. petro***; *Utricularia* spp. ***B. alismatis***; *Alisma plantago-aquatica*. ***Baris analis***; *Pulicaria dysenterica*. ***B. laticollis***; *Sisymbrium officinale, Brassica oleracea*. ***B. lepidii***; *Rorippa* spp.

and other Brassicaceae in marshy places. ***B. pilicornis***; *Reseda lutea*. ***B. scolopacea***; *Atriplex portulacoides*. ***Limnobaris dolorosa***; *Carex* spp. ***L. t-album***; *Carex* spp. ***Amalorrhynchus melanarius***; *Rorippa nasturtium-aquaticum, R. microphyllum*. ***Amalus scortillum***; *Polygonum aviculare* agg. ***Calosirus terminatus***; *Daucus carota*. ***Ceutorhynchus alliariae***; *Alliaria petiolata*. ***C. assimilis***; various Brassicaceae particularly *Brassica* spp. ***C. atomus***; *Arabidopsis thaliana*. ***C. cakilis***; *Cakile maritima, Crambe maritima*. ***C. cochleariae***; *Cardamine* spp. particularly *C. pratensis*. ***C. constrictus***; *Alliaria petiolata. C. halibaeus, Sisymbrium officinale*. ***C. erysimi***; *Capsella bursa-pastoris*. ***C. hepaticus***; *Sinapis* spp, *Brassica* spp. and other similar Brassicaceae. ***C. hirtulus***; *Erophila verna*. ***C. insularis***; *Cochlearia officinalis*. ***C. obstrictus***; various Brassicaceae particularly species of *Brassica, Sinapis, Sisymbrium* and *Lepidium*. ***C. pallidactylus***; various Brassicaceae particularly *Sisymbrium officinale*. ***C. pallipes***, *Sisymbrium* sp. ***C. parvulus***; *Lepidium* spp. particularly *L. heterophyllum*. ***C. pectoralis***; *Cardamine amara*, and sometimes otherspecies of Cardamine ***C. pervicax***; *Cardamine pratensis, C. amara*. ***C. picitarsis***; *Brassica napus* and other species of Brassicacea. ***C. pulvinatus***; *Descurainia sophia*. ***C. pumilio***; *Teesdalia nudicaulis*. ***C. pyrrhorhynchus*** ; and other species of *Brassicacea officinale*. ***C. querceti***; *Rorippa palustris*. ***C. rapae***; *Descurainia sophia, Sisymbrium officinale* and *Alliaria petiolata*. ***C. resedae***; *Reseda luteola*. ***C. sulcicollis***; various Brassicaceae particularly *Sisymbrium officinale*. ***C. thomsoni***; *Alliaria petiolata*. ***C. turbatus***; *Lepidium draba*. ***C. unguicularis***; *Arabis hirsuta*. ***Coeliodes rana***; *Quercus* spp. ***C. transversealbofasciatus***; *Quercus* spp. ***C. ruber***; *Quercus* spp. ***Coeliodinus nigritarsis***; *Betula* spp. ***C. rubicundus***; *Betula* spp. ***Datonychus angulosus***; *Stachys palustris, Geleopsis tetrahit* agg. ***D. arquatus***; *Lycopus europaeus*. ***D. melanostictus***; *Lycopus europaeus, Mentha aquatica*. ***D. urticae***; *Stachys sylvatica*. ***Drupenatus nasturtii***; *Rorippa nasturtium-aquaticum, R. microphylla*. ***Ethelcus verrucatus***; *Glaucium flavum*. ***Glocianus distinctus***; *Hypochaeris radicata*. ***G. moelleri***; *Hieracium* spp., *Leontodon* spp. ***G. pilosellus***; *Taraxacum* spp. ***G. punctiger***; *Taraxacum* spp. ***Hadroplontus litura***; *Cirsium arvense*. ***H. trimaculatus***; *Carduus nutans*. ***Micrelus ericae***; *Calluna vulgaris, Erica* spp. ***Microplontus campestris***; *Leucanthemum vulgare*. ***M. rugulosus***; *Matricaria recutita* and perhaps other species of Mayweed. ***M. triangulum***; *Achillea millefolium*. ***Mogulones asperifoliarum***; *Cynoglossum officinale, Echium vulgare* and perhaps other species of Boraginaceae. ***M. euphorbiae***; *Myosotis* spp. ***M. geographicus***; *Echium vulgare*. ***Nedyus quadrimaculatus***; *Urtica dioica*. ***Parethelcus pollinarius***; *Urtica dioica*. ***Poophagus sisymbrii***; *Rorippa nasturtium-aquaticum, R. microphylla, R. amphibia*. ***Sirocalodes depressicollis***; *Fumaria* spp. ***S. mixtus***; *Fumaria* spp., *Ceratocapnos claviculata*. ***S. quercicola***; *Fumaria* spp., particularly *F. officinalis*. ***Stenocarus ruficornis***; *Papaver rhoeas, P. somniferum*. ***Tapeinotus sellatus***; *Lysimachia vulgaris*. ***Thamiocolus viduatus***; *Stachys palustris*.

Trichosirocalus barnevillei; *Achillea millefolium*. *T. dawsoni*; *Plantago coronopus*, *P. maritima*. *T. horridus*; *Cirsium vulgare*, *Carduus nutans*. *T. rufulus*; *Plantago lanceolata*, *P. maritima*. *T. thalhammeri*; *Plantago maritima*, *P. coronopus*. *T. troglodytes*; *Plantago lanceolata*. *Zacladus exiguus*; small-flowered *Geranium* spp. *Z. geranii*; *Geranium pratense*, *G. sylvaticum*, *G. sanguineum*. **Mononychus punctumalbum**; *Iris foetidissima*, *I. pseudacorus*. **Eubrychius velutus**; *Myriophyllum* spp. **Neophytobius quadrinodosus**; *Persicaria amphibia*, *P. hydropiper*. **Pelonomus canaliculatus**; *Myriophyllum* spp. **P. comari**; *Potentilla palustris*, *Lythrum salicaria*. *P. olssoni*; *Lythrum portula*. *P. quadricorniger*; *Persicaria amphibia*, *P. lapathifolia*. *P. quadrituberculatus*; *Rumex crispus*, *Persicaria* spp. *P. waltoni*; *Persicaria hydopiper*. *P. zumpti*; *Glaux maritima*, *P. lapathifolia*, *P. hydropiper*. **Phytobius leucogaster**; *Myriophyllum* spp., *Potamogeton* spp. **Rhinoncus albicinctus**; *Persicaria amphibia*. *R. bruchoides*; *Persicaria maculosa*. *R. castor*; *Rumex acetosella*. *R. inconspectus*; *Persicaria amphibia*. *R. pericarpius*; *Rumex* spp. *R. perpendicularis*; *Persicaria amphibia*. **Rutidosoma globulus**; *Populus tremula*. **Cossonus linearis**; in dead wood of *Salix* and *Populus* spp. *C. parallelepipedus*; in dead wood of *Salix* and *Populus* spp. **Cryptorhynchus lapathi**; *Salix alba*, *S. triandra*, *S. viminalis*. **Gronops inaequalis**; *Atriplex prostrata*. *G. lunatus*; *Spergularia marina*, *S. media*. **Graptus triguttatus**; *Plantago lanceolata*. **Neliocarus sus**; *Calluna vulgaris*, *Erica* spp. **Strophosoma capitatum**; *Cytisus scoparius*, *Calluna vulgaris*, *Erica* spp. *S. fulvicorne*; *Calluna vulgaris*, *Erica* spp. *S. melanogrammum*; on trees particularly *Corylus avellana* and *Quercus* spp. **Barynotus moerens**; *Mercurialis perennis*.

  Otiorhynchus ligustici; *Anthyllis vulneraria*. **Caenopsis fissirostris**; *Calluna vulgaris*, *Erica* spp. *C. waltoni*; *Calluna vulgaris*, *Erica* spp. **Phyllobius pomaceus**; *Urtica dioica*. **Liophloeus tessulatus**; found on *Hedera helix*, but feeds on Apiaceae. **Polydrusus confluens**; *Ulex* spp., *Cytisus scoparius*. **Brachysomus hirtus**; *Primula vulgaris*. **Sitona ambiguus**; *Lathyrus pratensis*, *Vicia sepium*. *S. cylindricollis*; *Melilotus* spp. *S. hispidulus*; *Medicago* spp., *Trifolium* spp. *S. humeralis*; *Medicago* spp. partcularly *M. lupulina*. *S. lepidus*; *Trifolium pratense*, *T. repens*. *S. lineatus*; *Pisum sativum*, *Vicia faba* and other Fabaceae. *S. lineellus*; *Trifolium* spp. particularly *T. repens*, *Lotus corniculatus*, *Ornithopus perpusillus*. *S. ononidis*; *Ononis repens*. *S. puncticollis*; *Trifolium pratense*, *T. repens*. *S. regensteinensis*; *Ulex europaeus*, *Cytisus scoparius*. *S. sulcifrons*; *Trifolium pratense*, *Lotus corniculatus*. *S. suturalis*; *Lathyrus pratensis*, *Vicia sepium*. *S. waterhousei*; *Lotus* spp. particularly *L. corniculatus*. *S. griseus*; *Cytisus scoparius*, *Ononis repens*, *Anthyllis vulneraria*, *Ornithopus perpusillus*. *S. cambricus*; *Lotus pedunculatus*. *S. puberulus*; *Lotus* spp. **Hypera fuscocinerea**; *Medicago* spp. particularly *M. sativa*. *H. nigrirostris*; *Trifolium pratense*. *H. ononidis*; *Ononis* spp. particularly *O. repens*. *H. plantaginis*; *Lotus corniculatus*. *H. postica*; *Medicago* spp. particularly

Fig. 43 **a.** *Barynotus obscurus* (F.): adult feeding on African Violet. **b.** *Otiorhynchus porcatus* (Herbst): adult feeding on cultivated *Primula*. **c.** *Sitona lineatus* (L.): adult feeding on *Laburnum*.

M. lupulina and *M. sativa, Trifolium* spp. **H. suspiciosa**; *Vicia cracca* and other species of *Vicia, Lathyrus* and *Melilotus*. **H. venusta**; *Anthyllis vulneraria, Ulex minor*. **H. dauci**; *Erodium cicutarium*. **H. zoilus**; *Trifolium pratense, T. repens*. **H. diversipunctata**; *Cerastium arvense, Stellaria media, S. uliginosa, Myosoton aquaticum* and perhaps other Caryophyllaceae. **H. meles**; *Trifolium pratense, T. repens*. **H. arundinis**; *Sium latifolium*. **H. pollux**; on various Apiaceae in marshy places, particularly *Oenanthe crocata*. **H. rumicis**; *Rumex* spp., particularly *R. crispus, R. hydrolapathum, R. obtusifolius*. **H. pastinaceae**; *Daucus carota*. **Limobius borealis**; *Geranium* spp., particularly *G. pratense*. **L. mixtus**; *Erodium cicutarium*. **Larinus planus**; *Cirsium* spp., *Carduus* spp. **Lixus scabricollis**; *Beta vulgaris*. **Cleonis pigra**; *Cirsium arvense*. **Coniocleonus nebulosus**; *Calluna vulgaris*. **Rhinocyllus conicus**; *Cirsium* spp., *Carduus* spp. **Magdalis duplicata**; *Pinus sylvestris*. **M. memnonia**; *Pinus sylvestris*. **M. phlegmatica**; *Pinus sylvestris*. **M. ruficornis**; *Crataegus* spp., *Prunus* spp. **M. armiger**; *Ulmus* spp., particularly *U. procera*. **M. carbonaria**; *Betula* spp. **M. barbicornis**; *Malus* spp., *Pyrus* spp., *Prunus* spp., *Crataegus* spp. and other rosaceous trees. **M. cerasi**; *Quercus* spp. **Liparus coronatus**; *Anthriscus sylvestris*. **L. germanus**; *Heracleum sphondylium*. **Leiosoma deflexum**; *Anemone nemorosa, Ranunculus repens, Caltha palustris*. **L. oblongulum**; *Anemone nemorosa, Ranunculus* spp. **Hylobius abietis**; *Pinus* spp., particularly *P. sylvestris* and *P. contorta, Picea* spp. **H. transversovittatus**; *Lythrum salicaria*. **Syagrius intrudens**; *Pteridium aquilinum* and other ferns. **Pissodes castaneus**; *Pinus* spp. particularly *P. sylvestris*. **P. pini**; *Pinus* spp. particularly *P. sylvestris*. **P. validirostris**; *Pinus* spp., particularly *P. sylvestris*. **Orobitis cyaneus**; *Viola* spp.

    SCOLYTINAE. The larvae and adults of this subfamily are to be found beneath the bark, and for some species well into the sapwood, of various trees and shrubs. Only the major hosts are listed here, but most species will also occur in a range of similar or related hosts.

Scolytus intricatus; *Quercus* spp, and at times *Castanea sativa, Ulmus* spp., *Fagus sylvatica, Populus* spp. **S. laevis**; *Ulmus glabra*. **S. mali**; *Pyrus* spp., *Prunus* spp., *Crataegus* spp.. **S. multistriatus**; *Ulmus* spp. **S. pygmaeus**; *Ulmus* spp. **S. ratzeburgi**; *Betula* spp. **S. rugulosus**; *Prunus* spp., *Malus* spp., *Pyrus communis*. **S. scolytus**; *Ulmus* spp. **Acrantus vittatus**; *Ulmus* spp. **Hylastinus obscurus**; *Ulex* spp., *Cytisus scoparius*. **Hylesinus crenatus**; *Fraxinus excelsior*. **H. oleiperda**; *Fraxinus excelsior*. **Kissophagus hederae**; *Hedera helix*. **Leperisinus orni**; *Fraxinus excelsior*. **L. varius**; *Fraxinus excelsior*. **Hylastes angustatus**; *Pinus* spp., *Picea* spp. **H. ater**; *Pinus* spp. **H. attenuatus**; *Pinus* spp. **H. brunneus**; *Pinus* spp. **H. cunicularius**; *Picea* spp. **H. opacus**; *Pinus* spp. and occasionally in *Fraxinus excelsior* and *Ulmus* spp. **Hylurgops palliatus**; *Abies* spp., *Pinus* spp., *Picea* spp., *Larix* spp. **Phloeophthorus rhododactylus**; *Ulex* spp., *Cytisus scoparius*. **Polygraphus poligraphus**; *Picea* spp., *Pinus* spp. **Dendroctonus micans**; *Picea* spp. **Phloeosinus aubei**; *Thuja* spp. *P. thujae*; *Chamaecyparis* spp., *Cupressus* spp., *Thuja* spp. **Tomicus minor**; *Pinus* spp., occasionally in *Abies* spp. and *Picea* spp. *T. piniperda*; *Pinus* spp., and occasionally in *Picea, Abies* and *Larix* spp. **Xylechinus pilosus**; *Abies* spp., *Picea* spp. and occasionally *Larix* spp. **Ips acuminatus**; *Pinus* spp., *Larix* spp. *I. cembrae*; *Larix* spp., *Pinus* spp., *Picea* spp. *I. sexdentatus*; *Pinus* spp. *I. typographus*; *Picea* spp., *Abies* spp., *Pinus* spp. **Orthotomicus erosus**; *Pinus* spp. *O. laricis*; *Pinus* spp., *Larix* spp. *O. suturalis*; *Pinus* spp. **Pityophthorus lichtensteini**; *Pinus* spp. *P. pubescens*; *Pinus* spp., *Picea* spp. **Cryphalus abietis**; *Abies* spp., *Picea* spp. *C. piceae*; *Picea* spp. **Ernoporus caucasicus**; *Tilia* spp. *E. fagi*; *Fagus sylvatica, Quercus* spp., *Betula* spp. *E. tiliae*; *Tilia cordata*. **Trypophloeus asperatus**; *Populus* spp., *Salix* spp. *T. granulatus*; *Populus* spp. **Crypturgus subcribrosus**; *Picea abies*. **Dryocoetinus alni**; *Alnus glutinosa, Fagus sylvatica, Corylus avellana*. *D. villosus*; *Quercus* spp., *Fagus sylvatica, Castanea sativa*. **Lymantor coryli**; *Corylus avellana*. **Taphrorhychus bicolor**; *Fagus sylvatica, Quercus* spp. and other broad-leaved trees. **Xylocleptes bispinus**; *Clematis vitalba*. **Xyleborus dispar**; *Fagus sylvatica, Quercus* spp. and many other broad-leaved trees. *X. dryographus*; *Fagus sylvatica, Castanea sativa, Quercus* spp., *Ulmus* spp. *X. saxeseni*; *Quercus* spp., *Ulmus* spp., *Acer* spp., *Pinus* spp. **Xyloterus domesticus**; *Quercus* spp., *Fagus sylvatica*. *X. lineatus*; *Abies* spp., *Picea* spp., *Pinus* spp., *Larix* spp. *X. signatus*; *Quercus* spp., *Fagus sylvatica*.

PLATYPODIDAE

Platypus cylindrus; in felled *Quercus* spp., *Fagus sylvatica, Fraxinus excelsior*.

HOST PLANTS AND THEIR BEETLES

At times, particularly as a student new to the study of Coleoptera, one will find a plant with one or more species of beetle associated with it, and it can be helpful have a list of the more likely species that these might be. The following is a list of plants with the more frequent species that are associated with them.

EQUISETACEAE – horsetails

Equisetum fluviatile, Water Horsetail. *Bagous lutulentus*.

Equisetum arvense, Field Horsetail. *Grypus equiseti*.

Equisetum palustre, Marsh Horsetail. *Hippuriphila modeeri, Grypus equiseti*.

PTEROPSIDA – ferns

Pteridium aquilinum, Bracken. *Syagrius intrudens*.

Azola filiculoides, Water Fern. *Stenopelmus rufinasus*.

PINOPSIDA – conifers

***Abies* spp.**, Firs. *Laricobius erichsoni, Dryophilus pusillus, Zilora ferruginea, Stictoleptura rubra, Asemum striatum, Tetropium castaneum, Arhopalus rusticus, Hylurgops palliatus, Tomicus minor, T. piniperda, Xylechinus pilosus, Ips typographus, Cryphalus abietis, Xyloterus lineatus.*

Pinus sylvestris*,** Scots Pine, and other species of ***Pinus. *Plegaderus vulneratus, Melanophila acuminata,,Diacanthous undulatus, Paraphotistus impressus, Ampedus balteatus, A. elongantulus, A. sanguinolentus, Ernobius angusticollis, E. mollis, E. nigrinus, Hylecoetus dermestoides, Ostoma ferrugineum, Thanasimus formicarius, T. femoralis , Pityophagus ferrugineus, Scymnus nigrinus, Nephus quadrimaculatus, Exochomus quadripustulatus, Harmonia quadripunctata, Myzia oblongoguttata, Anatis ocellata, Abdera triguttata, Zilora ferruginea, Aulonium ruficorne, Corticeus fraxini, C. linearis, Chrysanthia nigricornis, Pytho depressus, Sphaeriestes castaneus, Lepturobosca virens, Anastrangalia sanguinolenta, Stictoleptura rubra, Strangalia attenuata, Asemum striatum, Tetropium castaneum, T. gabrieli, Arhopalus rusticus, A. ferus, Trinophylum cribratum, Pogonocherus fasciculatus, Acanthocinus aedilis, Cimberis attelaboides, Anthonomus varians, Brachonyx pineti, Magdalis duplicata, M. memnonia, M. phlegmatica, Hylobius abietis, Pissodes castaneus, P. pini, P. validirostris, Hylastes angustatus, H. ater, H. attenuatus, H. brunneus, H. opacus, Hylurgops palliatus, Polygraphus poligraphus, Tomicus minor, T. piniperda, Ips acuminatus, I. cembrae, I. sexdentatus, I. typographus, Orthotomicus erosus, O. laricis, O. suturalis, Pityophthorus lichtensteini, P. pubescens, Xyleborus saxeseni, Xyloterus lineatus.*

***Cupressus* spp.,** Cypresses. *Phloeosinus thujae.*

***Chamaecyparis* spp.,** Cypresses. *Phloeosinus thujae.*

***Thuja plicata*,** Western Red-cedar. *Phloeosinus aubei, P. thujae.*

NYMPHAEACEAE – Water-lily family

***Nymphaea alba*,** White Water-lily. *Donacia crassipes, Galerucella nymphaeae.*

***Nuphar lutea*,** Yellow Water-lily. *Donacia crassipes, Galerucella nymphaeae.*

CERATOPHYLLACEAE – Hornwort family

***Ceratophyllum submersum*,** Soft Hornwort. *Bagous subcarinatus.*

RANUNCULACEAE – Buttercup family

***Caltha palustris*,** Marsh-marigold. *Hydrothassa hannoveriana, H. marginella, Leiosoma deflexum.*

***Clematis vitalba*,** Traveller's-joy. *Leptophloeus clematidis, Xylocleptes bispinus.*

Anemone nemorosa, Wood Anemone. *Leiosoma deflexum, L. oblongulum.*
Ranunculus repens, Creeping Buttercup. *Leiosoma deflexum.*
Ranunculus flammula, Lesser Spearwort. *Bagous brevis.*
Ranunculus spp., Buttercups. *Hydrothassa glabra, H. marginella, Leiosoma oblongulum.*
Thalictrum spp., Meadow-rues. *Longitarsus brunneus.*

PAPAVERACEAE – Poppy family
Papaver somniferum, Opium Poppy. *Stenocarus ruficornis.*
Papaver rhoeas, Common Poppy. *Stenocarus ruficornis.*
Glaucium flavum, Yellow Horned-poppy. *Ethelcus verrucatus.*

FUMARIACEAE – Fumitory family
Ceratocapnos claviculata, Climbing Corydalis. *Procas granulicollis, Sirocalodes mixtus.*
Fumaria officinalis, Common Fumitory. *Sirocalodes depressicollis, S. mixtus, S. quercicola.*
Fumaria spp., Fumitories. *Sirocalodes depressicollis, S. mixtus, S. quercicola.*

ULMACEAE – Elm family
Ulmus glabra, Wych Elm. *Scolytus laevis.*
Ulmus spp., Elms. *Plegaderus dissectus, Lucanus cervus, Dorcus parallelipipedus, Ischnodes sanguinicollis, Megapenthes lugens, Elater ferrugineus, Xyletinus longitarsis, Nemozoma elongatum, Melandrya caraboides, Aulonium trisulcum, Prionychus ater, Scaphidema metallicum, Corticeus bicolor, Pyrochroa serraticornis, Anthonomus ulmi, Orchestes alni, Magdalis armiger, Scolytus intricatus, S. multistriatus, S. pygmaeus, S. scolytus, Acrantus vittatus, Hylastes opacus, Xyleborus dryographus, X. saxeseni.*

CANNABACEAE – Hop family
Humulus lupulus, Hop. *Psylliodes attenuata.*

URTICACEAE – Nettle family
Urtica dioica, Common Nettle. *Brachypterus glaber, B. urticae, Taeniapion urticarium, Nedyus quadrimaculatus, Parethelcus pollinarius, Phyllobius pomaceus.*

MYRICACEAE
Myrica gale, Bog-myrtle. *Orchestes iota.*

FAGACEAE – Beech family

Fagus sylvatica, Beech. *Paromalus flavicornis, Sinodendron cylindricum, Dorcus parallelipipedus, Microrhagus pygmaeus, Melasis buprestoides, Hylis sylvatica, Eucnemis capucina, Ampedus cinnabarinus, A. quercicolor, A. rufipennis, Procraerus tibialis, Elater ferrugineus, Hylecoetus dermestoides, Thymalus limbatus, Tillus elongatus, Laemophloeus monilis, Diplocoelus fagi, Anisoxya fuscula, Abdera quadrifasciata, Phloiotrya vaudoueri, Melandrya barbata, M. caraboides, Cicones variegata, Prionychus melanarius, Mycetochara humeralis, Corticeus unicolor, Tropideres sepicola, Orchestes fagi, Scolytus intricatus, Ernoporus fagi, Dryocoetinus alni, D. villosus, Taphrorhychus bicolor, Xyleborus dispar, X. dryographus, Xyloterus domesticus, X. signatus, Platypus cylindrus.*

Castanea sativa, Sweet Chestnut. *Sinodendron cylindricum, Pyrochroa serraticornis, Scolytus intricatus, Dryocoetinus villosus, Xyleborus dryographus.*

Quercus* spp.**, particularly ***Q. robur Pedunculate Oak. *Calosoma inquisitor, Paromalus flavicornis, Dendroxena quadrimaculata, Lucanus cervus, Gnorimus variabilis, Agrilus angustatus, A. laticornis, A. biguttatus, A. viridis, Microrhagus pygmaeus, Lacon querceus, Calambus bipustulatus, Ampedus balteatus, A. cardinalis, A. cinnabarinus, A. elongantulus, A. quercicola, Ischnodes sanguinicollis, Procraerus tibialis, Rhyzopertha dominica, Xestobium rufovillosum* (old trees), *Hadrobregmus denticollis, Xyletinus longitarsis, Anitys rubens, Hylecoetus dermestoides, Lymexylon navale, Opilio mollis, Hypebaeus flavipes, Sphinginus lobatus, Scymnus auritus, Abdera biflexuosa, Abdera quadrifasciata, Phloiotrya vaudoueri, Hypulus quercinus, Melandrya barbata, M. caraboides, Prionychus ater, Pseudocistela ceramboides, Mycetochara humeralis, Corticeus unicolor, Pyrochroa coccinea, P. serraticornis, Aderus oculatus, A. populneus, Prionus coriarius, Rhagium mordax, Grammoptera ustulata, G. abdominalis, Anoplodera sexguttata, Strangalia attenuata, Trinophylum cribratum, Pyrrhidium sanguineum, Phymatodes testaceus, Poecilium alni, Saperda scalaris, Cryptocephalus labiatus, C. querceti, Tropideres sepicola, Lasiorhynchites cavifrons, L. olivaceus, Neocoenorrhinus aeneovirens, N. interpunctatus, Attelabus nitens, Archarius pyrrhoceras, A. villosus, Curculio glandium, C. venosus, Orchestes pilosus, O. signifer, Coeliodes rana, C. transversealbofasciatus, C. ruber, Strophosoma melanogrammum, Magdalis cerasi, Scolytus intricatus, Ernoporus fagi, Dryocoetinus villosus, Taphrorhychus bicolor, Xyleborus dispar, X. dryographus, X. saxeseni, Xyloterus domesticus, X. signatus, Platypus cylindrus.*

BETULACEAE – Birch family

Betula pendula, Silver Birch, and ***B. pubescens,*** Downy Birch. *Sinodendron cylindricum, Diacanthous undulatus, Paraphotistus impressus, Ampedus* *balteatus, A. cinnabarinus, A. quercicola, A. pomorum, A. sanguinolentus, Hylecoetus dermestoides, Melandrya caraboides, Bolitophagus reticulatus, Diaperis boleti, Corticeus unicolor, Schizotus pectinicornis, Lepturobosca virens, Strangalia attenuata, Luperus flavipes, L. longicornis, Labidostomis tridentata, Cryptocephalus labiatus, C. nitidulus, C. parvulus, C. punctiger, C. pusillus, C. sexpunctatus, Temnocerus longiceps, T. nanus, Byctiscus betulae, Deporaus betulae, D. mannerheimi, Betulapion simile, Curculio betulae, C. rubidus, Anoplus plantaris, Orchestes rusci, Tachyerges pseudostigma, T. stigma, Ceeliodinus nigritarsis, C. rubicundus, Magdalis carbonaria, Scolytus ratzeburgi, Ernoporus fagi.*

Alnus glutinosa, Alder, and other species of **Alnus**. *Calambus bipustulatus, Chilocorus renipustulatus, Abdera flexuosa, Chrysomela aenea, Luperus flavipes, L. longicornis, Agelastica alni, Deporaus betulae, Anoplus roboris, Orchestes testaceus, Dryocoetinus alni.*

Carpinus betulus, Hornbeam. *Trachys minuta, Abdera quadrifasciata, Phloiotrya vaudoueri, Cicones variegata, Tropideres sepicola.*

Corylus avellana, Hazel. *Agrilus angustatus, A. laticornis, Hypulus quercinus, Gonioctena pallida, Luperus flavipes, Altica brevicollis, Smaragdina affinis, Cryptocephalus coryli, C. labiatus, C. nitidulus, C. sexpunctatus, Byctiscus betulae, Deporaus betulae, Apoderus coryli, Curculio nucum, Tachyerges pseudostigma, T. stigma, Strophosoma melanogrammum, Dryocoetinus alni, Lymantor coryli.*

CHENOPODIACEAE – Goosefoot family

Chenopodium spp., particularly **C. album,** Fat-hen. *Curtonotus convexiusculus, Cassida nebulosa.*

Atriplex prostrata, Spear-leaved Orache. *Gronops inaequalis.*

Atriplex portulacoides, Sea-purslane. *Baris scolopacea.*

Atriplex spp, Oraches. *Curtonotus convexiusculus, Cassida nebulosa.*

Beta vulgaris, Beet. *Cassida nebulosa.*

Beta vulgaris ssp. maritima, Sea Beet. *Lixus scabricollis,*

CARYOPHYLLACEAE – Pink family

Stellaria media, Common Chickweed. *Hypera diversipunctata.*

Stellaria uliginosa, Bog Stitchwort. *Hypera diversipunctata.*

Cerastium arvense, Field Mouse-ear. *Hypera diversipunctata.*

Myosoton aquaticum, Water Chickweed. *Hypera diversipunctata.*

Sagina spp., Pearlworts. *Sibinia primita.*

Spergula arvensis, Corn Spurrey. *Psylliodes cucullata, Sibinia pyrrhodactyla.*

Spergularia **spp.**, Sea-spurreys. *Cassida vittata, Sibinia arenariae, S. primita, Gronops lunatus.*

Silene vulgaris, Bladder Campion. *Cassida hemisphaerica.*

Silene uniflora, Sea Campion. *Cassida hemisphaerica.*

POLYGONACEAE – Knotweed family

Persicaria amphibia, Amphibious Bistort. *Pelonomus quadricorniger, Rhinoncus albicinctus, R. inconspectus, R. perpendicularis, Neophytobius quadrinodosus.*

Persicaria maculosa, Redshank. *Rhinoncus bruchoides.*

Persicaria lapathifolia, Pale Persicaria. *Pelonomus quadricorniger, Rhinoncus bruchoides.*

Persicaria hydropiper, Water-pepper. *Neophytobius quadrinodosus, Pelonomus waltoni, Rhinoncus bruchoides.*

Persicaria **spp.**, Knotweeds. *Gastrophysa polygoni, Chaetocnema concinna, Pelonomus quadrituberculatus, Rhinoncus perpendicularis.*

Polygonum aviculare **agg.**, *Perapion lemoroi, Amalus scortillum.*

Rumex acetosella, Sheep's Sorrel. *Kissister minimus, Mantura chrysanthemi, Apion haematodes, A. rubens, A. rubiginosum, Perapion curtirostre, P. marchicum, Rhinoncus castor.*

Rumex acetosa, Common Sorrel. *Apion cruentatum, Perapion affine, P. curtirostre.*

Rumex hydrolapathum, Water Dock. *Apion frumentarium, Perapion hydrolapathi, P. violaceum, Hypera rumicis.*

Rumex crispus, Curled Dock. *Hypera rumicis, Pelonomus quadriturberculatus.*

Rumex obtusifolius, Broad-leaved Dock. *Apion frumentarium, Perapion hydrolapathi, P. violaceum, Hypera rumicis.*

Rumex **spp.**, Docks. *Gastrophysa polygoni, G. viridula, Mantura obtusata, M. rustica, Chaetocnema concinna, Apion frumentarium, Rhinoncus pericarpius.*

PLUMBAGINACEAE – Thrift family

Limonium **spp.**, Sea-lavenders. *Neocrepidodera impressa, Pseudaplemonus limonii.*

Armeria maritima, Thrift. *Sibinia sodalis.*

CLUSIACEAE – St John's-wort family

Hypericum **spp.**, St John's-worts. *Chrysolina brunsvicensis, C. hyperici, C. varians, Cryptocephalus moraei.*

TILIACEAE – Lime family

***Tilia cordata*,** Small-leaved Lime. *Ernoporus tiliae.*

***Tilia* spp.,** Limes. *Sinodendron cylindricum, Aderus oculatus, A. populneus, Strangalia attenuata, Stenostola dubia, Ernoporus caucasicus.*

MALVACEAE – Mallow family

***Malva sylvestris*,** Common Mallow. *Podagrica fuscicornis, P. fuscipes, Aspidapion radiolus, A. aeneum, Malvapion malvae, Pseudapion rufirostre.*

***Althaea officinalis*,** Marsh-mallow. *Aspidapion soror.*

CISTACEAE – Rock-rose family

***Helianthemum nummularium*,** Common Rock-rose. *Meligethes brevis, M. solidus, Bruchidius cisti, Aphthona herbigrada, Altica helianthemi, Mantura matthewsi, Helianthemapion aciculare.*

VIOLACEAE – Violet family

***Viola* spp.,** Violets. *Orobitis cyaneus.*

CUCURBITACEAE – White bryony family

***Bryonia dioica*,** White Bryony. *Epilachna argus.*

SALICACEAE – Willow family

***Populus alba*,** White Poplar. *Dorytomus tremulae.*

***Populus tremula*,** Aspen. *Saperda populnea, Zeugophora flavicollis, Z. subspinosa, Z. turneri, Chrysomela tremula, Gonioctena decemnotata, Temnocerus tomentosus, Byctiscus populi, Dorytomus affinis, D. dejeani, D. tortrix, D. tremulae, Rutidosoma globulus.*

***Populus nigra* agg.,** Black-poplars. *Dorytomus filirostris, D. ictor, D. longimanus.*

***Populus* spp.,** Poplars. *Scymnus limbatus, Aromia moschata, Lamia textor, Saperda carcharias, Chrysomela populi, Phratora laticollis, P.vitellinae, P. vulgatissima, Crepidodera aurata, C. aurea, C. fulvicornis, C. nitidula, C. plutus, Rhampus pulicarius, Cossonus linearis, C. parallelepipedus, Scolytus intricatus, Trypophloeus asperatus, T. granulatus.*

***Salix fragilis*,** Crack-willow. *Plagiodera*

***Salix alba*,** White Willow. *Plagiodera versicolora, Dorytomus hirtipennis, Cryptorhynchus lapathi.*

***Salix triandra*,** Almond Willow. *Cryptorhynchus lapathi.*

***Salix purpurea*,** Purple Willow. *Tachyerges decoratus.*

Salix viminalis, Osier. *Plagiodera versicolora, Galerucella lineola, Cryptorhynchus lapathi.*

Salix caprea, Goat Willow. *Obera oculata, Lochmaea caprea, Dorytomus majalis, D. melanocephalus, D. rufatus, D. salicinus, D. taeniatus.*

Salix cinerea, Grey Willow. *Dorytomus majalis, D. melanocephalus, D. salicinus, D. taeniatus.*

Salix aurita, Eared Willow. *Dorytomus majalis, D. melanocephalus, D. rufatus, D. aurita, D. taeniatus.*

Salix repens, Creeping Willow. *Dorytomus melanocephalus, D. salicis,*

Salix spp., Willows. *Sinodrendron cylindricum, Dorcus parallelipipedus, Agrilus laticornis, A. viridis, Trachys minuta, Calambus bipustulatus, Paraphotistus nigricornis, Xestobium rufovillosum* (old trees), *Scymnus limbatus, Chilocorus renipustulatus, Anisoxya fuscula, Abdera flexuosa, Prionychus ater, Gracilia minuta, Aromia moschata, Lamia textor, Saperda carcharias, S. populnea, Chrysomela populi, C. tremula, Gonioctena decemnotata, Gonioctena viminalis, Phratora polaris* on dwarf mountain species, *P. vitellinae, P. vulgatissima, Luperus flavipes, L. longicornis, Crepidodera aurata, C. aurea, C. fulvicornis, C. nitidula, C. plutus, Cryptocephalus decemmaculatus, C. exiguus, C. punctiger, Temnocerus longiceps, T. tomentosus, Melanapion minimum, Archarius salicivorus, Acalyptus carpini, Ellescus bipunctatus, Isochnus foliorum, I. populicola, Rhampus pulicarius, Tachyerges pseudostigma, T. salicis, T. stigma, Cossonus linearis, C. parallelepipedus, Tryophloeus asperatus.*

BRASSICACEAE – Cabbage family

Sisymbrium officinale, Hedge Mustard. *Meligethes rotundicollis, Baris laticollis, Ceutorhynchus pallipes, C. obstrictus, C. pallidactylus, C. chalibaeus, C. pyrrhorhynchus, C. sulcicollis.*

Descurainia sophia, Flixweed. *Psylliodes sophiae, Ceutorhynchus pulvinatus, C. rapae.*

Alliaria petiolata, Garlic Mustard. *Phyllotreta ochripes, Ceutorhynchus alliariae, C. constrictus, C. thomsoni, C. sulcicollis.*

Arabidopsis thaliana, Thale Cress. *Ceutorhynchus atomus.*

Rorippa nasturtium-aquaticum, Water-cress *Amalorrhynchus melanarius, Drupenatus nasturtii, Poophagus sisymbrii.*

Rorippa microphylla, Narrow-fruited Water-cress. *Amalorrhynchus melanarius, Drupenatus nasturtii, Poophagus sisymbrii.*

Rorippa palustris, Marsh Yellow-cress. *Ceutorhynchus querceti.*

Rorippa amphibia, Great Yellow-cress. *Poophagus sisymbrii.*

Rorippa **spp.,** Water-cresses. *Phaedon cochlearia, Phyllotreta ochripes, P. tetrastigma, Psylliodes laticollis, Baris lepidii.*

Cardamine amara, Large Bittercress, *Ceutorhynchus pectoralis, C. pervicax.*

Cardamine pratensis, Cuckooflower. *Ceutorhynchus cochleariae, C. pervicax.*

Cardamine **spp.,** Bitter-cresses. *Phyllotreta diademata, P. ochripes, P. tetrastigma, C. pectoralis.*

Arabis hirsuta, Hairy Rock-cress. *Ceutorhynchus unguicularis.*

Erophila verna. Common Whitlowgrass. *Ceutorhynchus hirtulus.*

Cochlearia anglica, English Scurvygrass. *Phaedon concinnus.*

Cochlearia officinalis, Common Scurvygrass. *Ceutorhynchus insularis.*

Cochlearia **spp.,** Scurvygrasses. *Phyllotreta diademata.*

Capsella bursa-pastoris, Shepherd's-purse. *Ceutorhynchus erysimi.*

Teesdalia nudicaulis, Shepherd's-cress. *Ceutorhynchus pumilio.*

Lepidium heterophyllum, Smith's Pepperwort. *Ceutorhynchus parvulus.*

Lepidium draba, Hoary Cress. *Ceutorhynchus turbatus.*

Lepidium **spp.,** Pepperworts. *Ceutorhynchus obstrictus.*

Brassica oleracea, Cabbage. *Baris laticollis.*

Brassica napus, Oil-seed Rape. *Meligethes aeneus, Ceutorhynchus picitarsis.*

Brassica rapa, Turnip. *Helophorus porculus, H. rufipes.*

Brassica **spp.,** Cabbages. *Meligethes aeneus, M. coracinus, Phaedon cochleariae, Phyllotreta atra, P. consobrina, P. cruciferae, P. nigripes, P. punctulata, P. striolata, P. undulata, Psylliodes chrysocephala, P. cuprea, P. napi, Ceutorhynchus assimilis, C. hepaticus, C. obstrictus.*

Sinapis arvensis, Charlock. *Meligethes fulvipes, M. rotundicollis, M. viridescens, Phyllotreta consobrina, P. cruciferae, P. punctulata, Ceutorhynchus hepaticus, C. obstrictus.*

Coincya wrightii, Lundy Cabbage. *Psylliodes luridipennis.*

Cakile maritima, Sea Rocket. *Psylliodes marcida, Ceutorhynchus cakilis.*

Crambe maritima, Sea-kale. *Ceutorhynchus cakilis.*

Raphanus **spp.,** Radishes. *Phyllotreta striolata.*

RESEDACEAE – Mignonette family

Reseda luteola, Weld. *Ceutorhynchus resedae.*

Reseda lutea, Wild Mignonette. *Phyllotreta nodicornis, Bruchela rufipes, Baris picicornis.*

ERICACEAE – Heather family

Calluna vulgaris, Heather. *Amara infirma, Bradycellus caucasicus, Trichocellus cognatus, Ampedus sanguinolentus, Coccinella hieroglyphica,*

Lochmaea suturalis, Altica longicollis, A. oleracea, Micrelus ericae, Neliocarus sus, Strophosoma capitatum, A. fulvicorne, Caenopsis fissirostris, C. waltoni, Coniocleonus nebulosus.

Erica tetralix, Cross-leaved Heath. *Altica ericeti.*

Erica spp., Heaths. *Altica longicollis, A. oleracea, Micrelus ericae, Neliocarus sus, Strophosoma capitatum, S. fulvicorne, Caenopsis fissirostris, C. waltoni.*

PRIMULACEAE – Primrose family

Primula vulgaris, Primrose. *Eusphalerum primulae, Brachysomus hirtus.*

Lysimachia vulgaris, Yellow Loosestrife. *Tapeinotus sellatus.*

Glaux maritima, Sea-milkwort. *Pelonomus zumpti.*

CRASSULACEAE – Stonecrop family

Sedum acre, Biting Stonecrop. *Aizobius sedi.*

Sedum anglicum, English Stonecrop. *Aizobius sedi.*

ROSACEAE – Rose family

Filipendula ulmaria, Meadowsweet. *Galerucella tenella.*

Rubus idaeus, Raspberry. *Byturus tomentosus, Batophila rubi.*

Rubus spp., Brambles. *Dryophilus anobioides, Batophila aerata, B. rubi, Neocoenorrhinus germanicus, Anthonomus rubi.*

Potentilla palustris, Marsh Cinquefoil. *Pelonomus comari.*

Potentilla erecta, Tormentil. *Anthonomus brunnipennis.*

Fragaria vesca, Wild Strawberry. *Neocoenorrhinus germanicus.*

Fragaria x ananassa, Garden Strawberry. *Harpalus rufipes.*

Sanguisorba minor, Salad Burnet. *Altica helianthemi.*

Rosa spp., Roses. *Meligethes atratus, M. flavimanus, Anthonomus rubi.*

Prunus spinosa, Blackthorn. *Anthaxia nitidula, Neocoenorrhinus pauxillus, Anthonomus rufus.*

Prunus domestica, Plum. *Gnorimus nobilis.*

Prunus padus, Bird Cherry. *Anthonomus rectirostris.*

Prunus spp., Plums. *Prionychus ater, Mycetochara humeralis, Orsodacne cerasi, Involvulus caeruleus, Magdalis ruficornis, M. barbicornis, Scolytus mali, S. rugulosus.*

Pyrus spp., Pears. *Gnorimus nobilis, Temnocerus longiceps, Magdalis barbicornis, Scolytus mali, Scolytus rugulosus.*

Malus spp., Apples. *Sinodendron cylindricum, Gnorimus nobilis, Prionychus ater, Tetrops praeustus, Involvulus caeruleus, Anthonomus humeralis, A. piri, A. pomorum, Magdalis barbicornis, Scolytus rugulosus.*

Sorbus aucuparia, Rowan. *Gonioctena pallida, Involvulus cupreus, Anthonomus conspersus.*

***Crataegus* spp.,** Hawthorns. *Anthaxia nitidula, Agrilus sinuatus, Hadrobregmus denticollis, Opilio mollis, Tetrops praeustus, Orsodacne cerasi, O. lineola, Lochmaea crataegi, Involvulus caeruleus, Neocoenorrhinus aequatus, N. pauxillus, Anthonomus bituberculatus, A. chevrolati, A. pedicularius, Rhampus oxyacanthae, R. subaeneus, Magdalis ruficornis, M. barbicornis, Scolytus mali.*

FABACEAE – Pea family

Astragalus glycyphyllos, Wild Liquorice. *Pseudoprotapion astragali.*

Onobrychis viciifolia, Sainfoin. *Bruchidius olivaceus, Hemitrichapion reflexum, Stenopterapion intermedium.*

Anthyllis vulneraria, Kidney Vetch. *Tychius schneideri, Otiorhynchus ligustici, Sitona griseus, Hypera venusta.*

Lotus corniculatus, Common Bird's-foot-trefoil. *Meligethes carinulatus, Bruchidius cisti, Ischnopterapion loti, Pachytychius haematocephalus, Tychius squamulatus, Sitona lineellus, S. sulcifrons, S. waterhousei, S. puberulus, Hypera plantaginis.*

Lotus pedunculatus, Greater Bird's-foot-trefoil. *Ischnopterapion modestum, Synapion ebeninum, Sitona cambricus, S. puberulus.*

Ornithopus perpusillus, Bird's-foot. *Sitona lineellus, S. griseus.*

Hippocrepis comosa, Horseshoe Vetch. *Meligethes erichsoni, Hemitrichapion waltoni.*

Vicia cracca, Tufted Vetch. *Bruchus atomarius, Cyanapion spencii, Cyanapion gyllenhali, Eutrichapion viciae, Holotrichapion aethiops, Oxystoma cerdo, Oxystoma pomonae, Hypera suspiciosa.*

Vicia sepium, Bush Vetch. *Bruchus atomaria, Eutrichapion puntigerum, Holotrichapion aethiops, Oxystoma pomonae, Sitona ambiguus, S. suturalis.*

Vicia sativa, Common Vetch. *Bruchus atomaria, B. rufipes.*

Vicia faba, Broad Bean. *Bruchus rufimanus, Sitona lineatus.*

***Vicia* spp.,** Vetches. *Derocrepis rufipes, Eutrichapion vorax, Oxystoma craccae, Tychius quinquepunctatus.*

Lathyrus pratensis, Meadow Vetchling. *Bruchus loti, B. rufimanus, B. rufipes, Cyanapion afer, Eutrichapion ervi, E. viciae, Oxystoma pomonae, O. sabulatum, Tychius quinquepunctatus, Sitona ambiguus, S. suturalis, Hypera suspiciosa.*

Pisum sativum, Garden Pea. *Sitona lineatus.*

Ononis spinosa, Spiny Restharrow. *Holotrichapion ononis, Protapion ononidis.*

***Ononis repens*,** Common Restharrow. *Holotrichapion ononis, Protapion ononidis, Sitona ononidis, S. griseus, Hypera ononidis.*
***Melilotus alba*,** White Melilot. *Stenopterapion meliloti, Tychius crassirostris.*
***Melilotus* spp.,** Melilots. *Stenopterapion meliloti, Tychius meliloti, Sitona cylindricollis, Hypera suspiciosa.*
***Medicago lupulina*,** Black Medick. *Sitona humeralis, Hypera postica.*
***Medicago sativa*,** Lucerne. *Hypera fuscocinerea, Hypera postica.*
***Medicago* spp.,** Medicks. *Holotrichapion pisi, Stenopterapion tenue, Protapion filirostre, Tychius junceus, Sitona hispidulus.*
***Trifolium repens*,** White Clover. *Meligethes nigrescens, Protapion fulvipes, Sitona lepidus, S. lineellus, S. puncticollis, Hypera zoilus, H. meles.*
***Trifolium campestre*,** Hop Trefoil. *Protapion nigritarse.*
***Trifolium dubium*,** Lesser Trefoil. *Protapion nigritarse.*
***Trifolium pratense*,** Red Clover. *Bruchidius varius, Protapion apricans, P. ryei, P. trifolii, P. varipes, Sitona lepidus, S. puncticollis, S. sulcifrons, Hypera nigrirostris, H. zoilus, H. meles.*
***Trifolium medium*,** Zigzag Clover. *Protapion trifolii.*
***Trifolium arvense*,** Hare's-foot Clover. *Protapion dissimile.*
***Trifolium* spp.,** Clovers. *Longitarsus atricillus, Catapion curtisi, C. pubescens, C. seniculus, Ischnopterapion virens, Protapion apricans, P. assimile, P. difforme, P. fulvipes, P. laevicolle, P. schoenherri, Tychius lineatulus, T. picirostris, T. polylineatus, T. pusillus, T. stephensi, T. tibialis, Sitona hispidulus, Hypera postica.*
***Cytisus scoparius*,** Broom. *Dryophilus anobioides, Cryptolestes spartii, Bruchidius villosus, Gonioctena olivacea, Calomicrus circumfusus, Exapion fuscirostre, Pirapion immune, Protopirapion atratulum, Tychius parallelus, Strophosoma capitatum, Polydrusus confluens, Sitona regensteinensis, S. griseus, Hylastinus obscurus, Phloeophthorus rhododactylus.*
***Genista tinctoria*,** Dyer's Greenweed. *Meligethes bidentatus, Calomicrus circumfusus, Exapion difficile.*
***Genista anglica*,** Petty Whin. *Exapion genistae.*
***Ulex europaeus*,** Gorse. *Exapion ulicis, Sitona regensteinensis.*
***Ulex minor*,** Dwarf Gorse. *Hypera venusta.*
***Ulex* spp.,** Gorses. *Calomicrus circumfusus, Stenopterapion scutellare, Polydrusus confluens, Hylastinus obscurus, Phloeophthorus rhododactylus.*

HALORAGACEAE – Water-milfoil family
***Myriophyllum* spp.,** Water-milfoils. *Macroplea appendiculata, Bagous longitarsis, Eubrychius velutus, Pelonomus canaliculatus, Phytobius leucogaster*

LYTHRACEAE – Purple-loosestrife family

Lythrum salicaria, Purple-loosestrife. *Galerucella calmariensis, G. pusilla, Aphthona lutescens, Altica lythri, A. palustris, Lythraria salicariae, Psylliodes picina, Nanophyes marmoratus, Pelonomus comari, Hylobius transversovittatus.*

Lythrum portula, Water-purslane. *Dieckmanniellus gracilis, Pelonomus olssoni.*

ONAGRACEAE – Willowherb family

Epilobium hirsutum, Great Willowherb. *Altica lythri, A. palustris.*

Epilobium parviflorum, Hoary Willowherb. *Altica palustris.*

***Epilobium* spp.,** Willowherbs. *Altica oleracea.*

Chamerion angustifolium, Rosebay Willowherb. *Bromius obscurus.*

***Oenothera* spp.,** Evening-primroses. *Altica oleracea.*

VISCACEAE – Mistletoe family

Viscum album, Mistletoe. *Ixapion variegatum.*

EUPHORBIACEAE – Spurge family

Mercurialis perennis, Dog's Mercury. *Hermaeophaga mercurialis, Kalcapion pallipes, Barynotus moerens.*

Mercurialis annua, Annual Mercury. *Kalcapion semivittatum.*

***Euphorbia* spp.,** Spurges. *Aphthona atrocaerulea, A. euphorbiae, A. melancholica.*

LINACEAE – Flax family

***Linum* spp.,** Flaxes. *Chrysolina sanguinolenta, Longitarsus parvulus.*

ACERACEAE – Maple family

Acer campestre, Field Maple. *Gastrallus immarginatus, Anisoxya fuscula.*

Acer pseudoplatanus, Sycamore. *Exochomus quadripustulatus.*

***Acer* spp.,** Maples. *Ischnodes sanguinicollis, Mycetochara humeralis, Xyleborus saxeseni.*

GERANIACEAE – Crane's-bill family

Geranium sylvaticum, Wood Crane's-bill. *Zacladus geranii.*

Geranium pratense, Meadow Crane's-bill. *Aphthona nigriceps, Zacladus geranii, Limobius borealis.*

Geranium sanguineum, Bloody Crane's-bill. *Zacladus geranii.*

Geranium **spp.**, Crane's-bills. *Zacladus exiguus.*
Erodium cicutarium, Common Stork's-bill. *Hypera dauci, Limobius mixtus.*

ARALIACEAE – Ivy family

Hedera helix, Ivy. *Ochina ptinoides, Anobium inexpectatum, Clitostethus arcuatus, Nephus quadrimaculatus, Choragus sheppardi, Liophloeus tessulatus, Kissophagus hederae.*

APIACEAE – Carrot family

Anthriscus sylvestris, Cow Parsley. *Phytoecia cylindrica, Chrysolina oricalcia, Liparus coronatus.*

Anthriscus caucalis, Bur Chervil. *Phytoecia cylindrica.*

Sium latifolium, Greater Water-parsnip. *Hypera arundinis.*

Oenanthe crocata, Hemlock Water-dropwort. *Prasocuris phellandrii, Hypera pollux.*

Oenanthe aquatica, Fine-leaved Water-dropwort. *Prasocuris phellandri.*

Heracleum sphondylium, Hogweed. *Agapanthia villosoviridescens. Phaedon tumidulus, Liparus germanus.*

Daucus carota, Wild Carrot. *Phytoecia cylindrica, Calosirus terminatus, Hypera pastinaceae.*

GENTIANACEAE – Gentian family

Centaurium erythraea, Common Centaury. *Smicronyx reichi.*

SOLANACEAE – Nightshade family

Atropa belladonna, Deadly Nightshade. *Epitrix atropae.*

Hyoscyamus niger, Henbane. *Psylliodes hyoscyami.*

Solanum nigrum, Black Nightshade. *Pria dulcamarae.*

Solanum dulcamara, Bittersweet. *Pria dulcamarae, Ochrosis ventralis, Epitrix pubescens, Psylliodes affinis, P. dulcamarae, P. luteola.*

CONVOLVULACEAE – Bindweed family

Convolvulus arvensis, Field Bindweed. *Longitarsus pellucidus.*

Calystegia sepium, Hedge Bindweed. *Longitarsus rubiginosus.*

CUSCUTACEAE – Dodder family

Cuscuta epithymum, Dodder. *Smicronyx coecus, S. jungermanniae.*

MENYANTHACEAE – Bogbean family

Menyanthes trifoliata, Bogbean. *Bagous frit.*

BORAGINACEAE – Borage family

***Lithospermum officinale*,** Common Gromwell. *Longitarsus aenicollis.*

***Echium vulgare*,** Viper's-bugloss. *Meligethes planiusculus, Longitarsus anchusae, L. curtus, L. exoletus, L. quadriguttatus, Mogulones asperifoliarum, M. geographicus.*

***Pulmonaria* spp.,** Lungworts. *Longitarsus curtus.*

***Symphytum* spp.,** Comfreys. *Longitarsus curtus, L. nasturtii,*

***Myosotis* spp.,** Forget-me-nots. *Mogulones euphorbiae.*

***Cynoglossum officinale*,** Hound's-tongue. *Longitarsus anchusae, L. exoletus, L. quadriguttatus, Mogulones asperifoliarum.*

LAMIACEAE – Dead-nettle family

***Stachys officinalis*,** Betony. *Meligethes persicus.*

***Stachys sylvatica*,** Hedge Woundwort. *Meligethes brunnicornis, Dibolia cynoglossi, Datonychus urticae.*

***Stachys palustris*,** Marsh Woundwort. *Longitarsus reichi, Datonychus angulosus, Thamiocolus viduatus.*

***Ballota nigra*,** Black Horehound. *Meligethes ruficornis, Chrysolina banksi, Longitarus ballotae, Dibolia cynoglossi.*

***Lamiastrum galeobdolon*,** Yellow Archangel. *Meligethes atramentarius, M. kunzei.*

***Lamium album*,** White Dead-nettle. *Meligethes difficilis, M. haemorrhoidalis, M. morosus.*

***Galeopsis angustifolia*,** Red Hemp-nettle. *Meligethes serripes.*

***Galeopsis tetrahit*,** Common Hemp-nettle and ***G. bifida*,** Bifid Hemp-nettle. *Meligethes pedicularius, Chrysolina fastuosa, Dibolia cynoglossi, Datonychus angulosus.*

***Marrubium vulgare*,** White Horehound. *Meligethes nanus, Longitarsus ballotae.*

***Scutellaria galericulata*,** Skullcap. *Phyllobrotica quadrimaculata.*

***Teucrium scorodonia*,** Wood Sage. *Meligethes obscurus, Longitarsus membranaceus.*

***Nepeta cataria*,** Cat-mint. *Meligethes incanus, Longitarsus lycopi.*

***Glechoma hederacea*,** Ground-ivy. *Trachys scrobiculatus, Meligethes ovatus.*

***Prunella vulgaris*,** Selfheal. *Meligethes umbrosus, Squamapion cineraceum.*

***Clinopodium vulgare*,** Wild Basil. *Meligethes bidens, Longitarsus lycopi.*

***Origanum vulgare*,** Wild Marjoram. *Aphthona atrovirens, Longitarsus obliteratus, Squamapion flavimanum.*

***Lycopus europaeus*,** Gypsywort. *Longitarsus lycopi, Datonychus aequatus, D. melanostictus.*

***Thymus polytrichus*,** Wild Thyme. *Meligethes exilis, M. lugubris, Chrysolina cerealis, Longitarsus obliteratus, Squamapion atomarium.*

***Mentha arvensis*,** Corn Mint. *Meligethes gagathinus.*

***Mentha aquatica*,** Water Mint. *Cassida viridis, Chrysolina herbacea, Squamapion vicinum, Datonychus melanostictus.*

***Mentha* spp.,** Mints. *Longitarsus ferrugineus, L. lycopi.*

***Lavandula* x *intermedia*,** Garden Lavender. *Chrysolina americana.*

***Rosmarinus officinalis*,** Rosemary. *Chrysolina americana.*

PLANTAGINACEAE – Plantain family

***Plantago coronopus*,** Buck's-horn Plantain. *Mecinus collaris, Trichosirocalus dawsoni, T. thalhammeri.*

***Plantago maritima*,** Sea Plantain. *Longitarsus plantagomaritimus, Mecinus collaris, Trichosirocalus rufulus, T. thalhammeri, T. dawsoni.*

***Plantago lanceolata*,** Ribwort Plantain. *Chrysolina banksi, Mecinus circulatus, M. labile, M. pascuorum, M. pyraster, Trichosirocalus rufulus, T. troglodytes, Graptus triguttatus.*

***Plantago* spp.,** Plantains. *Chrysolina haemoptera, Longitarsus kutscherae, L. melanocephalus, L. pratensis.*

BUDDLEJACEAE – Butterfly-bush family

***Buddleja* spp.,** Butterfly-bushes. *Cionus hortulanus, C. scrophulariae.*

OLEACEAE – Ash family

***Fraxinus excelsior*,** Ash. *Sinodendron cylindricum, Dorcus parallelipipedus, Biphyllus lunatus, Chilocorus renipustulatus, Anisoxya fuscula, Abdera biflexuosa, Phloiotyra vaudoueri, Prionychus ater, Corticeus bicolor, Lytta vesicatoria, Stenostola dubia, Tetrops starkii, Platyrhinus resinosus, Hylesinus crenatus, H. oleiperda, Leperisinus orni, L. varius, Hylastes opacus, Platypus cylindrus.*

***Ligustrum* spp.,** particularly ***L. vulgare*,** Wild Privet. *Lytta vesicatoria.*

SCROPHULARIACEAE – Figwort family

***Verbascum thapsus*,** Great Mullein. *Cionus hortulanus, C. longicollis, Cleopus pulchellus.*

***Verbascum nigrum*,** Dark Mullein. *Cionus nigritarsis.*

***Verbascum* spp.,** Mulleins. *Longitarsus nigrofasciatus, L. tabidus, Cionus alauda.*

***Scrophularia nodosa*,** Common Figwort. *Longitarsus agilis, L. nigrofasciatus, Cionus alauda, C. hortulanus, C. scrophulariae, C. tuberculosus, Cleopus pulchellus.*

***Scrophularia auriculata*,** Water Figwort. *Longitarsus agilis, L. nigrofasciatus, L. rutilis, Cionus alauda, C. hortulanus, C. scrophulariae, C. tuberculosus, Cleopus pulchellus.*

***Antirrhinum majus*,** Snapdragon. *Brachypterolus antirrhini, B. vestitus.*

***Linaria vulgaris*,** Common Toadflax. *Brachypterolus linariae, B. pulicarius, Mecinus janthinus, Rhinusa antirrhini, R. collina, R. linariae.*

***Veronica officinalis*,** Heath Speedwell. *Longitarsus longiseta.*

***Veronica chamaedrys*,** Germander Speedwell. *Gymnetron melanarium, G. rostellum.*

***Veronica beccabunga*,** Brooklime. *Phaedon armoraciae, Prasocuris junci, Gymnetron beccabungae.*

***Veronica anagallis-aquatica*,** Blue Water-Speedwell. *Gymnetron villosulum.*

***Veronica catenata*,** Pink Water-Speedwell. *Gymnetron villosulum.*

***Pedicularis* spp.,** Louseworts. *Longitarsus holsaticus.*

LENTIBULARIACEAE – Bladderwort family

***Utricularia* spp.,** Bladderworts. *Longitarsus nigerrimus, Bagous petro.*

CAMPANULACEAE – Bellflower family

***Campanula glomerata*,** Clustered Bellflower. *Meligethes subrugosus.*

***Campanula trachelium*,** Nettle-leaved Bellflower. *Meligethes corvinus.*

***Campanula* spp.,** Bellflowers. *Cleopomiaris graminis, C. plantarum, Miarus campanulae.*

***Jasione montana*,** Sheep's-bit. *Meligethes subrugosus.*

RUBIACEAE – Bedstraw family

***Galium aparine*,** Cleavers. *Timarcha tenebricosa.*

***Galium* spp.,** Bedstraws. *Timarcha goettingensis, Sermylassa halensis.*

CAPRIFOLIACEAE – Honeysuckle family

***Viburnum opulus*,** Guelder-rose. *Anthaxia nitidula, Pyrrhalta viburni.*

***Viburnum lantana*,** Wayfaring-tree. *Pyrrhalta viburni.*

DIPSACACEAE – Teasel family

***Dipsacus fullonum*,** Wild Teasel. *Longitarsus fowleri.*

***Succisa pratensis*,** Devil's-bit Scabious. *Trachys troglodytes.*

ASTERACEAE – Daisy family

***Arctium* spp.,** Burdocks. *Ceratapion onopordi.*

***Carduus nutans*,** Musk Thistle. *Hadroplontus trimaculatus, Trichosirocalus horridus.*

***Carduus* .ssp.,** Thistles. *Cassida rubiginosa, C. vibex, Neocrepidodera ferruginea, Sphaeroderma rubidum, S. testaceum, Psylliodes chalcomera, Ceratapion onopordi, C. carduorum, C. gibbirostre, Larinus planus, Rhinocyllus conicus.*

***Cirsium arvense*,** Creeping Thistle. *Mordellistena acuticollis, Lema cyanella, Cassida rubiginosa, Ceratapion carduorum, Cleonis pigra, Hadroplontis litura.*

***Cirsium vulgare*,** Spear Thistle. *Cassida rubiginosa, Ceratapion carduorum, Trichosirocalus horridus.*

***Cirsium* spp.,** Thistles. *Mordellistena pumila, Cassida vibex, Neocrepidodera ferruginea, N. transversa, Sphaeroderma rubidum, S. testaceum, Ceratapion onopordi, Ceratapion gibbirostre, Larinus planus, Rhinocyllus conicus.*

***Centaurea nigra*,** Common Knapweed. *Pseudorchestes pratensis.*

***Centaurea* spp.,** Knapweeds. *Cassida vibex, Sphaeroderma testaceum, Ceratapion onopordi.*

***Hypochaeris radicata*,** Cat's-ear. *Olibrus affinis, Glocianus distinctus.*

***Leontodon* spp.,** particularly ***L. autumnalis*,** Autumn Hawkbit. *Olibrus flavicornis, O. pygmaeus, Glocianus moelleri.*

***Tragopogon pratensis*,** Goat's-beard. *Olibrus affinis.*

***Taraxacum* spp.,** Dandelions. *Glocianus pilosellus, G. punctiger.*

***Crepis* spp.,** Hawk's-beards. *Olibrus pygmaeus.*

***Pilosella officinarum*,** Mouse-ear-hawkweed. *Olibrus liquidus, Cryptocephalus aureolus, C. hypochaeridis.*

***Hieracium* spp.,** Hawkweeds. *Glocianus moelleri.*

***Filago vulgaris*,** Common Cudweed. *Olibrus pygmaeus, Acentrotypus brunnipes.*

***Gnaphalium* spp.,** Cudweeds. *Acentrotypus brunnipes.*

***Inula conyzae*,** Ploughman's-spikenard. *Pilemostoma fastuosa.*

***Inula helenium*,** Elecampane. *Cassida murraea.*

***Pulicaria dysenterica*,** Common Fleabane, *Cassida murraea, Baris analis.*

***Aster tripolium*,** Sea Aster. *Longitarsus brunneus, L. reichei.*

***Conyza canadensis*,** Canadian Fleabane. *Olibrus corticalis.*

***Tanacetum vulgare*,** Tansy. *Olibrus aeneus.*

***Seriphidium maritimum*,** Sea Wormwood. *Mordellistena nanuloides, Longitarsus absynthii.*

***Artemisia vulgaris*,** Mugwort. *Olibrus aeneus, Mordellistena parvula, Longitarsus succineus.*

***Achillea ptarmica*,** Sneezewort. *Cassida sanguinosa, Longitarsus succineus.*

***Achillea millefolium*,** Yarrow. *Olibrus millefolii, Mordellistena parvula, Longitarsus succineus, Microplontus triangulum, Trichosirocalus barnevillei.*

***Anthemis cotula*,** Stinking Chamomile. *Omphalapion beuthini, A. laevigatum.*

***Leucanthemum vulgare*,** Oxeye Daisy. *Longitarsus succineus, Diplapion stolidum, Microplontus campestris.*

***Matricaria recutita*,** Scented Mayweed. *Curtonotus aulicus, Olibrus aeneus, Ochrosis ventralis, Diplapion confluens, Omphalapion laevigatum, Pseudostyphlus pillumus, Microplontus rugulosus.*

***Tripleurospermum inodorum*,** Scentless Mayweed. *Curtonotus aulicus, Diplapion confluens, Omphalapion hookerorum.*

***Senecio jacobaea*,** Common Ragwort. *Longitarsus gracilis, L. jacobaeae.*

***Senecio vulgaris*,** Groundsel. *Longitarsus gracilis.*

***Senecio* spp.,** Ragworts. *Olibrus corticalis, Longitarsus dorsalis, L. flavicornis, L. ganglbaueri, L. ochroleucus, L. suturellus.*

***Eupatorium cannabinum*,** Hemp-agrimony. *Longitarsus aeruginosus, L. succineus.*

BUTOMACEAE – Flowering-rush family.
***Butomus umbellatus*,** Flowering-rush. *Bagous nodulosus.*

ALISMATACEAE – Water-plantain family.
***Sagittaria sagittifolia*,** Arrowhead. *Donacia dentata, Galerucella sagittariae.*
***Alisma plantago-aquatica*,** Water-plantain. *Bagous alismatis.*

JUNCAGINACEAE – Arrowgrass family
***Triglochin maritimum*,** Sea Arrowgrass. *Phaedon concinnus.*

POTAMOGETONACEAE – Pondweed family
***Potamogeton natans*,** Broad-leaved Pondweed. *Donacia versicolorea.*
***Potamogeton pectinatus*,** Fennel Pondweed. *Macroplea mutica.*
***Potamogeton* spp.,** Pondweeds. *Macroplea appendiculata, Bagous limosus, Phytobius leucogaster.*

RUPPIACEAE – Tasselweed family
***Ruppia* spp.,** Tasselweeds. *Macroplea mutica.*

ZOSTERACEAE – Eelgrass family
Zostera **spp.**, Eelgrass. *Macroplea mutica.*

LEMNACEAE – Duckweed family
Lemna **spp.**, Duckweeds. *Tanysphyrus lemnae.*

JUNCACEAE – Rush family
Juncus bufonius, Toad Rush. *Bagous lutulosus.*
Juncus sp., Rushes. *Aphanisticus emarginatus* (on *J. articulatus*), *A. pusillus.*

CYPERACEAE – Sedge family
Eleocharis palustris, Common Spike-rush. *Thryogenes nereis.*
Schoenoplectus lacustris, Common Club-rush. *Donacia impressa, D. vulgaris.*
Schoenus nigricans, Black Bog-rush. *Aphanisticus pusillus.*
Carex paniculata, Greater Tussock-sedge. *Donacia impressa.*
Carex acutiformis, Lesser Pond-sedge. *Donacia aquatica, D. impressa, D. thalassina.*
Carex rostrata, Bottle Sedge. *Donacia obscura, D. thalassina.*
Carex spp., Sedges. *Kateretes pedicularius, K. pusillus, K. rufilabris, Phalacrus caricis, P. substriatus, Plateumaris affinis, P. discolor, Notaris aethiops, N. scirpi, Thryogenes fiorii, Tournotaris bimaculatus, Limnobaris dolorosa, L. t-album.*

POACEAE – Grass family
Glyceria maxima, Reed Sweet-grass. *Donacia semicuprea, Notaris acridulus.*
Phalaris arundinacea, Reed Canary-grass. *Tournotaris bimaculatus.*
Ammophila arenaria, Marram. *Demetrias monostigma, Melanotus punctolineatus, Dicronychus equisetioides, Malachius barnevillei.*
Brachypodium pinnatum, Tor-grass. *Phalacrus fimetarius.*
Leymus arenarius, Lyme-grass. *Demetrias monostigma.*
Hordeum spp., Barleys. *Phalacrus corruscus.*
Secale cereale, Rye. *Zabrus tenebrioides.*
Triticum aestivum, Bread wheat. *Zabrus tenebrioides, Helophorus nubilus, Phalacrus corruscus.*
Phragmites australis, Common Reed. *Odacantha melanura, Donacia clavipes, Plateumaris braccata, Tournotaris bimaculatus.*

SPARGANIACEAE – Bur-reed family
Sparganium erectum, Branched Bur-reed. *Telmatophilus brevicollis, T. caricis, T. sparganii, Donacia bicolora, D. marginata, D. sparganii, D. vulgaris, Plateumaris sericea, Notaris aethiops, Thryogenes scirrhosus.*

Sparganium emersum, Unbranched Bur-reed. *Telmatophilus sparganii*.

TYPHACEAE – Bulrush family

Typha latifolia, Bulrush. *Stilbus oblongus, Telmatophilus typhae*.

Typha angustifolia, Lesser Bulrush. *Telmatophilus schoenherri, T. typhae*.

Typha **spp.**, Bulrushes. *Odacantha melanura, Demetrias imperialis, Coccidula scutellata, Donacia cinerea, D. vulgaris, Notaris scirpi, Tournotaris bimaculatus*.

LILIACEAE – Lily family

Narthecium ossifragum, Bog Asphodel. *Phalacrus substriatus*.

Lilium **spp.**, Lilies. *Lilioceris lilii*.

Asparagus officinalis, Asparagus. *Crioceris asparagi*.

IRIDACEAE – Iris family

Iris pseudacorus, Yellow Iris. *Plateumaris sericea, Aphthona nonstriata, Mononychus punctumalbum*.

Iris foetidissima, Stinking Iris. *Mononychus punctumalbum*.

Reference

Stace, C.A., 1997. *New Flora of the British Isles*, edn 2. Cambridge: Cambridge University Press.

Curatorial

INDOOR WORK

by J. Cooter

This section covers the procedures entailed in preparing the live beetle for the cabinet. I hasten to point out the purpose of studying insects is not simply to fill a cabinet or to tick off names on a check list in train-spotter/bird-watcher fashion. I hope this section is comprehensive; it has been compiled from many years' background as an amateur, combined with nearly 30 years professional experience in Museums. There is no such thing as the correct method; each person will have their own variation on the same theme. There are, however, INCORRECT methods which should be avoided at all costs.

A minority of coleopterists keep their material in alcohol or other liquid preservative, details of this are given later; here we deal with the preparation of a 'dry' collection.

Killing: The secret to good preparation lies in the killing method. When removed from the killing bottle, the beetle should be perfectly relaxed. The medium should be quick acting and present a minimum hazard to the user (it is good practice to regard any chemical used in entomology as potentially dangerous).

Ethyl acetate or 'acetic ether' is the most widely used killing agent today. It is cheap, freely available from large branches of dispensing chemists and reasonably safe to use, though it and its vapour are highly inflammable. It is such a good agent that others are not worth considering. When using ethyl acetate it is worth remembering that it will dissolve plastics and renders dimethyl hydantoin formaldehyde (D.M.H.F.) soft, prolonged exposure or immersion will dissolve the D.M.H.F.

Other organic fluids which have been used for killing insects include carbon tetrachloridae (a carcinogenic substance which leaves insects in a rigid inflexible state on death), ammonia (not domestic grade, but the concentrated 0.880 solution – can alter colours and renders insects difficult to prepare). They are mentioned here only to point out their unsuitability.

[*Cyanide*. Old textbooks on entomology recommend the use of potassium cyanide crystals under plaster in a glass bottle; it is best to regard this as inspired lunacy as cyanide, even in minute quantities, will kill healthy adult human beings. When exposed to moisture, even the moisture in air, it gives off an equally lethal colourless gas – hydrogen cyanide. Cyanide can only be rendered safe by chemical methods, burial or burning are quite ineffective and totally irresponsible. AVOID IT AT ALL TIMES; fortunately it is highly unlikely that you will be able to purchase cyanide. It is only mentioned here as a warning.]

Killing bottle: The term 'bottle' is here taken to be a jar or tube, any clear glass container with tight fitting lid will suffice but a larger tube of say 25-50mm diameter and between 50-75mm long is probably the ideal for most beetles (see for example plate 112). Larger species can be killed individually or in a large sized killing bottle; a ground glass stoppered spice or storage jar is preferable to a jam-jar or similar with a screw or metal lid, the plasticised liners and seals on these are destroyed by the vapour given off by the killing agent. Polypropylene bottles/tubes can be used but many of these are translucent, so it is not possible to see inside without removing the lid.

The bottle should be kept clean and dry, cotton wool should **NEVER** be used as the beetles will get irretrievably entangled in it. White blotting paper, filter paper, kitchen roll, 'soft' toilet tissue or cellulose wadding are all suitable and readily available, cheap and absorb the fluid any excess moisture forming within the bottle; these materials are cheap and readily replaced.

It is best to make up fresh bottles prior to each outing by pushing an amount of the absorbent material tightly in the bottom of the tube, adding a few drops of ethyl acetate followed by another layer of tissue. A piece of crumpled tissue is then added to give the beetles a large surface to run about on, and to help absorb excess moisture. The beetles are added and the stopper put in place, the whole being kept away from direct sunlight. The time taken for the beetles to die will depend upon the charge in the bottle, temperature, number and type of beetle. Histerids and other similarly hard-to-kill species should either be kept separate or put in first, the others added when these have stopped moving around. Experience is the only sure guide here; if in doubt, leave it another half-hour or overnight.

Points to remember are:
- Use a minimum of killing fluid.
- Keep the beetles away from direct contact with the killing fluid.
- Keep the killing bottle dry in use.
- Never over-fill the bottle with beetles.
- Keep large or voracious species separate (I still recall the agony of watching helplessly as my first specimen of *Hypebaeus flavipes* (F.) was devoured by a cantharid in the killing tube).
- Keep the killing bottle out of the sun.

Material that cannot be dealt with straight away: If the catch cannot be prepared directly there are several options available. If the delay is only a day, the beetles can be left in their tubes and placed in a refrigerator, or killed and left in an ethyl acetate killing bottle. Transfer to laurel will permit a longer period to elapse. For long periods, the material can either be left to dry or put in 70% solution of alcohol (ethanol, industrial methylated spirit (I.M.S.) or isopropyl alcohol) with 3-5% glacial acetic acid added. If alcohol is used it is advantageous to fill tubes two-thirds to one-half full and add a piece of tissue.

Live beetles should not be dropped in directly, as very often their elytra will open and wings extend during their death struggle. In addition the beetle will become stiff and brittle making it virtually impossible to manipulate the appendages during subsequent preparation and very difficult to dissect the genitalia without damage. Killing in ethyl acetate and then exposing the beetle to acetic acid vapour in a sealed container for 24hrs prior to transfer to alcohol will overcome this. A slip with the data must be added to each tube; use pencil or India ink, not biro or synthetic ink as these are taken into solution by the alcohol rendering the data slip totally blank.

Dry material. It is of prime importance to ensure the material is dead before packing. This may seem obvious. On removal from the killing bottle, the beetles are laid neatly in rows on cellulose wadding in a box (stout cardboard is perhaps preferable as its walls do help by absorbing excess moisture to a degree; this is not so with plastic or metal containers). If the appendages are roughly arranged there might be no need to relax the specimens later on. The layers are built up, and data kept with the beetles. Never mix material from different localities on the same layer. Keep the box full by placing spare wadding on top of the last layer as this will ensure the contents are not disturbed and mixed up when the box is transported. There is often a need to handle material in this way especially if collecting abroad where the aim is to amass as large a sample as possible in the time available. In the tropics the addition of silica gel or other moisture-absorbing compound and possibly a fungicide too is often necessary to minimise the risk of moulds. Some pesticide, say a drop or two of citronella oil soaked into the bottom and lid of the box, is also advantageous to ward off insect attack.

Wood shavings: A method used widely abroad and generally useful for larger species of beetles. It can be used at home for keeping material supple prior to preparation or whilst away on an expedition. Freshly killed material preserved in this way remains supple for two to three years and longer (depending upon size of the beetle). That said, any smaller species preserved by this method should be dealt with as quickly as possible upon return home as they do become brittle and difficult to prepare if left too long.

You will need a supply of clean wood shavings. The bedding sold for small rodent pets is suitable if first sieved to remove 'saw dust' and finer particles. After sieving leave it to dry thoroughly for two days on a paper sheet. A stock of small self-seal plastic bags and/or polypropylene screw-cap collecting bottles (50ml capacity is ideal but they are bulkier than the self-seal bags but the bags will need packing in a secure protective container, the bottles will not); see plate 110. Freshly killed material should be allowed to dry of all surface moisture by being placed on absorbent paper (newspaper will do if nothing else available) for about ten minutes. Care taken that the material does not desiccate too much or does not desiccate completely.

Mix the pre-dried beetles and shavings in a 1:1 ratio and place in the bottle or bag. Add 5-drops of ethyl acetate (this acts as an antimould agent), label each container. Place in a deep freeze (NOT refrigerator) at about -20°C. If in the field away from home and/or a freezer, keep the material in sawdust in collecting bottles treated with about ten drops of ethyl acetate. Take care to ensure the bottles are not exposed to direct sun or heat from other sources. If excess moisture develops within the bottle, empty the contents onto absorbent paper and allow to dry for about five minutes, return contents to the bottle and re-charge with 10 drops of ethyl acetate before sealing tightly. When back home, the material can be prepared for the freezer by drying slightly, then recharging the bottle with fresh ethyl acetate before storing in the freezer until needed for preparation or study.

Material removed from the freezer for preparation should be allowed to reach room temperature before work begins. Place the 'thawed' material in a small solvent proof container containing tissue to which 3-drops of ethyl acetate and one or two drops of water have been added. After an hour or so the beetles will be in optimum condition for preparing. You can of course remove only a few beetles from the freezer container at a time, returning the rest to the freezer for later.

Over-desiccation of material in the freezer can be avoided by adding a few more drops of water.

Laurel bottle: Like Joy's *'Practical Handbook of British Beetles'*, this is something coleopterists either swear by or swear at. Equally similarly, the laurel bottle, when used *correctly,* is a very handy aid that has stood the test of time. Although once recommended as a killing medium, it is very slow acting and should not be considered; ethyl acetate has superseded this use. For keeping freshly killed material relaxed it is very useful, though it should not be used for long periods as there is a tendency for its vapours to darken colours, and there can be a problem with excessive moisture.

Young leaves of the shrub cherry laurel (*Prunus laurocerasus*) are picked, cut into small pieces and then bruised and pressed down tightly in a large glass jar making a layer about one quarter the depth of the jar. Fresh laurel tends to be very moist so precautions must be taken. A layer of cellulose wadding is placed on top of the laurel layer and on top of this a blotting paper lining. The crushing of the leaves produces non-lethal quantities of hydrogen cyanide and benzaldehyde – the latter, smelling strongly of almonds – which keeps the insects relaxed indefinitely; indeed appendages may actually drop off on handling the specimens if they have been in direct contact with the laurel, or in the jar for excessively long periods. The beetles should thus be kept in batches in screws of absorbent paper or in small polythene self-seal bags which have been perforated with numerous pin-pricks. A lining of kitchen roll is inserted into the bag and the beetles placed in on one side of this and a data label inserted on the other. The bag is sealed and placed in the laurel jar (or in a plastic box containing crushed laurel).

Relaxing: The purpose of relaxing dry material is to render the musculature soft enough to permit manipulation of the appendages, and, if necessary, dissection of the genitalia. This goal is simply achieved by putting a lining of tissue, cellulose wadding or other absorbent material (not cotton wool) in a plastic or metal box with tight fitting lid. The multicompartment 'sorting trays', pl. 112, (see Bioquip website catalogue) are ideal for this purpose. Water is added soaking the lining, beetles added in rows, the lid replaced and the whole placed in an airing cupboard, on top of a radiator or similar warm place. After a few hours, often overnight, depending upon the size of the individual specimens, the beetles will be ready for preparation. It will become apparent when engaged in such work that it is good practice to relax at one time only sufficient beetles that can be mounted in one session and when possible make up each batch of beetles of similar size. The relaxing box should be cleaned out between use and a new lining used each time.

Some people advocate relaxing beetles by floating them on hot/warm water. Although I have not personally tried this, I would not recommend it as I can see beetles sinking, water being spilled and other potential disasters. Personal experience has shown the use of proprietary made up 'relaxing fluids' are a waste of money though many coleopterists use them.

Mounting: Whether the beetle has been removed from the killing bottle, alcohol tube, laurel or relaxing box, the subsequent preparation follows the same path. Because our fauna is well documented and composed mostly of small to medium-sized beetles, British coleopterists have traditionally carded all species. Some, however, prefer to pin larger species such as *Carabus, Nicrophorus* or *Lucanus*.

Before describing the methods of mounting, we should first consider the equipment and materials needed.

Equipment: The basic necessary items are few (see plate 112). Always buy the best quality available; quality does indeed increase with price.

Brushes. Two, three or more artist's good quality sable hair brushes, size 0, 00, or 000 being the most useful. Personal preference as to size will result from use. Brushes with synthetic hairs are much cheaper, but the hairs easily curl rendering continued use of the brush impossible. Sable hair is not affected by the range of organic solvents used in entomology.

Forceps. At least one pair of watchmaker's fine pointed straight forceps size 5 or 5A (off-set points). Two pairs are advantageous, and a third pair of less fine, say size 2, will be found very useful. If possible, these should be stainless (anticorrosive) and non-magnetic. As an illustration of buying best quality, a good pair of 5 or 5A, stainless, anti-magnetic will cost (2005) around £16. I have used the Swiss *Precista* make for many years and find them perfectly adapted for entomology, fine points, firm touch, perfect balance. I find 5A with the off set tips preferable to straight tipped 5. Dumont are another good quality make. Try not to drop them on the floor, when not in use replace the plastic cover over the points. Never take forceps of this quality into the field and never lend them to a friend.

Watchmaker's size 2 forceps are very useful when working with micro-pins and cabinet points. Their pointed ends are much more robust than 5 or 5A but unlike pinning forceps (next below) not knurled and much thinner making manipulation of short fine pins much easier.

Pinning forceps are used in Britain for handling mounting pins. Like all tools they vary greatly in quality and an inferior pair can easily lead to damage being caused when working the collection. They should close with gentle pressure, both ends meeting evenly over their length, those with a 'peg' will be less subject to 'shearing'. The addition of ribbing on the finger grips will minimise slipping, and 'knurling' on the ends will assist in gripping the pin, preventing slipping when pressing the pin into the cabinet. However, if the 'knurling' is too deep, the pin can slip sharply into the groove with the attendant likelihood of damage to the insect, even dislodging it from its card. Pinning forceps with magnetic qualities are a confounded nuisance when using steel pins (everyone should use steel pins). Ideally buy in person rather than through the post and test them before parting with your money.

Dissecting needles of the standard biological type are too coarse for the bulk of entomological work. A very useful tool is a micro-lepidopterist's headless stainless steel pin. These are too short to be hand held, and easily slip when held in forceps. A watchmaker's 'pin vice' comes in very handy here, being about 12cm long and having good balance (they are easily cut down to suit personal preference). The 'business end' is a three-jawed chuck activated by a screw collar. Thus pins of various lengths and diameters can be changed easily. Alternatively the individual pins can be glued to cocktail sticks or small diameter dowels cut to appropriate length. The thinnest micro-pins are too thin and will bend with frustrating ease. Undoubtedly the best dissecting pins are made from tungsten wire sharpened in molten sodium nitrite. They can be re-sharpened but this will only be required after many years use. I obtained five tungsten pins in the late 1960's and have three still in daily use, 2004; none have had to be re-sharpened, two somehow got lost.

Dissecting dish. Goldilocks would be good at selecting a dissecting dish, it should not be too large, too small, too deep or too shallow but just right; it must be heavy with a flat base. The spot testing palette (plate 112) used by biologists has advantages. They generally have a dozen or so shallow depressions and permit rapid transfer from one fluid to another, being of white porcelain the dissection and minute dissected parts are easily seen. The palette is sufficiently heavy and the individual depressions of just the right depth. An excavated glass block is very useful for different reasons being about 4cm square with a flat

Curatorial

base. The excavation is too deep and too small a diameter to allow easy dissection, but it is invaluable for sorting small beetles preserved in spirit, being deep the fluid will not evaporate quickly and the whole can be protected with a cover glass. Excavated glass blocks have other uses too, I use two such blocks for glue, one with gum arabic, the other with tragacanth.

Scissors. Your own pair of top quality scissors should be kept well hidden from other members of the family. A good pair will make the sharp clean cut necessary when trimming labels.

Instrument roll. The tie-up cloth dissecting instrument rolls are ideal for keeping the equipment safe when not in use or when in transit. Points of forceps and the heads of brushes are easily damaged if kept loose in boxes.

Pins: Pins have three main uses:

For manipulating the beetle during mounting, and for performing dissections.

For holding the card-mounted beetle and its labels.

For direct pinning of beetles.

The headless, stainless micro pins are available in a range of diameters (A-G with A the thinnest) and lengths. Diameters from (D) are suitably stout for manipulating beetles and dissecting. These pins are sold in lots of 100 of each size by Watkins & Doncaster; pins of other dimensions are marketed by other entomological suppliers. The size of pin used to hold a mounting card or to direct pin a beetle should be as long as possible. Alas, in Britain many old cabinets and store-boxes are too shallow to take Continental pins (38mm long) so English size 12 or stainless steel headless size G3 will have to be substituted. Brass (Japanned or plated) pins should NEVER be used, with time corrosion sets in at the point where the pin passes through the mounting card and into the cabinet/store-box bottom, eventually the pin will break at one or both of these points.

Continental pins are of two lengths, diameters 000 to 6 being 38mm long, and sizes 7 to 10 are 53mm long (intended for use with large tropical insects). Sizes 3 to 5 will be found adequate for British workers. They are made in stainless steel or black steel, with nylon or brass heads. Good makes include the Czech hand made Koštál brand (possibly the sharpest 'Continental' pin I have encountered and with heads that never seem to come off, alas not available in stainless steel), the German 'Karlsbader' and Austrian 'Double Elephant' and 'Austerlitz' brands.

In recent years some entomological dealers are now selling Japanese entomological pins. These are stainless steel with integral head and like 'Continental' length pins come in a range of diameters but at 41mm long they are likely to be too long for most traditional British *and* 'Continental depth' storage units.

Mounting cards: More and more British coleopterists are now using the die-stamped standard-sized cards sold by European dealers. They come in two types, with or without fine black lines at one end and in 69 sizes up to 80 x 35mm (including two triangular types), some dealers offering one or two sizes in black card. They are available as individual sizes in packs of 100, 250, 500 or 1000 depending upon size of card and dealer. The more elongate sizes are ideal for Staphylinidae, Elateridae etc, the more equi-dimensional ones for Coccinellidae, Histeridae etc. Invariably the coleopterist will settle on a few, perhaps, 6 to10, sizes through personal preference, the main families of interest or any other personal prejudice.

Whether to use cards with printed lines or not is purely personal preference. The unlined cards have more 'useable area' as mounting part of the beetle over the printed lines looks untidy and is generally avoided. Also the actual depth of the printed lines can very from batch to batch from the same supplier. It is not difficult to consistently place the mounting pin centrally and at an equal distance from the end of an unlined card. Indeed given the variation in depth of the printed lines and the reliance of using a line to pin through, the lined cards are far less versatile than unlined ones; the costs are identical.

Although a broad generalisation, different manufacturers use different quality card for their product. The majority use a matt surface card which seems to hold glue better and not show excess gum as a shiny film. The more 'egg-shell' surface card, although much whiter, is not so efficient with standard water-based gums used widely in entomology.

It is of course possible to cut one's own cards to whatever size is desired. A good quality Bristol Board (three sheet for small cards and four sheet for larger; now only obtainable from specialist artist supply shops) should be employed. A *very* sharp knife is needed for cutting as otherwise ragged edges or 'burring' will result. A range of standard sizes should be employed and a stock of each cut prior to use. Given the low cost and convenience of die-stamped cards, the onerous task of cutting one's own cards with the invariable result of no two exactly the same, holds little appeal and can be time consuming and more expensive.

Glue. There are many different types of glue in use by coleopterists. Each will tell you that theirs is, if not the best, then that it is very good. This might well be true, there is probably not one type of glue to recommend, there are, however, types to keep well away from and this includes any synthetic organic-solvent based glue. The glue you use should be strong, water soluble, clear, dry to a matt finish and should not stain the card or darken with age.

A straw poll on the <beetles-britishisles@yahoogroups.com> news group shows that the stationer's glue 'Gloy' (the undyed clear or translucent varieties) is the most popular, followed by Watkins & Doncaster's mounting glue. Others use gum tragacanth, gum arabic and various proprietary brands sold by entomological dealers, one or two even use wall-paper paste. Some museum-based entomologists now use specialist water soluble 'Evacon-R' a neutral pH reversible EVA, (water soluble, non-plasticised, pH7.5 ethylene-vinylacetate copolymer emulsion) available from Conservation By Design, Time Care Works, 5 Singer Way, Woburn Road Industrial Estate, Kempston, Bedford, MK42 7AW. It should be used diluted with water). It would seem most people try a few different types of glue before settling with one that they like.

Too much glue and glue in inappropriate places is a confounded nuisance obscuring microsculpture and ornament, matting pubescence and setae and distorting proportions of antennal and tarsal segments. I am not alone in returning beetles sent to me for identification unidentified with a diplomatic note basically saying 'I do not have the time to clean up your beetles to render them identifiable.'

Those beetles that require to be mounted on one side, for example *Meligethes*, some hydrophilidae, for mounting aedeagi 'end on' and at times of standard mounting such beetles as the larger histerids with smooth shiny integument, a very strong water based glue such as 'Secotine' (sold in British stationers) or 'Hercules' (widely used in central Europe) or 'fish glue' sold by many entomological dealers is required. The latter two are the glues I use for card mounting large beetles such as *Dytiscus, Carabus, Nicrophorus,* chafers etc. They are also a useful for repairing damage to larger beetles.

Pinning stage: This simple device (Plate 112) is a great help and the easiest way to ensure that mounting card, data and other labels are spaced evenly and at standard heights on the pin. Similarly, direct pinned insects will have their ventral surfaces at the same height.

The pin is pushed through the mounting card and then into the deepest hole in the stage, the data label in the next deepest and so on. Having every specimen at roughly the same height – the cards will be at identical heights, but the height of the individual beetles varies – is a great help and saves a lot of refocusing the microscope when making identifications or comparisons.

Dropping bottles: These are very useful and convenient for dispensing small quantities of liquids, for example alcohol for dissection work, clove oil and even killing fluid. The stoppers are vapour-tight and as delivery is accurate, very little chemical is wasted.

Pinning medium: A number of small blocks of Plastazote or similar material of various sizes and dimensions for pinning mounted beetles will be required. Mounted specimens are pinned into this for examination under the microscope, freshly prepared material pinned into it while being labelled; the uses are many.

Mounting a beetle

Card mounting: My first efforts at carding beetles were awful and doubtless I am not alone in this. The card was too thin and I certainly used far too much glue. The beetles had not been killed properly and/or were not relaxed enough to allow manipulation. Like all skills, the ability to mount improves with practice.

The beetle to be mounted should be perfectly relaxed, dry and dead (possibly all coleopterists have had beetles 'come alive' an hour or two after mounting). If the beetle has been stored in spirit or had its genitalia dissected in alcohol it should be dried off on filter paper or kitchen roll.

Once dry the beetle is placed on its dorsal surface and held gently but firmly with downward pressure from forceps. A pin and or forceps are then employed to tease out the legs, antennae and mouthparts. A small amount of gum is brushed centrally onto a mounting card and the beetle placed ventral side down onto the gum. Using a sable hair brush carrying a minimum of dilute glue the legs are arranged symmetrically, sometimes forceps will be require to facilitate this. Finally the mouthparts and antennae are arranged with the brush. See fig. 44.

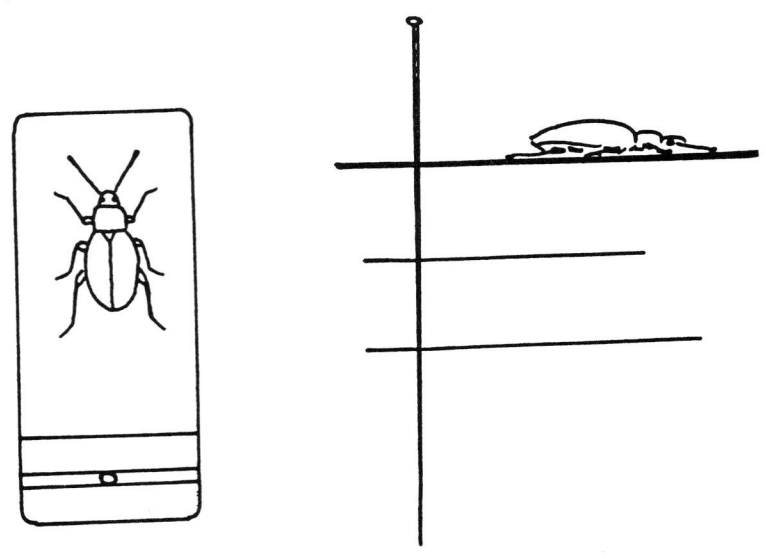

Fig. 44. A beetle mounted with gum directly onto a die-stamped mounting card.

Making an entire collection of 'perfectly set' specimen with all appendages symmetrically arranged, palpi included and mandibles open probably says more about the preparator's state of mind than entomological expertise. At the other end of the scale is the simple gumming of the beetle body to the card with little or no attempt to display the appendages. As with everything, if the reason for carrying out an action is understood, life becomes much easier. The purpose of mounting is to allow study of the beetle and to preserve it with a reduced risk of damage. Thus card mounting and preserving in alcohol are better alternatives to direct pinning (certainly for smaller beetles) and card pointing.

Care must be taken to ensure tarsi lie flat on the card and, ideally, with glue only on their undersurface. Antennal segments rarely have circular cross section, usually they are slightly dorso-ventrally flattened so any mis-alignment of the segments will not show true proportions of width. Antennal segments, generally being small and delicate, will also have their relative proportions distorted if gum adheres. If left un-mounted the antennae will generally twist and bend as the beetle dries out making measurement of or observation of relative lengths of individual segments very difficult. A small amount of adhesive on the terminal antennal segment only is the aim. This is true for virtually all types of antennae, lamellate and irregular types perhaps being exceptional. If it is necessary to take absolute measurements of individual antennal segments or to measure segments and calculate their ratio, it is a good idea to remove an antenna completely and mount it in Euparal, again ensuring all segments are in the same plane and not twisted and that the entire antenna is flat on the card or plastic mount.

With experience it will be known which features of each genus are required for study during identification. Thus Aleocharinae are likely to be dissected as routine prior to mounting, *Meligethes* mounted on their side, *Choleva* females to have their genital tergites extracted and so on. When preparing foreign *Agathidium* I make a shorthand note of relevant ventral features on a small label pinned under the mounting card (for example 'mc +; LL -; F/L comp' which translates to median carina present, lateral lines absent, femoral lines complete). The beetle is then mounted 'conventionally' dorsal side up after dissection of the genitalia; the ratio of pronotum to elytra can be measured, sutural stria and surface ornament to be studied.

If the beetle has been dissected see 'dissection' (pp 356-361) for treatment of the dissected parts.

Direct pinning: Beetles should be pinned through the right elytron at a point toward the base and about one-third elytral width from the suture. Push a suitable pin into the elytron and then make sure it continues its path vertically to both the longitudinal and transverse axis of the beetle. Take great care when pushing the pin through the ventral surface, do this gently and slowly and check the position of the point as soon as it pierces through. It is quite easy to incorrectly position the pin and it pushes out the coxa of the right middle leg, or passes right through the coxa making it impossible to manipulate the leg.

For larger direct pinned beetles, once the pin is successfully placed, put the beetle on the pinning stage and press right through. Next transfer the pinned beetle to a thick piece of Plastazote pushing the pin in until the ventral surface touches the Plastazote. The appendages are manipulated to their required position with watchmaker's forceps and held in place with micro-pins. If a dissection has been made, the dissected parts should be suitably mounted or stored and pinned adjacent to the beetle.

For smaller micropinned beetles that will next be staged (fig.45), it is advisable first to prepare the carrying pin – the standard large pin with a small length of polyporus or Plastazote 'stage', data labels beneath. Push the micro-pin about two-third through the beetle, tease out the appendages so underside characters are not obscured, then pin the beetle into the 'stage.'

If pinned beetles have been dissected, mount the dissection on a card or acetate strip and pin this immediately under the beetle above the data and determination labels.

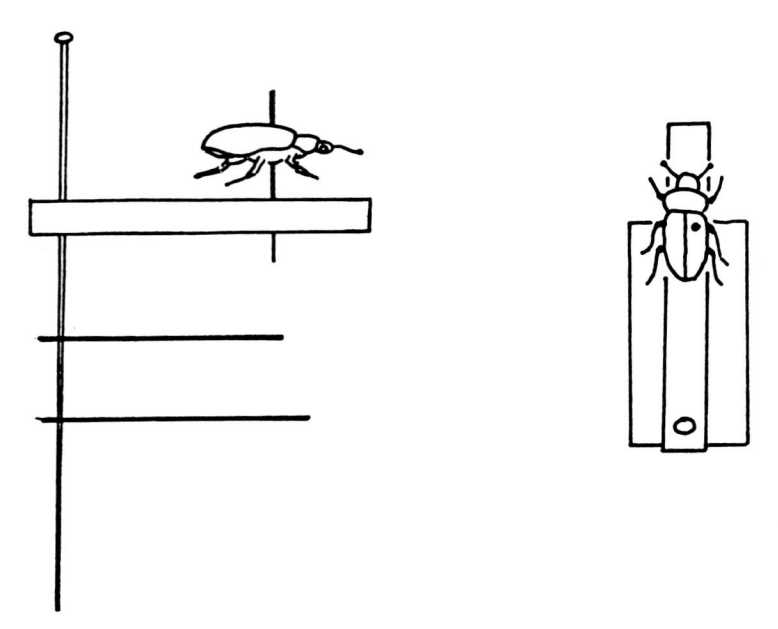

Fig. 45. A small beetle staged using polyporus or a modern synthetic alternative. The beetle is pinned with a stainless steel headless micro-pin.

Labelling

A collection of rare, perfectly mounted beetles without labels has very little use other than to add weight to the anti-collecting lobby's argument that we have no need to study actual insects but can make do with photographs.

During my lifetime working in museums I have seen many insect collections built up over a lifetime and representing a huge investment of time and money but with no data. The main use for such collections, assuming the museum would actually want to acquire anything of such low value, is museum display and 'education' both activities invariably resulting in destruction of the specimens.

Other collections have a little numbered label attached to each specimen. These possibly cross referenced to a written record in a catalogue or journal, but where that catalogue went to is anybody's guess.

There is a tendency today to computer generate labels. This is a contentious issue but one adopted through necessity. Hand writing does take time, photographically produced labels *will* turn black with time, traditional printing by comparison to computer generation is slow and expensive. If hand written labels are produced, we all need the odd one or two from chance collecting, these should be written with a top quality permanent drawing or Indian ink and fine nib pen. Pencil makes a permanent and handy alternative, but ball point and other synthetic water-based or solvent-based inks will fade with time or be affected by the substances in entomological use for deterring pests and mould.

The contentiousness of computer generated labels lies with two unknowns – permanency of ink and paper. It seems very likely that both will show no signs of deterioration for many years. It might be convenient to group computer printers into two broad categories, laser and non-laser. I have (currently) an HP deskjet959c and HP laserjet2200d printer, the latter is a wonderful machine and can print with needle-sharp clarity down to 2-point (requiring a hand lens to read). Both printers set at using a font such as Ariel Narrow at 4 or 4.5 point produce labels readable to the naked eye and of a size similar to or not much larger than a die-stamped mounting card (of course depending upon the card size and amount of data recorded). However, the ink jet printer has slight 'bleeding' around the print making of a rather fuzzy outline; older model ink jet printers show this fuzzyness to an almost unacceptable degree requiring the user to select a larger sized font.

A personal preference is to hand write at least the species name on Type labels, thus effectively adding my signature to each member of the Type Series (holotype and paratypes). A very sensible addition which I will adopt is to number the paratypes.

Data labels:

There is no justifiable excuse for not putting a data label on each and every specimen. The time involved in doing this is not great and needs only doing once.

If a lot of data is required, two labels or hand writing an extra line on the reverse of the computer generated label might be necessary. For most UK situations a single label will suffice, but for collecting abroad when latitude, longitude and altitude are required and habitat detail advantageous, two labels are preferable to one larger label.

Typical UK data should include the county or Watsonian-Praeger Vice County code (fig. 46) or the Praeger-Woodward 'Typomap' code (fig. 47). This should be followed by the location, often the nearest town or village or name of the nature reserve etc. A 10km or preferably 2km OS grid reference, date, captor's name and a brief habitat reference (for example 'red rotten oak', 'wet flush', '*Iris foetidissima*', 'sheep dung').

0	Channel Isles	42	Brecon	80	Roxburgh	3	Cork W.
1	West Cornwall	43	Radnor	81	Berwick	4	Cork Mid.
2	East Cornwall	44	Carmarthen	82	Haddington (East Lothian)	5	Cork E.
3	South Devon	45	Pembroke			6	Waterford
4	North Devon	46	Cardigan	83	Edinburgh (Midlothian)	7	Tipperary S.
5	South Somerset	47	Montgomery			8	Limerick
6	North Somerset	48	Merioneth	84	Linlithgow (West Lothian)	9	Clare
7	North Wilts.	49	Caernarvon			10	Tipperary N.
8	South Wilts.	50	Denbigh	85	Fife and Kinross	11	Kilkenny
9	Dorset	51	Flint	86	Stirling	12	Wexford
10	Isle of Wight	52	Anglesey	87	Perth S. and Clackmannan	13	Carlow
11	South Hampshire	53	Lincoln S.			14	Queen's County (Leix)
12	North Hampshire	54	Lincoln N.	88	Mid Perth	15	Galway S.E.
13	West Sussex	55	Leicester and Rutland	89	Perth N. (or E.)	16	Galway W.
14	East Sussex	56	Nottingham	90	Forfar	17	Galway N.E.
15	East Kent	57	Derby	91	Kincardine	18	King's County (Offaly)
16	West Kent	58	Cheshire	92	Aberdeen S.		
17	Surrey	59	Lancashire S.	93	Aberdeen N.	19	Kildare
18	South Essex	60	Lancashire mid.	94	Banff	20	Wicklow
19	North Essex	61	S.E.Yorks	95	Elgin (Moray)	21	Dublin
20	Herts.	62	N.E.Yorks	96	Easterness	22	Meath
21	Middlesex	63	S.W.Yorks	97	Westerness	23	Westmeath
22	Berkshire	64	Mid W.Yorks	98	Main Argyll	24	Longford
23	Oxfordshire	65	N.W.Yorks	99	Dunbarton	25	Rosscommon
24	Bucks.	66	Durham	100	Clyde Isles	26	Mayo E.
25	East Suffolk	67	Northumberland S.	101	Kintyre	27	Mayo W.
26	West Suffolk	68	Cheviotland	102	Ebudes S.	28	Sligo
27	East Norfolk	69	Westmorland and N.Lancashire	103	Ebudes Mid.	29	Leitrim
28	West Norfolk			104	Ebudes N.	30	Cavan
29	Cambridge	70	Cumberland	105	Ross W.	31	Louth
30	Bedford	71	Isle of Man	106	Ross E.	32	Monaghan
31	Huntingdon			107	Sutherland E.	33	Fermanagh
32	Northampton	Scotland		108	Sutherland W.	34	Donegal E. (or S.).
33	East Gloucester			109	Caithness	35	Donegal W. (or N.).
34	West Gloucester	72	Dumfries	110	Hebrides	36	Tyrone
35	Monmouth	73	Kirkcudbright	111	Orkneys	37	Armagh
36	Hereford	74	Wigtown	112	Shetlands	38	Down
37	Worcester	75	Ayr			39	Antrim
38	Warwick	76	Renfrew	Ireland		40	Derry (Londonderry)
39	Stafford	77	Lanark				
40	Shropshire	78	Peebles	1	Kerry S.		
41	Glamorgan	79	Selkirk	2	Kerry N.		

Key for Fig. 46.

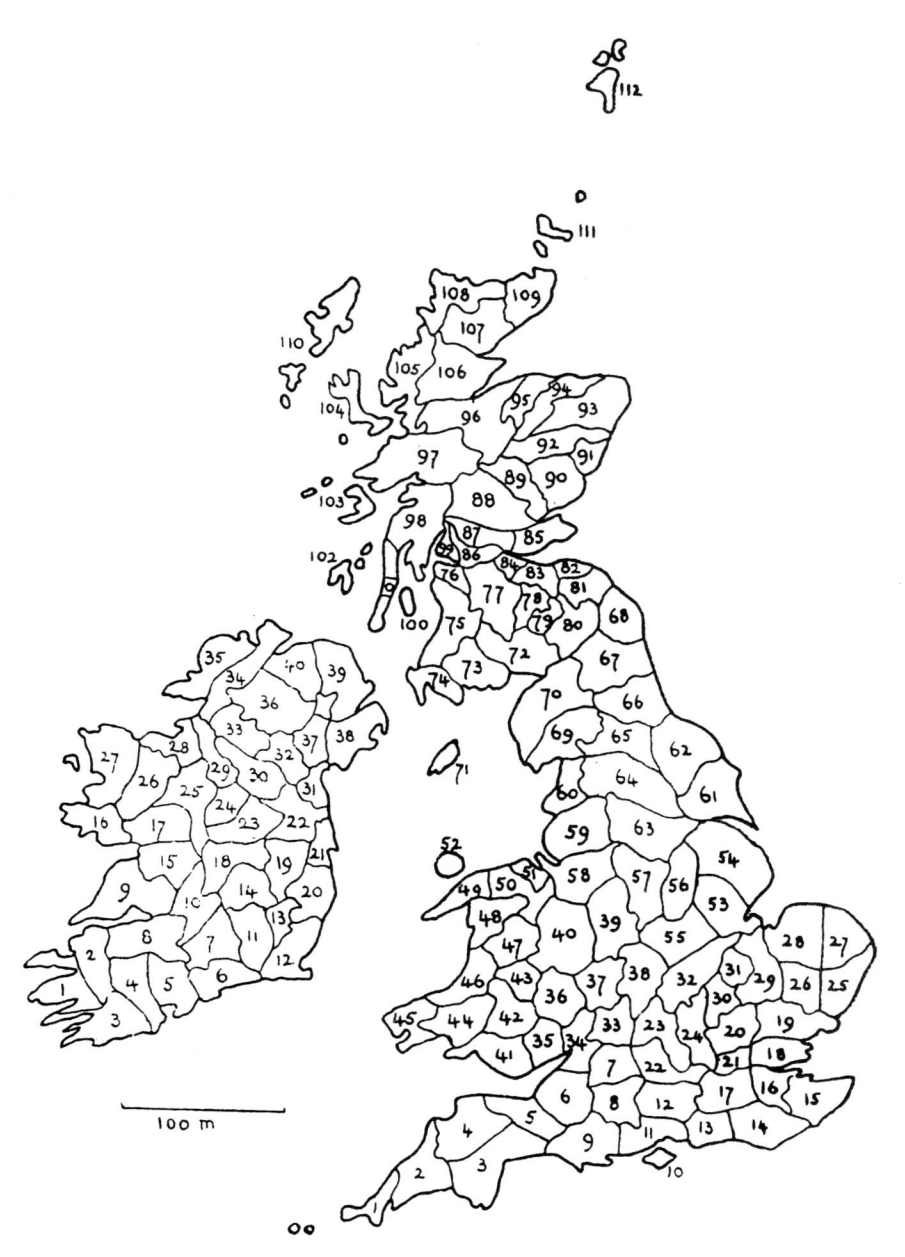

Fig. 46. The Counties and vice-Counties of the British Isles.

A	Anglesey	ES	East Suffolk	MO	Monaghan	SI	Shetland Isles
AM	Argyll, main	EX	East Sussex	MX	Middlesex	SK	Selkirk (Scotland)
AN	Aberdeen north (Scotland)	EY	North-east Yorkshire	MY	Mid west Yorkshire	SK	South Kerry (Ireland)
				ND	North Devonshire	SL	South Lancashire (England)
AN	Antrim (Ireland)	FE	Fermanagh	NE	North Essex	SL	Sligo (Ireland)
AR	Armagh	FF	Forfar	NG	North Galway	SN	South Northumberland
AS	Aberdeen south	FT	Flint	NH	North Hampshire	SP	Shropshire
AY	Ayr	GE	Gloucestershire east	NK	North Kerry	SR	Surrey
B	Bute, Arran and Clyde Isles	GM	Glamorganshire	NM	Nottinghamshire	SS	South Somerset (England)
		GW	Gloucestershire west	NN	North Northumberland	SS	South (or east) Sutherlandshire
BD	Bedfordshire			NO	Northamptonshire		
BF	Banff	HB	Hebrides	NS	North Somerset (England)	ST	Stafford (England)
BK	Berkshire	HD	Haddington			ST	South Tipperary (Ireland)
BR	Brecon	HF	Herefordshire	NS	North (or west) Sutherlandshire		
BW	Berwick	HT	Hertfordshire	NT	North Tipperary	SW	South Wiltshire
BX	Buckinghamshire	HU	Huntingdonshire	NW	North Wiltshire	SY	South-east Yorkshire
				NY	North-west Yorkshire		
CA	Caithness	I	Islay etc (Ebudes south)			TY	Tyrone
CB	Cambridgeshire			OI	Orkney Islands		
CD	Cardiganshire	IM	Isle of Man	OX	Oxfordshire	WA	Waterford
CH	Cheshire	IW	Isle of Wight			WC	West Cornwall (England)
CL	Clare			PB	Pembrokeshire		
CM	Carmarthenshire	KB	Kirkcudbright	PC	Perthshire south and Clackmannanshire	WC	West Cork (Ireland)
CR	Caernarvonshire	KC	King's County			WD	West Donegal
CT	Kintyre	KD	Kildare	PE	Peebles	WG	West Galway
CU	Cumberland	KF	Kinross and Fife	PM	Mid Perthshire	WH	Westmeath
CV	Cavan	KI	Kincardineshire	PN	Perthshire north	WI	West Inverness (Scotland)
CW	Carlow	KK	Kilkenny				
				QC	Queen's County	WI	Wicklow (Ireland)
DB	Denbighshire	L	London			WK	West Kent
DF	Dumfries	L	Lundy Island	RA	Radnorshire	WL	Westmorland and Lancashire north
DM	Durham	LA	Lanarkshire	RE	Ross-shire east		
DN	Dumbarton	LD	Londonderry	RF	Renfrewshire	WM	West Mayo
DO	Down	LE	Leitrim	RO	Rosscommon	WN	West Norfolk
DT	Dorset	LF	Longford	RW	Ross-shire west	WO	Worcestershire
DU	Dublin	LH	Louth	RX	Roxburghshire	WS	West Suffolk
DY	Derbyshire	LK	Limerick			WT	Wigtown
		LL	Linlithgow	S	Isle of Skye etc (north Ebudes)	WW	Warwickshire
EC	East Cornwall (England)	LN	Lincoln north			WX	West Sussex (England)
EC	East Cork (Ireland)	LR	Leicester and Rutland	SC	Scilly Isles	WX	Wexford (Ireland)
ED	Edinburgh (Scotland)	LS	Lincoln south	SD	South Devonshire	WY	South-west Yorkshire
ED	East Donegal (Ireland)	M	Mull etc (mid Ebudes)	SE	South Essex		
EI	East Inverness	MC	Mid Cork	SG	Stirlingshire (Scotland)		
EL	Elgin	M	Meath				
EK	East Kent	MG	Montgomeryshire	SG	South Galway (Ireland)		
EM	East Mayo	ML	Mid (or west) Lancashire)	SH	South Hampshire		
EN	East Norfolk	MM	Monmouthshire				
		MN	Merionethshire				

Key for Fig. 47.

Curatorial

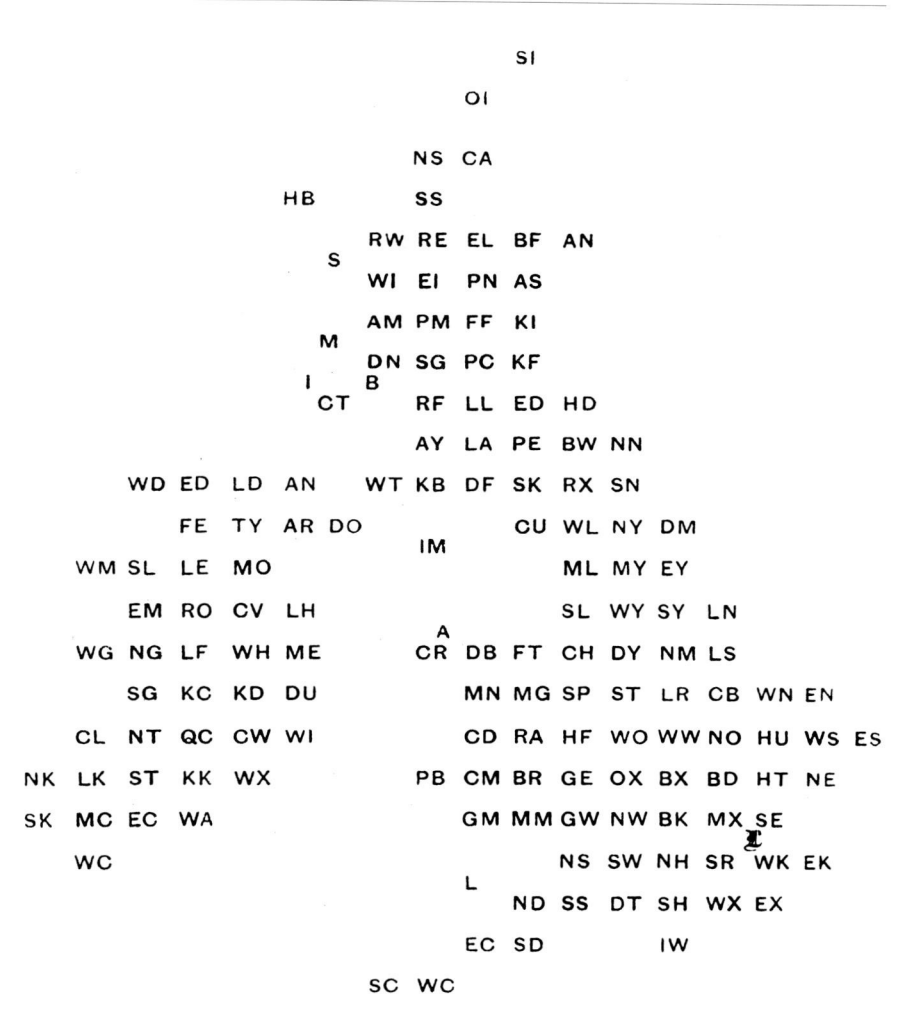

```
                              SI
                              OI
                         NS  CA
                         SS
                    HB
               RW  RE  EL  BF  AN
             S
               WI  EI  PN  AS
               AM  PM  FF  KI
           M
               DN  SG  PC  KF
         I   B
           CT  RF  LL  ED  HD
               AY  LA  PE  BW  NN
   WD  ED  LD  AN  WT  KB  DF  SK  RX  SN
       FE  TY  AR  DO           CU  WL  NY  DM
                         IM
 WM  SL  LE  MO                 ML  MY  EY
   EM  RO  CV  LH               SL  WY  SY  LN
                         A
 WG  NG  LF  WH  ME      CR  DB  FT  CH  DY  NM  LS
     SG  KC  KD  DU          MN  MG  SP  ST  LR  CB  WN  EN
   CL  NT  QC  CW  WI        CD  RA  HF  WO  WW  NO  HU  WS  ES
NK  LK  ST  KK  WX           PB  CM  BR  GE  OX  BX  BD  HT  NE
SK  MC  EC  WA               GM  MM  GW  NW  BK  MX  SE
   WC                             NS  SW  NH  SR  WK  EK
                         L
                             ND  SS  DT  SH  WX  EX
                         EC  SD             IW
                    SC  WC
```

Fig. 47. Typomap of the British Isles.

There is a seminal paper (Balfour-Browne, 1931:183-193) entitled '*A plea for uniformity in the method of recording insect captures*' which promotes the use of the 'Typomap' and another (Balfour-Browne, 1940:60-67) which includes explanation of the 'Vice County' system. Each system gives a abbreviated code to each county in the British Isles. The 'Typomap' uses letters, thus Herefordshire is HF, Isle of Wight is IW, Wexford is WX. The 'Vice County' system uses numbers, thus Herefordshire in VC36, Isle of Wight is VC10 and Wexford VC12. However, caution is required when using the 'Vice

County' numbering system as numbers 1 – 40 are duplicated, one set applying to England and Wales as far north as Shropshire (VC40, referred to in this system as Salop) the other 1 – 40 set numerating the counties of a seemingly united Ireland. Thus North Hampshire is 'mainland' Vice County 12, but 'Irish' Vice County 12 is Wexford). The Channel Isles were originally lumped together as VC0, but nowadays VC113, a geographical area not included in the 'Typomap' system.

Foreign data is usually a little more expansive and this is often due to the collector's unfamiliarity with the region, lack of reference points or maps. The labels should, as with UK ones, list the location in narrowing terms, thus country, province, county/prefecture, village, forest, marsh, mountain, nature reserve or distance and compass bearing along a road from a village, altitude, longitude and latitude, date, collector's name and habitat reference. A shorthand being adopted currently is to use universally understood abbreviations of country's names, for example by its capitalised e-mail address termination, thus CN = China, UA = Ukraine, TR = Turkey and so on.

Determination labels

The name of the beetle, who identified it and when the identification was made should be regarded as essential requirements, NB the beetle name includes the name of the author. Many people get by without determination labels thinking the very fact that the beetle mount is standing over its name in the cabinet will suffice. Generally people adopting this attitude do not use their collections much. A collection is a reference, specimens will be removed and replaced regularly and almost certainly a number will be required for examination when identifying fresh material, especially of 'more difficult' genera. Replacing each specimen accurately is easily achieved if each bears a determination label. Further, if a specimen is accidentally incorrectly replaced, the error – especially if the sole representative of that species in your collection – is easily spotted and rectified.

There are four parts to each name – genus, species, author and date. The first three should be recorded on the determination label. Confusion can arise in taxonomically unstable genera if the author's name or genus is omitted. A sex sign can be added to the label if required. The name of the identifier plus date (month plus year or simply year) the identification was made should also be recorded. Addition of the name is indication that the identifier is confident in correctly identifying the species. Inclusion of the date will show, for example, that the specimen was identified after or before a revision of the genus or species group was published.

Storage systems. The entomological cabinet is a storage system traditionally used in Britain generally with a number of store boxes as an adjunct for expansion and housing ephemeral material, unidentified specimens and duplicates. Many people use only store boxes. On mainland Europe a cabinet is a very rare item as are our preferred large double-sided heavy wooden store boxes. Generally collections are maintained in single sided 'cartons' with hinged or lift off lids. These come in a range of sizes but are of uniform depth. Modern 'cartons' are supplied with Plastazote or similar synthetic lining.

Cabinets: (See plates 118 and 119). My life time professional experience working with entomological collections in museums has led me to conclude that nobody makes an insect cabinet like they used to. Fortunately there are often good quality second-hand cabinets to be had, but make sure you are satisfied with the quality of the cabinet before parting with your money. Points to check are:

- All the drawers are sound, no gaps or splits in the woodwork.
- All drawer lids fit perfectly and (ideally) can be interchanged between drawers (this is rare and a sign of real quality). The wooden frame of the lid should sit tight on the top of the drawer all the way round, there should be no gaps and the opposite corners, when pressed, should not rock or lift.
- The drawers slide in and out of the carcase without interruption or jolt.
- The drawers have no free side to side play when in the carcase.
- The drawers are fully interchangeable with each other (ie any drawer can fit into any slot in the carcase). This is a feature seldom used when managing your collection, but it is very useful and indicates excellent quality, fit for the purpose.
- The carcase should be sound, free of gaps and splits. The doors, if present, should close and open freely.
- Remove all the drawers and examine the empty carcase inside and out and examine each drawer minutely inside and out.
- There should be no active woodworm, but if old woodworm is noted check that it is superficial and will not effect the structure of the cabinet.

Top quality manufacturers include Gurney, Hill, and Grange & Griffiths (a Middlesex based firm who purchased the Hill patents). Those made by Watkins & Doncaster are generally good quality and thoroughly serviceable, especially their older models.

It was traditional to line cabinet drawers with cork and paper but in recent years Plastazote has almost universally taken over as a drawer liner. Some, notably Hill cabinets, have removable slats running top to bottom in the drawers. These are handy as individual slats can be removed, this feature plus the universal substitution of Hill drawers anywhere in any Hill carcase of any age make for simplified expansion of the collection when the time comes (in addition the drawer lids and drawer trays of authentic Hill units are interchangeable, regardless of age). Beware of imitation Hill units, not made to the original Hill patent. Original Hill units are no longer made, but are available secondhand.

More recently people are using 'unit trays' in cabinets (see plate 118). These are small card trays lined with Plastazote and of dimensions such that the drawer is fully but not tightly packed with trays. One or two species can be housed in each tray (obviously depending upon size of the tray and insect). If a specimen is required for study under the microscope, the tray is removed, thus reducing the handling of the actual specimen. Pieces falling off specimens stay in the unit tray, so they are not lost, and it is clear which species they belong to. Again expansion of the collection is greatly simplified with the unit tray system.

If ordering your own stock of unit trays, it is very sensible indeed to send the dimensions of the cabinet drawer to the manufacturer, explain what size or sizes or trays you require and let them work out the actual size of trays required. I tried to do this myself and came unstuck. My intention was to have two sizes of the same width but with one half the length of the other but so that both sizes could be used in combination, filling the drawer with minimal gaps. The reality is that I can only use one size tray in each row, fortunately I got the row width right. Get it wrong yourself and you still have to pay for what you ordered. Always request unit trays to be made from acid free or neutral pH card and paper.

Plastazote is cut to size with a sharp craft knife of scalpel and glued into the bottom of each tray. If using a large number of unit trays the Plastazote can be purchased in bulk direct from manufacturers in sheets usually 1m x 1.5m and to any thickness required. Again speak with the suppliers representative as there are various ways to cut to thickness, some more economical than others. For general use 10mm thickness is adequate. The LD45 density Plastazote is almost universally used (other densities range from marshmallow-like to rock hard). If smaller quantities are required, entomological dealers will be able to supply.

Store boxes and cartons: The great advantage of store boxes or cartons is that they are much cheaper than a cabinet and you can start with one or two and expand as your collection increases. Store boxes (plate 115) have a disadvantage over cartons in that, being double sided, they take up a lot of desk space when working with your collection and they are much heavier than a carton. Store boxes are double sided, cartons single sided. Store boxes are stored standing up-right, cartons flat. There is another advantage with cartons, the one sided structure and storing them flat means the insects contained with are also stored flat, as if in a cabinet (unlike being stood on end as in a store box).

Cartons (plate116) are available in sizes 15cm x 18cm, 18cm x 23cm, 23cm x 30cm and 30cm x 40cm with standard depth of 5cm they are available with hinged or removable lid. Store boxes come in two depths, one to take pins up to 30mm long, the other 'continental depth' as they are often known accommodate pins up to 38mm. They are available in sizes 20cm x 25cm, 25cm x 36cm, 28cm x 40cm and 30cm x 44cm.

Cartons are much cheaper than store boxes (currently (2005) 23cm x 30cm carton £5.50p (useable area of 690 sq cm), 25cm x 20cm 'continental depth' store box £30.35p (useable area 1000sq cm). Both cartons and storeboxes these days are lined with Plastazote or similar material.

Marking rows: The drawer, carton or store box can easily be divided into the required number of rows. Consider first the species to go into the drawer (or carton, box) – obviously a row of six aleocharine Staphylinidae will take less width than a stag beetle. Rows can be marked by black thread tied and looped around short pins at the head of each row. This has the advantage of not marking the Plastazote, but with time the thread will invariably perish. A thin fibre tip pen or soft pencil are preferable to thread, pale or light lines look neater than bold dark ones. To ensure a straight line, cut a piece or cardboard or thin plywood to just fit inside the drawer (carton/box) make sure one edge is perfectly straight!

Pest control: Insect collections are readily attacked by a variety of insect pests, most notably the dermestid beetle *Anthrenus* (especially *A. verbasci* (L.) 'the varied carpet beetle' and *A. sarnicus* Mroczkowski 'the Guernsey carpet beetle') and psocids but 'clothes moth' and a variety of other dermestid beetles would also eat your prized specimens given half a chance. The speed with which *Anthrenus* can destroy a box or drawer full of dried insects is staggering; vigilance and precautionary measures must be rigorously enforced. See plates 114 and 114a.

Vigilance includes making at least a quarterly visual examination of each box or drawer containing one's insects, especially postal boxes and older store boxes as these invariably afford easy access to pests. Also, postal boxes, by their very nature, travel around and can arrive at your house with a fresh supply of live pests. 'House keeping' should include keeping one's 'study' clean and vacuuming the floor weekly to remove any insect debris that missed the waste bin or otherwise found its way off the desk. The signs to look for are:

- Dust like frass at the bottom of a store box or on and under a card mount in a drawer or flat-stored box/carton.
- Adults, larvae and larval skins (excellent illustrations in Peacock, 1993).

Precautions basically fall into three main categories:
- Chemical deterrents
- Protection
- Freezing

Traditional chemicals include paradichlorobenzene and naphthalene, both of which are banned under Health & Safety legislation in museums and other non-domestic situations. Vapona (dichlorvos) is now likewise banned. For about 15 years I have been using essential oils of camphor and citronella with success, both are natural insecticides, cedar and sandalwood have similar effect, and all four have pleasant odour and are to all intents and purposes 'safe' to use. Pin mounted glass fumigant cups are used to house the oil, one or two cups per box or drawer, keep the cups topped up – check every six months. With passage of time the oil loses its insecticide property and smell and reduces in volume to a thick viscous 'goo'. Every two or three years this 'goo' needs to be removed from the cup and fresh oil added. A little alcohol or other organic solvent and a cotton bud is useful in cleaning the cups and the task will be made easier by selecting cups without a deep internal 'anti-spill' lip.

A little oil spread around the rim of a store box is a good additional precaution.

Protection for boxes and cartons is readily afforded by putting each into a seal-seal polythene bag of suitable size. Check occasionally that no holes have developed through use. Protection for a cabinet in this way is not practically possible, instead, check the seal around the door(s) and replace if necessary.

Freezing is nowadays the standard museum method of pest control in dry insect collections. Medical freezers are generally used as these are more robust, maintain a much lower temperature more consistently and lack the food shelves/trays of domestic freezers. That said, the domestic freezer will do the job for a private collection. Whereas museums might cyclically run their entire collection through the freezer over a set time period, for 'home use' freezing is probably best done if pest attack is suspected or noticed.

Seal the box or drawer in polythene (large zip lock bags are helpful) and put into the freezer for two or three weeks. Remove and leave, still sealed in polythene, for at least 24 hours to attain room temperature. Freezing without polythene, or removal of polythene too early, will result in condensation forming on the mounted beetles and inside the box/drawer, which may lead to the specimens becoming mouldy or rotting. Remember also that frozen glass is far, far more brittle that glass at room temperature and the glass used in cabinet drawers is of a very thin grade.

For fully comprehensive up to date guidance and advice see Pinniger, 2001.

Spirit collections: very few people maintain spirit collections, most preferring carded or pinned specimens for study and comparison when identifying new material. Spirit collections are usually housed in 70 or 80% alcohol, and are of particular value to those with an interest in immature stages, which can only reliably be stored in spirit. Such collections require at least as much maintenance as dry collections, for although the problem of pests is removed, the potency and level of the spirit in the tubes must be carefully maintained, and evaporation can be a real problem especially in private houses where the temperature is usually high. The usual practice in museums is to store specimens in glass tubes full of alcohol, stoppered with cellulose wadding, and stored upside down in large sealed kilner-style jars, also full of alcohol. The larger jars allow some organisation of the tubes (ie one family or superfamily per jar), and also minimise the amount of topping up, or 'respiriting', that needs to be done. Jars are best stored in plastic trays in metal cabinets, protected from light and fluctuations in temperature. Labels should always be placed with the specimens inside the tubes, and need to be written, of course, in alcohol-resistant, indelible ink. Each tube can store all specimens of a given species from a given site on a given day (adults and larvae can even be stored together), or undetermined material from a given site on a given day. This system has the advantage that long series of a species can be stored with a minimum of effort (compared to mounting and labelling each one), and the material can be removed and prepared for a dry collection at any stage. On the other hand, it creates a logistic nightmare unless one has an excellent filing and numbering system worked out, preferably linked to a database, because the cylindrical tubes and jars cannot be stored in a fixed order. Furthermore, spirit collections are bulky, heavy and cannot be blended into the home furniture as seamlessly as a nice old fashioned insect cabinet. There may also be issues of fumes, and possibly a fire hazard. On the other hand, there is no better way to store larval beetles, so those with an interest in larvae need to have a small spirit collection, possibly as an adjunct to their dry collection of adult beetles. Spirit, although it will cause specimens to stiffen, is probably the best medium for very long-term storage of unmounted, undetermined material- laurel bottles and acetic acid are not advised in the long term as specimens over-soften and their limbs drop off, and pests may run rampage unnoticed through collections of papered or wadded material, with heartbreaking results.

Organising the collection

It makes sense to organise your collection in taxonomic order rather than adopting an eclectic personal system. The norm is to follow the current published check list, but as we all know, taxonomy is dynamic and changes to the systematic ordering of beetles occur regularly and in recent years more drastic changes have taken place than we have seen for a long time. Like volcanoes, often after a period of major activity there follows a long period of quiescence, with only the occasional burp to upset the illusion of stability.

Having written that, I really ought to point out that my own collection follows Kloet & Hincks 1977 check list, except the Leiodidae and *Ptinus fur* are missing. The latter being an oversight when pinning the names into the cabinet drawer, the former because I specialise in the Leiodidae and maintain this family as a separate world-wide collection with my British captures incorporated.

When starting a collection, the space required to house the accumulated specimens will not be great. They might all cram into one carton or store box. However, always avoid cramming; divide the sides of the store box or the carton into rows of equal width, the rows can be marked with soft pencil or with black thread tied to pins. This gives order to the collection and makes reference to it more efficient. In the early days of a collection, species can be grouped by family or genus. As the collection expands a box might be devoted to just one family. More boxes or cartons are added as space demands. You may or may not end up purchasing or otherwise acquiring a cabinet; many of our leading authorities housed their collections in store boxes and in many European countries cabinets are virtually unknown.

If you do set out your now large collection according to a revised check list or in your newly acquired cabinet, do remember to leave space every so often for additions to the British fauna. Plate 119 illustrates this point with specimens of *Ischnomera cyanea* in the bottom right corner – the other species of *Ishnomera* are in taxonomic sequence in this drawer, but *Ischnomera cyanea* was recognised as a British insect in 1990. Rather than re-lay the entire cabinet drawer, the additional species has been placed in space left for new additions to our fauna in he bottom of the drawer; always leave a space in each drawer or box for additions, 'laying-out' is a time consuming and very protracted business which most entomologists only ever do once. Additions occur in most families, this is a lesson I learned the hard way when I set out my 32-drawer cabinet and left space in each Staphylinidae drawer. I think the first addition to our fauna after I had set out the cabinet was *Pterostichus rhaeticus,* a relatively large beetle that would have fitted into the appropriate Carabidae drawer if I had only left some spare room.

How much room to leave for each species must be a matter of personal choice; do you really need more than one specimen of each sex of *Lucanus cervus,* or are you really a secret Lepidopterist by wanting all those colour variations of *Rutpela maculata*? Cabinets are to some extent a constraint on space. You have just acquired four Hill Units, you have 40 drawers to accommodate the British Coleoptera. It is odds-on that if you start in the top left corner of drawer number 1 of the first cabinet setting out labels all the way to the bottom right corner of drawer number 10 of cabinet four, you will be somewhere in the weevils with a lot of extra genera left over. Setting out needs planning and the occasional 'dummy run' during which you configure a few drawers knowing the number and sizes of the beetles but without actually pinning anything into the drawers. This exercise extends until you are satisfied 4000 species of various sizes will fit into your 40 drawers with a little room spare for additions to the British fauna. Alternatively, if when doing the 'dummy run' you realise the cabinet is too small, as happened with my own 32-drawer cabinet less Leiodidae, find a suitable break-off point and accommodate the remainder in store boxes. Thus my own cabinet ends with Kloet & Hincks' Hylobiinae, the Rhyparosominae and subsequent tribes filling three store boxes.

Once set out a cabinet is rarely if ever re-jigged to accommodate a revised classification. To do this is a logistic nightmare involving mass removals, which have to be safely accommodated elsewhere before being returned to the revised layout; most people like to spend their time collecting or studying beetle not filing them to a revised classification.

Degreasing

Mounted beetles frequently tend to produce lipids (body fat, grease) that appear on the body surface and if mounted on cards, soak into the mounting card. The grease attracts dust and colours the mounting cards. Removing the grease from the mounted insect cleans the body surface and makes taxonomic work easier, decolorizes the mounting cards and makes the specimen less attractive to *Anthrenus* attack.

The most effective way of degreasing the specimen employs a continuous extraction of the specimen with hot hexane using Soxhlet apparatus. This method however requires laboratory equipment and is therefore frequently difficult to perform.

As an easy to perform alternative to the continuous extraction with hot hexane a simple dip of the mounted specimen into a series of containers with organic solvents at room temperature can be used.

The solvent of choice for the room temperature is tetrachloroethylene $Cl_2C=CCl_2$ (also known as perchloroethylene). Unfortunately, the less toxic hexane does not produce satisfactory results when used at room temperature. Tetrachloroethylene is a fairly universal solvent that dissolves compounds of a range of polarities. It dissolves lipids very well too. Unfortunately, to some extent, this solvent (and all other chlorinated organic solvents) degrades the soft tissues holding insect sclerites together. Therefore, a prolonged exposure to this solvent (several hours) might result in the disintegration of the insect.

While working with tetrachloroethylene, general precautions (especially ensuring adequate ventilation) should be observed, this applies generally for organic solvents.

A row of about four jars or beakers filled with an organic solvent will be required. The specimen to be cleaned is pinned into a small piece of Plastazote which is then inverted thus submerging the mounted beetle in the first jar where it is left for a couple of minutes, then taken out and the solvent is allowed to drip back into this container. Once most of the solvent has evaporated the specimen is dipped into the fresh solvent in the second container. This step repeated until the fourth dish is reached. Any further specimens are always first dipped into the first dish (which becomes the most contaminated with fatty substances whereas the last is the least contaminated). This room temperature method is less effective than the continuous extraction with hot hexane and consumes much more solvent (the solvent has to be replaced frequently). It nevertheless gives good results and does not damage the specimen nor the mounting card or entomological pin. If you plan cleaning large quantities of specimens, however, disposing of the solvent might become an issue. The continuous method produces minimum waste and the solvent is recycled.

After washing your specimens in tetrachloroethylene if you observe that the body pubescence has become depressed or flattened onto the surface of the beetle dip it in hexane for a short while and allow it to evaporate. After this treatment the pubescence should assume the natural position.

References

Balfour-Browne, F., 1931. A plea for uniformity in the method of recording insect captures. *Entomologist's Monthly Magazine*, **67**:183-193.

Balfour-Browne, F., 1940. *British Water Beetles*, **1**. The Ray Society, London

Peacock, E.R., 1993. Adults and larvae of hide, larder and carpet beetles and their relatives (Coleoptera: Dermestidae) and of Derodontid beetles (Coleoptera: Derodontidae). *Handbk. Ident. Br. Insects.*, **5**(3):144pages. Royal Entomological Society, London.

Pinniger, D., 2001. *Pest management in museums, archives and historic houses*. Archetype Publications, London. ISBN 1-873132-86-7.

NOMENCLATURE

by J. Cooter

Since the publication of the last edition of this *Handbook* in 1991, there has been an attempt to give a large number of taxa 'English' names. As might be expected, these made-up names have had a mixed reception, and opinion quickly polarised to those in favour and those 'against'. I put 'against' in quotation marks because it is more a case of recognising that there is no purpose in creating another set of names for named organisms, so why bother trying to remember which species they refer to. Such systems were introduced, many years ago, in Germany and what was then Czechoslovakia, and in both countries the system received no interest and fell by the wayside. In Britain various attempts likewise failed; I recall in the 1970's being shown a copy of Sandar's *'Insect Book for the Pocket'* (Sandars, 1946) in which beetles were given new 'English' names, the one proposed for *Oedemera nobilis* (Scopoli) being 'the swollen-thighed swollen-thigh.' Need I say more?

The novice, if using the new made up names, will soon realise that there are no reference books using those names, and no way of cross-referencing. Worse still, if seeking advice from a more experienced coleopterist, it will be doubtful if that person has any idea which species the novice is referring to. For example 'the dark willow guest' (= *Melanapion minimum* (Herbst)), 'the Pashford pot beetle' (= *Cryptocephalus exiguus* Schneider).

Amongst the concocted names are a variety of extras, possibly made up over a glass of beer after a good day in the field. They are indistinguishable from the 'official' names and include 'the aquatic pile beetle' (= *Copelatus haemorrhoidalis* (F.), 'the artist' (= *Gyrinus urinator* Illiger) and the very aptly named 'hog-weed bonking beetle' (= *Rhagonycha fulva* (Scopoli), if explanation were needed, see plate 55.

It is never easy to gain acquaintance with a new jargon, it takes time, experience and the desire to learn a new subject. This applies to any subject freshly taken up for study. The extreme is learning a foreign language; nobody has suggested the student to use a set of made up words instead of the actual language being studied; a ridiculous idea of course. Scientific names are internationally understood, made up names are not widely understood even in their country of origin.

Seeing the family Histeridae referred to as 'carrion beetles' will compound the confusion of the novice. This common term has traditionally been applied to the Cholevinae and Silphidae. Few, if any Histeridae are carrion feeders, one British species, *Saprinus virescens* (Paykull), is a predator of chrysomelid larvae, others inhabit ant nests, *Paromalus, Abraeus* and *Plegaderus* are associated with rotten wood and so it goes on.

It is beyond the scope of this *Handbook* to describe the rules of the International Code of Zoological Nomenclature (ICZN, 1999); which are in places very complex. However, the rather simplistic notes given below will, it is hoped, help to explain some of the various conventions the student is likely to meet in publications, particularly Check Lists.

Typically a Scientific name consists of four parts which from left to right are generic name, species name, author and date (for example *Agathidium convexum* Sharp, 1866).

Example 1
Dytiscus marginalis Linnaeus, 1758
This shows *Dytiscus marginalis* was described and named by Linnaeus in 1758 and has remained in that genus ever since.

Example 2
Hydaticus transversalis (Pontoppidan, 1763)
Curved brackets around the author and date show that the species has been transferred from the genus in which it was originally placed by its author. In this case Pontoppidan placed *transversalis* in genus *Dytiscus* in 1763. Later (Leach in 1817 to be precise) erected the genus *Hydaticus* and Pontoppidan's

species was transferred to it from *Dytiscus*. Note, unlike botanical nomenclature there is no indication of who was responsible for the generic transfer or the actual date.

Example 3
Dytiscus semisulcatus Müller, 1776
 = *punctulatus* Fabricius, 1777

This is a case of synonymy, Müller and Fabricius both published descriptions of the same species, both placing it in genus *Dytiscus,* but each giving the same beetle (species) a different specific name. As Müller in 1776 was first to publish, Müller's name is used. This is an illustration of the Law of Priority. Alas, exceptions to this Law are numerous and various.

Example 4
A beetle might have been described and given a name already in use in that genus:
Aphodius fasciatus (Olivier, 1789)
 = *uliginosus* (Hardy, 1847)
 = *foetidus* (Fabricius, 1792) not (Herbst, 1783)
 = *putridus* (Herbst, 1789) not (Geoffroy in Fourcroy, 1785)
 = *tenellus* sensu auct. not Say, 1823

Here we can use the information already gained from the examples above. It is clear that Olivier described the species *fasciatus* in 1789. From reference to the published check list, we will see the genus *Aphodius* was created by Illiger in 1798 so Olivier must have placed his species in a genus from which it was subsequently transferred (in fact the genus *Scarabaeus*), hence the use of round brackets. Also clear to us from the examples above is that Hardy in 1847 described the same species Olivier had named *fasciatus* but called it *uliginosus*. However, because Olivier's name was earlier, the Law of Priority declares that *fasciatus* (Olivier, 1789) is the current name, and that *uliginosus* (Hardy, 1847) is a junior synonym.

The next line introduces a new, more complex, feature. Fabricius described the species *foetidus* (in another genus) in 1792, and in 1783 Herbst also described a different species but which he also called *foetidus*. (This happened again with Herbst's *putridus* of 1789 and Geoffroy's *putridus* of 1785; again these two *putridus* are in fact different species.) This illustrates the importance of always adding the author's name when referring to beetles (and indeed all insects). So, Fabricius and Herbst each described a different member of the same genus with the name '*foetidus*'; of course, there cannot be two different beetle species both called '*Aphodius foetidus*', so *foetidus* Fabricius, as it was named later (by nine years) than *foetidus* Herbst, becomes a JUNIOR HOMONYM, and is disregarded. It so happens that *foetidus* (Fabricius) is not just a junior homonym of *foetidus* (Herbst), it is also a junior synonym (i.e. the same species but described later) of *fasciatus* (Olivier)! So, when he described *foetidus* in 1792, Fabricius was describing a species that was already described, using a name that had already been used! Such an error would seem ridiculous today, but in the heady rush to describe the European fauna, in times of slow printing and poor communications, such errors were very common, and much of the work of taxonomists involves untangling them.

The last line means that the species *Aphodius fasciatus* (Olivier) was once confused, by various authors, with a different species called *Aphodius tenellus* described by Say in 1823, see example 6 for an explanation of the terms auct. and sensu auct.

Example 5

Philonthus succicola Thomson, 1860
= *proximus* Fowler, 1888 *non proximus* Kraatz, 1859

The interpretation is exactly the same as in example 4, the beetle we now know as *P. succicola* is synonymous with *proximus* Fowler, but not with *proximus* Kraatz. Therefore *proximus* Fowler is both a synonym and a homonym.

Example 6

Staphylinus melanarius Heer, 1839
= *globulifer* sensu auct. *non* Fourcroy, 1785

The addition of 'sensu auct' (also written auct.) means 'in the opinion of authors' (sensu auctorum) and shows Fourcroy's description of *globulifer* published in 1785 was misinterpreted by other coleopterists, in fact they supposed Heer's *melanarius* to be Fourcroy's species. ('auct. Brit' means 'of British authors' and shows the beetle known to British coleopterists is not the same as that referred to by workers in other countries.)

Example 7

Carabus monilis Fabricius, 1792
v. *gracilis* Kuster, 1846
= *consitus* auct. *non* Panzer, 1792

The species *monilis* and, because it is set flush with the name above, the variety *gracilis* both occur in Britain. However, the variety was by various authors (see example 6) once called *consitus* in mistake for Panzer's species.

Example 8

Carabus [*problematicus*] Herbst, 1796

ssp. *gallicus* Gehin, 1885

   = *catenulatus* Fabricius, 1801 *non* Scopoli, 1763

Square brackets are used to signify the nominal sub-species *problematicus* does not occur in Britain, but is represented by sub-species *gallicus* Gehin. The beetle used to be known as *catenulatus* of Fabricius not the true *catenulatus* of Scopoli.

There is no hard-and-fast rule for the uses of 'not', 'non' and 'nec' to mean 'not' when distinguishing homonyms and misidentifications; one checklist might write:

   = *catenulatus* Fabricius, 1801 *non* Scopoli, 1763

while another may prefer:

   = *catenulatus* Fabricius, 1801 *nec* Scopoli, 1763

and a third:

   = *catenulatus* Fabricius, 1801 not Scopoli, 1763

though in the last example the 'not', being an English word, should not be put in italics, while many people italicise the Latin 'non' and 'nec'. Purists may quibble, especially if the Latin 'sensu auct.' is combined in the same line with the English 'not', but for our purposes the three terms (not, non and nec) should be regarded as interchangeable.

References

ICZN, 1999. *International Code of Zoological Nomenclature*. 4th edn. 306pp. International Trust for Zoological Nonmenclature, London.

Sandars, E., 1946. *An insect book for the pocket*. Oxford University Press, London. 349pp + 36 col. pl.

IDENTIFICATION

by J. Cooter

It is an assumed objective that the student will want to know the names of the beetles collected. Many of the initial captures will be familiar from common knowledge (ladybirds, cock-chafer for example) and others from illustrations in text books (for example the majority of our Cerambycidae and most ladybirds). Others will look familiar, but seem somehow subtly different from their congeners. With increasing experience an increasing number of beetles will be identified in the field or at least placed in their correct genus or thereabouts. The smaller species, especially those in large genera such as *Cryptophagus, Longitarsus* and '*Apion*', will generally need careful examination under magnification, and very often will need dissection too.

By far the best method of identifying beetles is to use a reliably identified collection in conjunction with up-to-date identification keys and manuals. Identifying beetles in batches of the same genus is sound practice. The work will be quicker and the results more accurate; the key can be better understood, and more importantly a better understanding of the characters of the beetles and their differences from each other will be gained. Looking through a well laid out collection will increase one's knowledge and show clearly the subtle differences so difficult to express in a key or even in longer text. Thus after a short time, your own collection itself becomes a very useful aid for comparative work.

It is always worthwhile enquiring at the local museum to see if a collection is available for consultation; very often these will not be on public display but can be seen by appointment. The Natural History Museum in London keeps a separate 'British Collection' and makes a point of welcoming visitors wishing to identify beetles they have collected. In some museums, some of the identifications may be suspect, especially so in older collections and museums that do not have an entomologist on their staff. Indeed it seems a trend in local authority museums at the present time to employ 'generalists' good at putting on family events and addressing such issues as access for disabled and social inclusion; all very laudable in themselves, but mostly done at the expense of subject expertise and knowledge. Very often it is to the 'private sector' that one most profitably looks for help – local coleopterists if approached will generally be found to be very helpful and full of encouragement. The Advisory Panel of our own Society was established as a means to assist members with their identification and other entomological problems. Whenever sending material away for expert opinion it is good manners to INCLUDE SUFFICIENT STAMPS TO COVER RETURN POSTAGE, and good policy to write beforehand to enquire if it is convenient to send such material. Also, bear in mind that specialists may take some time to complete the identification of your material, and that sending overly frequent reminders is a good way of ensuring refusal of future requests. Do not be surprised if the odd specimen is retained by the person identifying your material. You might at the time feel a little aggrieved, but in reality this should not be the case; the process is one of mutual benefit, you have got all your material reliably determined, the identifier has a beetle which he needs. Identifications, as every coleopterist knows, are often onerous, and if dissection needs to be done, can be very time consuming. Most members of the Advisory Panel are not professional entomologists and so give freely of their limited spare time, (indeed many professionals through necessity carry out such identification work during their own time). The gift of the odd specimen is a small price to pay, and more often than not, the identifier will make good your loss with a few duplicates of his own.

Most coleopterists will, with time, amass a very useful library; a selection of some of the most useful books and journals is listed below and most of the 'beetle family' accounts (pages 1-199) include a useful bibliography. Although most of the systematic papers published in the journals of the Royal Entomological Society tend to deal with foreign faunas, the Society possesses an unrivalled entomological library for the free use of Fellows. Details of Fellowship can be obtained from The Registrar, Royal Entomological Society, 41 Queen's Gate, South Kensington, London SW7 5HR.

In order to facilitate identification some form of magnifying device is required. Many coleopterists make do, and indeed do good work with a simple x10 and/or x20 hand lens. Points in favour of a lens are low cost and being small can be carried on one's person. Points against are that being small it is easily lost, use puts a strain on one eye and there is a greater risk of damaging specimens. Further, with very small beetles, or those where dissection is required, even x20 magnification is insufficient.

The ideal is a stereoscopic wide-field binocular microscope with a range of magnifications of say, x10 to x80 or x100 (occasionally limited instances will require an even higher magnification), for example see plate 113. A rotating objective is preferable to the type that has to be slid in and out in order to change magnification. Zoom devices are also available and give a constant range of magnification between set limits. Although these instruments are not cheap, they are often no dearer than a good quality hi-fi separates system, family holiday abroad or a PC/laptop. A well made model will last several life times, will retain its value and can be readily sold at a later date or traded in for a higher specification model. The better models are sold by specialist dealers, either at their showrooms or via a regional representative who will make house calls. Microscope sales people really do know their business so explain fully your likely requirements and try out a range of models offered.

I always recommend people to contact the sales managers of the major manufacturers and flag up their needs and price range. Very often the company itself has microscopes to dispose of at greatly reduced prices. These might be last year's demonstration model, or one a customer returned as unsuitable; basically if you have the chance to buy a virtually new microscope from the manufacturer at a huge reduction you don't ask questions. Many microscopes these days are used for quality checking in the micro-electronic and computing industries. Often a company will want to replace all its microscopes in one go and will turn to a manufacturer for a 'good deal'. This results in several good quality microscopes being acquired by the manufacturer who will in turn sell these on; by flagging up your interest you stand a good chance of obtaining one before they are sold on to a microscope dealer and the price is increased. Remember, microscopes used in micro-electronic industry quality control have hardly ever had their focus mechanism used and checking is often carried out at fixed magnification, so not much wear there either. Bargains can be had by asking around and waiting, I know, this is how I obtained my own Leica MZ12.

It is also worth flagging up your requirement with a dealer that you know acquires industry used microscopes. He will give them a clean, a thorough check and a mark-up, but such microscopes are still very good value.

If pursuing purchase by other methods it is a good idea to take to the showroom a box of mounted beetles covering a range of sizes – for example *Amara, Bembidion* and *Apion* plus a very flat (say *Pediacus*) and a very convex species (*Cercyon* perhaps). Run these through an identification key to see if the described characters are easily seen under magnification. Some dealers will permit home trial upon receipt of a suitable deposit. Most microscopes are supplied with a pair of x10 eye-pieces and acquisition of a pair of x20 will greatly enhance the versatility of the instrument. Eyepieces should always be purchased in pairs. Remember, higher power eye-pieces only magnify the image captured by the objective. A higher magnification objective is preferable; relative costs are not too different.

Other useful accessories include an eye-piece micrometer for taking absolute measurements and for comparing dimensions and ratio. Rubber eye-cups are useful for those not wearing spectacles. These and other accessories can be purchased with the microscope or at a later date. A good heavy base of a decent size which permits all likely activities by the user – identification, dissecting, sample sorting etc – is a very handy feature.

With a microscope, a light source is needed. Some have a built-in light, others have none. Invariably a separate light is more useful and can be moved freely without having to move the subject and re-focus. Direct unfiltered light is too intense, causes too much reflection from the subject and obscures detail. Placing a filter between the light and beetles will help show up surface sculpture and other features. Moving the light source closer to or further away from the subject is also very beneficial, revealing subtle characters. Moving the light 'around the clock' so that the light shines from different directions will pick out characters such as pubescence. With practice the use of a filtered light source will be better understood; it is one of the secrets of determination.

Good light is a major key to identification. After 25+ years working with proper 'microscope lights' including swan-necked fibre optic types I found myself in the Natural History Museum, Geneva checking the identifications of all the Leiodinae from all Swiss museum and private collections. I was supplied with an adequate microscope, but my light source looked a bit odd. It consisted of a chemists retort stand holding a baked bean tin cut in half to act as a reflector for a low energy fluorescent tube which was wired in through the base of the tin. My first thought was along the lines 'if this is what the Swiss are reduced to, what hope for the rest of us?' However, when I turned the light on and looked at a beetle under the microscope the scales fell from my eyes, I could actually see circular punctures with their edges clearly visible, no high points of light obscuring features; it was truly amazing. Upon return home I purchased a desk light with single long arm and 20 watt low energy fluorescent tube. I took this to bits, cut the arm to about 180cm, re-assembled it, and have enjoyed seeing surface ornament in fine detail without any high-lights, shimmer or halo effects ever since (see plate 113). I do possess a top quality twin swan neck fibre optic microscope light but have never had need to use it.

Coleoptera keys are somewhat unusual in typically offering comparative choices in each couplet. Other Orders, such as the Diptera and Hymenoptera more often give a clearcut 'either/or' 'with/without' choice. Subtle differences 'less closely but more intensely' 'more closely but less intensely' are difficult to appreciate in isolation, but if authentic material is available so that direct comparison can be made the differences at once become very clear. Thus the importance of being able to refer to reliably identified material cannot be over-stressed and the value of determining generic batches becomes obvious.

Some identifications will run smoothly from the start to an obviously correct conclusion; in many works there are additional diagnostic characters listed under species headings. At other times an obviously incorrect conclusion will be reached, or indeed a point in the key where none of the choices seems to fit. At other times one will run through the whole key without success (commonly known as 'falling off the end' we all do it, do not worry). The best advice here is 'go back to the start', and this time take it more slowly and make sure *all* the characters in *each* couplet are considered. However, there is little to be gained, other than complete frustration and lack of confidence, in persisting. If continued lack of success results, it is better to put the problem beetle to one side, perhaps with a brief note of the difficulties, and try again another day or with a different, often older, key. Alternatively, put all such problematic beetles in a box ready for your next visit to the museum or a more experienced friend. By 'keeping at it' and not getting too dispirited, these problems WILL be overcome . . . only to be replaced by a different set of problems, but then life is like that.

Literature

As interest in the Coleoptera develops, there will be a need to obtain literature and build up a working library augmented with interesting, but out-of-date texts, back volumes of the entomological journals, foreign works and a variety of reprints.

The standard work on the British beetle fauna is *The Coleoptera of the British Islands* by Canon W.W. Fowler published in five volumes between 1887 and 1891 with a sixth, supplementary volume by Fowler and Donisthorpe appearing in 1913. The larger format illustrated volumes are very scarce and expensive, but the smaller format unillustrated edition, although again somewhat scarce, is generally reasonably priced. Although somewhat out of date, 'Fowler' is still a necessary reference and its continuing usefulness is a glowing testament to the thoroughness of its author.Joy's *'A Practical Handbook of British Beetles'* (two volumes published in 1932, reprinted 1976) is the most widely used identification manual although, with the passage of time, it too is somewhat out of date. Its keys are easy to follow, but great use is made of comparative differences and often these are only appreciated if the student has a series of reliably identified specimens to hand; it contains little information otherwise. The classification adopted by Joy is likely to be very confusing to the beginner. The best advice is simply ignore the higher classification, use the work to identify species.

The Royal Entomological Society's *'Handbooks or the Identification of British Insects'* - now published in conjunction with the Field Studies Council are very good, if somewhat technical and concise. The first volumes to deal with beetles appeared in 1953 and, to date (2004), well under half the British Coleoptera have been covered and all except the most recent parts are out of print. Very often the student is forced to turn to foreign publications. Reitter's *'Fauna Germanica'* (5 volumes, 1908 to 1916) is nowadays perhaps too out-of-date to be of much use to the British coleopterist, but Freude, Harde & Lohse's *'Die Käfer Mitteleuropas'* (16 volumes, 1965-2005) is widely used in Britain. Other very useful 'foreign' works include the *Fauna Entomologica Scandinavica*, *Fauna d'Italia* (the more recent volumes of which have keys in English), *Faune de France*, *Fauna Iberica* (again, some with English keys), the *Keys for the Identification of Polish Insects* and for Staphylinidae, the *Svensk Insektenfauna* series. The Czech publisher Kabourek, has two excellent volumes available – Carabidae of the Czech and Slovak Republics and the companion Elateridae volume. Both have English keys and contain a wealth of line drawings and colour illustrations of entire beetles. Another recent series of books produced in Germany, with alternate pages in German and English, provide very good illustrated keys to 'Carabidae and Cicindelidae', 'Cleridae' and 'Cerambycidae and Vesperidae'.

In recent years, popular beetle families have been the subject of excellent, readable well illustrated and reasonably prices books. These include Richmond Publishing Company's *'Common Ground Beetles'* by T. Forsythe and *'Ladybirds'* by Majerus and Kearns. The Field Studies Council's AIDGAP series *'A Key to the Adults of British Water Beetles'* is also a very useful and user-friendly book. It is always worth perusing bookstalls stocking new entomological titles whenever one is at an insect show, and keeping up with the catalogues of the several specialist book dealers, who advertise in the entomological journals or will e-mail you regular lists of new and secondhand books.

The *Coleopterist, Entomologist's Monthly Magazine, Entomologist's Gazette* and the *Entomologist's Record and Journal of Variation* all regularly carry papers and shorter articles devoted to the Coleoptera. The Balfour-Browne Club's journal '*Latissimus*' is devoted to aquatic beetles and is another essential read. Back numbers of the established journals contain a wealth of revisionary papers, keys, additions to and subtractions from our fauna as well as distributional, biological and ecological information.

Unfortunately for the beginner, although the information is all out there somewhere, it is at times out of date and at times difficult to locate. The ABE Books website is often an excellent source and worth visiting regularly in the hope of finding that long desired volume.

CHROMOSOME DIFFERENCES

by Robert Angus

In normal sexually reproducing organisms each individual has two sets of homologous chromosomes, one received from each parent. This is the diploid chromosome number (2N) Only the sex chromosomes show major differences. For egg and sperm production only one set of chromosomes (the haploid number, N) is permitted, so that at fertilisation the two-set complement is restored. This reduction to one set is achieved by a "reduction-division" (meiosis), which requires that the homologous chromosomes (one from each set) pair up. Any differences between homologous chromosomes within a species must still allow this pairing up.

To understand chromosome differences a brief account of chromosome structure may be helpful. Each chromosome consists of two strands – chromatids – which lie parallel to one another and join at a constriction called the centromere. The regions of the chromosome lying between the centromere and each end are the chromosome arms. During ordinary cell division (mitosis) the chromatids separate and the centromeres move them to opposite daughter nuclei. Centromeres may be positioned in the middle of the chromosome (metacentric chromosomes), or at or near one end (acrocentric or telocentric chromosomes). Intermdeiate centromere positions are called submetacentric or subtelocentric. The centromere position may be expressed as a Centromere Index (CI) – the length of the shorter arms of the chromosome as a percentage of the total length of the chromosome.

The chromosome pairs in a nucleus normally show differences in length. These can be measured from photographs, but because chromosomes condense (shorten) in the course of cell division, comparison of measured lengths from different nuclei is not helpful. The problem is avoided by using Relative Chromosome Length (RCL), the length of each chromosome expressed as a percentage of half the total lengths of all the autosomes (i.e. the chromosomes excluding the sex chromosomes and any B-chromosomes) – the total haploid autosome length. Comparison of chromosome complements is normally done after assembling the chromosomes as karyotypes. A karyotype in this sense is an arrangement with the chromosomes placed together in homologous pairs, with their centromeres lined up, the pairs in order of decreasing RCL from left to right, and with the sex chromosomes at the right, followed by any B-chromosomes. Note that while sex chromosomes and B-chromosomes are excluded from total haploid autosome length, they are assigned individual RCLs.

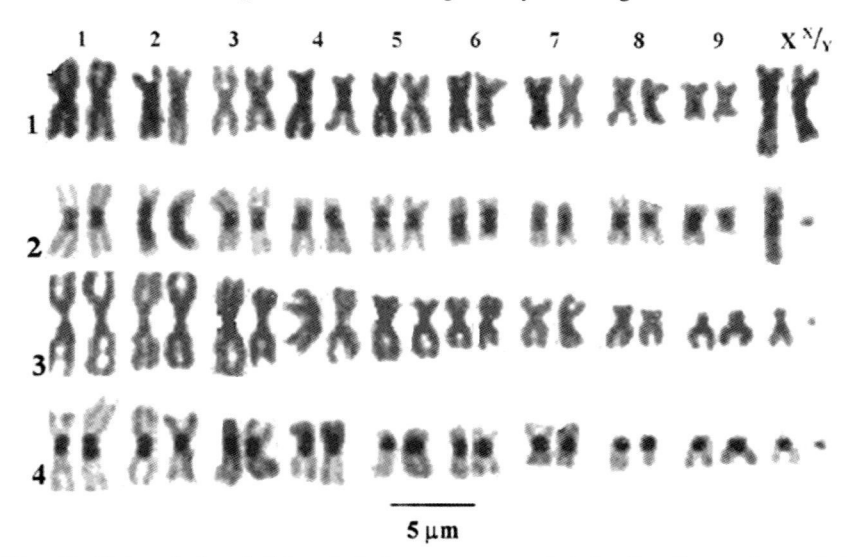

Fig. 48. Unbanded and C-banded chromosomes of *Aphodius fimetarius* and *A. pedellus*. The small acrocentric (more or less V-shaped) chromosomes 8, 9 and X of *A. pedellus* are very striking, as are the long C-bands on chromosomes 2 and X of *A. fimetarius*.

Row 1, *A. fimetarius*, ♂, plain; row 2, *A. fimetarius*, ♂, C-banded; row 3, *A. pedellus*, ♂, plain; row 4, *A. pedellus*, ♂, C-banded.

(Illustrated as figs 1 & 3 – 5 by Wilson (2001), (courtesy of Dr C. J. Wilson and with permission from the Tijdschrift voor Entomologie).

Chromosome differences result from a number of phenomena. Major ones are:

1. Inversion. This involves the breaking of a chromosome and repair with the broken fragment restored the other way round. This is normally detectable by a change in centromere position - e.g. from near the middle of the chromosome to at or near one end. Inversion does not preclude pairing of chromosomes because the uninverted sections can still pair.

2. Translocation. This involves breakage of a chromosome and attachment of the broken fragment to another chromosome. This causes mayhem at meiosis and is not normally present as a polymorphism within a species. When the karyotypes of different species are compared, the sequence of relative lengths of the chromosomes is often different. This results from translocation events in the course of the species' evolution.
3. Incorporation of new material. Typically this is "heterochromatin" - genetically inert DNA with the base-pairs more or less repetitive. This can be polymorphic within a species.
4. Chromosome fragmentation. This changes chromosome number. It can be polymorphic within a species, but this is unusual.
5. B-chromosomes. These are "optional extras", variable in number between and within individuals. They may be regarded as "parasitic DNA". Their occurrence means that differences in chromosome number do not necessarily mean more than one species is involved.
6. Polyploidy. Here individuals may have varying numbers of chromosome sets. This is normally associated with parthenogenesis. As a general rule, if chromosome differences are within a species, then individuals heterozygous for the characters (i.e. one chromosome showing one arrangement, its homologous partner the other one) will be found.

Methods

Chromosome preparations from adult beetles are often easy to obtain if suitable laboratory equipment is available. Such preparations may give surprising information, as with the unexpected discovery of *Aphodius pedellus* (De Geer) as a distinct species from *A. fimetarius* (L.) (Wilson, 2001). The protocol and list of chemicals given below show how this can be done.

Tissues available

Mid gut: The beetles should be healthy and should have been feeding for at least 24 hours. Mid-gut crypt cells undergo mitosis to replace cells lost carrying out digestion.

Testis: The age of the beetle is important. If too young only spermatogonial mitosis will be present, if too old, only spermatozoa. Details vary with different groups. Testis cells are often "sticky" with this schedule, and may form "rafts" in suspension – with both complete and damaged nuclei. Therefore apparent chromosome numbers may be misleading.

Ovary: Oogonial mitosis can be obtained in many beetles, though in some cases, such as *Helophorus* (Hydrophiloidea) I have never seen it.

Chemicals are listed separately. Stock solutions should be kept in darkness to avoid algal growth.

Procedure

1. Treat the living tissue with colchicine – either inject colchicine solution through the soft tissue at the apex of the abdomen (using a glass microelectrode over the end of a hypodermic syringe if necessary), or place the beetle in a solid watch glass of colchicine solution and partially detach the abdomen by inserting fine pins between it and the metasternum, at the side, and pulling the abdomen backwards. Cover the watch glass.
2. Leave for about 12.5 min. – the time depends on the size of the beetle – e.g. 10 min. for *Anacaena* but 15 min. for *Laccobius*.
3. Transfer the beetle to a watch glass containing _-isotonic Potassium Chloride (KCl). Remove the abdomen by breaking the lateral connections to the thorax (with fine needles or forceps, see above). Dissect out the mid-gut and gonads, and leave them in the KCl for the same time that they were in the colchicine (10–15 min.). Remove the rest of the specimen and keep it for reference. Keep the material in KCl out of strong sunlight.
4. Transfer the tissue to fixative (3 parts absolute ethanol to one part glacial acetic acid) in a fresh watch glass. Change the fixative three times. Keep it covered and leave for one hour. (Note: fatty tissue. Excess fat should be picked from the tissue while it is in KCl, and to some extent while it is in fixative. If fat remains it may spoil the preparation. If necessary, after the third change of fixative, change to a mixture of three parts fixative to one part chloroform, leave for 30 min., then go back to fixative and leave a further 30 min.)
5. At this stage the rest of the beetle may be killed (boiling water is quick!) and mounted, along with the abdomen and genitalia. Obviously, specimens from which chromosome preparations are obtained must be kept for future reference.
6. Take small pieces of tissue in fine forceps and place them (one at a time!) on clean dry slides. Immediately (before the tissue dries out) add a small drop of 45% acetic acid to disaggregate the cells. (I find a hypodermic necessary to get a small enough drop – too much and the preparation ends up round the edges of the slide). Examine the cell suspension under the microscope through transmitted light (close the condenser to see the cells). If the cells are not sufficiently disaggregated, tear the tissue apart with fine pins – usually necessary with gut, less so with testis. Next, with a pipette drop one drop of fixative on to the cell suspension. This will cause the drop to spread over the slide as a thin film. The spreading film can be guided by tilting the slide. Allow to dry horizontally – about five min. The slide may be either stained or set aside for C-banding. (Note: if you use "twinfrost" slides (with a ground glass area at one end, *on both sides*, you can label the preparation, in pencil, on the frosted area. If you use "single-frost" slides (ground glass area on one side only), you will find that this frosted area is generally on the bottom, so you cannot write on it without turning the preparation face downwards).

7. Stain the slides by putting them in 1–2% Giemsa stain at pH 6.8 (Sörensen) for about 10 min. Rinse briefly with unbuffered distilled water, shake off the excess water, and dry vertically. Freshly dry (or almost dry) slides may be checked under the microscope (x10 or x40). After 24 hours they may be examined under oil immersion (x100) (no coverslip) and photographed. Slides for C-banding should be kept for two days, unstained.

8. C-banding. Place the slide, at room temperature, in a Coplin jar of saturated barium hydroxide. Leave for about seven minutes. Then rinse in three changes of distilled water at pH 6.8. Transfer to a Coplin jar of 2 x SSC (salt-sodium citrate solution) at 60° C and leave for one hour. Rinse in three changes of distilled water at pH 6.8, and stain as in **7** above. C-banding picks out regions with "constitutive heterochromatin" (where the DNA tends to have the base-pair sequence repeated over and over again), which stain darkly, while the rest of the chromosome stains only weakly.

9. Photography. Slides may be photographed under oil-immersion, without cover-slips, the oil applied directly to the preparation. High-contrast film is best. Prints are normally done at a magnification of x3000. Chromosomes may be cut out and arranged to form karyotypes, as shown in Figs 1 – 4.

Chemicals for Chromosome Preparations

1. **Sörensen's Phosphate Buffer**
   Master solutions: $1/15$ Molar Na_2HPO_4 (Disodium hydrogen orthophosphate 'Na') and $1/15$ Molar KH_2PO_4 (Potassium dihydrogen orthophosphate 'K').
   'Na': One of these: Na_2HPO_4 anhydrous – molecluar. Weight = 141.96, so $1/15$ M = 9.46 grams per litre.
   $Na_2HPO_4.2H_2O$ (Sörensen's salt). Mol. Wt. = 177.99. $1/15$M = 11.87 g/l.
   $Na_2HPO_4.12H_2O$. Mol. Wt. = 358.15. $1/15$M = 23.88 g/l.
   The anhydrous salt is rather slow to dissolve and needs shaking till dissolved.
   'K': KH_2PO_4 anhydrous. Mol. Wt. = 136.09. $1/15$M = 9.07 g/l.
   Working solution: pH 6.8: 50 ml "Na" + 50 ml "K" + 900 ml distilled water. Nearly everything is done at pH 6.8.

2. Insect saline: 0.75% (7.5 g/l) NaCl in distilled water at pH 6.8.

3. Colchicine: 0.1% colchicine powder in insect saline at pH 6.8. (Note: Colchicine powder is very poisonous, and probably only obtainable in a laboratory).

4. 0.5-isotonic (hypotonic) KCl. 0.48% KCl in distilled water at pH6.8. (0.75 NaCl is 0.128 Molar. 0.064 M KCl is 0.64 x 74.55 = 4.77 g/l).

5. **Fixative**: 3 parts absolute ethanol to 1 part glacial acetic acid.
   (It is very important that fixative is anhydrous. 3 changes are enough for most pieces of tissue, but for larger ones a fourth change can be used. Fixation in final change is at least 30 min., and can usefully be 1 hour. For fatty tissue 1 part chloroform can be added to 3 parts fixative at the third change. Leave for 30 min, then replace with fixative without chloroform and leave a further 30 min.).

6. **For spreading cells on slides:** 45% acetic acid in distilled water.

7. **Stain.** Giemsa's stain as sold by laboratory suppliers. (Normal chromosome or nuclear stains, like haematoxylin and orcein, scarcely stain chromosomes prepared by this method).

For C-banding

8. Saturated Barium Hydroxide [$Ba(OH)_2$] in distilled water (unbuffered). Keep covered as it reacts with atmospheric CO_2 to form insoluble carbonate. The solubility of barium hydroxide varies with temperature. I normally work at between 25 and 28°C. As a guide, when rinsing in three changes of distilled water at pH6.8, the 3rd slide should make the first Coplin jar of rinse go cloudy.

9. 2 x SSC (Salt-Sodium Citrate) 0.3 M NaCl = 1.75 g/100 ml + 0.03 M trisodium citrate ($Na_3C_6H_5O_7.2H_2O$, mol. wt. = 294.1) = 0.882 g/100ml. (Made up in unbuffered distilled water). (This makes a total of 100 ml SSC.).

References

Angus, R.B., 1982. Separation of two species standing as *Helophorus aquaticus* (L.) (Coleoptera, Hydrophilidae) by banded chromosome analysis. *Systematic Entomology* **7**(3): 265-281.

Angus, R.B., 1988. A new sibling species of *Helophorus* F. (Coleoptera: Hydrophilidae), revealed by chromosome analysis and hybridisation experiments. *Aquatic Insects* **10**(3): 171-183.

Santiago-Blay, J.A. & Virkki, N., 1997. Evolutionary relationships within *Monoxia* (Coleoptera: Chrysomelidae: Galerucinae): chromosomal evidence for its intrageneric classification. *Caryologia* **49**(3-4): 257-265.

Shaarawi, F.A. & Angus, R.B., 1990. A chromosomal investigation of five European species of *Anacaena* Thomson (Coleoptera: Hydrophilidae). *Entomologica Scandinavica* **21**(4): 415-426.

Wilson, C. J., 2001. *Aphodius pedellus* (De Geer), a species distinct from *A. fimetarius* (Linnaeus) (Coleoptera: Aphodiidae). *Tijdschrift voor Entomologie* **144**: 137–143.

THE SYSTEMATIC IMPORTANCE OF THE MALE GENITALIA IN COLEOPTERA

by R.A. Crowson

In beetles, as in other insects, it has long been established that in adult males the most striking differences between closely related species are often to be found in the aedeagus. Many systematists, indeed, have taken the male genitalia by themselves to provide a sufficient basis for the distinction and classification of species in their own particular groups. Quite apart from its practical disadvantages when dealing with female specimens, this approach is unsound in principle. If we define a species as an assemblage of potentially inter-breeding populations, then there are many species known in which the aedeagus shows a great deal of variation, in respects which would elsewhere be relied on to distinguish species, and others where 'good' species show little or no detectable difference in the male genitalia. A further practical difficulty arises from the fact that the male genitalia are 3-dimensional and somewhat flexible structures, whose appearance may be considerably influenced by the processes of preparation and by the precise angle from which the finished preparation is viewed. When all these qualifications are allowed, it remains true that the examination of the aedeagus is a very useful, if not indispensable, procedure for the coleopterist seeking reliable species determinations in 'different' genera. The external genitalia of the females as a rule show far less species-variability than does the aedeagus, but a wholly internal structure, the spermatheca, not infrequently differs notably between related species and has been used for their discrimination.

The structure of the aedeagus may also provide important evidence for classificatory relations at levels much higher than the specific, though this aspect of it is rarely likely to be of much concern to the average beetle collector. In the suborder Adephaga, the absence of a true basal piece to the tegmen seems to be a basic feature; in the vast group Cucujiformia, including the Cleroidea, Lymexyloidea, Cucujoidea, Chrysomeloidea and Curculionoidea, a special 'ring' or 'Cucujoid' form of the tegmen is a basic feature – though the basic pattern is often so much modified as to be hardly recognisable. The division of the penis (median lobe) into separate dorsal and ventral sclerites characterises the superfamily Dascilloidea, and the incompleteness of the 'ring' of the tegmen on the ventral side (morphologically) is the most reliable single feature by which to distinguish those Heteromera with 4-4-4 tarsi from the Clavicornia of the Cerylonid group. The orientation of the aedeagus, both in the retracted and the extruded position, may be of systematic importance. An aedeagus which is normally oriented when extruded very often lies on one side when retracted, for example in many Cerambycidae, Erotylidae, Chrysomelidae etc.; in a number of families, the entire structure is inverted, both in the retracted and the extruded condition, as for example in many Cucujidae, many Heteromera such as

Pyrochroidae, certain Staphylinidae etc. In beetles with an inverted aedeagus, the normal copulatory position is for the male and female to be facing in opposite directions and only in contact with each other at the end of the abdomen, whereas with a normally oriented aedaegus, the male climbs onto the back of the female.

Systematically useful characters of the aedeagus may be found in the tegmen (basal piece plus parameres), the median lobe or the internal sac. The last named structure is apt to be overlooked, unless the aedeagus is cleared and mounted in canada balsam or a similar medium, permitting its study by transmitted light. The internal sac is a more or less membranous tubular structure, usually invaginated into the penis and not easily everted from it; it often has more or less complex and characteristic structures on its walls, and may be extended at its apex (in the everted condition) into a long thin tube known as a flagellum, for example in Cerambycidae and Cucujidae.

Besides the aedeagus proper, the last two abdominal segments (nos. 8 and 9) normally partly or wholly withdrawn inside segment 7, show more or less marked modifications in the sexes which are liable to be of systematic importance, particularly in forms like Staphylinidae and Cantharidae where these segments are more or less exposed. The last visible sternite (ventrite) in most beetles is that of segment 7, and this too often shows distinct sexual differences; in some beetles, notably in the scraptiid *Anaspis,* various species of *Stenus,* and some Heteromera, there may be species-characteristic sexual modifications of some of the more anterior abdominal sternites.

EXAMINATION OF COLEOPTEROUS GENITALIA

by J. Cooter

Equipment

Most instruments necessary for the extraction and dissection of the required parts of beetle genitalia should already be in the possession of the coleopterist (see plate 112).

1. *Fine brush* similar to the type used for mounting, ideally an artist's good quality sable hair brush, with fine point; size 000 or 00 are suitable. Brushes with synthetic hairs are cheaper, but the hairs often easily and permanently distort rendering the brush useless; this has happened to me after the first use of a cheaper brush bought in a sale: so much for bargains!

2. *Needles* – undoubtedly the best are made from fine gauge tungsten wire. However, their manufacture may be beyond the capabilities of the average coleopterist as it requires special equipment. The wire is cut to length and held in a metal clamp fixed to a rotating eccentric. To sharpen them they

must be dipped into molten sodium nitrite (not a solution) with an electric potential between the wire and electrolyte. Acceptable results can be had by hand dipping the wire into the molten sodium nitrite without electricity involved. Care must be taken to hold the wire in an insulated carrier and to take other sensible precautions such as wearing safety glasses and doing the job while your wife is out of the house. The degree of point can be regulated by time and depth of immersion. Too long immersion will result in flat cut ends. I was lucky enough to obtain a small stock of these needles and used them for over 30 years; they do not blunt or bend in use, but can split or be damaged accidentally. Re-sharpening is as described above. Tungsten wire is very expensive. Without the necessary equipment my small stock diminished I turned to using microlepidopterist's stainless steel headless pins of 10mm or 12.5mm length. These can be held in forceps, attached with a strong synthetic or epoxy glue to a match- or cocktail stick or similar, or better still held by a pin vice with screw-activated chuck.

3. *Forceps*. One or preferably two pairs of watchmaker's very fine pointed stainless steel antimagnetic forceps (No. 5A or 5) with a pair of less fine-pointed (No. 2) and pinning forceps close to hand. The points of dissecting forceps should be protected when not in use. Good quality forceps are not cheap (£15 - £20+ per pair (2005)) but if looked after they will last many years. My experience is that natural history and laboratory suppliers generally do not stock fine pointed forceps of sufficiently good quality; I always purchased my own and those for museum use from a watchmakers supply house; the Swiss 'Precista' or 'A. Dumont & Fils' brands are superb, light weight with perfect balance and strong but fine points. Dumont forceps are stocked by Lydie Rigout.

4. *Microscope*. A stereoscopic microscope is the ideal for entomological use, but genitalia preparations can easily be made with monocular or non-stereo microscopes. A hand lens can be substituted, but needs to be mounted on a stand as both hands are required whilst dissecting.

5. *Lighting*. The microscope light is very useful and the filter can be left out to give an intense beam. A desk light with low energy fluorescent tube is useful, other bulbs generate too much heat which, apart from making the worker uncomfortable, quickly dries up the dissection.

6. *Dissecting dish*. Some form of shallow vessel is needed. The best I have found is the white porcelain palette used by biologists for spot-testing. This is heavy, white and opaque with shallow depressions. Too much water or alcohol in the dissecting vessel will cause everything to float about and possibly move out of the field of view; too little water or alcohol leads to drying out and dissected parts being accidentally flicked out of the dish. The spot-testing palette has another great advantage in having usually twelve depressions. As these are set out in rows of three, top to bottom, four

dissections can be undertaken in one session. The top row may have the beetle plus water, the next, the genitalia plus alcohol and the third, if needed, may be used for xylene, clove oil or another reagent. Care must of course be taken not to mix up the dissections, writing letters in permanent marker to donate the rows and numbers for the columns is advisable, but beware some 'permanent' markers are affected by alcohol and other solvents.

7. *Boiling vessel.* This should not be of too small a diameter or too deep, thus a test-tube is quite unsuitable. A stainless steel egg-cup or a chemist's porcelain micro-crucible are ideal, the latter more so as they are supplied with lids. Both are heat and chemical resistant. If boiling, two or three pin-heads can be added to prevent 'bumping' - the formation of large bubbles which can, and sometimes do, eject the contents of the vessel.

8. *Heat.* For boiling, a spirit lamp is suitable. To soften the tissue of a dry beetle boiling is not strictly necessary; warming and keeping warm for a time will suffice; cold water will do but a much longer soaking time is needed. Laboratory electric 'hot plates' with thermostatic control are ideal, but very expensive.

9. *Hydroxide.* Potassium hydroxide solution is more potent than sodium hydroxide solution of comparable concentration. It acts to break down the musculature enabling the component parts to be dissected from the surrounding tissue. It must be remembered that prolonged immersion will eventually dissolve everything - chitin included. Thus maceration in hydroxide should never be left unattended. This latter property is useful when the internal structure of the median lobe of the aedeagus is to be studied and the chitin of the median lobe is very thick and obscures fine detail. The chitin of the lobe can be rendered thinner and thus more transparent by immersion in hydroxide. That said, in 35+ years studying Coleoptera I have never personally had to resort to hydroxide immersion.

Any dissection exposed to hydroxide solution must be washed in dilute acetic acid and then in several changes of water in order to expel any hydroxide residue. IMPORTANT – both sodium and potassium hydroxide in solution and as solids (which are deliquescent) cause burns to skin and will ruin clothing. Great care in the use of such solutions must be exercised.

10. *Plastazote.* One or two postcard-sized or smaller pieces of Plastazote should always be to hand.

11. *Bent pin.* An entomological pin with head removed and the cut end heated and then hammered flat makes an ideal tool for picking up larger genitalia from solution. Smaller genitalia or dissected parts get 'sucked' into the hairs of a brush, a fine pipette is better, but delivers a small quantity of solvent with the dissected part. As the object of transfer is to remove the dissection from that very solvent a pipette is not ideal.

A steady hand and fine forceps might suffice, but GREAT care is required when picking up delicate tubular or globular spermathecae.

Dissecting the Genitalia

The student should become familiar with those species and genera that need to be dissected, the degree of dissection involved, and the orientation of the parts when finally mounted. These last two points vary greatly; there is not, and indeed cannot be, a hard and fast rule. It may be possible to determine the sex of a beetle from external characters; again these vary from family to family, but where present, generally a reference to the differences is given in the literature ('Fowler' being very good in this respect). As an example, I take the genus *Choleva* – the males have the segments of the anterior tarsi dilated, in females they are more linear. The male aedeagus shows useful specific differences, and with the females the tergites of the last segments of the abdomen are characteristic, females of the majority of the Catopinae have a poorly chitinised spermatheca. If the sex of the beetle is known, care can be taken and the required parts only dissected.

The component parts of the genitalia are very delicate, often tubular, structures so extreme care must be taken throughout dissection and mounting.

With dry material it is necessary to soften the tissues before attempting to remove the genitalia, if this is not done, damage is almost certain to occur. The tissues may be softened quite easily in water, the abdomen or part of it can be removed and immersed. The length of time required for the softening to occur will vary according to the nature and size of the abdomen but it can be speeded up if the water is heated. The addition of hydroxide will also help, but if used, the parts must be thoroughly washed in dilute acetic acid followed by several changes of clean water in order to remove the caustic residue, which will continue to attack the chitin.

The standard method I employ when dissecting dry material is to relax the whole beetle. The beetle is removed from its card and placed on damp tissue in one section of a laboratory 'sorting tray' (see plate 112). A little extra water is added to any empty compartments, the lid fitted and the box placed in the airing cupboard over night. The compartments of the 'sorting tray' are identified by an embossed number/letter code, A1-6, through to D1-6. I have marked a rectangle of Plastazote marked with a grid of 24 squares A1 – D6. After removing the beetle from its card and placing it in a 'sorting tray' compartment, I pin the vacated mounting card with its labels into the corresponding square on the Plastazote rectangle; one de-0mounted beetle per compartment, one pin per square. This minimises mixing up specimens and data during the dissection process.

If dealing with newly acquired unmounted dry material a number of specimens with the same field data can be placed in the same compartment. Remember here we are preparing dried hitherto unmounted or previously mounted beetles for dissection, after dissection the beetle will require mounting and the dissected parts will require clearing and mounting in Euparal or other suitable medium. This represents a lot of time consuming work so never relax more material than can be dealt with in one session, it can go mouldy if left too long in a humid atmosphere.

Any freshly killed beetles or those removed from fluid preservative will not need softening; it is always a more straightforward task to dissect fresh material is best carried out prior to mounting. If the beetle is not mounted it should be held gently but firmly with forceps or a needle. With small beetles, say 3mm and less, the whole beetle can be placed in alcohol in a depression in the porcelain palette and the last two segments of the abdomen carefully removed with a fine pin, this usually exposes the genitalia. I have on occasion carded small Staphylinidae and dissected them the next day while the tissues are still soft, but the beetle is firmly held by the mounting gum.

If the abdomen has been removed, the genitalia can be extracted by pulling out the tissues and genitalia through the opening with a pair of fine forceps. In many cases all that is required is to tear the pleurites or tergites open and pull out the genitalia surrounded by a mass of tissue. The more heavily sclerotised beetles such as histerids and most weevils generally permit removal of the genitalia via the raised pygidium – this being held open with a pin or forceps. However, it must be borne in mind that very often the apical segments themselves have characteristically shaped sclerites (especially useful in some small Staphylinidae). In such cases a needle inserted under the posterior edge of tergite of abdominal segment 7 can be moved posteriorly so that the anterior edge of abdominal tergite 8 is pulled clear. In such a way segments 8 and 9 will be removed from the abdomen. These are transferred to the dissecting dish, the genitalia removed with pins and forceps and the sclerites carefully separated and great care taken not to confuse tergites and sternites. The occasional beetle will die with the aedeagus partly extruded, this is especially true of material collected in flight interception or pitfall traps, and with careful use of pin and forceps the whole armature can be removed.

Having removed the genitalia and their associated tissues, the next stage can begin. This involves the removal of the unwanted musculature and fatty tissue. If the genitalia are to be used solely for confirming an identification, there may be no need to progress further, but if the genitalia is small or weakly chitinised, it will be necessary to clean off the unwanted tissues that will otherwise obscure the 'hard parts'. The genitalia and tissue are placed in water which may be heated or have hydroxide added. The dish is then placed under the microscope and the sclerotised genitalia teased out from the tissue, some of which might have to be carefully pulled away with the forceps.

Although potassium hydroxide solution is universally recommended to macerate the soft tissues and render their removal easier, I have never found its use necessary even with dried beetles which have been relaxed prior to dissection (but see also 'clove oil preparation' below). Potassium hydroxide will also dissolve the chitin of the aedeagus itself and it is recommended for 'thinning' the chitin to render the internal detail of the median lobe more readily visible for study under transmitted light. Likewise, and I specialise in a group of beetles where study of the internal sac is often a necessity for correct identification, I have never had to resort to hydroxide solution to make the internal detail of the median lobe visible. My advice is to try dissection and preparation without hydroxide and see how you get on. Replacing the dissected genitalia and mass of adhering tissue into the 'sorting dish' in a warm atmosphere generally enables removal of tissue and a clean dissection. Potassium hydroxide is a very strong alkali and in its solid form is deliquescent; it is a dangerous substance capable of buring skin and clothing, request Health & Safety 'COSHH' safety notes from your supplier when ordering.

With many genera, for example *Stenus* and *Catops,* the relative lengths and attitudes of the parameres and median lobe are of great importance and so should not be separated from each other. With others, for example *Olibrus*, most small aleocharine staphs, *Philonthus* and *Quedius,* the dissection should be taken further as the diagnostic points are at this stage not revealed. This will generally consist of separating the median lobe and paramere(s) and any other genital sclerite and great care must be taken not to damage the individual structures. In some genera (*Philonthus, Quedius*) the single paramere has diagnostic value, in others (most aleocharine staphs) the paired parameres do not.

The dissected parts can be removed from the alcohol or water with the 'bent pin' or forceps. If a brush is to be used, a useful tip is to soak up the bulk of the liquid with blotting paper leaving a thin film in the dish. With this minimal fluid, the brush will retain its point and the risk of minute genital sclerites being drawn into the hairs by capillary action is avoided.

Mounting

In a few cases it will be necessary to treat the dissected aedeagus further in order to reveal its internal structure, as is necessary with xantholine staphylinids and most *Leiodes* species as well as some Carabidae. Details of this are given below. Alternatively it is sometimes necessary to view the dissected parts in more than one orientation, genus *Colon* for example, which effectively rules out mounting, and in such cases the van Doesburg method should be employed, see p.366.

With the bulk of species, the genitalia can be lightly glued to the mounting card behind the beetle from which they were dissected, along with any sclerites dissected during the process, see figure 49. Tottenham (1954) recommended gluing dissected tergites on the left and sternites on the right; it is sound practice to adopt a systematic standard method such as this as it gives a degree of uniformity and avoids unsightly labelling.

Mounting in a medium

A growing number of coleopterists use the synthetic water-soluble transparent mountant *dimethyl hydantoin formaldehyde* (D.M.H.F.). It is crystal clear and permits observation of minute detail whilst setting rock-hard and thus affording maximum protection. Being water soluble, it permits removal of the genitalia for reorientation or re-carding should the need arise. Alas, it sometimes does not attach too well to the mounting card and can break away with the included dissection; this habit should be overcome if the card is scratched minutely with a pin prior to adding the D.M.H.F.

In order to study or observe the internal structure of dissected genitalia, examination in transmitted light is usually necessary. For this purpose, the synthetic mountant Euparal is almost universally employed; D.M.H.F. is generally used to affix the dissected genitalia to the mounting card and to protect the delicate parts whilst allowing easy visibility.

Whilst checking the proofs for this edition of the *Handbook* a very interesting and useful paper (Liberti, 2005) appeared. The characteristic and properties of several mountants are discussed in detail and it is highly likely that this paper will have much influence on the type of mountant coleopterists will use in the future. The student should carefully read this paper and if able, experiment with the compounds and methods described.

Given the date of Liberti's paper, this section continues with traditional methodolody which is widely used and has given time-proven results and is thus perfectly acceptable.

It is good practice to keep all dissected parts, data and other labels *with* the beetle they relate to. The practice of preparing genitalia for transmitted light as a microscope slide which is stored elsewhere is not to be recommended. Other authors describe how to make a slide by punching a circular hole in mounting card and gluing a fragment of microscope cover slip on one side - the 'slide' is then pinned under the mounting card before addition of data labels; to view, the 'slide' is moved to one side. This method is acceptable, but in light of experience, perhaps too complicated and time consuming to become popular but made easier if a stock of 'slides' is prepared in advance. The most widely used type of transmitted light final mount is described below.

Clove oil preparation. With a few genera, or certain species within certain genera, the internal sac of the aedeagus displays diagnostic characters. It is not necessary to evert the sac, but to observe the fine detail the chitin of the median lobe has to be rendered more transparent. **A note of caution:** *very often, if not always, the internal sac is clearly visible through the chitin of the median lobe. If this is so, then there is no need to render the chitin of the median lobe more transparent by immersing in hydroxide solution and rinsing in dilute acetic acid to neutralise the alkali then rinsing repeatedly in changes of water to dispel the dilute acid then* *passing through organic reagents to remove the water.* Once the aedeagus is dissected and still in the dissecting dish, check the clarity of the internal sac.

NEVER LET A DISSECTION DRY OUT AT ANY STAGE AS AN AIR BUBBLE WILL INVARIABLY FORM WITHIN THE MEDIAN LOBE. The presence of an air bubble will obscure or distort the internal sac. The bubble can be removed by immersing the aedeagus in alcohol or Euparal solvent and applying VERY GENTLE pressure with fine forceps forcing the bubble towards the basal orifice. too much pressure will break the median lobe and/or distort the internal sac.

If the chitin of the median lobe requires clearing, the dissected aedeagus is immersed in cold 10% potassium hydroxide solution for several hours during which time it must be examined under the microscope regularly in order to assess the degree of detail being rendered visible. When the internal detail is sufficiently clearly visible, the median lobe is removed and washed in dilute acetic acid to neutralise the caustic solution, after this it is washed in two or more changes of water and then dehydrated in 70% alcohol and, ideally, then absolute alcohol. Next it is transferred to clove oil for clearing. After about one hour or when the internal detail is clearly visible, it can be transferred back to Euparal solvent or 'essence' then to Euparal, alternatively it can be transferred directly from clove oil to Euparal. Experience is a good guide here. I have made good genital preparations with internal sac perfectly clear simply by dissecting the genitalia in 70% alcohol, transferring to clove oil, then to Euparal essence and finally to Euparal mountant.

For mounting, the recommended method is to spread a small quantity of mountant on an acetate* strip cut a little longer and a little narrower than the mounting card. This mountant should be as shallow as possible to cover the aedeagus but form a flat-topped 'blob'; a hemispherical 'blob' will act as a lens and give a distorted image. Orientation is checked and if necessary adjusted as the mountant dries (after 24 hours is sensible time) and finally the acetate strip is pinned under the beetle mounting card, followed by the data and determination labels, see figure 50. Strips of the acetate (or plastic) can be cut before hand and kept as a stock with one's mounting cards and setting equipment.

---

* the term 'acetate' is used here in a purely vernacular non-scientific manner. The material I use is a clear 'plastic' sheet sold by model shops for a range of uses (doll's house windows, model aircraft cockpits, model car windscreens etc). Although this method is used throughout Europe and I have employed it for years, a strong notice of caution must be added here. The 'acetate' or 'plastic', whatever it actually is, it is an organic compound and as such is vulnerable to attack by organic solvents. Exposure to ethyl acetate for example will make the 'acetate' or 'plastic' soft and probably turn it into a blob in the middle of which is the dissected aedeagus. Some, including myself, pour a little ethyl acetate into a store box if pest attack is suspected and yes, I once did this to a box that contained some beetles actually mounted on acetate strips. A few days later when opening the box the beetles, staphylinids I recall, were embedded within an elongate blob of yellowed acetate like mini-spring rolls. Fortunately these were specimens without data, so no great loss, but beware and be aware of the properties of the materials you use and how they can interact with each other.

Presentation

The final mount should be kept on the same pin as the beetle, usually underneath it (Fig.48). With small beetles, e.g. Ptiliidae, aleocharines, *Longitarsus*, apionid and ceutorhynchine weevils etc., and other genera where examination under transmitted light is not necessary, the aedeagus or spermatheca can be placed in a patch of Euparal, D.M.H.F. or even a minimum of mounting glue on the mounting card alongside the beetle, making sure that enough mountant is added so that the genitalia is covered or sufficient glue to secure the dissection without obscuring it. Do not build the mountant up into a semi-spherical blob, this will act as a lens and distort the image of the genitalia contained within.

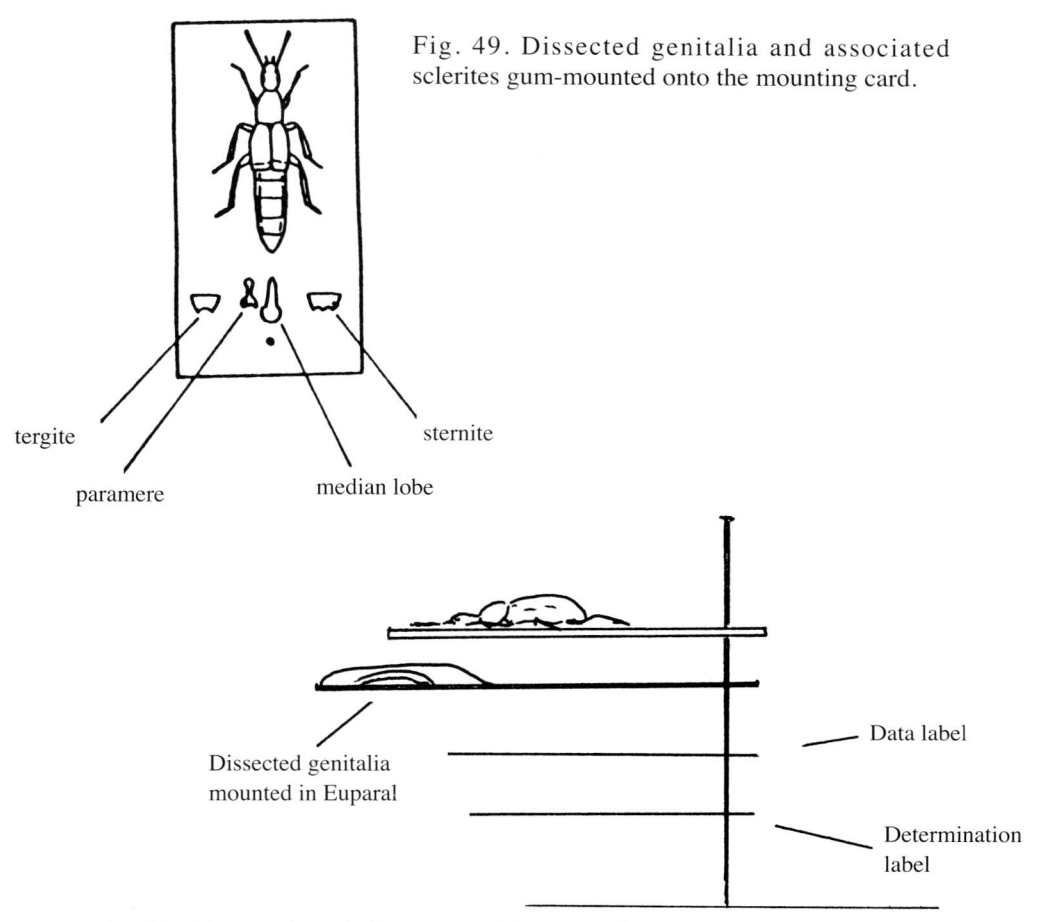

Fig. 49. Dissected genitalia and associated sclerites gum-mounted onto the mounting card.

tergite   paramere   median lobe   sternite

Fig. 50. Dissected genitalia mounted in Euparal on an acetate mount pinned under the beetle mounting card but above the data and determination labels.

MALE GENITALIA

AEDEAGAL TYPES

Carabid beetle (dorsal view) Fig. A

Tenenbrio molitor L. Fig. B

Phyllobius glaucus Scop. Fig. D

Hydrophilus piceus L. Fig. C

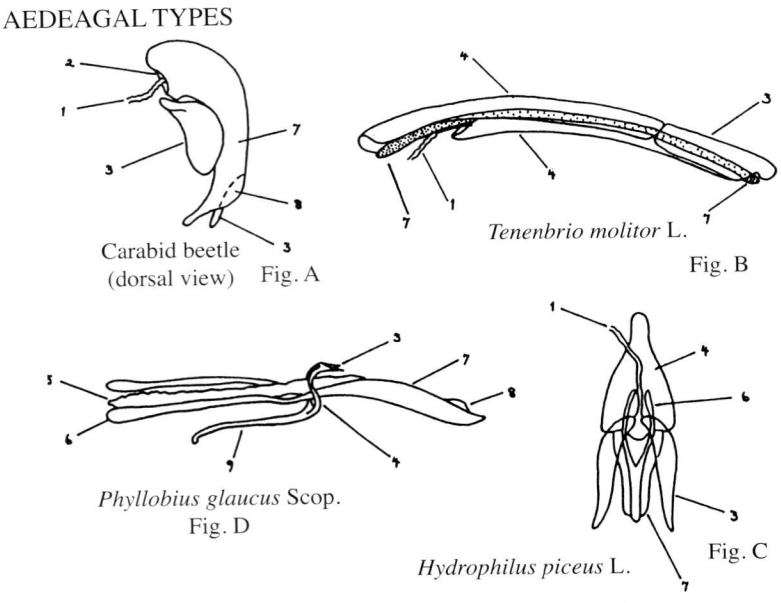

Fig. 51. Four major types of male genitalia found in the Coleoptera.
(1) ductus ejaculatoris (2) basal oriface; (3) paramere; (4) basal piece;
(5) internal sac; (6) apophysis; (7) median lobe; (8) ostium; (9) manubrium.

Fig. 51a. The *Articulate type* illustrated here with the aedeagus of a carabid beetle, dorsal view. Usually asymmetrical with basal piece often unsclerotised and thus seemingly absent. Parameres articulating to the median lobe by a true condyle. This type of aedeagus is found in most Adephaga and Staphylinoidea.

Fig. 51b. The *Vaginate type* as exhibited by *Tenebrio molitor* L., show here in lateral view. Basal piece and parameres forming a pipe or a (dorsal or ventral) channel through which the median lobe moves; the parameres are unable to move much. Sometimes the median lobe is so reduced that the tegmen assumes a large part of its copulatory function. Found in many Tenebrionoid families for example.

Fig.51c. The *Trilobate type* illustrated here by the aedeagus of *Hydrophilus piceus* L., dorsal view. The structure is symmetrical, with basal piece well sclerotised, the parameres articulate to the basal piece. The first connecting membrane allows very little independent movement of the median lobe and basal piece. Sharp & Muir (1912) refer to this as the *Byrrhoid type* and it is found in many families. It is regarded as the most primitive form within the Coleoptera.

Fig. 51d. The aedeagus of *Phyllobius glaucus* (Scopoli) illustrates the *Annulate type* of aedeagus. The basal piece forms a complete ring around the median lobe and as the two are only loosely connected, considerable movement of the median lobe is possible. Parameres are usually reduced, in some instances to a pair of processes firmly fixed to the basal piece. Found in the Phytophaga and some other families.

Transitional types also occur; aberrant types of aedeagus are found in the families Scirtidae and Oedemeridae for example.

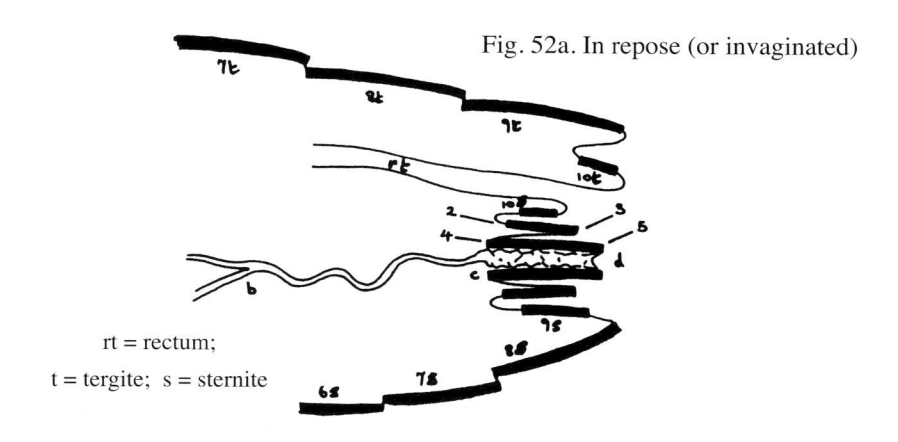

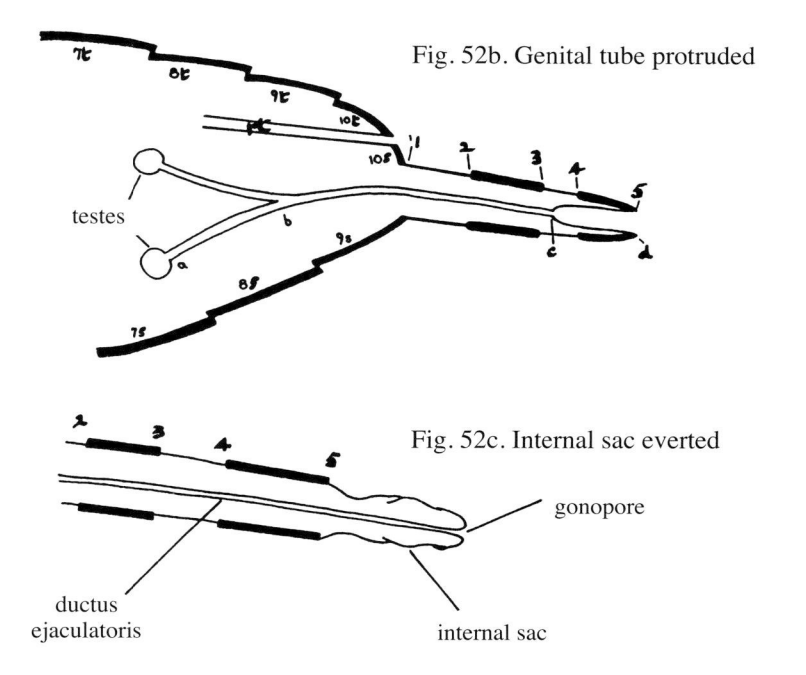

Fig. 52a. In repose (or invaginated)

rt = rectum;
t = tergite; s = sternite

Fig. 52b. Genital tube protruded

testes

Fig. 52c. Internal sac everted

gonopore

ductus ejaculatoris

internal sac

OUTLINE MORPHOLOGY OF THE MALE GENITAL TUBE
(diagammatic, after Sharp & Muir, 1912)

The features described below relate to a general case, the accompanying illustrations are diagrammatic and do not represent a particular beetle. The Society is most grateful to the Royal Entomological Society for permission to use these diagrams, taken from Sharp & Muir (1912).

A pair of SEMINAL DUCTS lead from the TESTES and form what is known as the ZYGOTIC PORTION, (a – b). These ducts join to give a long, single, highly irregular tube that is joined to the body wall – the AZYGOTIC PROTION, (b – d). The azygotic protion is divided into a long slender STENAZYGOTIC PORTION, (b – c) beyond which the tube enlarges to form the EURAZYGOTIC PORTION (c – d and 5 – 1). That part of the eurazygos that is not external (c – d) is the INTERNAL SAC. In all cases observed by Sharp and Muir, the internal sac is evaginated during copulation (thus approximating to the mammalian 'penis') and forms a continuation of the external parts of the genital tube. In many cases there is no demarcation between the stenazygotic and eurazygotic portions, in such a case the internal sac is said to be undifferentiated. The portion of the tube that is external is termed 'phallic.'

The sclerites situated on the 'phallic' portion fall into two groups:
  i. Those on the distal portion are known as the MEDIAN LOBE (5 – 4).
  ii. Those nearest to the base are known as the TEGMEN (3 – 2).

The membranes between these two groups of sclerites are the FIRST CONNECTING MEMBRANE (4 – 3) and the SECOND CONNECTING MEMBRANE (2 – 1). The median lobe plus the tegmen forms the aedeagus. The MEDIAN ORIFICE is the opening on the median lobe through which the internal sac is evaginated. The MEDIAN FORAMEN is the aperture at the base of the median lobe through which the ejaculatory duct passes. The LATERAL LOBES are the distal extensions of the tegmen, more usually referred to as PARAMERES (and increasingly in certain families simply as tegmen).

FEMALE GENITALIA

There are two main groups, the *Saccular Type* without separate bursa copulatrix (Fig. 52) and the *Tubular Type* with a separate bursa copulatrix (Fig. 53) which is generally combined with the development of an ovipositor.

An abberent type occurs in the genus *Cyphon* Paykull, where the female plays the active part during coitus and possesses special copulatory apparatus known as the *prehensor* which develops from the ventral wall of the vagina.

Different names have been applied to the same parts of Coleopteran genitalia by various authors, Tuxen (1970) contains a glossary with 5,400 terms giving synonyms used by other authors.

Curatorial

Fig. 53

Fig. 54

FEMALE GENITALIA

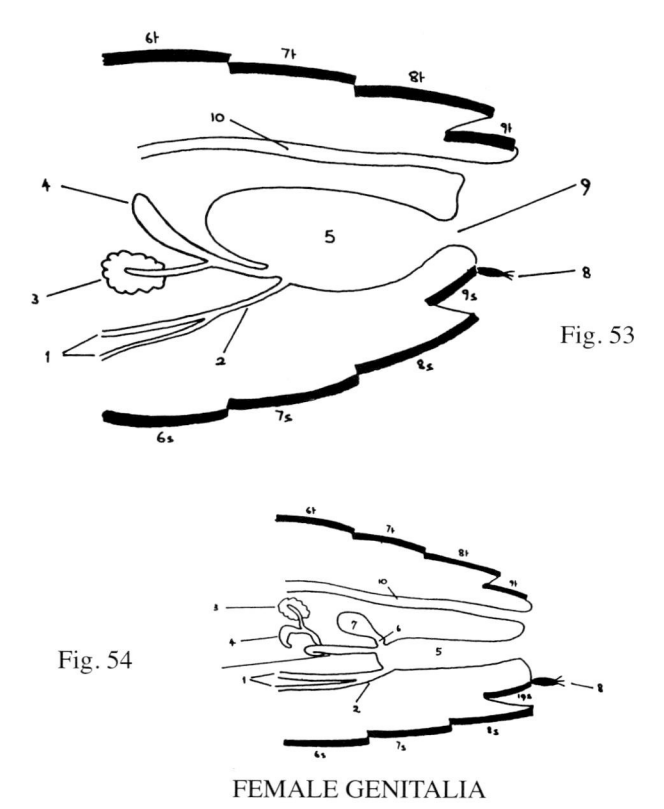

Fig. 53. Saccular type without separate bursa copulatrix.

Fig. 54. Tubular type with separate bursa copulatrix.

(1) lateral oviducts; (2) common oviduct (oviductus communis); (3) spermathecal gland; (4) spermatheca; (5) vagina; (6) ductus bursae; (7) bursa copulatrix; (8) genital style; (9) vulva; (10) rectum.

References

Liberti, G., 2005. Improved solutions of two water-soluble media for mounting beetle genitalia. *The Coleopterist* **14**(1): 29-35.

Sharp, D. and Muir, F.A.G., 1912. The comparative anatomy of the male genital tube in Coleoptera. *Transactions of the Entomological Society of London,* 1912 (part3): 477 - 642, pls 42-78. Reprinted without change (1969) by the Entomological Society of America.

Tottenham, C.E., 1954. Staphylinidae. *Handbook. Ident. Br. Insects.,* **4**(8a). Royal Entomological Society, London, 79pp.

Tuxen, S.L., (ed.) 1970. *Taxonomist's glossary of genitalia in insects.* Copenhagen (1956) (Coleoptera section by C.H.Lindroth and e.Palmer) (Second revised edition 1970).

THE VAN DOESBURG METHOD
by Ashley Kirk-Spriggs

In the case of poorly sclerotised or complicated genital structures which require examination at different orientations, it is desirable to store genitalia in a wet state. Several storage methods have been used in the past which ensure genitalia are always associated with the specimen from which they came, for example glass 'Durham' tubes. These, however, prove expensive to buy, and the cork stoppers become impregnated with glycerol which eventually leaks down the pin and stains the cabinet lining. The Dutch entomologist, P.H. van Doesburg, developed a method which is now used extensively by dipterists and hemipterists and increasingly by coleopterists in U.K. institutions and abroad.

Procedure: Polythene tubing of 3mm bore and 5mm outside diameter is heated gently over a spirit lamp until the polythene becomes soft and sticky, this should be done in a well-ventilated room and care taken not to heat the polythene too much or it will ignite. Having softened the tubing, place the heated end between the forceps a little way from their tips and squeeze the forceps together and hold for a few seconds to give the polythene chance to bond. With the tube now sealed at one end, cut off approximately 15mm. Transparent tubing is best suited for entomological use.

After cooling, check the tube is properly sealed before going on to the next stage. Fill a pipette or syringe with glycerol; then insert its point in the tube as far as it will go and fill half the tube. Continue to hold the tube vertically to ensure the glycerol settles into the base of the tube. The polythene caps for these tubes are made from polythene rod. A 4mm length of rod is cut with a scalpel or razor blade (scissors will burr the rod). The cut rod is then inserted into the open end of the tube to seal. After a dissection, the genitalia are placed in the tube with glycerol and the rod cap replaced. The tube is pinned through the flattened heat-bonded end under the dissected beetle's mounting card.

If genitalia are mounted on microscope slides or otherwise kept separate from their beetle, the two must be cross-referenced and both should carry full data. Slides can be stored in proprietary microscope-slide boxes or cabinets. A mounting method that separates the dissected parts from the parent beetle is not to be recommended. The two can become irretrievably separated, for examples the slides lost after death of their owner. With large numbers of slides the system can become unwieldy.

Materials
David Henshaw, 34 Rounton Road, Waltham Abbey, Essex, EN9 3AR supplies both the tubing and rod.

Breeding

REARING LARVAE

by Dr M.L. Luff

Introduction

Whereas a Lepidopterist may spend as much time searching for the immature as for the mature stages of butterflies and moths, the collector of beetles, rarely carries out similar methods of collecting, for the following reasons:

1. The feeding habits of beetle larvae are very varied, some being carnivorous, actively seeking living prey; some feed upon carrion or upon dung; many burrow in the soil or in the solid wood of trees, and others are parasites or inquilines in the nests of ants and other creatures.
2. Many beetles require several years in which to complete their metamorphoses.
3. In general, the larvae of beetles are unattractive aesthetically, devoid of the beauty of form and coloration shown so conspicuously by the caterpillars and, frequently, by the pupae of Lepidoptera.
4. Much more care, attention and elaboration of apparatus (and, consequently much more time) is, in general, required to rear beetles successfully.

Nevertheless, there is a greater feeling of self-satisfaction in bringing successfully to maturity some interesting beetle than in the rearing butterflies and moths. It happens, too, not infrequently, that the coleopterist in search of a definite species finds on searching its known habitat that he is too early, the season perhaps being late and that it is represented by fully, or almost fully, grown larvae. It is pity to neglect these, for from them it may be possible with due care to rear the beetles. Many Coleoptera have phytophagous larvae, to rear which, even from the egg-stage, is almost as simple as rearing caterpillars. While the careful and detailed working-out of the life histories of beetle species is rather the province of the specialist or research student and requires incessant application and much time, the conscientious working coleopterist may still add not a little to the knowledge of the bionomics of beetles.

The whole essence of rearing larvae is to reproduce as accurately as possible the actual conditions under which they normally exist. The chief drawback to rearing them is the duration of the larval period, for the majority take at least a year to attain maturity, while some take three or more years.

Such annual or extended life-cycles are usually synchronised with the time of year by an obligatory resting phase or 'diapause' at some stage in the (often) immature adult stage of the life cycle. As the timing of this phase is often linked to natural changes in outdoor environmental features such as day length and temperature, it is important to subject larvae being reared to natural conditions wherever possible, rather than keeping them indoors 'in the warm' to hasten their development.

A further very important point to observe is the exact nature of the food. This is not always at first apparent; for instance, a larva might be found under bark, but this does not necessarily mean that it is lignivorous; it may be predaceous on other larvae or feed on moulds, hyphae of fungi, dead insects, or even spider webs. Hence, until the beginner has become better acquainted with their habits, it is advisable to collect a reasonably large supply of the wood, fungus, or whatever the larvae are found in, so that their exact feeding habits may be observed in captivity. The more experienced the collector becomes, the less this precaution will prove necessary. Another equally important point to observe is the actual *condition* of the food, especially with regard to its degree of moisture and decay. Generally speaking, it is better to have the food too moist rather than too dry; moreover, it is fatal to give larvae dry rotten wood when they have been living in sound wood. Breeding experiments have proved that the successful development of lignivorous larvae depends to a greater extent on the nature and condition of the wood (i.e., the thickness of bark, presence of sap, degree of moisture, stage of deterioration, etc.) than on the species of tree from which they were obtained, since the majority of these larvae are undoubtedly polyphagous.

Two fatal conditions in rearing are drought and mould, and continuous inspection is necessary if these are to be avoided. Food should be moistened and examined for signs of mould periodically. The maintenance of good ventilation is the surest way of inhibiting mildews; nonporous rearing containers should never be exposed to sunlight as this causes excessive condensation. The growth of mould in soil may be inhibited by occasionally sprinkling the surface with water containing five parts to the hundred of common salt.

No special breeding cages or other apparatus are required for general rearing purposes, because they can all be improvised from jam-jars, flower-pots, etc.; but a few dozen glass-topped tins roughly 6-8cm diameter and 3-4cm deep are extremely useful: the bottoms should be covered with pieces of damp blotting paper to maintain a humid atmosphere or to absorb excess moisture as the case may be.

One can also use plastic pots or beakers, covered by cling-film which is perforated in a few places to aid ventilation. As an alternative to blotting paper, the base can be filled to a depth of 1cm with plaster of Paris, which is then moistened with water. The remainder of this chapter will outline rearing techniques which can be tried for larvae living in a range of habitats.

Remember, however, that there is no sure guide to successful rearing of any beetle species. Be prepared to experiment, and adjust the basic methods to each individual species. The satisfaction of rearing a difficult species will be all the greater if you have had to devise your own scheme in order to be succeed. If rearing of a rare species is achieved, consider returning the adult(s) to their natural habitat, so that the species is conserved. But be wary about releasing reared species away from their normal environments, as outlined in the 'Code for Insect Introductions,' published by the Joint Committee for Conservation of British Insects (JCCBI).

During the course of rearing, a number of larvae may be found to be parasitised: do not throw them away. Preserve some of the parasite larvae and place the host and its contents in a separate tin. The resulting imagines, together with their larvae and full data, should be sent to an authority on the group. Such material is badly needed.

Finally every rearing cage or container should bear a label with the number of the species as recorded in the catalogue; self-adhesive labels are suitable for this purpose.

Terrestrial predatory larvae

This description applies to many species of Carabidae and Staphylinidae, as well as Cantharidae and Lampyridae and most Coccinellidae. Larvae may be obtained either by collecting in the field, or by extraction from the substrate in a Tullgren funnel or similar device. If gravid female beetles are found (recognisable by their distended abdomens), they may lay eggs if kept on a moist medium such as moist sieved soil or peat (as supplied by garden centres for use as seedling compost). When any larvae hatch, they can be treated similarly to those collected in the field.

The problem in rearing such larvae is that they must be kept individually in order to avoid cannibalism. Each can be maintained in a small container with a moistened absorbent base (blotting paper, plaster of Paris, soil or peat). The addition of a further layer or piece of moss provides convenient shelter for the larva. Food, either live prey such as Collembola, small dipterous larvae, or chopped-up pieces of larger food such as mealworm larvae, earthworms, etc. should be added regularly. Be careful, however, to remove any dead, uneaten food before it decomposes. Some larvae will eat unnatural food such as small pieces of raw or cooked meat. It is always preferable, however, to find an apparently 'natural' food if possible. Cool and moist conditions must be ensured, and the food changed daily. Lampyrid (glow-worm) larvae feed on snails such as *Helix* and *Limnaea* species, and will display their light if gently handled. They pupate, like many terrestrial larvae, among moss and leaf litter on, or near the surface of the soil. Cocinellids may also be reared as above, fed with leaves bearing colonies of aphids. Detailed advice on rearing ladybirds is given by Majerus & Kearns (1989).

Larvae of tiger beetles (*Cicindela*) can be found in vertical burrows on dry, exposed sandy or peaty soil. In order to rear the larvae, blocks of soil about 25cm deep and 15cm square, with a burrow in the centre should be dug up, and held together by wrapping in polythene bags. Place the soil in a south facing, sunny position, and feed the larvae with small caterpillars or similar food, held in forceps. Keep the soil surface slightly dampened. When the larvae pupate in the burrows, cover the surface with polythene or glass to prevent drying out.

Aquatic predatory larvae

This includes mainly larvae of Dytiscidae, but also other Adephaga such as Gyrinidae and Hygrobiidae, and the aquatic species of Hydrophilidae.

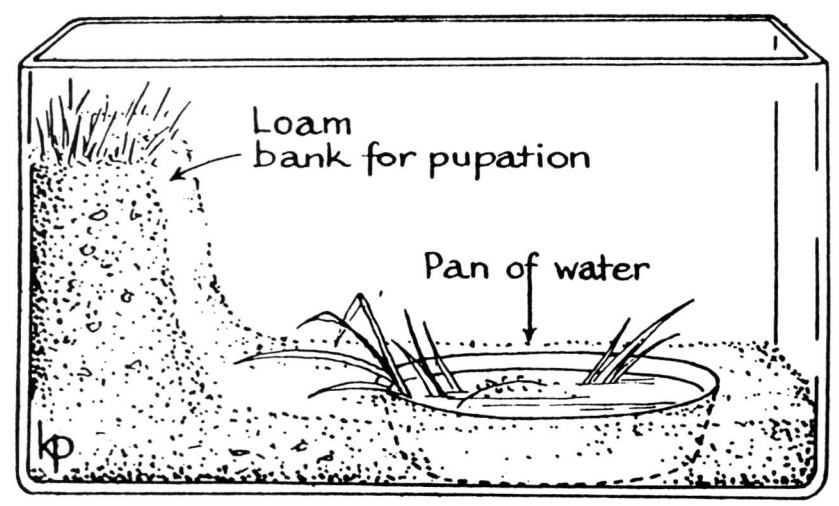

Fig. 55. An aquarium terrarium.

Dytiscid larvae are best obtained in the late summer and reared by placing them individually – as they are cannibalistic – in a shallow glass or earthenware dish almost filled with water. It is advisable to use a quantity of the water from which they were originally taken as this would contain plenty of minute crustacea to support them during their early stages. Some small aquatic plants should be thrown in to provide anchorage for the larva and to enable it to crawl out on to *terra firma* (a piece of sheet cork is suitable for the latter purpose). The dish and its contents should then be lowered into a larger tank (preferably of glass so that the metamorphoses may be followed more closely), the bottom of which has been covered with about 2.5cm of moist loam and has a steep bank of loam at one end (Fig. 54). Some dytiscid larvae, particularly those of

Dytiscus, pupate more readily when confronted by a vertical 'wall'. Larvae will feed on freshwater crustacea, worms and even on small tadpoles. Always procure and preserve the larval exuvium as soon as the pupa is revealed. When it is desired to rear several aquatic species for general interest and observation a shallow pool should be constructed (see Macan, 1982). Straight-sided glass beakers have proved quite suitable for rearing species up to the size of *Ilybius ater* De Geer. These are provided with a basal depth of soil, fine silt, gravel or mud from the bed of a pond. Do not let this mud be much deeper than 2cm and plant in it a sprig of some small aquatic plant such as *Ranunculus fluitans, Luronium (Alisma) natans* or *Elodea,* and also a small fragment of some aquatic moss. Care must be taken to avoid too much vegetation or it will be very difficult to keep the life history under observation. After the plants have been introduced, the beaker is filled with water from a pond up to within a half-inch of the top and is covered with a square of glass or piece of cling film held down with a rubber band. The beaker is then set aside for at least a week for the silt to settle and for the plant to take root. Pond-water should always be used since it introduces an abundance of small Crustacea and other suitable food for the young larvae.

It is possible, by varying the plant introduced into the beaker and by the type of pond from which the water is taken, to vary the conditions from acid (sphagnum bog) to alkaline (chalk pond), consequently widening the range of species it is possible to rear.

With these small aquaria two separate courses are open. Either they may be used exclusively for breeding *ex ovo;* or larvae taken in the field may be reared. In the former case adult beetles, one species only to each beaker, are introduced and experience has shown that it is best to use two females to each male to allow selection of mate. It is advisable to note the number of specimens introduced, since after a sufficient number of eggs have been observed, the imagines must be removed or they will devour either the eggs or the young larvae. The particular season at which a breeding aquarium is begun is of no importance, since it is possible to keep the imagines alive for up to two years under these conditions and they will breed in their due season. All that is required is that the adults be fed at intervals: the larvae of *Chironomus* spp. are excellent for this purpose. For a beaker with three specimens of an *Agabus,* fifteen to twenty *Chironomus* larvae should be an adequate supply for perhaps ten days. Very little is known of the egg-laying habits of the carnivorous water beetles and it is always necessary to make regular observations to ascertain the periods and sites of oviposition. The eggs are laid either at random on the plants, frequently in the axils of the leaves, or in the mud at the bottom, either just on the surface or even about half an inch below it. Once oviposition has been observed, constant watch must be kept until sufficient eggs are judged to have been laid: the adults are then removed. Emergence usually takes place in five to ten days, but it has been recorded that eggs of *Agabus montanus* Stephens. have overwintered, emergence taking place more then 240 days after laying. The young larvae will usually be satisfied with Entomostraca and Cladocera and this will suffice to rear nearly all the Hydroporines to full size, more food being introduced from cultures as may be required. Cannibalism among the larvae is of very varying prevalence and only experience will show the optimum number of larvae that can safely be allowed. It is worthwhile to allow for more eggs being laid than this optimum number as cannibalism will make use of the excess as valuable food supply. With Agabines and Colymbetines feeding with *Chironomus* larvae becomes necessary when the larvae increase in size. Up to a limit, the more food offered the more rapid is growth.

All aquatic Adephaga pass through three larval instars and pupation takes place out of the water. With the technique outlined above, it is necessary to remove the larvae from the aquaria to terraria for pupation. When they are ready for this transfer they become restless and will be noticed trying to escape from the water. They are then plump and the fat bodies may be seen to be well developed.

For a terrarium one can use either a smaller beaker or fairly shallow glass bowl. Ordinary soil is fairly tightly pressed into it. Divide the bowls into about four compartments, either with glass plates or stiff board, and place one larva in each compartment. The dividing plates must reach right up to the covering glass or the larvae will wander around and this may lead to confusion and mistakes in identification when dealing with field-caught material, unless care is taken to ensure only pure cultures in both aquaria and pupation terraria. It is best to have a few bits of dead grass or moss lying on the surface of the soil in the terraria, for some species seem to prefer to build their pupal cells on the surface, using the dead grass as rafters in the mud roof. Others will excavate a fairly deep burrow before enlarging the end to form a pupal cell.

It is worthwhile to mention that certain species appear normally to breed during the winter months, so that it is possible to maintain the interest during the whole year.

With species of a larger size, such as *Hydaticus, Acilius* and *Dytiscus*, breeding from the egg requires larger containers to serve as aquaria. Rather stronger plants will be required, since these genera oviposit within the stem tissues.

Rearing larvae captured in the field may all be done in beaker aquaria, even with the largest *Dytiscus*. As when rearing *ex ovo,* care must be taken that the individuals placed in tumblers are a pure 'culture' or otherwise serious mistakes in identification of larvae will be inevitable. For this reason this method is not so satisfactory in the results achieved, since without a lot of experience it is very difficult to compare and identify an active living larva with preserved material. In genera such as *Hydroporus* the specific characters are so fine that it is often very difficult to be sure of the specific identity or of the difference between two preserved specimens, let alone a living and a dead specimen.

Adult Gyrinidae are more difficult to keep alive in aquaria for any lenth of time, but gravid females may be kept long enough for them to lay their eggs. Larvae can then be reared as with Dytiscidae; when fully grown the third instar becomes dark brown, and needs to climb out of the water onto a vertical surface on which to pupate. Either stones at the edge of the aquarium, or stems such as those of *Phragmites,* sticking vertically out of the water, will suffice. The larvae pupate in cocoons formed from debris in the water, and adults will emerge after a few weeks.

Adult Hydrophilidae such as *Hydrochara* and *Hydrophilus* require the same technique as is used for the larger dytiscids; but they are very much more cannibalistic. *Helochares, Enochrus, Cymbiodyta* and *Hydrobius* may be fairly satisfactorily reared in the beaker aquaria; but it has been found that there is a high rate of cannibalism among them and that their restlessness brings them rapidly into dangerous contact with one another.

Many of the remaining genera of the Palpicornia have the larvae either terrestrial or semi-aquatic; only a few are truly aquatic. For such genera as *Limnebius, Ochthebius, Helophorus, Hydrochus, Anacaena, Paracymus,* aquaria-terraria provide the best means of rearing them successfully. Bowls may be filled with soil from the edge of a pond, shaped in such a way as to leave a depression occupying one-third or half of the bowl. The depression is filled with water to a varying depth, depending on the genus intended to be reared. It may be from one or two millimetres up to two or three centimetres. The 'bank' portion is provided with small plants and a sprig of moss or another small aquatic plant and a young plant of *Glyceria* is planted in the 'pool'. As always, the aquarium-terrarium must be allowed to settle down for a few days. Adult beetles may then be introduced. The position in which the egg-masses are deposited varies enormously in this family. Some place the cases direct in the water, others partly in the water and others in or on the drier portions of the bank. Both *Limnebius* and *Ochthebius* are, like *Hydraena,* feeders on Confervae and the larvae are as easily drowned as are most *Hydraena* larvae. The aquaria-terraria intended for them are therefore nearly without water - wet mud suits them very well. Most of the other genera mentioned above prefer a wetter situation. Feeding is occasionally required and very small *Lumbricus* earthworm, freshly killed, are quite suitable. Cannibalism is advantageous in that it allows one to avoid some of the tedious time spend 'feeding' the aquaria-terraria.

According to Angus (1973), larvae of *Helophorus* may be reared on damp blotting paper in pill boxes, in a similar way to some Carabids, but ensuring that the paper is saturated rather than flooded. They should be fed on *Tubifex* worms; when fully grown they are provided with a bank of earth in which to pupate.

Aquatic phytophagous larvae

This includes most Haliplidae, Dryopidae, Elmidae and probably Scirtidae. They are best reared in aquaria-terraria as suggested for the semiaquatic Hydrophilidae above, but with plant material and fresh-water algae as food. Some, such as *Hydraena*, will survive on encrusting algae found on decaying leaves at the water's edge. The water should be changed regularly, and replaced with water similar to that from where they were collected. An alternative to soil at the edge of the container, is to use a slightly tilted box or tank, and to wedge a few layers of bark in at the shallow end. This provides a damp, but not submerged, pupation site.

A further, interesting group of plant feeding but aquatic larvae, are those of the chrysomelid genera *Donacia* and *Plateumaris*. These may be collected from late summer until mid winter, although the latter season involves dipping one's hands into icy-cold water and the handling of cold slimy mud. By means of a strong hook at the end of a stout handle, or by the use of a grappling iron (if the water is deep) or by the use of the hands alone where it is sufficiently shallow, masses of tangled rhizomes and roots of *Typha, Nuphar, Potamogeton, Phragmites*, etc., may be torn up. Portions should be washed carefully and not too violently in the water and searched over. If the locality is a favourable one, glabrous brown cocoons, oval in form, will probably be found attached by one end to the sides or under surfaces of the rhizomes or occasionally to the roots growing from these, together with small fat white larvae attached by a posterior pair of hollow respiratory spines to the epidermis and cortex of the same plant structures. The cocoons will be found to contain pupae or newly-emerged adult beetles. During the summer and autumn, larvae of *D. vulgaris* Zschach and *P. sericea* (L.) may be collected easily and if the portions of rhizome are placed in water in shallow receptacles, the mode of feeding adopted by the larva and the making of the cocoon can be observed. The period of pupal life is short and within a period of two or three weeks, adult beetles, prepared for hibernation, can be extracted from the cocoons. In coastal areas *Plateumaris braccata* (Scopoli) should be sought for in the rhizomes of *Phragmites*.

Dung-inhabiting larvae

Dung-feeding larvae are by no means easy to rear, as the dung is very susceptible to mould, especially when the earth beneath it is damp. If larvae are reared by the following method, however, this complication should be overcome. Fill a medium-sized flower-pot to two-thirds with damp loamy soil and press it well down (Fig. 55). Next fill the pot up to just below the rim with fresh moist dung and introduce the larvae, making small holes for them if necessary. Then stand the pot in a shallow dish of water and place it in a well-ventilated situation such as a shady corner of the garden. Do not moisten the dung from above. Some dung-living larvae (e.g., many Histeridae,

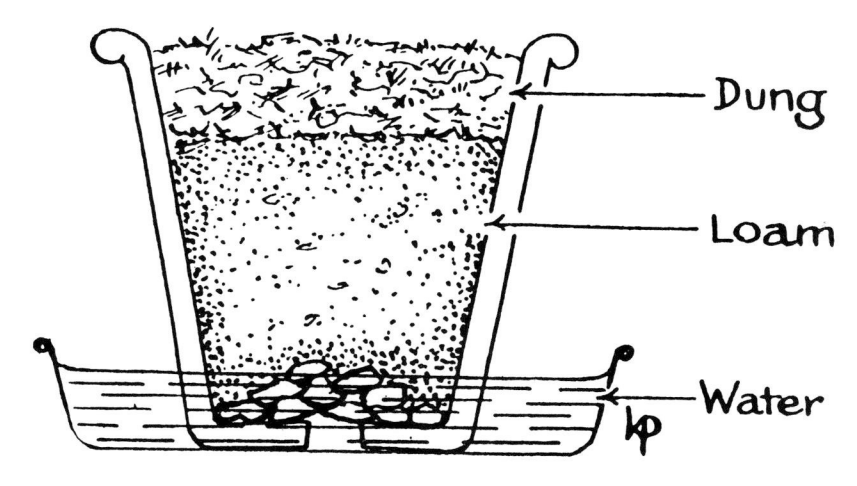

Fig. 56. A pot for dung beetle larvae.

Staphylinidae, etc.) are predaceous on other larvae (especially of Diptera), so the dung should be well stocked with the latter before the former are introduced.

Soil-inhabiting root feeders

This includes many larvae of Elateridae, as well as some chafers (Scarabaeidae) and Byrrhidae, Dascillidae, etc. Scarabaeid larvae that feed on fibrous roots should be reared in jam-jars filled with earth and root fibres or, better still, in a flower-pot containing a living plant.

An alternative, intensive method that has been used successfully to study the development of such larvae individually, is to make a plaster of Paris block, 20cm square by 5cm deep. The surface of the block is then drilled out with a cork borer or similar tool, making a series of chambers, about 2cm diameter by 1cm deep, in the surface of the block. Larvae are kept separately in each chamber; the whole block is kept moist by standing in a tray of water, and covering the block with a glass sheet. Each larva is fed daily.

Larvae of Elateridae (click beetles) are found in a wide range of soil types and conditions. The breeder must simulate very different kinds of natural habitats if he is to be successful with click beetles. Since so many species turn cannibal on the least provocation, it is important to keep individuals apart. Corked vials, with a small runnel cut down one side of the cork to allow aeration, filled to at least an inch below the cork with soil or wood, are very suitable. Small jam jars, tins with a few holes in the lid, or any other closed receptacle will do equally well. The wood or soil should be damp but not wet.

Larvae will eat their way through a cork or escape through the nick in the cork if they can at all reach it - and they can achieve the seemingly impossible in this respect. For food, living turf or roots of the plants beneath which the larvae were found may be supplied to the soil feeders, or grains of corn may be substituted for nearly all species. This is particularly useful when breeding in vials; each time the soil is changed, fresh grains are provided; it is then simple to see if the larvae have been feeding since the last meal. Mouldy grains must be replaced quickly. In spring, the larvae of many species feed at a great pace, and one larva may consume a dozen sprouting grains in a few days; if more are not supplied, the breeder is asking for trouble in the way of bored-cork escapes. Some species, like *Oedostethus quadripustulatus* (F.), appear to have very short feeding periods, whereas others, like *Agriotes* spp., feed during many months of the year. *Ctenicera cuprea* (F.) feeds strongly in spring and summer, but rarely in autumn and winter; *Athous campyloides* Newman, on the other hand, feeds nearly as much in autumn and winter (if mild) as during the other seasons.

Terrestrial phytophagous larvae

This accounts for most Chrysomelidae and Curculionidae. Surface-feeding larvae, found on the leaves of their host plant, include many chrysomelids (hence their name 'leaf-beetles'), but few weevils except the genus *Cionus*, whose larvae are covered with a glutinous layer of slime. Rearing such species seldom presents any difficulty as their larval period is comparatively short. They develop satisfactorily in a glass-topped tin containing a layer of damp blotting paper and a sprig of their food plant, which should be replenished at the first signs of wilting. Sometimes it will be found convenient to introduce a few plants into the garden so that a regular supply of the right foliage may be at hand. Alternatively, the host plant may be kept with its stem in water, but care must be taken to prevent larvae leaving the plant and drowning in the water, especially when they are searching for somewhere to pupate. Rearing phytophagous beetle larvae is in many ways similar to rearing Lepidoptera larvae and methods used by lepidopterists may usefully be copied.

Species whose larvae live internally within plant tissue (most weevils, many Alticinae - flea beetles) are more difficult. Plants thought to be containing larvae can be transplanted into pots in the house and enclosed so as to capture any emergent adults. Leaf-mining species of tribe Rhamphini can be reared on cut shoots of their host tree stood in water; virtually any beech tree will yield the larval mines of *Orchestes fagi* (L.) if young shoots are collected in May. Leaves rolled by rhynchitid weevils, and seed pods infested by Apionidae, can be collected and maintained until emergence of the adult beetles.

Beetles that breed in flowers, such as the abundant *Meligethes,* and species of *Anthonomus* can be reared similarly by collecting infested plant material.

Wood and bark inhabitants

There is a great variety of beetle habitats associated with wood, from standing trees in varying stages of health, through recently cut logs, and their gradual stages of decay to thoroughly rotted stumps filled with wood mould. The immature beetles associated with these vary accordingly: Cerambycidae, Buprestidae (if lucky), Curculionidae: Scolytinae in fresh or relatively-recent dead wood; Lucanidae, some Elateridae and Scarabaeidae (for example *Gnorimus nobilis* (L.)), Melandryidae etc. in well decomposed but dry conditions. The general rule mentioned earlier applies here: always attempt to rear wood-inhabiting larvae in as near natural conditions as is possible. Remember that many larvae in such habitats are predatory, so a variety of prey organisms also need to be present. For sub-cortical species such as 'scolytids', if pieces of bark of suitable size are placed with their inner faces clamped towards one another it is usually possible thus to obtain pupae and later the beetles. Owing to the curvature of the strips of bark, it is of course necessary to use some pressure to put the two faces in sufficiently close apposition and therefore the portions used must be of limited size. Desiccation is here as fatal as excess of moisture, the latter condition being conducive to a deadly growth of moulds. Fitting the curving pieces of bark against household fire-logs has not proven successful, but others might try it nevertheless.

In order to rear cerambycid larvae, a comparatively sound section of the infested branch should be sawn off and brought back with the larvae. Holes of a diameter slightly greater than the maximum breadth of the larva should be drilled longitudinally at one end: they should be of sufficient depth to accommodate individual larvae, which should be inserted head first. When the larvae are collected, it should be observed whether they are feeding in the heart wood or under or in the bark and the holes drilled accordingly. The section should then be inserted to a depth of 6-8cm in a tray of moist sand. A day or two later – by which time the larvae will be secure in their burrows – the section should be inverted so that they are feeding head-upwards unless they have been observed to feed head-downwards in the field, as a small number of species do. The wood should be lightly sprayed with slightly saline water periodically, and the sand moistened with a weak aqueous solution of potassium permanganate to prevent the formation of mildew. To prevent the adults subsequently escaping, the wood and the tray should be enclosed in a cylinder of wire gauze or perforated zinc. Alternatively, one may simply keep sections of wood without introducing extra larvae into them, in the hope that larvae already in the wood will complete their development and emerge as adults. Be patient though; drying out of the wood in particular extends the developmental period, and beetles have been known to take years to emerge from cut timber (the maximum noted is 30 years in the case of an American buprestid!). See also the chapter 'Cerambycidae' by Martin Rejzek, pp 121-142.

Larvae such as Lucanidae and Elateridae, living in crumbling, decayed wood, should be placed with sizeable portions of the decayed wood in plastic boxes or similar containers. Spaces between larger fragments of wood should be filled with smaller pieces and wood powder. The lid should be perforated to prevent condensation, but the wood should be kept slightly moistened but not wet.

Detritus and fungal feeders

Very many Staphylinidae, Latridiidae, Mycetophagidae etc., are found in a range of habitats including fungal fruiting bodies, litter and refuse, bird nests, straw and compost heaps. A container should be filled with the material from which it is hoped to rear larvae, and treated as for larvae in thoroughly rotten wood (see above). Larvae found by sieving the material can be added if required. If the material is placed in a darkened box, with a small hole on top to which is taped an inverted clear glass or plastic vial, this functions as a crude 'emergence trap'. The larvae in the material will avoid the light, and remain in the damp conditions of their rearing medium; emergent adults, however, are usually attracted to the light, and will be found crawling around in the vial, trying to escape.

References

Angus, R. B., 1973. The habitats, life histories and immature stages of *Helophorus* F. (Coleoptera, Hydrophilidae). *Transactions of the Royal Entomological Society of London*, 125:1-26.

Macan, T.T., 1982. The Study of Stoneflies, Mayflies and Caddis Flies. *The Amateur Entomologist* No. 17. Amateur Entomologists' Society, Feltham.

Majerus, M.E.N. & Kearns, P., 1989. *Ladybirds* (Naturalists' Handbooks 10) 103pp. Richmond Publishing, Slough.

Conservation

CONSERVATION

by Jonathan R Webb

(The opinions expressed in this chapter are that of the author and not necessarily shared with English Nature).

1. The Changing State of the British Countryside

The invertebrate fauna of Britain is one of the best recorded in the world. We have the enviable situation, not shared by many other countries, of having a long history of biological recording and collecting. This has allowed us to draw up a comparatively good picture of our fauna, including beetles.

We are also at an advantage because we live on an island. This 'island perspective' makes our fauna more distinct and more easily definable as it has clear boundaries (i.e. the sea) that are barriers for colonisers. The disadvantage is that the British fauna is more impoverished when compared to many other countries of similar size, latitude and land-use on the continent.

Although our countryside is still changing, over the last two decades the outright destruction of many of our semi-natural sites has slowed down. This is mainly due to the fact that a lot of the last remaining sites are either protected by law, impossible to improve agriculturally or they receive some other form of protection under the planning system. What we are left with throughout much of the rural countryside is a patchwork of agricultural land with widely dispersed little 'island' sites of semi-natural wilderness. These islands may be woodlands, wetlands, heathlands or unimproved grassland, but they all too often share one thing in common – they are isolated from each other. Isolation is a big problem: the distance between sites makes it is very difficult for many species to colonise other suitable areas. There is also a greater possibility of local extinctions, as such smaller sites invariably contain smaller numbers of individuals within any given colony. This makes them more susceptible to catastrophic events such as disease, weather, pollution or inappropriate management. For these areas, site management alone will inevitably fail, and our main aim for the survival of invertebrate species diversity should be to re-create the links between the isolated sites.

There are some larger areas of land that are less intensively managed, such as our National Parks. These tend to have a much larger mosaic of habitats and are often the mainstay of many of our important areas for beetles. Not only is isolation less of a factor in these large sites, but they also provide more 'choice' for species, in particular a rich variety of ecotones (habitat edges), which many of our rarer species tend to be associated with.

The rate of decline of many of our semi-natural habitats is much reduced from what it was twenty years ago. The importance of heathlands and woodlands, in particular, is now much better highlighted. The total area of heathland has more or less stabilised and may even be increasing in some areas mainly due to restoration projects. The total area of woodland is increasing, but ancient woodland is still being destroyed, mainly by development.

It is now fair to say that a lot, but not all, of the current habitat destruction and degradation tends to be focused on particular habitats. These habitats are most often linked with old agricultural systems but also include wetlands drained for new agricultural land. Those habitats of particular importance to beetles and that are decreasing at a relatively fast rate, include traditional orchards and unimproved grassland. The losses of both these habitats have been catastrophic in the past, with 97% of unimproved grassland being lost between 1930 and 1984 (Fuller 1987). There is still a lot of anecdotal evidence that such areas are still decreasing at a high rate and evidence from specific counties have shown that losses between the mid-1980s and the early 1990s is still high (Jefferson & Robertson 1996). Losses of traditional orchards in England have also been severe in recent decades, with estimates by Common Ground (2000) ranging from 40% loss (in south Buckinghamshire between 1945 and 1975) to 95% (in Wiltshire since 1945). An estimate based on Defra census data and pesticide usage figures suggests that less than 7,000 ha of traditional orchard habitat remain in England. It is highly probable that this habitat is still in strong decline.

Site quality is another issue but one which is more difficult to measure. At one stage the UK was losing many hedgerows. Between 1984 and 1990 the net loss of hedgerow length in England was 21% (UKBAP 1995) but now it seems that this has more or less evened out as new hedgerows have been created. It is difficult to tell what the quality of these new hedgerows may be, however. Parklands, often a coleopterist's heaven, may also have declined in ecological quality. Many such sites, although they still contain a good number of veteran trees, tend to be in intensive farmland where the grassland is either over-grazed or mown a number of times a year. Trees in such sites often show bark damage, from stock or machines, and excessive tree surgery, including removal of dead wood. A survey of one Midland county suggested that over 70% of all parkland sites were in intensive grazing agricultural management (Webb and Bowler, 2001). The age structure of the trees within such habitats can also be of great concern; many parklands lack new cohorts of younger trees to replace them. There has been no comprehensive survey of sites to ascertain the size of this problem but anecdotal evidence suggests that it is the case with the majority of parklands.

Pollution of one type or another is also affecting the remaining semi-natural areas. This includes point source pollution, such as sewage and industrial waste, and diffuse pollution, such as agricultural drift and eutrophication (nutrient enrichment) of our rivers. The effect of pollution is often more difficult to quantify and measure than direct destruction, but it is certainly a major factor. Wetland and water systems are particularly susceptible to this form of degradation.

Another highly threatened area of land which often contains beetle interest are 'brownfield' sites. Brownfield sites are those areas that have previously been affected by human development. This can include anything from derelict housing, old industrial workings to a large mining operation, such as a clay pit or sand pit. Many of the more interesting brownfield sites, such as those along the Thames corridor, are of exceptional quality for invertebrates because of their early successional nature, i.e. they contain plenty of bare areas and a variety of flower-rich grassland and scrub that provides warmth, food source a shelter for a host of insects. The government has set targets for a minimum of 60% of new houses to be developed on brownfield sites. Together with targets for increasing household densities, the preferential use of previously developed land is intended to control urban expansion and avoid the destruction of land perceived to be more natural. Previously developed land available for redevelopment amounts to approximately 65,000 ha in England (about 0.5% of the total land area). This land is not confined to urban areas but is also found in urban fringe and rural situations, e.g. China clay tips in Cornwall. Needless to say, the government target for brownfield development presents an obvious threat to a large number of sites that are potentially very interesting for all insects, as well as beetles.

In summary:

- Nature Conservation should be about biological enrichment of the land in which we live, not about micromanagement of small, isolated sites. Therefore, the remaining islands of habitat must be properly managed and efforts should be made to link up such sites.
- Although the rate of loss of many semi-natural habitats has declined and even halted in some instances, we must remain aware that particular habitats are still under threat.
- The quality of many sites may also be decreasing, although this is more difficult to measure.

2. Collecting and Recording

For many coleopterists reading this book, collecting and recording species will be the main focus for most work. Your activities as a recorder are of paramount importance to the conservation of beetles but **only when you report what you have found**. It should be an important aim and goal of site managers and members of nature conservation bodies to encourage and allow you onto a site when they know you will provide them with information. Reassure site managers that you will not be causing any long-term effects on their site. It is my personal view that for the Coleoptera, the collecting of all but a few species protected by law is a positive, rather than a negative, issue when your findings are recorded. Collecting has no substantial effect on the invertebrate populations of most species. Far more animals are killed through habitat destruction and predation, particularly from birds. As a strapline to one talk on insect conservation once said, 'conserve insects – shoot a blue tit'.

Always gain permission to collect and record from a site. In the case of private land this may be difficult but spending the time and effort to determine the landowner and gain their permission will outweigh the problem of getting caught. When working on any land, but in particular sites managed by nature conservation organisations, you may have to be aware of various health and safety issues. Public liability is another issue that collectors and surveyors should be aware of. It is best to make an initial visit to a site with a site manager.

When you have finished your work and carried out your identifications and analysis, send a report to the site manager. Try to steer away from just a list or even an annotated list of species – it will generally mean very little to a site manager. If you are only looking for a specific species or community (e.g. dead-wood beetles), then tell the site manager. It also very important to record the habitat and locality of the animal on site, not just the site location. This may help the manager make better judgements about site management. Give them some feedback;

- State where you think the main importance of the site is and why. Perhaps the site contains a good dead-wood assemblage within hedgerow trees along one of its site boundaries or, perhaps the water beetles in the brackish ditches are of high quality; both these pieces of information have may effect future site management.
- State what you feel about the site's importance based on the species you have found: little or no importance, local importance or national importance? This may, of course, change in time but site managers like to know how their site is rated. If you are not sure, then say so.
- If there are problems on the site that are easily rectifiable and do not conflict too much with other overriding interest, then tell them. Is it too shady? Too tidy? Too homogenous? Too dry?

They may be surprised to learn what you think the best part of their site is. Perhaps it was the old sand pit and band of ruderal plants growing along the eroding path within their heathland. If you don't tell them there is a good chance that nobody will and these sort of features may be seen as a potential problem or an eyesore. Also, if you get chance, show them an animal or two. If it's alive then all the better, but a carded specimen will also be of use. This adds no end to a site manager's interest. All of sudden you have put a 'face' to a latin name and this may steer them towards better conservation of the beetles in question.

Remember that many sites (the majority of SSSIs and County Wildlife Sites and reserves) are selected for their botanical or ornithological interest. There may be times where such issues conflict and, in such cases, compromises can be sought. Perhaps an area of the site can be managed for the species in question that will not stifle the site's other interests. Entomological site management is often about compromise.

Being good at entomology and the ability to find rare species does not necessarily mean that he or she is good at prescriptions for site management. To do this, experience is needed in determining what effect specific management tools (such as grazing) may have on a site. All too often entomologist's say to a site manager 'this bit here is great so don't do anything with it' without realising that sites are always in a state of change and succession. Often one must put the clock back to earlier parts of succession by what may initially look like quite radical kinds of management. Using mechanical diggers may raise some alarm when clearing vegetation to expose bare soil but it does not take long to regenerate. If done in the correct area, this will have little effect on other site features of interest.

If in doubt about site management, then it is better to make no comments about it. The best thing to do may be to point out what is good on a site and then let the site manager determine how it got to that point and how it should be re-created.

2.1 Protected Species

There are a few species that are fully protected by law and these must not be collected without a licence. If you do wish to carry out work on these species and your efforts are for its conservation and/or scientific study, then you may apply for a licence by contacting the appropriate country agency. The table below shows the beetle species that are protected by law.

Where a beetle species receives full protection it is an offence to:

- Intentionally kill, injure or take such a wild animal
- Possess such an animal
- Intentionally or recklessly damage, destroy or obstruct access to any structure or place which such an animal uses for shelter or protection.
- Intentionally or recklessly disturb such an animal while it is occupying a structure or place used for shelter or protection.

(This is a general guide to the provisions of the Wildlife & Countryside Act (as amended). Consult the relevant legislation for further details)

*Beetles			
Chrysolina cerealis	Rainbow leaf beetle	Full Protection	1981
Curimopsis nigrita	Mire pill beetle	Damage/destruction of place of shelter/protection S.9(4)(a) only	1992
Graphoderus zonatus	Water beetle	Full Protection	1992
Hydrochara caraboides	Lesser silver water beetle	Full Protection	1992
Hypebaeus flavipes	Moccas Beetle	Full Protection	1992
Limoniscus violaceus	Violet click beetle	Full Protection	1988
Lucanus cervus	Stag beetle	Sale only S.9(5)	1998
Paracymus aeneus	Water beetle	Full Protection	1992

*Such lists are reviewed on a five-yearly basis and this list may change in the future.

3 A who's who of users of entomologists' information

The **National Biological Records Centre (BRC)**, part of the Institute of Terrestrial Ecology at Monks Wood, runs quite a large number of Coleoptera recording schemes. When individuals who are prepared to give time to such endeavours run these, they are invaluable in giving an overview of a species' national distribution and also collating ecological information on the species themselves.

A large number of **Local Biological Records Centres** exist throughout the country. These may be housed in a variety of locations; local museums; county or district planning departments; and County Wildlife Trusts. The degree of beetle and other entomological information that they store is variable although most systems now tend to be fully computerised. It is not uncommon that the backlog of old information can still only be found on card index systems. These records centres play an important role in supplying information to local authorities, consultants, conservation bodies, students and the interested public.

The **four country agencies** were created in 1991 out of the former Nature Conservancy Council to advise government on nature and geology within each of their specific countries. These are **English Nature, Countryside Council for Wales, Scottish Natural Heritage** and the **Environment & Heritage**

Service (Northern Ireland). All four agencies have local offices and are supported by a national office of scientists and experts. The set-up of each country agency differs slightly but each one has staff dealing purely with insect conservation. Each country agency manages, and help others to manage, Sites of Special Scientific Interest (SSSIs) or Areas of Special Scientific Interest (ASSIs) and National Nature Reserves (NNRs). It is the main aim of each country agency to make sure that these areas stay in good condition and supports the features for which it was designated. The agencies also run research programmes on issues that affect wildlife as well as taking a lead role in helping the government implement the UK Biodiversity Action Plan.

The **Joint Nature Conservation Committee (JNCC)** is the UK Government's wildlife adviser, undertaking national and international conservation work on behalf of the four country nature conservation agencies. JNCC provides an overview of the work of the country agencies as well as giving advice on UK policy and legislation regarding species. It also commissions and supports surveillance and monitoring schemes to assess and report upon the changing status of species in the UK. As part of the latter programme, JNCC has produced two Red Data Books (Shirt 1987, Bratton 1991) and several species reviews to assess the conservation status of invertebrates in the UK. Unfortunately these are now out of date and need revising.

The **Environment Agency** in England and Wales and **Scottish Environmental Protection Agency (SEPA)** has similar roles. They are the leading public bodies responsible for protecting and improving the air, land and water environment. The Environment Agency has a vested interest in watercourse quality and structure and they have carried out some interesting studies on river shingle and on water beetles over the last few years. They may have records of species for some stretches of rivers and will also be interested in receiving information from such areas.

The **County Wildlife Trusts**, of which there are forty-seven, are co-ordinated through The Wildlife Trusts whose headquarters is based at Newark. Each of the local Trusts cover an individual county or area of Britain and are responsible for the care and management of a very large number of nature reserves. They also act in an independent advisory capacity on matters of wildlife importance in the county, including planning matters and sustainable development. They are particularly interested in information on the insect life within their county or area. Information on Wildlife Trust reserves will be of particular importance to their site managers.

Invertebrate Link is another organisation that has been set up to further invertebrate conservation. It was originally called the Joint Committee for the Conservation of British Invertebrates (JCCBI) and its main functions are to co-ordinate the activities of existing societies and organisation. These include the country agencies, National Trust, The Wildlife Trusts as well as professional and amateur entomological societies, such as the Royal Entomological Society and the British Entomological and Natural History Society. Invertebrate Link has produced codes on insect collecting as well as lists of important and threatened habitats.

Buglife - The Invertebrate Conservation Trust, has also recently been set up. It is the first organisation in Europe committed solely to the conservation of all invertebrates. Its main aim is to prevent invertebrate extinctions and maintain sustainable populations of invertebrates in the UK. This new charity sees its role as undertaking and promoting crucial study and research, promoting sound management, ensuring that more resources are put into invertebrate conservation, supporting the work of other conservation organisations and promoting education and publicising invertebrates and their conservation.

3.1 The Changing Nature of Entomology

There are a small number of professional entomologists employed by the country agencies who work in the fields of conservation, but by far the main body of people interested in insect conservation are amateur recorders and collectors. There are also a number of entomologists working within the National Museums, although many of these tend to cover world taxonomy rather than concentrating on British species. There are also many curators working within natural history sections of local museums who have a keen interest in entomology.

There has been an interesting development in recent times mainly due to increased environmental awareness such as changes in the planning process and the advent of the UK Biodiversity Action Plan. For example, there is now often the requirement for biological survey for and environmental impact assessment as part of a developers' planning application. This has meant that there is a greater need for environmental consultants and these must often take entomology into account. Therefore, there are now several entomological consultants now operating within Britain.

4 Nature Conservation and Entomology
4.1 Where we were

Nature conservation and its science is still a comparatively new concept. Although many individuals have been making efforts since the 19th Century to conserve the wilder parts of Britain, it is only in the last 50 years that the principles of this concept have become much better understood. In comparison, invertebrate conservation is still in its infancy, with the main concepts only really being born two decades ago.

The increase in our knowledge of beetles and insect conservation over these past two decades has been immense. This has mainly been due to two groups of people: the pioneers of invertebrate conservation within government agencies and other nature conservation bodies; and the amateur entomologists who have spent so much of their valuable time in recording beetles. From the early '80s to the beginning of the '90s there were a number of dedicated invertebrate surveys and projects undertaken on an unprecedented scale by the statutory agencies. This included the setting up of the Invertebrate Site Register, major surveys of particular habitat types, the activities of various field units and the production of the Red Data Books and scarce species reviews. At one stage there were over fifteen entomologists working for what was then the Nature Conservancy Council. These were the halcyon years of entomology for the government agencies and many thoughts and ideas were pioneered at this time. Many of the projects that were set up over this time period often did not answer the questions that they originally posed, or never really asked the right question in the first place. It cannot, however, be overstated how much these projects helped to build up a picture of our invertebrate fauna and laid the basic concepts for future invertebrate conservation. This included the promotion of the importance of habitat mosaics, vegetation structure, bare exposed substrate, scrub and mild disturbance were all promoted as important features rather than problematical issues on site. The skill and experience gained by the agency staff employed at the time has meant that a high degree of knowledge still exists within the conservation world which can be drawn upon. Relative newcomers to invertebrate nature conservation find advice from such people invaluable (as I have). Unfortunately, a lot of this information is not readily accessible to a wider audience and, in many cases, not published at all.

In summary:

- Invertebrate conservation is still a relatively new subject
- The overall understanding of entomology is now far greater than it was twenty years ago. This allows conservationists to make a better-informed judgement on the conservation future and how we should proceed.

4.2 Where we are

At a national level, a large proportion of beetle conservation is now heavily influenced by the UK Biodiversity Action Plan (UK BAP). Below is a short explanation of the UK BAP, in how it came about and what it includes. The UK BAP website can be found at www.ukbap.com.

4.2.1 The UK BAP

In June 1992, the Convention of Biological Diversity was signed by 159 governments at the Earth Summit, which took place in Rio de Janeiro. It entered into force on 29 December 1993 and it was the first treaty to provide a legal framework for biodiversity conservation. It called for the creation and enforcement of national strategies and action plans to conserve, protect and enhance biological diversity.

In 1995 the UK Biodiversity Steering Group published *Biodiversity: The UK Steering Group Report. Meeting the Rio Challenge*. This contained costed action plans to conserve 116 species and 14 habitats together with recommendations for future biodiversity action plans in the UK (Tranche 1). Since then, a further six volumes of species and habitat action plans have been published (Tranche 2). There are now 436 biodiversity action plans, 391 species action plans (SAPs) and 45 habitat action plans.

Of these, a total of 85 beetles species have been included in either costed action plans, grouped action plans (such as river shingle beetles) or as uncosted statements (see appendix 1). Each species has a Lead Partner, one of their main roles being to report on the progress of the action plans and the current achievements made. In many instances, the Lead Partner is one of the statutory bodies such as English Nature, Countryside Council for Wales, Environment & Heritage Services (NI), Scottish Natural Heritage or the Environment Agency. There are, however, a number of non-governmental organisations (NGOs) which lead on many of the beetle species. The Balfour Browne club is the lead partner for most of the UK BAP water beetles, the People's Trust for Endangered Species leads on the Stag Beetle (*Lucanus cervus*) and the Noble Chafer (*Gnorimus nobilis*). The various wildlife trusts also lead on some species, e.g. Suffolk Wildilfe Trust lead on *Cryptocephalus exiguus* and Hampshire and Isle of Wight Trust on *Pachytychius haematocephalus*)

The UK BAP also recognised the importance of work at the local level and recommended the setting-up of various local action plans. Many of these are now in operation and most of them operate at the county level. They vary significantly in style and content although most are very similar to the UKBAP layout. Species tend to include those present in the UK BAP and they also often include further invertebrates considered to be of local interest.

The success of the UK BAP has been to bring the plight of many very rare species to a larger audience. This has worked well with big 'striking' animals such as the stag beetle. It has also highlighted some wider changes to our countryside, such as the loss of orchards in Worcestershire, Gloucestershire and Herefordshire, which is the mainstay for *Gnorimus nobilis*. It has also helped to highlight the importance of particular sites, such as Windsor Park which holds nine UK BAP species, five of which are confined to this site.

Certain issues however, have still yet to be addressed properly. Many of the species chosen are only found on one or a few sites. Hence, action has been limited to these sites. Only where species have been more widespread (such as *G. nobilis* in orchards) have they helped broaden interest to a wider perspective.

We are also still at the early stages for many of the Species Action Plans, with information still being drawn together about the animals' ecology and distribution, and it is fair to say that, in many instances, their ecology has often been difficult to determine. To some extent the gap has widened between species and habitat conservation and it is also often harder to get action for a species that does not have an Action Plan in the UK BAP.

There is a further round of BAP planning in 2005 and it is anticipated that this will be repeated every five years. This gives all of those concerned an opportunity to put the case forward for inclusion of further species, as well as exclusion of those species not deemed to be of importance or that have achieved the targets. The real test will be to make the Habitat Action Plans more invertebrate-friendly as these will affect far more localised and rare species than the current Species Action Plans.

4.2.2 Monitoring our National Sites (SSSIs)

There are now a large number of SSSIs where invertebrates are included in the list of features of interest, although very few sites have been notified on the grounds of invertebrates alone. Since notification, however, little work has been undertaken to monitor the invertebrate interest on these sites. A greater amount of entomological study on these sites is needed to determine the current quality of the fauna. This is a huge task and the methods used will include both proxy habitat tests as well as survey work to ensure the assemblages present are still of a suitable quality, this is often termed 'favourable condition'.

It is English Nature's hope that amateur entomologists will become involved with this process at a local level but this approach is still in the early stages of development. In the meantime, you are encouraged to make your data/records etc available to the local/county wildlife trust or biological records centre so that they can be incorporated into a larger picture.

4.2.3 Information flow

Over the last two decades recording effort has also increased and the computerisation of data has meant that our understanding of what is common and rare has changed. This is complicated by the fact that certain species have declined or increased their range over this timescale. A lot of the literature concerning rarity is now out of date and is in need of revision. It may not, however, be a particularly practical and useful idea to update such reviews in a paper format as these, like their predecessors, will date quickly. It may, therefore, be of more use to produce future such documents on the web where they are more easily updated.

There has also been a number of changes concerning the storage and access of biological data throughout the UK. English Nature runs the Invertebrate Site Register, a paper database set up in the early 1980s, which has since been computerised. It contains nearly quarter of a million records of species, mostly from designated sites. This database is maintained at English Nature headquarters in Peterborough.

The Countryside Council for Wales collates records of all invertebrates in Wales but with the emphasis on the less common species. These are currently held on Recorder 3.3 but will eventually be transferred to Recorder 2002. It is the intention to make data available through the NBN as far as data ownership rules and other confidentiality issues allow. The database currently holds about 450,000 records, of which 122,000 are of beetles. These records are used daily in such issues as SSSI selection, assessment of planning applications, scientific research, advising on habitat management and helping local authorities develop Biodiversity Action Plans. CCW would be pleased to receive further records of invertebrates from Wales. Please send these to Adrian Fowles, Senior Invertebrate Ecologist, CCW, Maes y Ffynnon, Ffordd Penrhos, Penrhosgarnedd, Bangor, Gwynedd, LL57 2DN.

By far and away one of the best sources of information in many parts of the country are the Local Records Centres. Although these can be very variable, a good number of them hold a great deal of invertebrate information.

4.2.4 Local Beetle Conservation

There has also been a lot of progress with defining sites of local interest. In many instances these are called County Wildlife Sites, Sites of Biological Importance or something similar, but they generally tend to mean the same thing. Such sites do not have any legal protection but they are often recognised by the local planning authority and as such have some degree of protection from development, unless the importance of the development is deemed to outweigh the importance of the site. Criteria for County Wildlife Sites vary considerably between counties. The vast majority of these sites are judged on habitat, bird or botanical criteria, although a fair few also use butterflies and dragonflies. Less than ten county wildlife site systems use criteria based on other invertebrates, such as deadwood beetles or the presence of Red Data Book species. Although these sites are not protected by law as SSSIs are, they should not necessarily be seen as second-tier sites. It must be remembered that SSSIs are a representative series rather than an exhaustive list of all nationally important sites. Therefore, many County Wildlife Sites, as well as many non-specified areas, may have extremely important beetle fauna.

The information on County Wildlife Sites is often held by the local planning authority, the county wildlife trust, the local records centre or a combination of them. In many counties, criteria have been designed to help select such sites and these will often include criteria for selection on its invertebrate interest. An unspecified but high number of these sites will have beetle importance. County wildlife Sites represent a perfect opportunity for coleopterists to get their sites recognised for local or eventually national interest.

4.2.5 Site Designation and Quality

Guidelines for the selection of biological SSSIs was published in 1989 and drew heavily on the Invertebrate Site Register, which was begun in 1980. At that time, there was a strong focus on butterflies and dragonflies, the best-known groups, and it was recognised that an improved understanding and more complete knowledge was needed for consideration of invertebrate assemblages. For example, the guidelines conclude that 'The process of analysing species assemblages, combined with assessing the presence of rare species at localities within a major habitat type, is likely to provide a sound basis for selecting important invertebrate sites in the future'.

In the past, many sites have been notified purely on the presence of rare species. Rarity in itself is not always a reliable factor for determining site quality. Some species have probably always been rare, others lead a very cryptic lifestyle and, although are seemingly rarely encountered, they can actually be quite widespread. For many species we still have yet to determine what their lifecycle and habitat requirements are. The presence of such an animal on a site is meaningless if we do not know what it does. It may have flown in from the nearest barley field! We have also seen over the past decade, that the rarity of species is variable over time. Some collectors become experienced at finding specific species and the species themselves can vary tremendously with climatic and other factors. A good example is the flax beetle *Longitarsus parvulus* (Paykull) and the jewel beetle *Agrilus biguttatus* (F.), both of which have expanded their ranges but are still regarded as nationally scarce.

4.2.6 Use of Assemblages

An assemblage is a community of invertebrates typically associated with the given habitat. The most useful measures of assemblages should be a mixture of both typicalness and rarity.

One should be aware of the differences between assemblages and lists of animals. Many sites have lists of animals that are referred to as if they are assemblages. Assemblages come from one habitat (and often from sub-habitats) and have some basic ecological relationship with each other.

Knowledge of invertebrate assemblages has significantly increased over the last 20 years and particular habitats such as dead wood (saproxylic species), bare ground and exposed riverine sediment (ERS) have all now been much more thoroughly investigated. They are especially rich for various invertebrates, in particular beetles. A whole community or assemblage of species with a high affinity to a particular semi-natural habitat or feature, which itself has a limited distribution, should bear considerable weight in assessing the importance of a site. Such stenotypic species are said to have a high fidelity for the habitat in question.

A number of Indices have been developed over the last decade; many are based on either rarity and/or fidelity. They include the Saproxylic Species Quality Indices, or SQI for short, (Fowles et al, 1991), the index of Ecological Continuity (IEC) (Alexander 1988), which also deals with saproxylic species, the classification and ranking of water beetles (Foster 1992) and a number of papers and reports dealing with exposed river sediments (Eyre and Lott 1990, Eyre & Luff 2002, Sadler & Bell 2000, Sadler & Bell 2002, Fowles et al in prep).

There have recently been a number of publications dealing with definitions of further communities. These include seepages and flush communities (Boyce 2002), wetland beetles (Lott 2003), calcareous grassland assemblages (Alexander 2003), woody debris (Godfrey 2003) and grazing marshes (Drake in prep). These reports represent initial work done in determining species assemblages and should be seen as the first few steps to gaining a greater hold on species assemblages.

5 Summary

Invertebrate conservation is a relatively new subject and, as such, there is still much to learn. A lot has changed over the past twenty years, including the advent of the UK Biodiversity Action Plan and increased knowledge on the ecology and location of many of our beetles. Many of the very rare species have benefited from the UK BAP but the importance of assemblages is becoming more apparent. The amateur coleopterist has a very important part to play in invertebrate conservation, particularly species recording and ecological study.

Finally, much of the conservation section in the last edition of the Coleopterist's Handbook is now out of date, hence the drafting of this new section. Invertebrate conservation will have failed if this chapter is not also out of date in twenty years' time.

APPENDIX 1

Ampedus nigerrimus	a cardinal click beetle	GS
Ampedus ruficeps	a cardinal click beetle	GS
Ampedus rufipennis	a cardinal click beetle	GS
Anisodactylus nemorivagus	a ground beetle	S
Anisodactylus poeciloides	Saltmarsh longspur	P
Anostirus castaneus	Chestnut coloured click beetle	P
Aphodius niger	Beaulieu dung beetle	P
Badister collaris	a ground beetle	S
Badister peltatus	a ground beetle	S
Bembidion argenteolum	a ground beetle	P
Bembidion humerale	a ground beetle	S

Conservation

Bembidion nigropiceum	a ground beetle	S
Bembidion testaceum	a ground beetle	GP
Bidessus minutissimus	a diving beetle	P
Bidessus unistriatus	a diving beetle	P
Byctiscus populi	Leaf-rolling weevil	P
Carabus intricatus	Blue ground beetle	P
Cathormiocerus britannicus	Lizard weevil	P
Ceutorhynchus insularis	a weevil	S
Chrysolina cerealis	Rainbow leaf beetle	S
Cicindela germanica	Cliff tiger beetle	P
Cicindela hybrida	Northern dune tiger beetle	P
Cicindela maritima	Dune tiger beetle	S
Cicindela sylvatica	Heath tiger beetle	P
Cryptocephalus coryli	Hazel pot beetle	P
Cryptocephalus decemmaculatus	10 spotted pot beetle	S
Cryptocephalus exiguus	Pashford pot beetle	P
Cryptocephalus nitidulus	a leaf beetle	P
Cryptocephalus primarius	a leaf beetle	P
Cryptocephalus sexpunctatus	6 spotted pot beetle	P
Curimopsis nigrita	Mire pill-beetle	P
Donacia aquatica	a reed beetle	P
Donacia bicolora	a reed beetle	P
Dromius quadrisignatus	a ground beetle	S
Dromius sigma	a ground beetle	S
Dryophthorus corticalis	Wood-boring weevil	GS
Dyschirius angustatus	a ground beetle	S
Elater ferrugineus	a click beetle	GS
Ernoporus tiliae	Bast bark beetle	P
Eucnemis capucina	False click beetle	GS
Gastrallus immarginatus	Maple wood-boring beetle	P
Gnorimus nobilis	Noble chafer	P
Gnorimus variabilis	Chafer	GS
Graphoderus zonatus	Spangled water beetle	P
Grouped plan for river shingle beetles		GP
Grouped statement for *Harpalus sp.*		GS
Grouped statement for saproxylic beetles		GS
Harpalus cordatus	a ground beetle	GS
Harpalus dimidiatus	a ground beetle	S
Harpalus froelichi	a ground beetle	P
Harpalus obscurus	a ground beetle	S

Harpalus parallelus	a ground beetle	GS
Harpalus punctatulus	a ground beetle	S
Helophorus laticollis	a water beetle	P
Helophorus laticollis	a water beetle	P
Hydrochara caraboides	Lesser silver water beetle	P
Hydrochus nitidicollis	a beetle	GP
Hydroporus cantabricus	a diving beetle	S
Hydroporus rufifrons	a diving beetle	P
Hypebaeus flavipes	Moccas beetle	GS
Laccophilus poecilus	a diving beetle	P
Lacon querceus	a click beetle	GS
Limoniscus violaceus	Violet click beetle	P
Lionychus quadrillum	a ground beetle	GP
Lucanus cervus	Stag beetle	P
Malachius aeneus	Scarlet malachite beetle	P
Megapenthes lugens	a click beetle	GS
Melanapion minimum	a weevil	P
Melanotus punctolineatus	a click beetle	P
Meotica anglica	a beetle	GP
Oberea oculata	Eyed longhorn beetle	P
Ochthebius poweri	a water beetle	S
Orchestes testaceus	Jumping weevil	P
Pachytychius haematocephalus	Gilkickers weevil	P
Panagaeus cruxmajor	Crucifix ground beetle	P
Paracymus aeneus	Bembridge beetle	P
Perileptus areolatus	a ground beetle	GP
Procas granulicollis	a weevil	P
Protapion ryei	a weevil	S
Psylliodes luridipennis	Lundy cabbage flea beetle	S
Psylliodes sophiae	Flixweed leaf beetle	P
Pterostichus aterrimus	a ground beetle	P
Pterostichus kugelanni	a ground beetle	P
Stenus palposus	a rove beetle	P
Synaptus filiformis	a click beetle	P
Tachys edmondsi	Edmond`s ground beetle	P
Tachys micros	a ground beetle	S
Thinobius newberyi	a rove beetle	GP

P= plan   S = statement   GP = group plan   GS = group statement

APPENDIX 2: GLOSSARY

Stenotypic
Species with very exacting and narrow ecological requirements as opposed to eurytypic which are species which can tolerate a wide range of ecological variables.

Saproxylic
A species that requires dead or decaying wood to complete its life cycle.

Cohort
Formally a unit in the Roman Army but it is also often used in referral to stands of trees all of a similar age.

Eutrophication
Nutrient enrichment of water bodies.

Ruderal
A habitat consisting of recently disturbed earths which have been recolonised by plant species which proliferate in such areas e.g. willowherbs, hogweed, dandelion, cow parsley and nettles.

References

Alexander, K.N.A., 1988. The development of an index of ecological continuity for deadwood associated beetles *In*: R.C. Welch (Ed.), Insect indicators of ancient woodland. *Antenna* 12: 96-71

Alexander, K. N. A., 2003. *English Nature Research Report No. 512.* A review of the invertebrates associated with lowland calcareous grassland. English Nature

Anon., 1995. Biodiversity: The UK Steering Group Report. Volume 2: Action Plans. London: HMSO.

Anon., 1981. *Wildlife and Countryside Act* HMSO, London.128pp.

Boyce, D.C., 2002. *English Nature Research Report* No. 452. A review of seepage invertebrates in England. English Nature.

Bratton, J. H., 1991. British Red Data Books: 3. Invertebrates other than insects. Joint Nature Conservation Committee.

Common Ground. 2000. *The Common Ground book of orchards.* London: Common Ground.

Eyre, M.D. and Lott, D.A., 1997. Invertebrates of Exposed Riverine Sediments. Marlow: *Environment Agency R& D Technical Report* W11

Eyre, M.D. and Luf, M.L., 2002. The use of ground beetles (Coleoptera: carabidae) in conservation assessments of exposed riverine sediments in Scotland and northern England. *Journal of Insect Conservation*, **6**, pp 25-38

Foster, G.N., 1992. *UK Nature Conservation No.1: Classification and Ranking of Water Beetle Communities.* Joint Nature Conservation Committee.

Fowles, A.P., Alexander, K.N.A. and Key, R.S., 1990. The Saproxylic Quality Index: evaluating wooded habitats for the conservation of dead-wood Coleoptera. *Coloepterist* 8(3): 121-141, November 1999.

Fowles, A.P., 1997. The Saproxylic Quality Index: an evaluation of dead wood habitats based on rarity scores, with examples from Wales. *Coleopterist* 6(1): 57-60.

Godfrey, A., 2003. *English Nature Research Report* No. 513. A review of the invertebrate interest of coarse woody debris in England. English Nature

Harding, P.T. and Rose, F., 1986. *Pasture-woodlands in Lowland Britain: A review of their importance for wildlife conservation.* Huntingdon. Institute of Terrestrial Ecology.

Hyman, P.S., (revised Parsons, M.S.) 1992. *A review of the scarce and threatened Coleopetera of Great Britain*. Part 1. UK Nature Conservation: 3. Peterborough: Joint Nature Conservation Committee.

Hyman, P.S., 1994. *A review of the scarce and threatened Coleoptera of Great Britain*. Part 2. UK Nature Conservation: 12. Peterborough: Joint Nature Conservation Committee.

Jefferson R. G., and Robertson H. J., 1996. *English Nature Research Report No. 169*. Lowland Grassland: Wildlife Value and Conservation status. English Nature

Lott, D. A., 2003. *English Nature Research Report* No. 488. An annotated list of wetland ground beetles (Carabidae) and rove beetles (Staphylinidae) found in the British Isles including a literature review of their ecology. English Nature.

Sadler, J. P. and Bell, D., 2000. *English Nature Research Report, No. 383*. A comparative site assessment of exposed riverine sediment beetles in South-West England. English Nature, Peterbrough

Sadler, J.P. and Bell, D., 2002. *Invertebrates of Exposed Riverine Sediments – Phase 3: Baseline Faunas*. Environemnt Agency R&D Technical Report W1-034/TR. Bristol.

Shirt, D. B., 1987. British Red Data Books: 2. Insects. Joint Nature Conservation Committee, Peterborough.

Webb, J. R. and Bowler, J., 2001. English Nature Research Report No.416. County surveys of parkland: The Staffordshire Experience. English Nature.

Creating a database

CREATING A DATABASE FOR YOUR RECORDS

by D.R. Nash

A beetle collection with data and determination labels is one form of database. In order to become a serious coleopterist it is necessary to form a collection of beetles because:

- the majority of British species are too small to identify accurately in the field with many also needing dissection of the genitalia to effect accurate identification.

- much identification of beetles relies on microscopical examination of comparative features between species – puncturation, microsculpture, relative proportions of body parts etc.

- a collection enables specimens to be re-examined, for example when a single established species is later found to actually consist of two species or when new information throws doubt upon the accuracy of the original identification.

Some coleopterists suggest, however, that the only written information needed with a collection of beetles is that provided by the labels on the pin or underside of the mounting card of each specimen. I do not subscribe to this view because:

- using just a properly laid out collection (i.e. in latest checklist order) the only opportunity one has to begin compiling a full species list for a site or a particular date, is before the individual specimens from each collecting trip are assigned to their places in the collection. Once this has been done, any valuable or interesting associations between the individual species are likely to be lost, as well as the ability to analyse the assemblage according to RDB status etc.

- a typical personal collection consists of a series of specimens (traditionally a maximum of between 6-12, often less with large obvious beetles) to represent each species. In some cases, moreover, a series of a species may originate from the same place and same date. These series only represent a tiny fraction of that person's field encounters with those species over what may be over half a century of collecting.

- the data label with a specimen needs the following: date, locality, county, at least 4-figure National Grid reference, collector's name and field data. (see p. 322). Although computer generated labels can be produced with very tiny print enabling small labels in proportion to the beetle mounts, precise field data can still take up far too much space and this, together with other information, is often omitted. The only solution, as I see it, is to record and store this valuable information elsewhere.

The 1954 first edition and 1975 revised second edition of this 'Handbook' contained extensive sections on keeping a journal and a separate determination book as well as suggesting the need for a field note-book in one's haversack.

The journal or 'Coleopterist's Day Book' was considered *(op. cit.* 1954:47) *'an adjunct of special importance for those prepared to make detailed studies of the Coleoptera of the areas they investigate, for within are recorded not only lists of species and numbers of examples collected on specific dates and at described places, but also those valuable pieces of extended data which comprise the observations made in the field, hastily jotted down in the field note-book.'*

It was recommended that the journal should contain dates of determination against each species recorded or captured which, together with a number on the specimen label, would enable cross-reference of specimen and journal entry to a second fast-bound book - the determination book. This second book would not only duplicate all of the information in the journal (except habitat and field data relating to an individual specimen) but would supply exact dates upon which specimens were named, provide details of the current location of specimens e.g. determiner's/collector's collection, donation to national/provincial museum, donation to a private collector, as well as giving a reference to the transfer of the record to a card index, county/local list or a publication.

This summary should serve to indicate the importance placed upon these recording books by the authors of the first edition. As a tyro, the writer followed their advice and maintained an impressive-looking journal/determination book for some fifteen years. During that time I gradually came to question more and more the value of recording in this very time-consuming and cumbersome way. I also discovered that only one coleopterist known to me kept a journal and separate determination book as recommended in the 'Handbook'. As collecting beetles can often result in very mucky hands, I personally don't take a notebook into the field, preferring instead to rough pencil my field data and other observations onto the self-adhesive labels encircling each of my specimen tubes and then transfer this information to my notebook at home.

Reflecting upon the reasons for the great emphasis placed upon such books by the highly competent coleopterists who produced that first handbook over fifty years ago, one wonders if they deliberately recommended that an excess of written, cross-referenced information be associated with a collection to ensure that the collections of novices following their advice were more scientifically valuable than many of those in the past. These writers, moreover, were unaware of the great upsurge of national recording of precise data on all living organisms which was to occur in this country during the next few decades and which was to result in a more immediate sharing of basic information among naturalists. At the present time, much data (particularly that relating to less-notable species) which in the past would probably have remained buried in a collector's journal, is being sent to the national recording schemes organised by the Biological Records Centre at Monks Wood (see 'Recording Coleoptera' pages 403-412), local recording scheme organisers and county recorders or used to compile site lists for English Nature, county conservation trusts, etc.

Although only brief mention is made of a card-index in the first two editions of the 'Handbook', the writer considers that this is a practical alternative (in conjunction with a notebook kept at home) to the journal and determination book as originally advocated and that it is probably the most efficient recording system for those without the knowledge or inclination to computerise their records.

A separate record card (at least the 6 x 4 inches size) should be headed up for each species on the British List which the coleopterist intends to record. Whilst these cards can be arranged exactly as in the latest check list to provide a system which facilitates the compilation of locality lists in systematic order, it is far quicker to extract or insert data if the families have their genera arranged alphabetically.

The front of the card can be used for recording all personal captures and field data relating to the species. As fine a pen as possible should be used (0.3 mm nib or less) with the writing as small and clear as can be conveniently managed, so that as much information as possible can be compressed onto each card. It is suggested that a single line be used for each entry, as shown below, except where additional field or rearing data needs to be included:

DATE	FIELD DATA	LOCALITY and VC	GRID REF.
10. v. 1997	Many beaten from oak	Brantham; 25	TM10.33

The reverse of each card can be used to list such things as all known county records, data from other sources (e.g. museum collections), information from other coleopterists, published references to the species, and so on.

When the front or back of a card is filled, additional cards can be headed with the species name and affixed with a rustless paper-clip.

The notebook/diary/journal kept at home contains dates and places etc. of collecting trips as well as listing the species recorded. Several years ago, I abandoned my original huge and unwieldy A4 fastbound tome (covering a couple of decades) for small (10 x 17 cm) 100 page 'single cash' books, each of which contains details of a one year's fieldwork. I chose such a book because it was small enough to keep at hand when determining beetles and it had four red pre-ruled columns which are of exactly the width required for listing my captures (wide column for 'species', narrow ones for 'date/grid reference', 'field data' and, in my case, 'computer database number'). Species based solely on field determinations are entered immediately after the collecting trip; those needing critical examination for determination are entered as they are named from collected material. I estimate on the generous side the number of pages required for entering all the species recorded on a particular date before heading up a new page for entering details of the next collecting trip. If you do decide to follow my example, do be sure to put clear labels on the front so the cash books are not discarded at your decease because someone thought they merely detailed your housekeeping finances!

This is not the place to discuss computer databases in detail but a few personal remarks on the topic may help those considering computerising their records.

In 1987, thanks to Alan Sugar and Amstrad the personal computer (PC) had started to become affordable. I decided computerising my records was the way forward so I invested in an Amstrad computer and printer and began experimenting with computerising the thousands of records on my card index. Cheap early PCs had very little memory (RAM) compared to the huge amounts available now, so the number of records which could be stored was limited, they functioned very slowly by today's standards and their disk operating system (DOS) was not at all user-friendly or easy to learn as is the Microsoft Windows system which has replaced it. At that time, there were simple database programmes available which allowed data to be entered into what was virtually an index card on the screen. Most of these were so-called 'flat file' databases which did not allow data on one of your databases to be searched for use by one of your others so I soon switched to a relational database which permitted individual databases to be linked so that their contents could be shared. The advantages of using a computer for storing your coleoptera records are:

- Speed of access to data. Individual records can be viewed almost instantly – very much quicker than flicking through a card index or journal.

- Individual records or groups of records can be called up on screen or printed at the touch of a button. Thus, reports for individual localities, counties etc. can be produced at the touch of a button – extremely useful for sending to landowners, conservation bodies etc.

- Coleopterist computer users can usually (but not always) exchange data files enabling one person's records to be dropped straight into another's database. Sophisticated searches can be carried out, for example you can ask the programme to find and print to the screen or printer, all records of beetles beaten from oak, all records of beetles you've found in June, or in a particular year, or all these search factors in combination. Imagine how long it would take to get this information from a card index with only a few hundred records let alone from the many thousands on an established one.

- As a bonus, you can also print your own data and determination labels thereby saving much time (see 'Indoor Work', p. 322).

Most databases are designed for use within the business environment so examples in their manuals are aimed at this market, resulting in the coleopterist having to extrapolate from these in order to customise the database for his or her use. To-day there are computer database programmes available which have been designed specifically for recording living organisms. The two most frequently used by entomologists are English Nature's 'Recorder' package and 'MapMate', with the latter enabling one to also produce dot maps showing the distribution of your records for each species. (see section on 'Recording' – pp 403-412 for more detail.)

In my experience, coleopterists of long-standing are, understandably, daunted by the task of inputting many thousands of records from their handwritten data. If you are just starting your study of beetles, however, or currently have only a relatively small number of records, I suggest you give serious thought now to computerising your data.

Useful addresses

MapMate is available from: Teknica Ltd., The White House, Montacute Road, Stoke Sub Hamdon, Somerset TA14 6UQ (price currently ca £24.99 plus annual licence fee). < http://www.mapmate.co.uk >

The JNCC's 'Recorder' software is available from the national Biodiversity Network < http://www.nbn.org.uk > (price currently £30.00p)

Recording

RECORDING COLEOPTERA

by Trevor J. James

Introduction

With an increasing demand for more and more detailed information about our natural environment, the work of even casual students of beetles becomes potentially valuable. This chapter therefore aims to show what we mean by 'biological recording' as it relates to beetles, and what use we might make of the information.

What is a biological record?

At its simplest, a 'biological record' is made when anyone cares to write down (or 'record') the occurrence of a species at a location on a date. On its own, this information might only be of curiosity value for the person making the original observation. Gathered together, such records begin to be a powerful tool for research, conservation and so on.

In practice, a really useful biological record needs more than this bare minimum. It might need the name of the person who identified the specimen as well as the field recorder. It could well do with having notes on the habitat – both the broad habitat type and the particularly circumstances in which the insect occurred. It might also have details of the weather, temperature, associated species, the names of food-plants and the method of capture etc. All of these details increase the long-term value of the record.

Why record beetles?

We might just want to record them for our own interest. In fact, this will be the first reason why most 'amateurs' (perhaps as opposed to paid 'professionals') might want to record. But there may be other reasons.

For conservation action, or to defend important areas from adverse impacts of development, changes in land use etc., we increasingly need good quality, detailed information on both habitats and species. Such information is also increasingly needed to be up-to-date in order to qualify for use, especially if used at public inquiries into planning, for example. Obviously recording such a large group as beetles is impossible across the board on a very regular time-scale, but the better we can make our information, the more strength there is for the conservation case.

We may need to support research activities. The biology of individual beetle species is mostly poorly known. We may have a fair idea that they are, for example, saproxylic species, but what might this mean in real terms for this species, and how might it relate to others in the same dead wood habitat? Without more, and more detailed, observations, we may never know.

Who can record beetles?

The answer to this is brief: – anyone. Good quality records come from careful collection, accurate identification, and attention to detail. These are not just attributes of the professional entomologist, but they do need application and determination. The records should also improve with experience.

Going about recording

Any casual observation of a beetle can make a useful record, but we get the best out of the activity if it is approached with some planning. This is where 'recording' as a subject overlaps with 'field survey'.

Planning the recording effort.

It helps if we have some clear objectives. The most useful 'recording' comes from a methodical approach. We may, for example, plan to produce an inventory of a particular site, which will require a recognition of the different habitat types available (grassland, forest, dead wood, flowers, water bodies, etc.) and a programme of re-visiting different habitats at different times of the year, with different collecting methods (see the chapter on field collecting). Or we may be focusing on a specific group of insects (e.g. water beetles) and planning our field work to examine a proportion of available relevant habitat in a geographical area.

Either way, the planning process for carrying out the field work should ideally also take into account the way we might want to record the results. For example: if we are recording water beetles, we may need to plan to record the nature of the water body, the substrate, and the vegetation present at each sample point. We may, therefore, need to arrange for tailored recording forms, etc. before the event (see below for recording methods etc.)

Localities, sites and habitats

It is easy to lose valuable information by being unclear about where a specimen came from. To get the best out of the recording activity, we need to be clear about the difference between 'locality', 'site' and 'habitat' (and perhaps 'micro-habitat'). Large numbers of important, historic specimens languish in museum collections with paltry information like 'New Forest' written on their mounts, and are almost worthless as a result.

Localities. The locality should be recorded by name from a published Ordnance Survey map, so that it can be traced by someone else in the future. It should be a recognisable, individual place-name, not just 'Home Farm' or 'Fox Covert'. It should also not just be a vast area, like 'New Forest'!

Sites. This may be more precisely defined. Very often other organisations, such as English Nature, the local biological records centre or wildlife trust will have defined 'sites' for recording purposes already, and these could be requested beforehand for details of site boundaries. Different people approach 'site' recording differently, but a useful way of thinking of a site is as a defined area of land recognisable as an individual management unit. This may be easier in some areas than others, because sometimes a 'site' just has to be a bit of recognisable ground in an otherwise open landscape. In any event, if there are already-defined 'sites' available in your area which you can use, then make use of them, because it helps enormously with defining and making use of data later. Local records centres can often provide digitised outlines of such sites at large scale on an OS map.

The grid reference. **This should not be an optional extra**. For modern recording, it is **vital**. The more accurate it is, the better: at least 4-figures for general recording activities, but preferably 6-figures for specific sites or parts of larger sites (and precision is best), or even 8- or 10-figure for highly localised features like veteran trees or small ponds. Grid references are also not a substitute for recording the 'site' or the habitat, as each is complementary in terms of the information it gives, and one can be used to cross-check the other.

An important aspect of the use of grid references is to remember that any grid-reference defines an intersection of the OS grid **and** a defined square of land to the north-east of the intersection. The more detailed the grid reference, the smaller the square of land (and therefore the more precise the reference). If your recording point is within a grid square, make sure you record the right grid reference for the square (the bottom left-hand corner) and avoid 'centred' grid references for sites, especially if these are large sites. This may seem academic, but it is essential for the effective later use of records.

Habitats. We live in a world of classification, and nowadays there are classification systems which try and standardise the way we approach definitions of things such as 'habitat' so that information can be comparable. However, it is better to try and record something rather than nothing at all. Again, precision helps. For example, it is better to record beetles from 'damp pasture' than just 'grassland', or from 'beech woodland' than just 'woodland' or even 'deciduous woodland'.

Micro-habitats are specially important for most beetle species, and experienced collectors will recognise them, even if there is currently no recognised classification system to accommodate the data: red-rotten heartwood of oaks; dead reed litter; flower heads of umbellifers; the gills of fungi; wet mud by a river; shingle and so on. Very often great precision in recording these features is highly valuable: for example the fine distinctions between silt, clay and fine sand as a water margin habitat.

Field recording methods

Field sampling methods are dealt with elsewhere. Here we are considering the means of collecting the record. The most basic (and still very effective) is the straightforward field notebook and a pen or pencil! Pencils are especially useful in the rain, and, if you can find one, a waterproof notebook is also handy!

Field **record cards** have been used for a considerable time, and are useful especially for intensive surveys of specific groups. The Biological Records Centre at CEH Monks Wood has issued a number of standard record cards on behalf of specific recording schemes (see below). The drawback of using them is often that we may not be just restricting our interest to one taxonomic group, and even if we are, we may only make a note of a very few species on one card for a locality, which tends to get through rather a lot of cards for a very few records. They may however be especially useful for compiling records for a locality for a defined time period (although the temptation to use such a system as an open-ended list should be avoided, as this loses the detail of dates, micro-habitats and precise localities). The BRC has also issued general recording cards, which can be useful for miscellaneous records.

Even if we do not want to use BRC recording cards, we may have it in mind to carry out some kind of systematic survey, and if so, a standard recording form is useful, in order to try and prompt for what detail is required and to standardise the way that data are recorded. These can nowadays be produced readily enough with word processors, although card versions might need special printing, perhaps on waterproof material like Tyvek for use with indelible markers.

With modern technology, however, there are developing a range of other methods of field recording. **Data loggers** are becoming more practicable, and can be linked to a **GPS** (global positioning system) to give reasonably accurate grid references automatically. Ordinary hand-held computers (palm-tops) can also be used, and are becoming more capable of accepting hand-written notes using a stylus. The drawbacks of any such system are the cost and the danger of losing data if the equipment gets damaged. A special advantage is the reduction in data handling, which can cut down error rates as well as speed up data management, especially if data are recorded or transferred directly into a copy of one of the standard database systems. The ability to record details of habitat etc. directly in the field is also a strong advantage.

Whichever way data are captured in the field, an important aspect is the consideration given to the way that records are made in order to be able to extract the most from them in the future. While much recording will no doubt always be casual, if more systematic recording is undertaken, it will be important to ensure that standard details of habitat etc. for a specific location and collection of species occurrences are collected at the same time. This process can be summarised under the term sampling. A **sample**, therefore is a collection of species occurrences which is recorded along with the same associated data at a place at a time. Using modern database systems, the importance of the idea of the 'sample' as a means of linking related data is growing.

Names and naming

The process of identification and naming a species is obviously central to a useful record. The business of identification is dealt with elsewhere, but it is worth reviewing here the importance of understanding how taxonomic names work, in order for records to be made securely.

The scientific binomial is the basis of every identification. At any one time, this is referable to a 'standard' for that taxon (or species) – the type specimen. The actual name used will be determined by what is recognised as the relevant name when any specimen is identified. The relevant name may change over time, or a species may be 'split' into two component species when further study is carried out by taxonomists. The results can be problematic for recording.

To get round this as much as possible, there are two principal ways of 'quality checking' the name. The first is to ensure that the proper 'authority' (the name of the person who devised the binomial) is given with any record, which then ensures that the meaning of the name is clear. The second, is to make reference to a published source for the name used in an identification.

The most recent standard 'complete' checklist of British Coleoptera is Pope (1977). Earlier checklists by Kloet & Hincks (1945), or the names used in Joy (1932) or Fowler (1887-1913) will be especially out-of-date. Experienced coleopterists may attempt to keep up to date with changes in nomenclature, but each group of beetles may have numerous changes (e.g. recent checklists of weevils in Morris (1997, 2002)), and some of these may not be apparent in the British literature. There is currently (2003) a programme of work going on to produce a new 'standard' checklist (see *The Coleopterist* website – details below). Some sections of this revised checklist are already in use by some coleopterists (e.g. Staphylinidae). Privately published 'complete' lists of British Coleoptera are also in circulation.

Whatever the source of names used, in making a record it is important that a published source can be quoted for them, and that the 'authority' is given, to avoid potential confusion. This is especially important in groups which are undergoing rapid changes in nomenclature.

Verification, validation and vouchers

Identifying beetles, as with many insect groups, will require a certain amount of expertise and experience. Some groups are more difficult than others. Many beetles need to be identified by a specialist for certainty.

The business of making accurate records depends heavily on correct identification, and it is therefore essential that anyone going about systematic recording should understand the need to try and ensure that identifications are accurate. For some other groups of organisms, there is a well-established system of species **verification** in place, with panels of 'experts' as referees identified for difficult groups. For beetles, this is more problematic, as many groups have few specialists available, and there is currently no organised society under which a panel of referees could be established. Assistance in identification may be found from one or more of the following:

- The County Recorder for your area, if there is one (see *The Coleopterist* website, or contact your local natural history society or local biological records centre – see below).
- A national Recording Scheme organiser (if there is one) (see below).
- A regional or national museum with good collections and professional natural history curators.

Whatever course is taken, or if you feel that your own skills are good enough, it is important in any record that the name of the determiner of the specimen is recorded, and the date of determination, so that this can be cross-checked with any subsequent changes of names.

The **validation** of records concerns ways of ensuring the accuracy of the other detail of the record. With modern electronic data processing (see below), there are often automatic checks built in to recording systems, but keeping a log of what is done to records also helps.

Approaching recording in a methodical way will help. Giving locality and habitat information in records in a standard and uniform way is important. Using a standard approach to the definition of dates (with as precise a date as possible) is also highly important, especially in producing records which might be computerised by others. Likewise, using a single form for individuals' names is also important (and ensuring that each record is identifiable with a person's name) (see below for Data Protection Act etc.). Finally, especially if records are being compiled from other sources, the origin of a record is also essential (see below for information on intellectual property rights etc.).

Compiling data, even in a manual form, should be done with a view to minimising the likelihood of errors being created. Minimising the number of transcriptions of records is therefore important, as is double-checking things like grid references or dates, which can be especially prone to errors.

The importance of keeping **vouchers** needs to be mentioned here. Vouchers are specimens which have been identified in the process of making a record, and which are retained as evidence for the accuracy of that record. Vouchers need to have details of the locality, date, habitat, collection method and collector associated with them. If someone is maintaining their own collection, some system of cross-referring specimens with records is recommended (perhaps specimen numbers). In the longer term, especially for important vouchers such as specimens of 'difficult' species, a field recorder needs to try and ensure that such material finds its way into a good, well-supported public reference collection somewhere, such as a National Museum, regional museum or the reference collections of organisations like the British Entomological & Natural History Society. Making a will to try and ensure this happens is recommended. This may seem a bit over-the-top, but far too many important collections of insects have been destroyed by unsympathetic surviving relatives of the collector!

The structure and operation of Coleoptera recording in Britain

Coleoptera studies in Britain are not co-ordinated under one organisation. However, there have been some attempts at formalising the way records are gathered. The website of *The Coleopterist* magazine (www.coleopterist.org.uk) is the principal means of bringing together information about Coleoptera recording in the country, and there is a related internet discussion group (beetles-britishisles@yahoogroups.com).

Recording nationally

The principal mechanism for collating records nationally for many years has been the formal **Recording Scheme.** The Biological Records Centre at CEH Monks Wood, Abbots Ripton, Huntingdon, has co-ordinated these schemes for about 40 years, although most schemes have not been running for this time. The currently listed schemes under the BRC are:

   Cantharoidea and Buprestoidea (Soldier Beetles, Jewel Beetles, Glow-worms etc.)

   Carabidae (Ground Beetles)

   Cerambycidae (Longhorn Beetles)

   Chrysomelidae and Bruchidae (Leaf and Seed Beetles)

   Coccinellidae (Ladybirds and allies)

   Cryptophagidae: Atomariinae (Atomariine Beetles)

   Curculionidae (Orthocerous species) (Orthocerous Weevils)

   Dermestoidea and Bostrichoidea (Larder Beetles and allies)

   Dytiscidae, Hydrophilidae etc. (Water Beetles)

   Elateroidea (Click Beetles and allies)

Ptiliidae (Ptiliid Beetles)

Scarabaeoidea (Dung Beetles and Chafers)

Scolytidae (Bark Beetles)

Staphylinidae (Rove Beetles) (general scheme)

Staphylinidae: Stenini (Rove Beetles of genera *Stenus* and *Dianous*)

Scirtidae (Scirtid Beetles)

Basic details of these can be obtained from the BRC's website (www.brc.ac.uk). They are also listed on *The Coleopterist* website. Some of these Recording Schemes are more currently active than others. Other families do not currently have active national recording schemes for them.

The basic function of the BRC Recording Schemes over the years has been to compile data which can subsequently be used especially to produce national atlases of groups. Atlases for a number of groups have appeared recently, or are in press (2003), including the Carabidae, Cerambycidae, Cantharoidea and Buprestidae, Chrysomelidae and Bruchidae. Others are actively being compiled for future publication (e.g. Water Beetles). All of the schemes depend almost entirely on voluntary support from their organisers and others. The records supplied to these schemes have become more detailed over time, so that data may be increasingly available for more detailed analysis, using the BRC's Oracle database.

Recording locally

Very often local activity in field recording is organised by a local natural history society or local records centre. Many counties have locally-designated '**county recorders**', operating under the auspices of these bodies. Their functions vary widely, but they may act to collate records locally from local volunteers, validate and process these records, and pass them down to the national recording schemes. They often communicate records (especially those of importance) to local records centres, which use them for conservation advisory purposes and local Biodiversity Action Plans etc.

A list of County Recorders with contact details is maintained on *The Coleopterist* website. Full details of local records centres are on the website of the National Federation for Biological Recording (www.nfbr.org.uk), which also has a link from *The Coleopterist* website.

A particular feature of recording at the local level is the adoption by many (but not all) county recorders of the '**Vice-County**' recording system. Originally adopted by botanists in the mid-19th century in order to standardise recording areas and mutual responsibilities of local recorders, the system has been adopted by other taxonomic groups. The areas covered in 'Vice-Counties' are sometimes similar to those of old administrative counties, others are part-counties. The aim was to try and render areas of recording more comparable (therefore counties like Norfolk and Essex are divided into two Vice-Counties).

With modern recording using 10km squares of the Ordnance Survey grid (or smaller subsets), this function is less important, but they still serve to define local areas of responsibility, and also have the advantage of being stable in their boundaries. Details of these Vice-County boundaries can be obtained either from small-scale maps available from the BRC, or are available via the National Biodiversity Network (www.nbn.org.uk).

The management of records

The days of card indexes are numbered. While such methods are relatively straightforward, cheap, and locally effective, they cannot compete with modern computer data management.

Data capture software is becoming increasingly sophisticated. Some people still use their own customised database systems, such as locally-tailored versions of Microsoft Access, or Excel spreadsheets, but increasingly others are using one or other of the proprietary databases designed specifically for biological recording. These include especially Aditsite, BioBase, MapMate and Recorder 2002. The latter three in particular have the advantage of operating with in-built and relatively up-to-date species checklists and the ability to link with geographical information systems (GIS) etc. The advent of such software is making it easier for data to be managed effectively by those actually collecting it, which further reduces errors in data collations, as long as those doing the initial data capture understand the principles of running computer databases, and the need to ensure that data standards are upheld (such as correct use of names, standard habitat descriptions, standard date forms).

Data sharing is also becoming more practicable through the use of standardised data systems. The days of passing floppy discs around with ASCII files are becoming numbered. Both MapMate and Recorder 2002 have in-built facilities to enable sharing of records through e-mail. MapMate's system in particular is predicated on this use.

The Recorder 2002 database system is important because it embodies the 'data model' of the National Biodiversity Network. This data model not only enables data to be transferred between copies of the database in the same way as MapMate's system, without losing the identity of the original record (and its originator), but also will enable data to be shared through the NBN's internet 'Gateway' (see below). Because of this, it embodies a range of internal standards which require users to enter certain basic kinds of data (e.g. the names of determiners of records). Its data model is also based on the 'sample', as described above, so that it can handle quite sophisticated information about habitats etc. in association with the species records themselves. Recorder 2002 will accept input of data from Excel spreadsheets, and is also designed to inter-communicate with MapMate, BioBase and other databases.

The National Biodiversity Network

This developing initiative is aimed at enabling data to be shared remotely between partner organisations through an internet Gateway. The internet functions enable data from disparate data sets to be merged and overlaid with site and habitat information. The NBN data model used in Recorder 2002 is at the centre of this data sharing capability.

An important partner in the NBN is the BRC at Monks Wood, which has 'mobilised' many data sets from national recording schemes on the Gateway already from its Oracle database. An example is the entire database of the Ground Beetle Recording Scheme. This enables precise data to be linked with habitat information from elsewhere. The plan is for more data sets to be mobilised in this way, enabling not only localised sharing but the use of data by other partners in the NBN and elsewhere.

Making data available in this way has highlighted the need for those carrying out recording to be aware of things like copyright, the Data Protection Act, and the need to be sure that those whose records are being used are happy for them to be available within the controls of the NBN Gateway (which are administered for each data set by those supplying it). The NBN Trust has therefore developed a range of advice and information which is available through its website (www.nbn.org.uk). Some of the more important issues which need to be considered by those carrying out recording are:

The Data Protection Act. This requires anybody holding information relating to people to be registered and for the use of this information to be within agreed limits. It applies to both manual and electronic information, and covers any information which can be associated with a person through the way the data are held (such as addresses associated with names). For the purposes of ordinary biological recording a mere name associated with a record is not generally considered by the Data Protection Registrar as being relevant, but details of addresses, recording abilities, membership of organisations etc. is. Therefore if data are passed down to others with such information, this may only be done with the knowledge and agreement of the person concerned, and in a way already registered under the DPA.

Copyright and Intellectual Property Rights. These have not normally figured in recording before, but with electronic transfer of data, are becoming important. Essentially any record will have rights associated with it belonging to the person who made the record. So, its use must acknowledge these, or take account of them through, for example a waiver. Modern recording scheme cards have such a waiver reproduced on them so that data submitted on them is covered for specified uses. Historic records can present some problems, but advice on this can be obtained from the NBN.

Glossary

GLOSSARY

by J. Cooter

As the reader might notice, the majority of terms and definitions listed below have been taken from Fowler (1887: ix-xv). No attempt has been made to list or define all the various anatomical terms the student will encounter when attempting identifications or more formal studies of the Coleoptera. These are fully explained in the best (but expensive) entomological dictionary in print Nichols and Schuh (1989). Other recommended works include Crowson (1954 and 1956). In addition some of the more recent *RES Handbooks* carry very useful glossaires. Crowson's (1981) *The Biology of The Coleoptera* is a seminal work in which numerous technical terms are described.

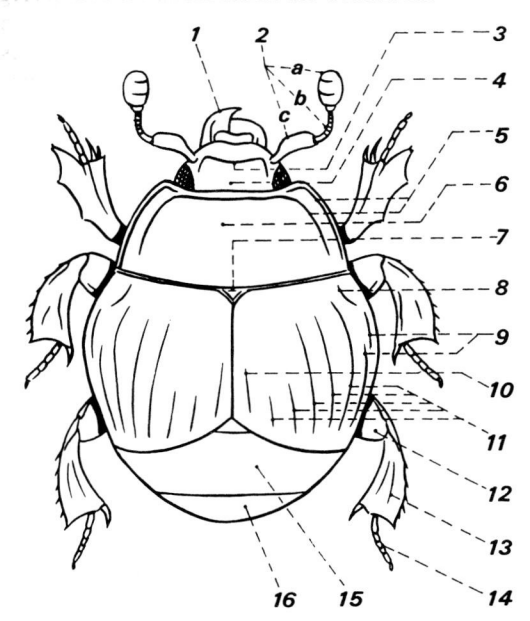

Fig. 57. *Hister unicolor* L., dorsal view (after Halstead, D.G.H., 1963, *Handbk, Ident. Br. Insects,* **4**(10):5).

(1) mandibles; (2) antenna, (a, club; b, funiculus; c, scape); (3) frontal stria; (4) frons; (5) pronotal striae; (6) pronotum; (7) scutellum; (8) humeral stria; (9) sub-humeral striae; (10) sutural stria; (11) dorsal striae; (12) posterior femur; (13) posterior tibia; (14) posterior tarsus; (15) propygidium; (16) pygidium.

Abdomen. The hindermost principle division of the body.

Aborted. Incomplete, undeveloped.

Acicular. Terminating in a sharp point like a needle (*acus*).

Aciculate. Covered with small scratches which appear as if made by a needle.

Aculeate. Produced to a point; as applied to one group of *Hymenoptera* it means furnished with a sting.

Acuminate. Terminated in a point.

Agglutinate. Fastened closely together so as to form one piece.

Aedeagus. (Etymologically correct *aedoeagus*). The *median lobe* plus *tegmen*. The term 'penis' should not be used; the two organs are not analogous. Despite this, Torre-Bueno and Tuxen in their entomological dictionaries refer to 'penis' when speaking of 'aedeagus.'

Aeneus. Bright brassy or metallic golden-green colour.

Alutaceous. Covered with minute cracks like the human skin.

Ambulatory. Relating to walking, see also *cursorial* and *natatorial*.

Anal. Pertaining to the apex or extreme end of the *abdomen*.

Annulate. Ringed (of colour).

Ante-. A prefix signifying in front of; e.g. *anteocular*, in front of the eye.

Anterior. Foremost; in front; of the part nearest the head.

Apex. In Coleoptera the parts of the body are described in relation to an imaginary central point (approximating to the centre of the scutellum), between the *pronotum* and *elytra;* the part furthest from this is the *apex*, the part nearest the base.

Apical. Relating to the *apex*.

Apneustic. Breathing through the tissues, not by means of special respiratory organs.

Apodous. Without legs; e.g. larvae of *Cercyon*.

Apodeme. Any cuticular growth of the body wall.

Apophysis. Any elongate process of the body wall, internal or external.

Appendage. Any part or organ attached by a joint to the body or to any other structure.

Appendiculate. Furnished with appendices, e.g. extra lines or furrows at the end of other lines or furrows.

Approximate. Brought near to one another.

Apterous. Wingless.

Articulated. Jointed.

Asperate. Roughness of the surface.

Glossary

Asymmetrical. One side of the body different from the other side.

Attenuated. Gradually diminished or narrowed.

Azygos. (not *azygous*) Unpaired portion of the *genitalia,* (i.e. *genital tube* beyond the junction of the *seminal ducts*) (Male)

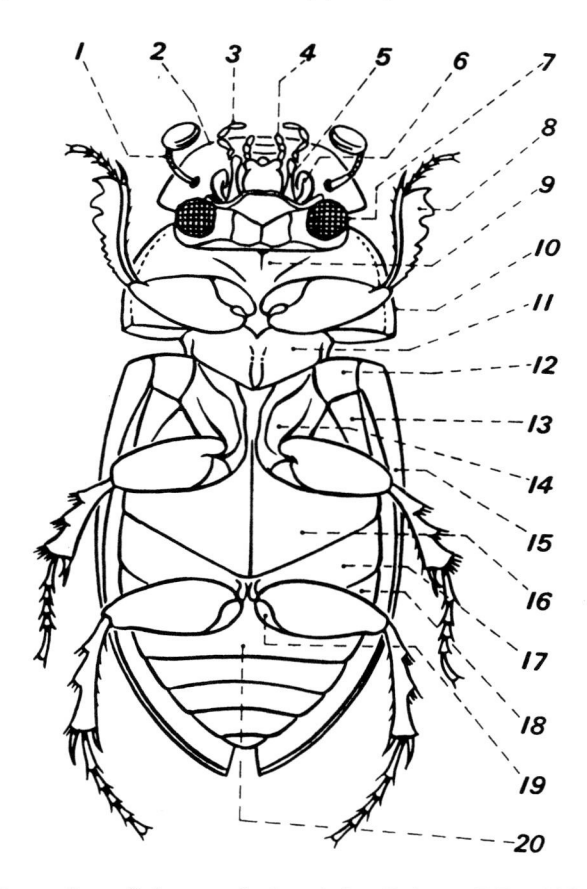

Fig. 58 *Aphodius rufipes* (L.) ventral view (after Britton, E.B., 1956, *Handbk. Ident. Br. Insects,* **5**(11):4).

(1) mandibles; (2) clypeus; (3) maxillary palp; (4) labial palp; (5) maxilla; (6) mandible; (7) eye; (8) anterior leg; (9) prosternum; (10) pronotum; (11) mesosternum; (12) mesepimeron; (13) metepisternum; (14) middle coxa; (15) elytral epipleuron; (16) metasternum; (17) posterior coxa; (18) first visible abdominal segment; (19) trochanter; (20) second visible abdominal segment.

Base. see *apex*.

Basal. Pertaining to the *base*.

Basal Orifice. Basal (proximal) opening of the *aedeagus* through which the *ejaculatory duct* enters. Often displaced to the ventral side.

Basal Piece. The unpaired basal (proximal) part of the *tegmen,* usually *sclerotised* and may form a complete ring or tube around the *median lobe*. Absent in some cases.

Bead. A fine elevated line generally at the perimeter or centre of a *sclerite* or structure.

Example: a basal bead of the pronotum. See also '*carina*' and '*keel*'

Bi-. A prefix to signify something which is composed of two parts, e.g. *bifid,* two-pronged; *bilobed,* divided into two lobes, *biforous,* having two apertures.

Border, Bordered. When a margin has a raised edge.

Buccal. Relating to the mouth or sides of the mouth.

Bursa Copulatrix. The proximal blind end of the vagina, with which it is narrowly connected. It receives the *internal sac* during copulation (Female).

Calcar. A spur, strong spine or pointed process.

Callosity, Callus. A slight projection or elevation, usually rounded.

Callose. Furnished with such a projection.

Campodeiform. Of a larva; the three principle divisions of the body are defined, and possessing three pairs of legs.

Canaliculate. Furnished with one or more channelled furrows.

Capillary. Slender and hair-like, applied to antennae.

Capitate. The term used to describe an abruptly clubbed antenna.

Carina. A keel or longitudinal elevated line.

Carinate, Carinated. Furnished with a *carina*.

Castaneous. Chestnut-coloured.

Catenulate. Chain-like.

Caudal marginal line. The line of junction which runs parallel to the hind coxae on the first abdominal sternite, which may either follow the hind coxal edge exactly or diverge from it to a varying degree posteriorly. Useful in specific diagnosis of the genera *Meligethes* and *Carpophilus*.

Cercus. (Plural, *cerci*). Usually a paired, attenuated – sometimes segmented – process projecting from the ninth abdominal tergite. The *cerci* of many coleopterous larvae are long, and directed posteriorly.

Chitin. Chemically, a nitrogeneous polysaccharide of very complex structure, often admixed with other substances of doubtful composition. Pure Chitin is colourless. Chitin forms the essential constituent of the insect exoskeleton.

Cicatrix. A large deep, scar-like impression.

Ciliate. Furnished with cilia or fringes of hairs more or less parallel.

Clavate. Clubbed or club-shaped, especially of the antennae (Fig. 58).

Clypeus. A cranial *sclerite* anterior of and generally fused to the *frons* see Fig. 42 (larva).

Common. Extending over two neighbouring portions of the body, e.g. elytra with a common spot.

Compressed. Flattened laterally.

Concolorous. Uniform in colour.

Confluent. Running into one another, applied to coloration and puncturation.

Connate. Soldered together.

Connecting Membranes. Sharp and Muir (1912, p. 485) state that this term cannot be commended. The *genital tube* may exist without *sclerites*, in such a case are these *connecting membranes*? However the term is widely used and is very useful; there are two *connecting membranes*. First Connecting Membrane – connects the *median lobe* to the *tegmen*. Second Connecting Membrane – connects the *tegmen* to the *apex* of the *abdomen*. (Male).

Coprophagous. Feeding on excrement.

Cordate, Cordiform. Heart-shaped, usually applied to the *pronotum*.

Coriaceous. Leathery.

Cornea. Lens of *ocellus*.

Corneous. Horny.

Costate. Furnished with elevated longitudinal ribs (*costae*).

Costiform. In the shape of a *costa* or raised rib.

Coxal. Relating to the *coxae;* the *coxal* or *cotyloid cavities* are the cavities in which the *coxae* articulate.

Crenate, Crenulate. Furnished with a series of very blunt teeth which take the form of segments of small circles.

Crepuscular. Occurring at twilight, relates to habits.

Cretaceous. Chalky.

Cribriform. Perforated, like a sieve.

Cruciform. Cross-shaped.

Cupule. Small cup-shaped organs with which the anterior *tarsi* of certain males (especially among the *Dytiscidae*) are furnished; they are used as suckers for adhering.

Cursorial. Adapted for running.

Cuspidate. Sharply pointed.

Cyathiform. Cup-shaped

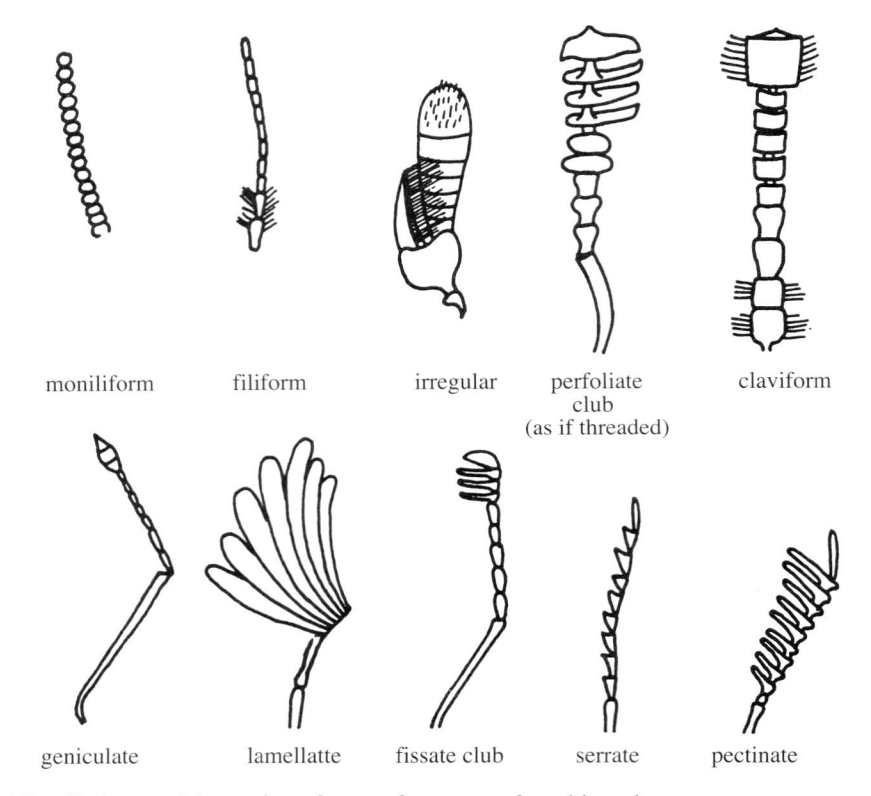

moniliform   filiform   irregular   perfoliate club (as if threaded)   claviform

geniculate   lamellatte   fissate club   serrate   pectinate

Fig. 59 Some of the various forms of antennae found in coleoptera.

Decumbent. Lying down.

Deflexed. Bent down, compare *reflexed.*

Dehiscent. Gaping apart towards the *apex.*

Dentate. Toothed, furnished with small teeth or tooth-like prominences.

Denticulate. With a row of small teeth.

Depressed. Flattened as if by pressure from above. (Compare *compressed*).

Dichotomy. A division into pairs.

Dichotomous Table. A scheme for identification which gives an alternative choice at each step.

Digitate. see *Palmate.*

Dimorphic, Dimorphous. Presenting two distinct types in the *same sex,* e.g. females of *Dytiscus circumcinctus* Ahrens may have smooth or sulcate elytra.

Glossary

Dimorphism, sexual. Differences in form *between the two sexes* of a species.

Disc. The middle, central portion.

Distal. Away from the centre of the body or point of attachment.

Distinct. Of spots, punctures, etc.; not touching or running into each other.

Divaricate. Used of two parts that are contiguous at the base and very strongly *dehiscent* at the apex (forked).

Dorsal. The upper surface.

Ductus ejaculatorius. The ectodermal, mostly median and unpaired, exit tube of the efferent system, opening by the *gonopore* at the tip of the *median lobe*. Its distal part sometimes called the *bulbus ejaculatoris* when enlarged. (Male).

Ductus spermathecae. The primary canal through which the sperms enter the *spermatheca* from the *vagina* or *bursa copulatrix*. (Female).

Ecdysis. Moulting of the external skeleton.

Elytra. The chitinous anterior pair of wings; the wing cases (singular = *elytron*).

Emarginate. Notched, with a piece cut out of the margin.

Entire. Without excision, emargination, or projection.

Entirely. Sometimes used when describing colour, meaning the whole insect or whole part being referred to.

Epigeal. Fungi maturing above the soil.

Eruciform. Of larvae when the three major divisions are not evident, the legs may be degenerate or wanting.

Excised. Cut away.

Exoskeleton. The hard chitinous integument of the Arthropods, fulfills the same function as an (endo-) skeleton, but is external.

Explanate. Widened out or expanded. Joy (1932) defines it as 'A slight hollowing out, close to margin, most commonly used for the sides of the thorax'. This definition should not be confused with 'emarginate' or 'excised' (above).

External Lobe. A synonym for *basal piece*.

Facets. The lenses or divisions of the compound eye.

Facies. General aspects of a species, genera or group of insects.

Farinose. Presenting a mealy appearance.

Fascia. A coloured band.

Fascicules. Tufts of bristles or hairs, resembling small brushes.

Ferruginous. Rust-red colour.

Filiform. Generally applied to antennae. Thread-like; elongate and of about the same thickness throughout. (Fig. 58).

Fissate. A type of lamelliform antennae, the terminal segments are only poorly laminate. Fig. 58. e.g. *Lucanus cervus.*

Flabellate, Flabelliforme. Fan-shaped; of antennae: having the upper segments prolonged into branches.

Flagellum. The sclerotised terminal prolongation of the *ductus ejaculatoris* usually concealed within the *internal sac* when in repose, but sometimes very long and constantly protruding through the *ostium* (male). In more general terms a flagellum is any long thin process attached to a larger (often globular) organ, e.g. the spermatheca in *Longitarsus* and some *Agathidium* species; this term is also used for the combined *funiculus* and *club* of the antennae, especially in Curculionidae.

Fossorial. Adapted for digging or burrowing.

Fovea. A large round or elongate depression on the surface.

Foveate. Furnished with such depressions.

Free. Of the head: visible, not hidden by the thorax; of a part of the body: movable, not fused to the adjoining part.

Frontal Suture. Paired suture between the frons and one or other of epicranial halves. See Fig. 42 (*larva*).

Frons. An unpaired cranial sclerite, bears the median ocellus (where present).

Funiculus. The segments of the antennae between the *scape* (first elongate segment) and the club; especially applied to the Curculionidae.

Fuscous. Brown or tawny-brown.

Fusiform. Spindle-shaped, broadest in the middle and gradually narrowed in front and behind to a more or less pronounced point.

Geniculate. Elbowed or kneed, abruptly bent upwards or downwards (Fig. 58).

Genitalia. The reproductive organs as a whole. Both sclerotised and unsclerotised parts, male or female.

Genital Opening. The opening of the *ductus ejaculatorius* into the *ductus cornmunis* on ventral surface.

Genital Segments. Those segments principally, but not exclusively, partaking in the formation of the copulatory organs; in male abdominal segment *ix,* in the female abdominal segments *viii* and *ix.* The segments are usually reduced or modified.

Genital tube. The *median lobe* plus *tegmen* plus connecting membranes. Compare '*aedeagus*' above.

Gibbous, Gibbose, Hump-backed, very convex.

Glabrous. Smooth, hairless and without punctures or raised sculpture; quite glabrous surfaces in Coleoptera are usually shining.

Gonopore. The external opening of the genital duct (in male and female). In the male, it is often situated on an intromittent organ. When a genital chamber or an *endophallus* is formed the opening of this cavity may be narrowed so as to form a second *gonopore* in which case the mouth of the *ductus ejaculatoris* (in males) or *oviductus communis* (in females) is called the *primary gonopore,* the latter opening the *secondary gonopore* or *gonotreme;* in males *phallotreme*, in females *vulva* or *oviporus*. Opens between abdominal segments *viii* or *ix*, or on abdominal segment *viii*. This is an ambiguous term and it should be deduced whether primary or secondary gonopore is meant. The term should be reserved for the primary gonopore.

Granulate. With small, rounded-off elevations.

Granulation. Applied to the eyes, the granulation of which is said to be fine or coarse accordingly as the facets are more or less numerous and pronounced.

Gressorial. Adapted for walking.

Gular. Pertaining to the throat.

Habitat. The natural environment where the beetle normally lives.

Habitus, Habitus figure. Generalised view of the whole specimen as seen from above.

Halteres. Or balancers: two small knobbed appendages attached to the thorax. They are modified wings which vibrate very rapidly and act as balancers during flight. They are characteristic of Diptera, though also occur in the Strepsiptera.

Heteromerous. With the posterior tarsi composed of less joints than the anterior and intermediate ones.

Hirsute. Set with thick long hairs.

Hispid. Set with short erect bristles.

Humeral. Relating to the shoulder, *(humerus)*.

Hybrid. The offspring of two different species.

Hypermetamorphosis. A metamorphosis in which the insect passes through two or more markedly different larval instars, usually accompanied by marked change in larval life, e.g. Meloidae.

Hypogeal (Hypogeous). Subterranean fungi, developing and becoming mature entirely in the soil; this includes the truffles *(Tuber* spp.).

Hypha. The filament of a fungus.

Imago. (Plural *imagines*). The perfect, completed state of an insect; the adult.

Imbricate. Overlapping one another like tiles on a roof.

Impunctate. Without punctuation.

Incrassate. Thickened.

Infuscate. Darkened; more or less fuscous in colour.

Inquiline. Living within another organism or in its nest but not as a parasite.

Insertion. Point of attachment of movable parts, e.g. antennae.

Instar. The progressive stages in the history of an insect constitute complete and distinct periods each of which is known as an *instar;* the new stage after each *ecdysis*.

Integument. The body wall.

Internal sac. Invaginated cavity at distal end of the median lobe of the *aedeagus* into which opens the *ductus ejaculatoris*. Everted during copulation.

Interspace. The space between punctures. Some authors refer to the space between the rows of punctures (*striae*) on the elytra as 'elytral interspaces.' Nowadays the spaces between striae are generally referred to as 'interstices', and those between individual punctures as 'space(s) between punctures'

Interstices. The spaces between *striae* or rows of punctures: the term is properly applied to the *elytra* only.

Iridescent. Exhibiting prismatic colours, changing in different lights.

Juxta. In composition means near, as juxta-ocular, situated near the eye.

Keel. A fine raised line. See also 'carina' above; the two terms are interchangeable.

Labrum. An unpaired cranial sclerite, the 'upper lip' covers the base of the mandibles. See Fig. 42 (*larva*).

Lamina. A flat plate or scale.

Lamellate, Lamelliform. Of antennae: having the apical segments like leaves of a book; as in the Lamellicornia (see Fig. 58.).

Lateral. Pertaining to the sides.

Lateral Lobes. Of the *aedeagus*. Often used in a purely descriptive manner; a *paramere*, (not in all cases are *parameres* lateral or paired).

Lignicolous. Dwelling in wood.

Lignivorous. Wood-feeding.

Linear. Narrow, elongate, parallel sided; applied to the whole insect or a particular portion.

Lineated, Lineate. With longitudinal stripes.

Lobes. Parts of an organ separated from one another by a more or less deep division.

Lunulate. Crescent-shaped.

Maculate. Spotted.

Mandibles. The biting jaws. (See Fig. 42. *larva*).

Margin. The outer edge.

Margined. Furnished with a more or less distinctly produced or carinate outer edge.

Maxillae. The lower jaws, always smaller than the *mandibles*.

Median. Central.

Median Suture. (Median epicranial suture), the suture between the frons and the two epicranial halves. (See Fig. 42 *larva*).

Membraneous. Of the consistancy of a membrane.

Mesonotum. The upper surface of the *mesothorax*.

Mesothorax. The middle segment of the *thorax*.

Metanotum. The upper surface of the *metathorax*.

Metasternum. The under surface of the *metathorax*. (Fig.57)

Metathorax. The third segment of the *thorax*.

Molar. The grinding surface of the mandible.

Moniliform. Of antennae; as if formed of beads. (Fig. 58).

Mucronate. Terminating in a sharp point.

Mutic. Without point or spine.

Myrmecophilous. Living with ants.

Myxomycetes. The 'slime moulds'. Not true fungi as they never form hyphae but spend most of their life as a naked mass of protoplasm (*plasmodium*).

Natatorial. Adapted for swimming.

Necrophagous. Feeding on dead and decaying matter.

Normal. Usual or natural; this term is used very loosely, but it is often very useful, and its meaning in comparison is always easily understood from the context.

Obconical. A reversed cone, with the thickest part in front often used to describe antennal segments which become thicker towards their apex.

Obsolete. Almost effaced or only slightly marked.

Ocellated, Ocellate. Furnished with round spots surrounding by a ring of a lighter colour.

Ocelli. Small extra simple eyes usually situated on the top of the head (e.g. in adults of many Dermestidae). Ocelli (as distinct from compound eyes) are the only ones present in the larval stages (except in third instar larvae of some non-British Strepsiptera).

Ochraceous. Brownish-yellow.

Onisciform. Shaped like a woodlouse.

Onychium. The last segment of the tarsi which bears two *onychia* or claws.

Operculum. A lid.

Orbital. Relating to the upper border of the eye, as supra-orbital, situated above this upper border.

Ostium. Opening through which the internal sac is everted during copulation. Usually situated dorsally and distally on the *median lobe* of the *aedeagus*.

Oval, Ovate, Ovoid. Egg-shaped or elliptical - these are technically different geometric shapes, but in entomology the term is loosely applied.

Ovipositor. The female organ by means of which eggs (*ova*) are deposited. Only one type is to be found in the Coleoptera, this is often called the oviscapt type. Formed by the prolongation or modification of the posterior abdominal segments. Absent in some Coleoptera.

Palmate. Widened and divided like a hand: if the divisions are long and slender, the term *digitate* is used.

Palpus. (Plural *palpi*) Auxiliary organ of the mouth-parts.

Papillae. Small rounded tubercles.

Paramere. A pair of appendages (sometimes coalescent or even completely joined) forming the distal (apical) part of the *tegmen* and usually protruding on either side of the *median lobe* of the *aedeagus*.

Parthenogenesis. The development of an embryo from an egg without fertilisation.

Patella. A little bowl or cup.

Patelliform. Cup or bowl shaped.

Pectinate. Toothed like a comb. (Fig. 58).

Peduncle. A piece supporting an organ, or joining one organ to another like a neck.

Penis. Originally meaning the intromittent organ, the term has been used interchangeably with *aedeagus* or with *phallus*. Thus it is not homologous throughout the insect class and should be replaced by *aedeagus*. Also, see 'aedeagus' above.

Pentamerous. With five joints.

Penultimate. The last but one.

Perfoliate. Formed of joints separated as it were strung together by a common thread or narrow support running through them (Fig. 58).

Glossary

Phytophagous. Feeding on plants.

Piceous. A somewhat dark to very dark colour sometimes with a green or yellow sheen.

Pilose. Hairy, covered with hair-like pubescence. *Verticillate-pilose,* of antennae, with hairs set round the vortex of each segment.

Plasmodia. In *myxomycetes.* the multinucleate motile mass of protoplasm, characteristic of the growth phase.

Pitchy. Blackish-brown: a somewhat loose colour term (see *piceous*).

Pleural. The lateral surfaces of the segments.

Pleurite. Lateral plates of the segments.

Plicate. Furnished with a fold or folds.

Polymorphous. Of various forms.

Polyphagous. Feeding on many kinds of food.

Pores. Large isolated punctures.

Process. A projection: any prolongation.

Productile. Capable of being lengthened out or produced.

Pronotum. The upper surface of the *thorax.*

Propygidium. Penultimate dorsal segment of the abdomen (visible in certain Histeridae, etc., to which the term is applied: it is not used of the Staphylinidae).

Prosternum. The under surface of the *thorax.*

Prothorax. The first segment of the *thorax.*

Pseudopod. A 'false foot' – a fleshy protuberance on the ventral surface of the terminal segment of the larval abdomen.

Pubescence. Shiny hairiness or down.

Puncture. A small depression on the surface, usually round.

Punctate (Puncturation). Covered with punctures. The terms '*punctae*' and '*punctation*' are American and seldom, if ever, feature in European texts.

Punctate-striate. With rows of punctures imitating and taking the place of *striae,* opposed to *striate-punctate,* with punctured *striae;* however, the terms have been loosely used often interchangeably so.

Pupa. In Coleoptera, the stage preceding the imago.

Puparium. The *integument* or 'skin' of the pupa.

Pygidium. Last dorsal segment of the *abdomen.*

Pygopodium. Terminal flat sclerite of abdomen (characteristic of *Elateridae*).

Pyriform. Pear-shaped.

Quadrate. Square.

Quadri-. In composition indicates four times, e.g. *quadrimaculate,* with four spots.

Raptorial. Adapted for seizing and devouring prey.

Receptaculum seminalis. The *spermatheca,* often including the *ductus spermathecae.*

Reflexed. Bent up.

Remiform. Oar-shaped.

Reniform. Kidney-shaped.

Reticulate. Covered with a network of scratches or cross *striae.*

Retinaculum. A produced tooth-like process, usually arising at the inner margin of the mandible.

Rostrum. Prolongation of the head between the eyes, especially applied to the Curculionidae.

Rostrate – in the form of a beak or *rostrum.*

Rufous. Reddish.

Rugose. Wrinkled, roughened.

Rugulose. Slightly wrinkled.

Salient. Extended, jutting out.

Saltatorial. Adapted for leaping.

Scansorial. Adapted for climbing.

Scape. The term applied to the first segment of the antennae when it is considerably developed and usually held at an angle to the rest of the antenna (the *funiculus* and *club*). Chiefly applied to the Curculionidae.

Sclerites. The chitinous plates which collectively make up the exoskeleton. They may be very hard, or quite soft.

Scrobe. Lateral furrow of the *rostrum,* holding the base of the antennae when at rest; chiefly applied to the Curculionidae. The *scape* fits into this.

Sculpture. Modifications of the surface in the way of punctuation, striae, elevations, etc., as opposed to structure, which has reference to the shape and construction of the various parts of the body.

Scutellum. A dorsal plate of the *mesonotum.* (Fig. 57).

Securiform. Hatchet-shaped.

Secretion. Matter produced by glands of the body.

Sensillae. A simple sense organ; a body hair connected to the nervous system.

Serrate, Serriform. With teeth like a saw. (Fig. 58).

Seta. A long outstanding bristle or stiff hair.

Setaceous. Gradually tapering to the tip, like a bristle.

Setiform. Shaped like a bristle.

Setose, (Setigerous, Setiferous). Set with or bearing *setae*.

Shagreened. A surface divided into microscopically equal areas, very fine sculpture with no punctuation (like shark skin).

Simple. With no unusual addition or modification, e.g. without spines, dilation, emargination, etc.

Sinuate. Slightly waved.

Spatulate. Narrowed at base and enlarged towards extremity.

Spermatheca. The receptable of the sperm during coition. An ectodermal invagination ventrally and posteriorly at the end of the abdominal segment VIII. In Coleoptera sometimes connected with the *bursa copulatrix,* sometimes opening by *seminal canal* (female).

Spiracle. Respiratory openings on the surface of the body, the external orifice of the *tracheae*.

Sporophore. The spore-producing or supporting structure, the 'fruiting-body' in myxomycetes (the 'powdery fungi' of Fowler and other authors).

Spur. Spike-like projection, occurs on the legs of many beetles and is a useful character in diagnosing, for example, some *Longitarsus* species. *Spurs* are generally on the tibiae; projections on the femora are referred to as *teeth*.

Squamose. Covered with larger or smaller scales *(squamae)*.

Stadium. The interval between *ecdyses*.

Sternite. The ventral plates of a segment.

Sternum. The ventral surface.

Striae. Impressed lines.

Stridulation. The sound produced by the friction of scraping of one surface against another. Many beetles also produce sound by expelling air through the *spiracles*.

Strigose. Scratched.

Striole. An abridged or rudimentary *stria*. *Striolate* - furnished with such small *striae*.

Style. A pointed process.

Sub-. In composition, indicates almost or slightly, as sublinear, subquadrate, etc.

Subulate. Tapering, terminating in a fine and sharp point, like an awl.

Sulcate. Furrowed.

Sulciform. Shaped like a furrow.

Sutural. Pertaining to the *suture*.

Suture. The line on which the elytra join (= elytral suture); the line of junction of any two adjacent parts.

Synanthropic. Living with humans.

Tegmen. The single divided sclerite situated basally (proximally) of the *median lobe* and often surrounding it when in repose. Usually divided into *basal piece* and *parameres*. The *tegmenite* is an isolated *basal sclerite* of the *tegmen* situated on the second *connecting membrane*.

Temple. The part of the head behind the eye.

Tergite. The dorsal plate or *sclerite* of a segment.

Terminal. The last, the end of a series of segments.

Testaceous. Yellowish, usually with a dusky tinge; not a bright yellow, although the term is loosely used, it is applied to almost all yellowish or reddish-yellow shades. Also, but rather infrequently 'testaceous' means having a hard outer cover, a 'test' as in tortoises or sea urchins, for example.

Tetramerous. With four joints.

Tomentose. Cottony.

Transverse. Broader than long.

Trivial name. The name of the species.

Tooth. A pointed projection, usually on the inner surface of the femur.

Trophi. Parts of the mouth used when feeding.

Truncate. Abruptly cut right across in a straight line.

Tubercle. A small abrupt elevation of varying form.

Unicolorous. Of the same colour throughout.

Unisetose. Bearing one *seta*.

Variolose. Covered with small impressions or pits, a pock-marked appearance.

Vellum. The thin membrane forming part of the apical and marginal portions of a *paramere*.

Vellum aedeagus. The thin membraneous covering of the intromittent organ.

Vermiculate. Covered with irregular, sinuate, worm-like *striae*.

Versicolorous. Of various colours.

Vertex. Upper flattened surface of the head (site of ocelli when present).

Vesicant. Raising a blister, (as applied to *Lytta* etc.).

Villose. Covered with long loosely-set hairs.

Xylophagous. Wood-feeding.

References

Britton, E.B., 1956. Coleoptera. Scarabaeoidea (Lucanidae, Trogidae, Geotrupidae, Scarabaeidae). *Handbooks for the Identification of British Insects,* **5**(11): 29 pages. (Revised edition by Jessop, L. 1986). Royal Entomological Society, London.

Crowson, R.A., 1955. *Natural Classification of the Families of Coleoptera.* Nathaniel Lloyd & Co., London. Reprinted 1967 with additional papers by E.W. Classey, Hampton, Middlesex.

Crowson, R.A., 1956. Coleoptera. Introduction and keys to families. *Handbooks for the Identification of British Insects,* **4**(1): 59 pages. Royal Entomological Society, London.

Fowler, W.W., 1887. *The Coleoptera of the British Islands.* Volume 1. Reeve & Co., London.

Halstead, D.G.H., 1963. Coleoptera. Histeroidea. *Handbooks for the Identification of British Insects,* **4**(10): 16 pages. Royal Entomological Society, London.

Joy, N.H., 1932. *A practical handbook of British beetles.* Volume 1. (622 pages). Witherby, London.

Nichols, S.W. and Schuh, R.T., 1989. *The Torre-Bueno glossary of entomology.* 840 pages. Hardback. ISBN 0-9133424-13-7. Distributed in Europe by Apollo Books, Kirkeby Sand 19, DK 5771 Stenstrup, Denmark (http://www.apollobooks.com).

Sharp, D. and Muir, F.A.G., 1912. The comparative anatomy of the male genital tube in coleoptera. *Transactions of the Entomological Society of London.* Reprinted 1969 without change by the Entomological Society of America.

Index

Index of Beetle Genera referred to in the text

Abax	19	*Amalus*	286, 295
Abdera	115, 274, 291, 293, 294, 297, 305	*Amara*	19, 20, 21, 232, 270, 298, 343
Abraeus	24, 337	*Amarochara*	252
Acalles	171	*Amauronyx*	39, 216, 252, 253, 254
Acalyptus	167, 284, 297	*Amidobia*	251
Acanthocinus	126, 127, 128, 137, 143, 266, Pl. 87a, Pl. 87b	*Amischa*	251, 252, 253, 254, 255, 257
Acanthoscelides	143, 145	*Ampedus*	67, 68, 272, 291, 293, 298, 392, Pl. 51, Pl. 52
Acentrotypus	281, 307	*Amphicyllis*	29
Acilius	372	*Amphimallon*	51, 54, 56
Aclypea	37	*Amphotis*	83, 86, 253
Acrantus	78, 289, 293	*Anacaena*	13, 349, 351, 373
Acritus	24, 234	*Anaglyptus*	125, 128, 132, 136, 142
Acrotona	253, 256	*Anaspis*	112, 113, 227, 276, 353
Acrotrichis	25, 27, 251, 254, 255, 256	*Anastrangalia*	128, 135, 276, 291
Actinopteryx	27	*Anatis*	94, 274, 291, Pl. 67
Actocharis	230	*Anchonidium*	175
Acupalpus	19, 21	*Ancistronycha*	72, 73
Adalia	96, 98, 99	*Anisodactylus*	236, 392
Aderus	119, 227, 276, 293, 296	*Anisosticta*	98
Adistemia	106	*Anisotoma*	29, 30, 34, 221, 234
Adota	230	*Anisoxya*	115, 274, 293, 297, 302, 305
Aegialia	56, 57	*Anitys*	77, 272, 293
Aeletes	23	*Anobium*	75, 76, 77, 79, 191, 272, 303
Aepus	19, 224, 230, Pl.22	*Anomala*	51, 56
Afromorgus	48	*Anommatus*	95, 96, 206, 234
Agabus	9, 11, 371, 372	*Anoplodera*	124, 128, 135, 276, 293
Agapanthia	126, 129, 132, 136, 138, 142, 276, 303	*Anoplotrupes*	52, 56
Agaricochara	237	*Anoplus*	167, 284, 294
Agaricophagus	29, 31	*Anostirus*	392, Pl. 53
Agathidium	29, 30, 31, 33, 34, 221, 234, 321, 337, 420, Pl. 118	*Anotylus*	44, 230, 234
Agelastica	143, 151, 152, 279, 294	*Anthaxia*	59, 60, 271, 299, 300, 307
Aglenus	115, 119	*Antherophagus*	91, 238
Aglyptinus	29	*Anthicus*	119, 229, 230, 237
Agonum	19, 20, 21	*Anthocomus*	81, Pl. 62
Agrilus	59, 60, 189, 271, 293, 294, 297, 300, 391, Pl. 49	*Anthonomus*	166, 167, 221, 283, 284, 291, 292, 299, 300, 376
Agriotes	66, 376	*Anthrenus*	74, 191, 236, 260, 269, 331, 335, Pl. 114a
Agrypnus	65	*Anthribus*	162
Ahasverus	259, 268, 269	*Apalus*	118
Aizobius	164, 281, 299, Pl. 105	*Aphanisticus*	60, 189, 225, 271, 309
Alaobia	230, 251, 256	*Aphidecta*	98, 100, 274
Aleochara	186, 223, 230, 251, 252, 253, 254, 256,	*Aphodius*	49, 51, 52, 53, 54, 55, 56, 57,58, 186, 229, 234, 239, 338, 339, 347, 348, 351, 392, 415, Pl. 42
Alianta	221	*Aphthona*	143, 144, 153, 154, 279, 296, 302, 304, 310
Alophus	172		
Alosterna	128, 135, 142	*Apion*	161, 196, 281, 341, 343
Alphitobius	116, 269	*Aplocnemus*	80
Alphitophagus	116	*Aplotarsus*	66
Altica	143, 144, 153, 154, 279, 294, 296, 299, 302	*Apoderus*	163, 196, 281, 294, Pl. 104
		Apteropeda	143, 153, 154
Amalorrhynchus	170, 286, 297	*Araecerus*	162

Archarius	186, 284, 293, 297	Caenocara	77, 191
Arena	230	Caenopsis	171, 216, 287, 299, Pl. 105
Arhopalus	125, 127, 136, 276, 291	Caenoscelis	91, 224
Aromia	124, 126, 129, 136, 142, 276, 296, 297	Cafius	46, 230
		Calambus	272, 293, 294, 297
Asaphidion	69, 231, Pl. 24	Calathus	19
Asemum	125, 127, 128, 136, 276, 291	Callidium	124, 127, 128, 136, 139, Pl. 83, Pl. 119
Aspidapion	164, 281, 296		
Aspidiphorus	234	Callosobruchus	143, 145, 146
Atheta	223, 230	Calomicrus	143, 151, 152, 279, 301
Athous	66, 376	Calosirus	286, 303
Atomaria	91, 92, 93, 223, 224, 229, 237	Calosoma	270, 293
Attagenus	191, 236, 269	Calvia	98, 101
Attelabus	163, 196, 281, 293	Cantharis	71, 72, 73, 189, 230
Aulonothroscus	64	Carabus	18, 19, 20, 21, 22, 182, 232, 315, 319, 339, 340, 393, Pl. 4 - 13
Badister	19, 392		
Baeocara	27	Carcinops	24
Bagous	16, 169, 216, 285, 290, 291, 292, 301, 303, 306, 308, 309	Cardiophorus	63, 66, 67, 189
		Carpelimus	45, 233, 236
Balanobius	166	Carpophilus	83, 84, 87, 88, 89, 221, 267, 416
Baris	169, 216, 226, 285, 294, 297, 298, 307	Cartodere	105, 106, 234
		Cassida	79, 143, 144, 148, 195, 230, 277, 294, 295, 305, 307, 308, Pl. 97- 99
Barynotus	172, 287, 288, 302		
Batophila	143, 153, 154, 280, 299	Catapion	225, 282, 301
Batrisodes	39, 251, 252, 254, 255	Cathormiocerus	171, 172, 216, 393, Pl. 105
Batristilbus	257	Catopidius	32, 34, 221, 239
Batrisus	39	Catops	31, 34, 358
Bembidion	19, 21, 22, 231, 232, 233, 235, 236, 343, 392, 393, Pl. 25	Cephennium	35
		Cerambyx	122, 123, 128, 129, 136, 205
Berosus	13	Cerapheles	80
Betulapion	164, 282, 294	Ceratapion	281, 307
Biblopectus	38	Cercyon	13, 184, 230, 237, 343, 414
Bidessus	232, 393	Cerophytum	63
Biphyllus	95, 114, 234, 274, 305	Cerylon	95, 221
Bisnius	46, 235	Cetonia	50, 51, 56, 186, 251, Pl. 45
Bitoma	116, Pl. 63	Ceutorhynchus	169, 170, 216, 225, 286, 297, 298, 393
Blaps	20, 116, 192, 239		
Bledius	19, 44, 231, 232, 233, 235, 236	Chaetocnema	143, 144, 154, 280, 295
Blethisa	Pl. 20	Chaetophora	61
Bolitophagus	116, 275, 294	Chilocorus	98, 274, 294, 297, 305
Brachinus	18	Chlaenius	19, 20, 231
Brachonyx	166, 284	Choleva	32, 34, 321, 356
Brachygluta	39	Choragus	162, 280, 303
Brachygonus	67	Chrysanthia	117, 120, 275, 291
Brachypterolus	81, 83, 273, 306	Chrysolina	143, 149, 150, 151, 160, 277, 295, 302, 303, 304, 305, 384, 393, Pl. 93
Brachypterus	81, 82, 273, 292		
Brachysomus	287, 299		
Bradycellus	19, 20, 224, 270, 298	Chrysomela	143, 149, 150, 278, 294, 296, 297
Brindalus	56	Cicindela	18, 182, 225, 231, 370, 393, Pl. 19
Bromius	143, 158, 280, 302	Cicones	116, 275, 293, 294
Broscus	19, 230	Cidnopus	67
Bruchela	162, 280, 298	Cillenus	Pl. 27
Bruchidius	143, 145, 146, 159, 216, 217, 277, 296, 300, 301	Cimberis	162, 228, 280, 291
		Cionus	167, 196, 283, 284, 305, 306, 376, Pl. 103
Bruchus	143, 145, 277, 300		
Bryaxis	39	Cis	107, 108, 109, 110, 221
Buprestis	59	Clambus	58
Byctiscus	163, 281, 294, 296, 393	Claviger	40, 253, 255, 256
Byrrhus	61, 62, 234, Pl. 50	Cleonis	174, 196, 289, 307
Bythinus	39	Cleopomiarus	285, 306
Byturus	95, 274, 299	Cleopus	167, 283, 285, 305, 306

Index

Clitostethus	98, 100, 101, 274, 303	*Denticollis*	67, 189
Clivina	233	*Deporaus*	162, 163, 196, 281, 294
Clytra	143, 156, 159, 197, 250, 251	*Dermestes*	191, 236, 260, 264, 267, 269
Clytus	125, 128, 136, 142, Pl. 85	*Derocrepis*	143, 153, 280, 300
Coccidula	98, 99, 274, 310	*Diacanthous*	272, 291, 293
Coccinella	98, 100, 101, 181, 251, 274, 298, Pl. 66	*Dianous*	224, 232, 410
		Diaperus	116
Coelambus	11	*Diastictus*	56
Coeliodes	169, 286, 293	*Dibolia*	143, 280, 304
Coeliodinus	169, 286	*Dicheirotrichus*	19, 236
Coelostoma	13	*Dicronychus*	65, 67, 69, 225, 272, 309
Colenis	29, 31	*Dictyoptera*	69, 235
Colon	28, 32, 33, 34, 228	*Dieckmanniellus*	165, 283, 302
Colydium	116	*Dienerella*	105, 106, 221
Coniocleonus	289, 299	*Diglotta*	230
Conopalpus	193	*Dinarda*	251, 252, 254
Copelatus	11, 337	*Dinoptera*	123, 125, 128, 135
Copris	48, 52, 56, 229	*Diplapion*	281, 308
Corticaria	105, 106, 107, 230	*Diplocoelus*	95, 274, 293
Corticarina	105, 230	*Dolichosoma*	80
Corticeus	116, 275, 291, 292, 293, 294, 305	*Donacia*	17, 143, 147, 195, 228, 277, 291, 308, 209, 310, 374, 393, Pl. 89
Cortinicara	105		
Corylophus	103, 104, 230	*Dorcatoma*	77, 78, 227
Cossonus	287, 296, 297	*Dorcus*	24, 50, 56, 186, 271, 292, 293, 297, 305
Creophilus	186		
Crepidodera	143, 153, 154, 280, 296, 297	*Dorytomus*	167, 196, 223, 285, 296, 297
Crioceris	143, 147, 277, 310	*Dromius*	20, 234, 393
Cryphalus	290, 291	*Drupenatus*	170, 286, 297
Cryptarcha	83, 86, 192	*Drusilla*	251, 253, 254, 255, 256, 257
Crypticus	116, 236	*Dryocoetinus*	290, 293, 294
Cryptocephalus	143, 155, 156, 157, 158, 280, 293, 294, 295, 297, 307, 337, 388, 393, Pl. 90, Pl. 91	*Dryophilus*	75, 272, 291, 299, 301
		Dryophthorus	165, 256, 393
		Dryops	15, 66, 187, 233
Cryptophagus	91, 92, 93, 229, 237, 238, 341	*Drypta*	231
Cryptorhynchus	171, 196, 287, 296, 297	*Dyschirius*	19, 231, 233, 235, 236, 393, P. 21
Crypturgus	290	*Dytiscus*	11, 12, 319, 337, 338, 371, 372, 373, 418, Pl. 2
Ctenicera	66, 376		
Cteniopus	Pl. 72	*Elaphrus*	233, Pl. 18
Ctesias	191, 233, 237, Pl. 58	*Elater*	67, 272, 292, 293, 393
Curculio	166, 196, 284, 293, 294, Pl.100	*Eledona*	192
Curimopsis	61, 62, 384, 393	*Ellescus*	223, 285, 297
Curtonotus	20, 270, 294, 308	*Elodes*	16, 187
Cyanapion	282, 300	*Emus*	46
Cyanostolus	89, 233, 236	*Enalodroma*	253
Cybocephalus	83, 87, 88	*Endomychus*	96, 97, Pl. 65
Cychramus	83, 86, 192, 237	*Enicmus*	105, 106, 116, 234
Cychrus	19, Pl. 14	*Ennearthron*	107
Cylindrinotus	116	*Enochrus*	13, 373
Cymbiodyta	373	*Ephistemus*	91
Cypha	221	*Epierus*	23
Cyphon	16, 17, 364	*Epilachna*	98, 100, 274, 296
Cyrtusa	29	*Epiphanis*	62, 63
Cytilus	61	*Epitrix*	143, 154, 280, 303
Dacne	94	*Epuraea*	83, 84, 85, 87, 88, 192, 235
Dascillus	187	*Ernobius*	76, 77, 78, 228, 234, 272, 291
Dasytes	80	*Ernoporus*	290, 293, 294, 296, 393
Datonychus	169, 286, 304, 305	*Ethelcus*	286, 292
Demetrias	20, 270, 309, 310, Pl.30	*Eubria*	16
Dendroctonus	89, 176, 290	*Eubrychius*	170, 287, 301
Dendrophilus	24, 184, 251, 252, 254, 255,	*Eucnemis*	63, 271, 293, 393
Dendroxena	37, 185, 271, 293, Pl. 35	*Euconnus*	36, 237, 255

Eudectus	42	Helochares	13, 373
Euheptaulacus	54, 56, 229	Helophorus	13, 17, 184, 270, 298, 309, 348, 351, 373, 378, 394
Eulagius	114, 120		
Euophryum	171	Helops	235
Euplectus	38, 39, 237	Hemicoelus	76, 77, 78
Eurycolon	33	Hemitrichapion	282, 300
Euryptilium	26	Henoticus	91
Euryusa	256	Heptaulacus	54, 56, 229
Eusphalerum	42, 271, 299	Hermaeophaga	143, 153, 280, 302
Eutheia	35, 255	Hetaerius	25, 251, 253, 254, 255
Euthiconus	35	Heterocerus	15, 232, 233, 236
Eutrichapion	282, 300	Heterothops	46, 230, 251, 253
Exapion	161, 281, 301	Hippodamia	98, 101
Exochomus	98, 102, 274, 291, 302	Hippuriphila	143, 153, 280, 290
Falagria	257	Hister	24, 25, 184, 413
Ferreria	166, 207	Holoparamecus	97
Fleutiauxellus	67, 232	Holotrichapion	282, 300, 301
Furcipus	166	Homoeusa	253, 254, 255, 256, 257
Gabrius	46	Hoplia	56
Galeruca	143, 151, 152, 160	Hydaticus	12, 337, 372
Galerucella	143, 144, 151, 152, 153, 195, 278, 291, 297, 299, 302, 308, Pl. 94	Hydnobius	29, 31, 238
		Hydraena	14, 232, 373, 374
Gastrallus	76, 77, 78, 272, 302, 393	Hydrobius	373
Gastrophysa	143, 149, 277, 295	Hydrochara	373, 384, 394
Geodromicus	234	Hydrochus	13, 373, 394
Georissus	14, 231	Hydrocyphon	16
Geotrupes	49, 52, 54, 56, 57	Hydrophilus	13, 362, 373, Pl. 32
Gibbium	75	Hydroporus	11, 12, 373, 394,
Glaphyra	128, 136	Hydrosmecta	232
Glischrochilus	83, 86, 87, 192	Hydrothassa	143, 149, 150, 278, 292
Glocianus	225, 286, 307	Hygrobia	10, 182
Gnathoncus	223	Hygrotus	11
Gnatocerus	116, 164, 269	Hylastes	76, 176, 289, 291, 292, 305
Gnorimus	50, 56, 186, 271, 293, 299, 377, 388, 393, Pl. 46- 48	Hylastinus	289, 301
		hylecoetus	272, 291, 293, 294
Gonioctena	143, 149, 150, 278, 294, 296, 297, 300, 301	Hylesinus	289, 305
		Hylis	63, 271, 293
Gonodera	117	Hylobius	175, 196, 227, 229, 289, 291, 302
Gracilia	124, 128, 136, 142, 276, 297	Hylotrupes	124, 128, 136, 142, Pl. 82
Grammoptera	125, 128, 135, 142, 276, 293	Hylurgops	289, 291
Graphoderus	12, 384, 393, Pl. 3	Hypebaeus	79, 80, 212, 227, 273, 293, 312, 384, 394
Graptus	172, 287, 305, Pl. 105		
Gronops	171, 287, 294, 295	Hypera	167, 173, 174, 196, 225, 287, 294, 295, 300, 301, 303
Grynobius	75, 191		
Grypus	166, 283, 290	Hyperaspis	98, 100
Gymnetron	168, 285, 306	Hypocaccus	24, 225, 236
Gyrinus	9, 10, 337	Hypocassida	143
Gyrohypnus	251, 253	Hypocoprus	90
Gyrophaena	42, 237	Hypulus	115, 275, 293, 294
Habrocerus	43	Ilybius	9, 11, 182, 197, 371
Hadrobregmus	70, 80, 272, 293, 300	Ilyobates	252, 255
Hadroplontus	286, 307	Involvulus	163, 281, 299, 300
Halacritus	23	Ips	176, 290, 291
Haliplus	10	Ischnodes	272, 292, 293, 302
Hallomenus	114, 193	Ischnomera	117, 120, 334
Halobrecta	230	Ischnopterapion	165, 282, 300, 301
Halyzia	98, 100, 101, Pl. 68	Isochnus	285, 297
Haploglossa	223, 251, 252, 255	Isomira	117
Harmonia	98, 101, 102, 274, 291	Ixapion	164, 165, 202, 281, 302
Harpalus	19, 20, 21, 182, 270, 299, 393, 394	Judolia	124, 128, 135, 142, Pl. 80
Helianthemapion	165, 281, 296	Kalcapion	164, 237, 281, 302

Kateretes	81, 82, 273, 309	*Longitarsus*	143, 144, 153, 154, 155, 159, 216, 226, 279, 292, 301, 302, 303, 304, 305, 306, 307, 308, 341, 361, 391, 420, 427
Kheper	47		
Kibunea	189		
Kissister	24, 271, 295		
Kissophagus	289, 303	*Lophocateres*	78
Korynetes	79, 236	*Lordithon*	43, 237, Pl. 38
Labidostomis	143, 156, 159, 280, 294	*Loricera*	18
Laccobius	13, 349	*Lucanus*	48, 50
Laccophilus	11, 394	*Luperus*	48, 50, 56, 186, 271, 292, 293, 315, 334, 384, 388, 394, 420
Lacon	65, 67, 272, 293, 394		
Laemophloeus	273, 293	*Lycoperdina*	97, 237
Laemostenus	20, 239, Pl. 28	*Lyctus*	23
Lagria	117	*Lymantor*	290, 294
Lamia	126, 129, 137, 142, 276, 296, 297	*Lymexylon*	272, 293
Lamprinodes	253, 254, 255, 256	*Lyprocorrhe*	251, 254
Lamprohiza	70	*Lythraria*	143, 154, 280, 302
Lampyris	70, 191, Pl. 54	*Lytta*	117, 275, 305, 429
Langelandia	116, 235	*Macronychus*	15, 236
Laricobius	73, 272, 291	*Macroplea*	16, 143, 147, 277, 301, 308, 309
Larinus	174, 289, 307	*Magdalis*	76, 161, 174, 196, 289, 291, 292, 293, 294, 299, 300, Pl. 101
Lasioderma	75, 77, 221, 259, 269		
Lasiorhynchites	281, 293	*Malachius*	80, 273, 309, 394
Lasiotrechus	Pl. 23	*Malthinus*	72, Pl. 56
Latheticus	116, 269	*Malthodes*	71, 72, 73, 215, 217, Pl. 57
Latridius	105	*Malvapion*	282, 296
Lebia	20	*Mantura*	143, 153, 154, 280, 295, 296
Leiodes	28, 29, 30, 33, 34, 213, 238, 358	*Margarinotus*	25, Pl. 33
Leiopus	125, 128, 133, 137, 142	*Mecinus*	167, 168, 196, 285, 305, 306
Leiosoma	175, 196, 285, 291, 292	*Medon*	45
Leistus	19, 234	*Megapenthes*	67, 272, 292, 394
Lema	143, 147, 277, 307	*Megarthrus*	42
Leperisinus	176, 289, 305	*Megatoma*	191, 236
Leptacinus	251, 254, 256	*Melanapion*	164, 281, 297, 337, 394
Leptinotarsa	143	*Melandrya*	115, 193, 275, 292, 293, 294, Pl. 70
Leptinus	33, 223, 238, 253		
Leptophloeus	273, 291	*Melanimon*	117, 236
Leptura	125, 128, 135, 142, Pl. 78, Pl. 119	*Melanophila*	59, 234, 271, 291
		Melanotus	67, 189, 272, 309, 394
Lepturobosca	122, 128, 135, 276, 291, 294	*Melasis*	62, 271, 293
Lesteva	186, 232	*Meligethes*	83, 84, 87, 88, 161, 192, 228, 273, 296, 297, 298, 299, 300, 301, 304, 305, 306, 319, 321, 376, 416
Licinus	19, 225		
Lilioceris	143, 147, 148, 277, 310		
Limnebius	14, 373	*Melinopterus*	51
Limnichus	15, 16	*Meloe*	117, 118, 238, Pl. 73
Limnobaris	169, 286, 309	*Melolontha*	51, 54, 56, 57, 186, Pl. 43
Limobius	173, 174, 289, 302, 303	*Meotica*	394
Limoniscus	67, 68, 384, 394	*Mesosa*	125, 128, 132, 136, 142
Liocyrtusa	31	*Metabletus*	19
Liogluta	252, 256	*Metoecus*	115, 194, 238
Lionychus	232, 394	*Metopsia*	42
Liophloeus	172, 287, 303	*Miarus*	167, 285, 306
Liparus	175, 289, 303	*Micralymma*	19, 224
Lissodema	119	*Micrambe*	91, 92
Litargus	114	*Micrelus*	170, 286, 299
Lithocharis	237	*Micridium*	26, 27
Lithostygnus	106	*Microcara*	16
Lixus	174, 289, 294	*Microglotta*	186
Lochmaea	143, 151, 152, 153, 279, 297, 299, 300	*Microlomalus*	24
		Micropeplus	42, 45, 219, 234
Lomechusa	254, 256, 257	*Microplontus*	169, 286, 308
Lomechusoides	255	*Microptilium*	27

Microrhagus	63, 271, 293	Ochthebius	14, 373, 394
Microscydmus	36	Octotemnus	107
Millidium	26, 27	Ocypus	46, Pl. 39
Miscodera	232	Ocys	20, 224
Mniophila	143, 153, 154, 230	Odacantha	8, 270, 310
Mogulones	286, 304	Odonteus	49, 52, 56, 237
Molorchus	125, 127, 128, 136, 142, 276, Pl. 81	Oedemera	117, 336
		Oedostethus	376
Mononychus	169, 232, 287, 310	Oenopia	101
Monotoma	89, 219, 229, 237, 249	Oiceoptoma	37
Mordellistena	110, 111, 114, 275	Olibrus	90, 274, 307, 308, 358
Mordellochroa	110	Oligella	26
Morychus	61	Omalium	42, 230
Murmidius	95	Omaloplia	56
Mycetaea	97	Omophron	18
Mycetochara	117, 275, 293, 299, 302	Omorgus	48
Mycetophagus	77, 114, Pl. 69	Omosita	83, 85, 86, 88, 192
Mycetoporus	43	Omphalapion	281, 208
Mycterus	118	Oncomera	117
Myllaena	44	Ontholestes	184
Myrmechixenus	115, 116, 120, 229	Onthophagus	52, 53, 54, 55, 56, 229
Myrmetes	24, 251	Onthophilus	24
Myrrha	98	Oodes	8, 19
Myzia	98, 274, 291	Oodescelis	246
Nacerdes	117, 230	Oomorphus	143, 155, 159
Nalassus	116	Ootypus	91
Nanophyes	165, 216, 283, 302	Opatrum	116, 236, 254, 255, Pl. 105
Nanoptilium	26	Ophonus	19, 21
Nargus	31, 34	Opilo	80, 273, Pl. 61
Nathrius	124, 136, 142	Orchesia	115, 193
Nebria	18, 19, 21, 22, 182, 230, 231, 232, 234, Pl.15	Orchestes	168, 284, 285, 292, 293, 294, 376, 394
Necrobia	79, 191, 236	Orectochilus	9, 233
Necrodes	37	Orobitis	175, 216, 217, 235, 289, 296
Nedyus	286, 292	Orsodacne	122, 143, 144, 145, 159, 160, 277, 299, 300
Negastrius	67, 69, 232		
Neliocarus	287, 299, Pl. 105	Orthocerus	116
Nemadus	31, 34	Orthochaetes	168, 216
Nemozoma	79, 273, 292	Orthoperus	103, 104, 105
Neocoenorrhinus	163, 181, 293, 299, 300	Orthotomicus	290, 291
Neocrepidodera	143, 154, 280, 295, 307	Oryzaephilus	259, 264, 267, 269
Neophytobius	287, 295	Osphya	Pl. 75
Nephanes	27	Ostoma	79, 272, 291
Nephus	98, 99, 100, 101, 102, 274, 291, 303	Othius	251, 253, 254, 255, 256, 257
		Otiorhynchus	171, 224, 235, 287, 288, 300
Nestus	45	Oulema	143, 147, 148
Neuraphes	36, 254	Oulimnius	14
Nicrophorus	32, 37, 184, 315, 319, Pl. 36	Oxylaemus	95, 115
Niptus	75, Pl. 59	Oxyomus	51, 53, 56
Nitidula	192	Oxypoda	234, 251, 252, 254, 255
Nossidium	26	Oxyporus	45
Notaris	166, 283, 309, 310	Oxystoma	282, 300
Noterus	10	Oxythyrea	57
Notiophilus	18, 19, 20, 21, 22, 182, Pl. 17	Pachytodes	129, 135
Notothecta	251, 252, 254, 255	Pachytychius	285, 300, 388, 394
Notoxus	119	Paederus	46
Oberea	123, 125, 127, 128, 137, 142, 276	Palorus	116, 192, 269
Obrium	122, 127, 128, 129, 133, 136, 142, 276, 394	Panagaeus	233, 294, Pl. 31
		Parabathyscia	32, 34, 206
Ochina	76, 272, 303	Paracorymbia	128, 134, 135
Ochrosis	143, 280, 303, 308	Paracymus	373, 384, 394

Index

Paralister	24	*Pocadius*	83, 86, 237
Paramecosoma	91	*Podabrus*	72
Paraphotistus	272, 291, 293, 297	*Podagrica*	143, 153, 154, 280, 296
Paratillus	79	*Poecilium*	125, 128, 136, 229, 293
Parethelcus	286, 292	*Pogonocherus*	125, 126, 127, 128, 133, 137, 142, 229, 276, 291, Pl. 86
Paromalus	24, 271, 293, 337,		
Patrobus	234	*Pogonus*	19, 236
Pedostrangalia	124, 128, 134, 135, 137	*Polistichus*	236
Pella	257	*Polydrusus*	172, 287, 301
Pelonomus	170, 287, 295, 299, 301, 302	*Polygraphus*	289, 291
Pelophila	Pl. 16	*Polyphylla*	56
Pentaphyllus	116	*Pomatinus*	15, 233, 236
Peranus	223	*Poophagus*	170, 286, 297
Perapion	281, 295	*Porcinolus*	61
Perileptus	394	*Prasocuris*	143, 149, 150, 278, 303, 306
Phaedon	143, 144, 149, 150, 160, 278, 298, 303, 306, 308	*Pria*	83, 84, 88, 273, 303
		Priobium	76, 77
Phalacrus	98, 228, 274, 309, 310	*Prionocyphon*	16
Phaleria	117, 230, 236	*Prionus*	125, 127, 135, 140, 142, 197, 212, 276, 293
Philonthus	46, 186, 216, 229, 230, 233, 237, 339, 358,		
		Prionychus	117, 275, 292, 293
Philopedon	172	*Procas*	166, 205, 283, 292, 394
Phloeocharis	42	*Procraerus*	67, 272, 293
Phloeophthorus	176, 289, 301	*Propylea*	98
Phloeosinus	290, 291	*Protaetia*	51, 56, 250, 251, 254
Phloeostiba	235	*Protapion*	165, 282, 300, 301, 394
Phloiophilus	78, 80, 221	*Proteinus*	42
Phloiotrya	275, 293	*Protopirapion*	282, 301
Phosphaenus	70	*Psammodius*	56, 225
Phratora	143, 144, 149, 150, 278, 296, 297	*Psammoporus*	51, 56
Phylan	116, 236	*Pselactus*	170, 230
Phyllobius	172, 287, 292, 362	*Pselaphaulax*	39
Phyllobrotica	143, 151, 152, 279, 304	*Pselaphus*	39
Phyllodrepa	42	*Pseudapion*	282, 296
Phyllopertha	51, 57	*Pseudaplemonus*	281, 295
Phyllotreta	143, 144, 153, 154, 155, 279, 297, 298	*Pseudocistela*	117, 275, 293
		Pseudoprotapion	282, 300
Phymatodes	124, 128, 136, 276, 293	*Pseudopsis*	45, 219, 220
Phytobius	170, 284, 287, 301, 308	*Pseudorchestes*	168, 285, 307
Phytoecia	129, 137, 138, 142, 276, 303	*Pseudostyphlus*	168, 285, 308
Phytosus	225, 230	*Pseudotriphyllus*	114
Pilemostoma	143, 148, 277, 307, Pl. 96	*Pseudovadonia*	128, 135, 136
Pirapion	282, 301	*Psilothrix*	80
Pissodes	229, 289, 291	*Psylliodes*	143, 144, 153, 154, 159, 160, 280, 292, 294, 297, 298, 302, 303, 307, 394, Pl. 95
Pityophagus	83, 86, 192, 273, 291,		
Pityophthorus	290, 291		
Plagiodera	143, 149, 278, 296, 297	*Psyllobora*	98, 100
Plagionotus	122, 123, 128, 132, 136	*Ptenidium*	26, 230, 251, 252, 255
Plataraea	252	*Pterostichus*	19, 20, 21, 182, 232, 234, 334, 394, Pl. 26
Plateumaris	16, 17, 143, 147, 277, 309, 310, 374		
		Pteryx	27
Platycerus	56	*Ptilinus*	76, 79, 191, Pl. 60
Platycis	69	*Ptiliola*	26
Platydracus	251, 253, 256, 257	*Ptiliolum*	26
Platynaspis	98, 100	*Ptilium*	26, 27, 251
Platynus	19	*Ptilodactyla*	62
Platypus	116, 176, 196, 290, 293, 305	*Ptinella*	27
Platyrhinus	114, 162, 280, 306	*Ptinomorphus*	75
Plectophloeus	38	*Ptinus*	75, 223, 238, 256, 260, 269, 333
Plegaderus	23, 24, 271, 291, 292, 337,	*Ptomophagus*	31, 34
Pleurophorus	56	*Pulion*	81

Pycnomerus	116	Sermylassa	143, 151, 152, 235, 279, 306
Pyrochroa	118, 119, 192, 275, 292, 293, Pl. 74, Pl. 76	Siagonium	44, 186
		Sibinia	168, 216, 285, 294, 295
Pyropterus	69	Silis	72
Pyrrhalta	143, 151, 152, 195, 278, 279, 306	Silpha	37, 184, Pl. 34
Pyrrhidium	124, 128, 136, 276, 293, Pl. 84	Silusa	235
Pytho	118, 275, 291	Simplocaria	61
Quedius	46, 223, 224, 232, 250, 251, 253, 255,256	Sinodendron	50, 57, 186, 271, 293, 296, 299, 305
		Sirocalodes	286, 292
Rabocerus	119	Sitaris	118
Reesa	260	Sitona	172, 173, 196, 232, 287, 288, 300, 301
Reichenbachia	39		
Rhagium	125, 127, 128, 135, 142, 235, 276, 293, Pl. 77	Sitophilus	165, 259, 264, 265, 267, 269, Pl. 102
Rhagonycha	71, 72, 337, Pl. 55	Smaragdina	143, 156, 280, 294
Rhamphus	168	Smicronyx	168, 216, 285, 303
Rhantus	Pl. 1	Smicrus	27, 237
Rhinocyllus	174, 289, 307	Sogda	30, 238
Rhinoncus	170, 287, 295	Soronia	83, 86, 87, 192, 235
Rhinosimus	119	Spercheus	14, 184
Rhinusa	285, 306	Sphaeridium	13
Rhizophagus	89, 207, 221, 235	Sphaeriestes	119, 229, 234, 276, 291
Rhopalodontus	107	Sphaerites	23
Rhopalomesites	170	Sphaerius	7, 14, 231
Rhynchaenus	168	Sphaeroderma	143, 153, 154, 280, 307
Rhyssemus	56	Sphaerosoma	96, 97
Rhyzobius	98, 99	Sphindus	234
Rhyzopertha	259, 269, 272, 293	Sphinginus	273, 293
Riolus	15	Sphodrus	20, 239
Rugilus	45	Squamapion	164, 225, 281, 304, 305
Rutidosoma	169, 287, 296	Staphylinus	46, Pl. 40
Rutpela	125, 128, 136, 334, Pl. 117a	Stegobium	75, 76, 260, 269
Rybaxis	39	Stenagostus	67, 189
Rypobius	103, 104, 105	Stenelmis	14, 15
Salpingus	119, 221	Stenichnus	36, 251
Saperda	124, 125, 127, 128, 129, 132, 137, 138, 140, 142, 276, 293, 296, 297, Pl. 88	Stenocarus	169, 286, 292
		Stenocorus	125, 128, 135, 142
		Stenolophus	19, Pl. 29
Saprinus	24, 337	Stenopelmus	166, 176, 283, 290
Saprosites	50, 55, 57, 186	Stenopterapion	282, 300, 301
Scaphidema	275, 292	Stenoria	118
Scaphidium	40	Stenostola	128, 137, 142, 276, 296, 305
Scaphisoma	40, 41	Stenurella	128, 136
Scaphium	40	Stenus	45, 224, 232, 233, 353, 358, 394
Scarabaeus	47, 48	Stethorus	98, 100
Schizotus	119, 275, 294	Sticticomus	119
Sciodrepoides	31	Stictoleptura	124, 125, 127, 128, 135, 276, 291, Pl. 79
Scirtes	16		
Scolytus	176, 196, 289, 292, 293, 294, 296, 299, 300	Stilbus	90, 274, 310
		Strangalia	122, 128, 136, 276, 291, 293, 294, 296
Scopaeus	45		
Scraptia	112, 113, 227	Strophosoma	287, 293, 294, 299, 301
Scydmaenus	36	Subcoccinella	98, 100
Scydmoraphes	36	Sulcacis	107, 110
Scymnus	98, 99, 100, 102, 274, 291, 293, 296, 297	Sunius	253
		Syagrius	175, 289, 290
Selatosomus	66, 69	Symbiotes	97, 234, 256
Sepedophilus	43, 46, 220, 221	Synapion	282, 300
Serica	51, 57, Pl. 44	Synaptus	394
Sericoda	234	Syncalypta	61
Sericoderus	103, 104		

Index

Synchita	116	*Trechus*	20, 236
Tachinus	42, 43	*Tribolium*	116, 192, 259, 263, 264, 268, 269
Tachyerges	285, 294, 296, 297	*Trichius*	50, 56, 57
Tachyporus	43, 46, 220, 221, Pl. 37	*Trichocellus*	19, 224, 270, 298
Tachys	231, 233, 394,	*Trichodes*	79, 118
Tachyusa	233	*Trichonyx*	38, 39
Tachyusida	256	*Trichophya*	43
Taeniapion	164, 281, 292	*Trichosirocalus*	169, 287, 305, 307, 308, Pl. 105
Tanymecus	172	*Trigonogenius*	260
Tanysphyrus	176, 283, 309	*Trimium*	38
Tapeinotus	170, 286, 299	*Trinophylum*	136, 137, 138, 142, 276, 291, 293
Taphrorhychus	290, 293	*Triplax*	93, 94, 221, Pl. 64
Tarsostenus	79	*Trissemus*	39
Tasgius	236	*Tritoma*	94
Telmatophilus	91, 221, 274, 309, 310	*Trixagus*	64, 65, 227
Temnocerus	281, 294, 296, 297, 299	*Tropideres*	280, 293, 294
Tenebrio	116, 192, 260, 264, 267, 269, 362	*Tropiphorus*	172
Tenebriodes	78, 264	*Trox*	48, 52, 56, 186, 223, 236
Teredus	95, 115	*Trypocopris*	52, 56
Teretrius	23	*Trypophloeus*	290, 296
Tesarius	57	*Tychius*	168, 216, 285, 300, 301
Tetratoma	114, 221	*Tychobythinus*	39, 256
Tetropium	124, 127, 128, 136, 276, 291	*Tychus*	39
Tetrops	125, 127, 128, 129, 133, 137, 142, 276, 299, 300, 305	*Typhaea*	114, 259, 268
		Typhaeus	52, 54, 56, 186, Pl. 41
Thaiosophila	257	*Tytthaspis*	98, 100
Thalassophilus	232	*Urophorus*	84
Thalycra	83, 86, 235	*Velleius*	46, 186, 218, 238
Thamiocolus	284, 286, 304	*Vibidia*	98
Thanasimus	79, 191, 273, 291	*Vincenzellus*	119
Thanatophilus	37	*Xanthogaleruca*	143
Thaneroclerus	78, 79	*Xantholinus*	256
Thes	106	*Xanthomus*	116
Thiasophila	251, 252	*Xestobium*	75, 76, 80, 191, 272, 293, 297
Thinobaena	230	*Xyleborus*	176, 290, 291, 292, 293, 302
Thinobius	7, 45, 231, 394	*Xylechinus*	290, 291
Throscus	64	*Xyletinus*	76, 77, 80, 272, 292, 293
Thryogenes	166, 283, 309	*Xylita*	115
Thymalus	78, 272, 293	*Xylocleptes*	176, 290, 291
Tilloidea	79	*Xyloterus*	85, 176, 290, 291, 293
Tillus	79, 273, 293	*Zabrotes*	143, 145, 146
Timarcha	143, 149, 277, 306, Pl. 92	*Zabrus*	20, 21, 270, 309
Tmesiphorus	257	*Zacladus*	169, 207, 302, 303
Tomicus	290, 291	*Zeugophora*	142, 143, 144, 277, 296
Tomoxia	110, Pl. 71	*Zilora*	115, 275, 291
Tournotaris	166, 283, 309, 310	*Zorochros*	66, 67, 68, 69
Trachyphloeus	171, 216	*Zyras*	250, 251, 253, 254, 255, 256, 257
Trachys	60, 189, 271, 294, 297, 304, 306		